W9-AGF-950

FOURTH EDITION

OPERATIONAL AMPLIFIERS AND LINEAR INTEGRATED CIRCUITS

Robert F. Coughlin

Frederick F. Driscoll

PRENTICE HALL, Englewood Cliffs, New Jersey 07632

Library of Congress Cataloging-in-Publication Data

Coughlin, Robert F.
Operational amplifiers and linear integrated circuits / Robert F.
Coughlin, Frederick F. Driscoll. -- 4th ed.
p. cm.
Includes bibliographical references and index.
ISBN 0-13-639923-1
1. Operational amplifiers. 2. Linear integrated circuits.
I. Driscoll, Frederick F., 1943- . II. Title.
TK7871.58.O6C68 1991
621.39'5--dc20 90-43585
CIP

Editorial/production supervision: *Ellen Denning*
Interior design: *Jayne Conte*
Cover design: *Bruce Kenselaar*
Manufacturing buyer: *Ed O'Dougherty*
Cover photograph: Gary Gladstone/The Image Bank

Printed in the United States of America

10 9 8 7 6 5 4 3

ISBN 0-13-639923-1

PRENTICE-HALL INTERNATIONAL (UK) LIMITED, *London*
PRENTICE-HALL OF AUSTRALIA PTY. LIMITED, *Sydney*
PRENTICE-HALL CANADA INC., *Toronto*
PRENTICE-HALL HISPANOAMERICANA, S.A., *Mexico*
PRENTICE-HALL OF INDIA PRIVATE LIMITED, *New Delhi*
PRENTICE-HALL OF JAPAN, INC., *Tokyo*
SIMON & SCHUSTER ASIA PTE. LTD., *Singapore*
EDITORA PRENTICE-HALL DO BRASIL, LTDA., *Rio de Janeiro*

To Our Partners in Ballroom Dancing
and
Our Lifetime Partners,
Barbara and *Jean*

As
We Grow Older
We Grow Closer

Contents

10 AC PERFORMANCE: BANDWIDTH, SLEW RATE, NOISE, AND FREQUENCY COMPENSATION 250

11 ACTIVE FILTERS 272

Preface

It has been our purpose in the first three editions, and again now in the fourth edition of this text, to show that operational amplifiers and other linear integrated circuits are both fun and easy to use, especially if the application does not require the devices to operate near their design limits. It is the purpose of this book to show just how easy they are to use in a variety of applications involving instrumentation, signal generation, filter, and control circuits.

When first learning how to use an op amp, one should *not* be presented with a myriad of op amps and asked to make an informed selection. For this reason, our introduction begins with an inexpensive, reliable op amp that forgives most mistakes in wiring, ignores long lead capacitance, and does not burn out easily. Such an op amp is the 741, whose characteristics are documented in Appendix 1 and whose applications are sprinkled throughout the text.

If a slightly faster op amp is needed for a wider bandwidth, another inexpensive and widely used op amp is the 301. See Appendix 2 for its electrical characteristics and Chapter 10 to learn when one might prefer the 301 over the 741. We also use the stable CA3140 and TL081, which are pin-for-pin higher-frequency replacements for the 741.

Where appropriate, we have added specialized op amps: the LM339 single supply comparator in Chapter 2 and the LM311 high-performance comparator in Chapter 4. Where better dc performance is required, we have added BiFET op amps in Chapter 7. The excellent low-bias-current and low-offset-voltage OP-07 is also employed for instrumentation applications.

The fourth edition is organized into a set of core chapters that should be read first. They are Chapters 1 through 6 and proceed in a logical teaching sequence to show how the op amp can be used to solve a variety of application problems.

The limitations of op amps are not discussed in Chapters 1 to 6 because it is very important to gain confidence in using op amps before pushing performance to its limits. When studying transistors or other devices, we do not begin with their limitations. Regrettably, much of the early literature on integrated circuits begins with their limitations and thus obscures the inherent simplicity and overwhelming advantages of basic integrated circuits over basic transistor circuits. For these reasons, op amp limitations are not presented until Chapters 9 and 10 for those readers who need to understand some of their limitations with respect to dc and ac performance. Furthermore, not all op amp limitations apply to every op amp circuit. For example, dc op amp limitations such as offset voltages are usually not important if the op amp is used in an ac amplifier circuit. Thus dc limitations (Chapter 9) are treated separately from ac limitations (Chapter 10). The remaining chapters have been written to stand alone. They can be studied in any order after completion of Chapters 1 through 6.

The servoamplifier and subtractor circuits have been added to Chapter 5. Chapter 6 has been extensively rewritten to show the latest techniques using a modulator/demodulator IC and trigonometric function generator IC to make a *precision* triangle/square/sine-wave generator whose frequency can be adjusted over a wide range by a *single* resistor and whose amplitude can be adjusted without affecting frequency, or vice versa.

Chapter 7 deals with the specialized applications that can best be accomplished by op amps combined with diodes.

Chapter 8 is concerned with problems of measuring physical variables such as force, pressure weight, and temperature. Bridge and instrumentation amplifiers are ideal for these measurements.

Chapter 11 simplifies the design of active filters. The four basic types of active filters are shown: low-pass, high-pass, band-pass, and band-reject filters. Butterworth filters were selected because they are often used and easy to design. If you want to design a three-pole (60-dB/decade) Butterworth low- or high-pass filter,

Chapter 11 tells you how to do it in four steps with a pencil and paper. No calculator or computer program is required. Basic algebra is the only mathematics that is required throughout the text. The sections on bandpass and notch filters have been completely rewritten to simplify both their design procedures and fine tuning.

A fascinating integrated circuit, the multiplier, is presented in Chapter 12 because it makes analysis and design of communication circuits very easy. Modulators, demodulators, frequency shifters, a universal AM radio receiver, and a host of other applications are performed by the multiplier, an op amp, and a few resistors. This chapter has been retained because numerous instructors have written to say how useful it is as a hardware-oriented teaching tool that introduces the principles of single-sideband, suppressed-carrier, and standard amplitude-modulation principles used in radio communication.

Chapter 13 is included for those who need to use the ubiquitous 555 IC timer and/or the XR2240 counter timer.

A chapter (14) has been added (because of requests of many instructors) on digital-to-analog and analog-to-digital converters. More specifically, instructors have requested information on converters that (1) are microprocessor compatible, (2) simple to select via a data bus, (3) low in cost and easy to use, and (here comes the hard one) (4) stand-alone circuits for both a microprocessor-compatible DAC and an ADC that can be (a) breadboarded by students in a laboratory and (b) do *not* require a microprocessor. This has been done!

Since almost all linear integrated circuits require a regulated power supply, we have, because of requests from readers, retained material on power supply design and analysis and moved it to Chapter 15. The latest IC regulators are used to show how you can make excellent linear regulated supplies at low cost for (a) 5-V digital logic ICs, (b) ± 15-V linear ICs, (c) combined (a) and (b) for microprocessor supplies, and (d) either positive or negative adjustable supplies.

In this fourth edition we have incorporated suggestions from students, instructors, and practicing engineers and technicians from all parts of the United States and from Indonesia, Poland, Japan, and the USSR.

This edition contains more than enough material for a single-semester course. All circuits have been personally lab tested by the authors and their recommendations for laboratory work have been added to the end of each chapter. The material is suitable for both nonelectronic specialists who just want to learn something about linear ICs and for electronic majors who wish to use linear ICs. In addition, we have added Learning Objectives to the beginning of each chapter, again in response to requests from readers.

We thank Mrs. Phyllis Wolff for the preparation of the original manuscript and continued support in this fourth edition. A special thanks to our colleague Robert S. Villanucci who has been particularly helpful with his suggestions and for testing of ideas and to Dean Alexander Avtgis for his continued support.

We also thank two highly respected analog engineers, Dan Sheingold of

Analog Devices and Bob Pease of National Semiconductors, for their constructive criticism, technical corrections, and guidance in areas we found difficult.

Finally, we thank our students for their insistence on relevant instruction that is immediately useful, and our readers for both their enthusiastic reception of this book and their perceptive comments.

ROBERT F. COUGHLIN
FREDERICK F. DRISCOLL

Boston, Mass.

CHAPTER 1

Introduction to Op Amps

LEARNING OBJECTIVES

Upon completing this introductory chapter on op amps, you will be able to:

- Draw the circuit symbol for a 741 op amp and show the pin numbers for each terminal.
- Name and identify at least three types of package styles that house a general-purpose op amp.
- Identify the manufacturer, op amp, and package style from the PIN.
- Correctly place an order for an op amp.
- Count the pins of an op amp from the top or bottom view.
- Identify the power supply common on a circuit schematic, and state why you must do so.
- Breadboard an op amp circuit properly.

1-0 INTRODUCTION

One of the most versatile and widely used electronic devices in linear applications is the *operational amplifier*, most often referred to as the op amp. Op amps are popular because they are low in cost, easy to use, and fun to work with. They allow you to build useful circuits without the need to know about their complex internal circuitry. The op amp is also very forgiving of wiring errors because of its self-protecting internal circuitry.

1-1 A SHORT HISTORY

1-1.1 The Early Days

George Philbrick is one of the people credited with developing and popularizing the op amp. He worked first at Huntington Engineering Labs, and then with his own company, Philbrick Associates. He was instrumental in designing a single vacuum-tube op amp and introduced it in 1948. These early op amps and subsequent improved versions were intended primarily for use in analog computers. The word "operational" in operational amplifiers, at that time, stood for *mathematical operations*. The early op amps were used to make circuits that could add, subtract, multiply, and even solve differential equations.

Analog computers had limited accuracy, no more than three significant figures. Hence analog computers were replaced by the faster, more accurate, and more versatile digital computers. However, digital computers were not the demise of the op amp.

1-1.2 Birth and Growth of the IC Op Amp

Fairchild brought out the 702, 709, and 741 integrated circuit op amps between 1964 and 1968, while National Semiconductor introduced the 101/301. These integrated circuit op amps revolutionized certain areas of electronics because of their small size and low cost. Even more important, they reduced drastically the task of circuit design. For example, instead of the tedious and difficult task of designing an amplifier with transistors, designers could now use one op amp and a few resistors to make a superb amplifier.

The design time for an amplifier made with an op amp is about 10 seconds. Moreover, IC op amps are inexpensive and take up less space and power than discrete components. Circuit functions that can be made with one or two op amps and a few components include signal generation (oscillators), signal conditioning, timers, voltage-level detection, and modulation. This list is just about endless.

1-1.3 Further Progress in Op Amp Development

Major improvements were made in op amps in two respects as their fabrication technology became more precise. First, field-effect transistors were substituted for certain bipolar junction transistors within the op amp. JFETs, at the op amp's input, draw very small currents and allow the input voltages to be varied between the power supply limits. MOS transistors in the output circuitry allow the output terminal to go within millivolts of the power supply limits.

The first BiFET op amp was the LF356. The CA3130 features bipolar inputs and a complementary MOS output. It is appropriately named BiMOS. These amplifiers also are faster and have a higher-frequency response than the 741.

The second major improvement was the invention of dual and quad op amp packages. In the same 14-pin package occupied by a single op amp, designers fabricated *four* separate op amps. All four op amps in the package share the same power supply. The LM324 is a popular example of the quad op amp and the LM358 is a popular dual op amp.

1-1.4 Op Amps Become Specialized

Inevitably, general-purpose op amps were redesigned to optimize or add certain features. Special function ICs that contain more than a single op amp were then developed to perform complex functions.

You need only to look at linear data books to appreciate their variety. Only a few examples are

1. High current and/or high voltage capability
2. Sonar send/receive modules
3. Multiplexed amplifiers
4. Programmable gain amplifiers
5. Automotive instrumentation and control
6. Communication ICs
7. Radio/audio/video ICs

General-purpose op amps will be around for a long time. However, you should expect to see the development of more complex integrated circuits on a single chip that combine many op amps with digital circuitry. In fact, with improved very large scale integrated (VLSI) technology, it is inevitable that entire systems will be fabricated on a single large chip.

A single chip computer is today's reality. A single-chip TV set will happen eventually. Before learning how to use op amps, it is wise to learn what they look like and how to buy them. The op amps' greatest use will be as a part in a system that interfaces the real world of analog voltage with the digital world of the computer as will be shown in Chapter 14. If you want to understand the system you must understand the workings of one of its most important components.

1-2 741 GENERAL-PURPOSE OP AMP

1-2.1 Circuit Symbol and Terminals

The op-amp symbol in Fig. 1-1 is a triangle that points in the direction of signal flow. This component has a *part identification number* (PIN) placed within the triangular symbol. The PIN refers to a particular op amp with specific characteristics. The 741C op amp illustrated here is a general-purpose op amp that is used throughout the book for illustrative purposes.

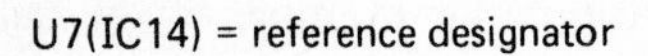

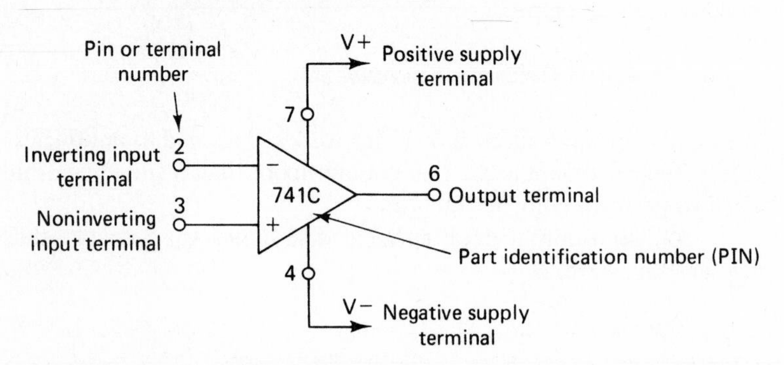

FIGURE 1-1 Circuit symbol for the general-purpose op amp. Pin numbering is for an 8-pin mini-DIP package.

The op amp may also be coded on a circuit schematic with a *reference designator* such as U1, IC 101, and so on. Its PIN is then placed beside the reference designator in the parts list of the circuit schematic. All op amps have at least five terminals: (1) The positive power supply terminal V_{CC} or $+V$ at pin 7, (2) the negative power supply terminal V_{EE} or $-V$ at pin 4, (3) output pin 6, (4) the inverting (−) input terminal at pin 2, and (5) the noninverting (+) input terminal at pin 3. Some general-purpose op amps have additional specialized terminals. (The pins above refer to the 8-pin mini-DIP case discussed in the following section.)

1-2.2 Circuit Schematic

The equivalent circuit of a 741 op amp is presented in Fig. 1-2. It is a complex, third-generation design made of one capacitor, 11 resistors, and 27 transistors.

Transistors Q_1 and Q_2, with their supporting circuitry, make up a high-gain differential *input stage*. Q_{14} and Q_{20} are the complementary *output stage*. Q_{15} and

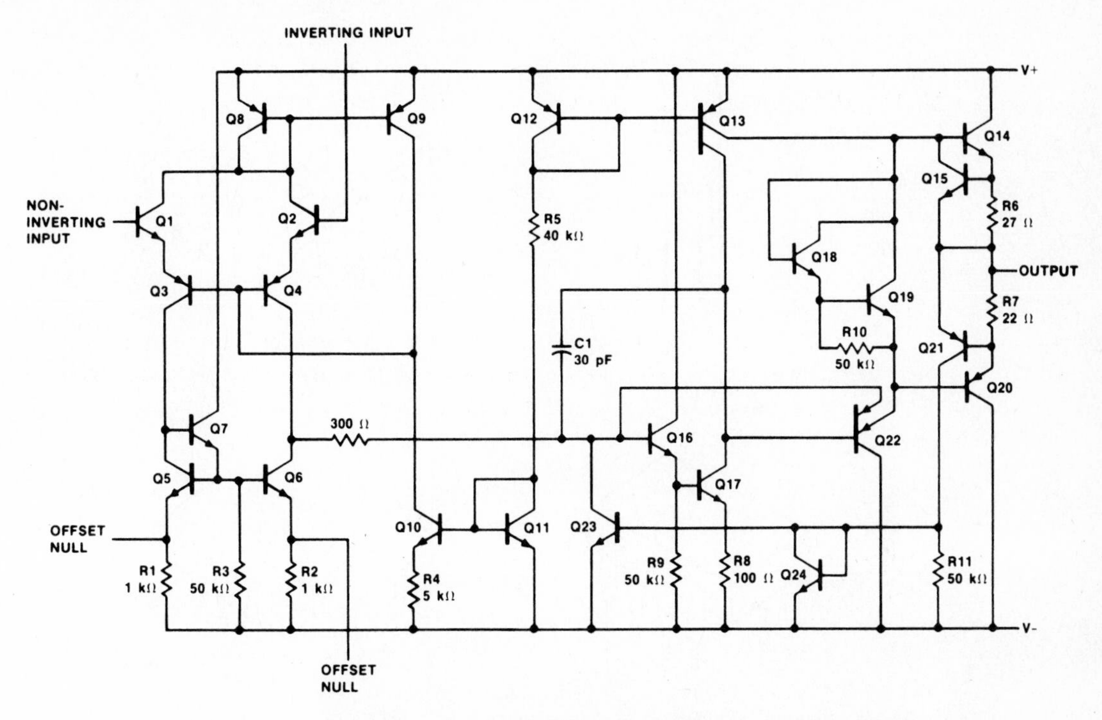

FIGURE 1-2 Equivalent circuit of a 741 op amp. (Courtesy of Fairchild Semiconductor, a Division of Fairchild Camera and Instrument Corporation.)

Q_{21} sense the output current and provide short-circuit protection. The remaining transistors comprise the *level shifter* stage, which interfaces the input stage with the output stage.

1-3 PACKAGING AND PINOUTS

1-3.1 Packaging

The op amp is fabricated on a tiny silicon chip, and packaged in a suitable case. Fine-gage wires connect the chip to external leads extending from a metal, plastic, or ceramic package. Common op amp packages are shown in Fig. 1-3(a) to (c).

The metal can package shown in Fig. 1-3(a) are available with 3, 5, 8, 10, and 12 leads. The silicon chip is bonded to the bottom metal sealing plane to expedite the dissipation of heat. The tab identifies pin 8 and the pins are numbered counterclockwise when you view the metal can from the *top*.

The popular 14-pin and 8-pin dual-in-line packages (DIPs) are shown in Fig. 1-3(b) and (c). Either plastic or ceramic cases are available. As viewed from the *top,* a notch or dot identifies pin 1 and terminals are numbered counterclockwise.

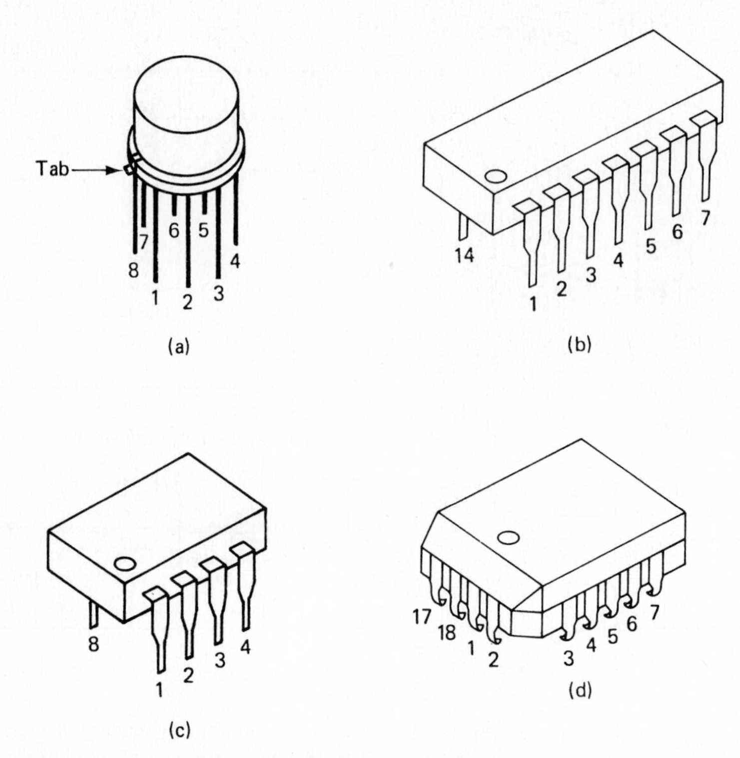

FIGURE 1-3 The three most popular op amp packages are the metal can in (a) and the 14- or 8-pin dual-in-line packages in (b) and (c). For high-density ICs the latest surface-mounted technology (SMT) package is shown in (d).

Complex integrated circuits involving many op amps and other ICs can now be fabricated on a single large chip or by interconnecting many large chips and placing them in a single package. For ease of manufacture and assembly, pads replace the leads. The resulting structure is called surface-mounted technology (SMT), shown in Fig. 1-3(d). These packages provide a higher circuit density for a package of a given size. Additionally, SMTs have lower noise and improved frequency-response characteristics. SMT components are available in (1) *plastic lead chip carriers* (PLCCs), (2) *small outline integrated circuits* (SOICs), and (3) *leadless ceramic chip carriers* (LCCCs).

1-3.2 Combining Symbol and Pinout

Manufacturers are now combining the circuit symbol for an op amp together with the package view into a single drawing. For example, the four most common types of packages that house a 741 chip are shown in Fig. 1-4. Compare Fig. 1-4(a) and

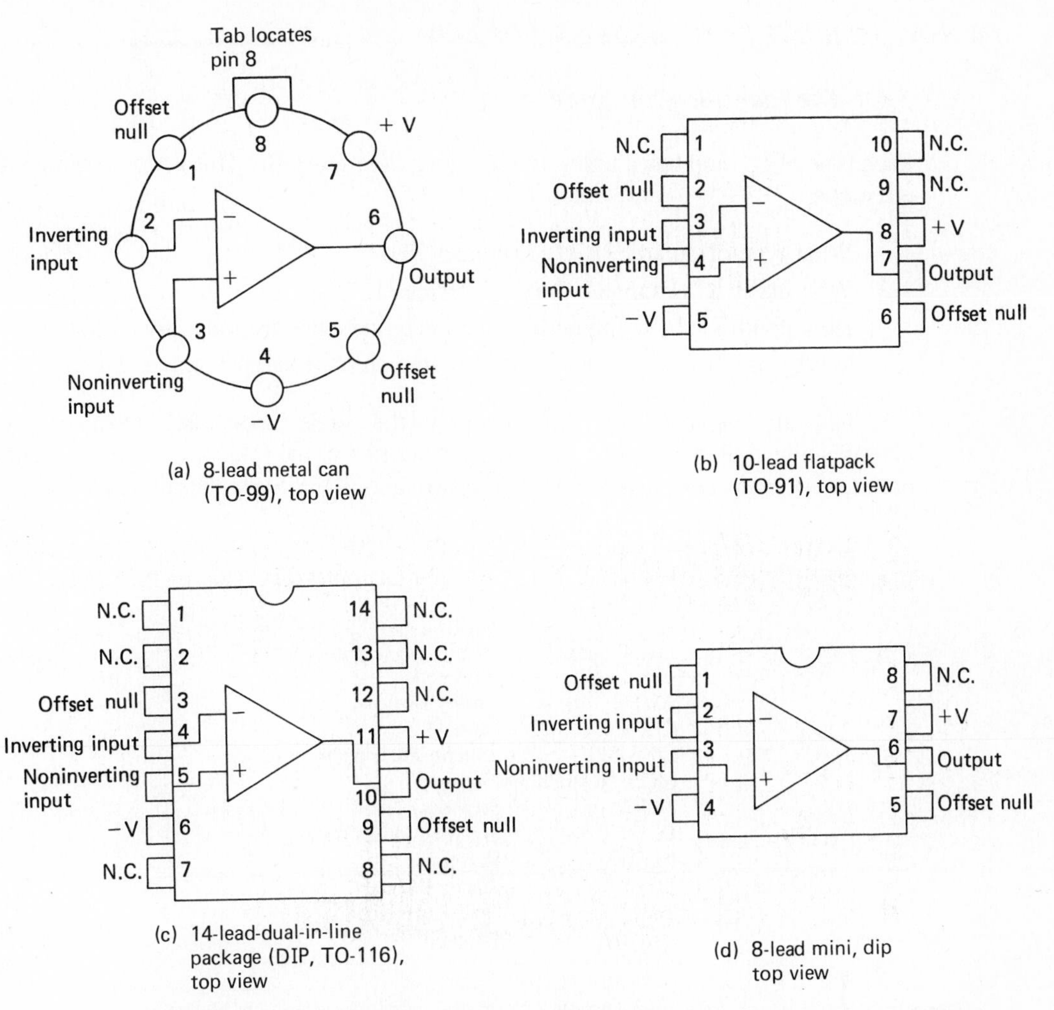

(a) 8-lead metal can (TO-99), top view

(b) 10-lead flatpack (TO-91), top view

(c) 14-lead-dual-in-line package (DIP, TO-116), top view

(d) 8-lead mini, dip top view

FIGURE 1-4 Connection diagrams for typical op amp packages. The abbreviation NC stands for "no connection." That is, these pins have no internal connection, and the op amp's terminals can be used for spare junction terminals.

(d) to see that the numbering schemes are identical for 8-pin can and 8-pin DIP. A notch or dot identifies pin 1 on the DIPs, and a tab identifies pin 8 on the TO-5 (or the similar TO-99) package. From a top view, the pin count proceeds counterclockwise.

The final tasks in this chapter are to learn how to buy a specific type of op amp and to present advice on basic breadboarding techniques.

1-4 HOW TO IDENTIFY OR ORDER AN OP AMP

1-4.1 The Identification Code

Each type of op amp has a letter–number identification code. This code answers four questions:

1. What type of op amp is it? (Example: 741.)
2. Who made it? (Example: Analog Devices.)
3. How good is it? (Example: the guaranteed temperature range for operation.)
4. What kind of package houses the op amp chip? (Example: plastic DIP.)

Not all manufacturers use precisely the same code. But most use an identification code that consists of four parts written in the following order: (1) letter prefix, (2) circuit designator, (3) letter suffix, and (4) military specification code.

Letter prefix. The letter prefix code usually consists of two or three letters that identify the manufacturer. The following examples list some of the codes.

Letter prefix	Manufacturer
AD	Analog Devices
CA	RCA
LM	National Semiconductor Corp.
MC	Motorola
NE/SE	Signetics
OP	Precision Monolithics
RC/RM	Raytheon
SG	Silicon General
TL	Texas Instruments
UA(μA)	Fairchild

Circuit designation. The circuit designator consists of three to seven numbers and letters. They identify the type of op amp and its temperature range. For example:

062C

J-FET input op amp → 062 ; "C" identifies commercial temperature range

The three *temperature-range codes* are as follows:

1. C: commercial, 0 to 70°C
2. I: industrial, −25 to 85°C
3. M: military, −55 to 125°C

Letter suffix. A one- or two-letter suffix identifies the package style that houses the op amp chip. You need the package style to get the correct pin connections from the data sheet (see Appendix 1). Three of the most common package suffix codes are:

Package code	Description
D	Plastic dual-in-line for surface mounting on a pc board
J	Ceramic dual-in-line
N, P	Plastic dual-in-line for insertion into sockets. (Leads extend through the top surface of a pc board and are soldered to the bottom surface.)

Military specification code. The military specification code is used only when the part is for high-reliability applications.

1-4.2 Order Number Example

A 741 general-purpose op amp would be completely identified in the following way:

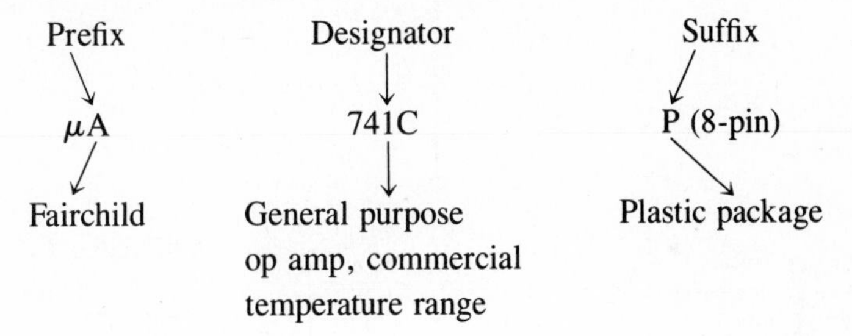

1-5 SECOND SOURCES

Some op amps are so widely used that they are made by more than one manufacturer. This is called *second sourcing*. People employed by Fairchild designed and made the original 741. Fairchild then contracted for licenses with other manufacturers to make 741s in exchange for a license to make op amps or other devices.

As time went on, the original 741 design was modified and improved by all manufacturers. The present 741 is in its fifth or sixth generation of evolution. Thus, if you order a 741 8-pin DIP from a supplier, it may have been built by Fairchild (μA741), Analog Devices (AD741), National Semiconductor (LM741), or others. Therefore, always check the manufacturer's data sheets that correspond to the device you have. You will then have information on its exact performance and key to the identification codes on the device.

1-6 BREADBOARDING OP AMP CIRCUITS

1-6.1 The Power Supply

Power supplies for general-purpose op amps are bipolar. As shown in Fig. 1-5(a), the typical commercially available power supply outputs ±15 V. The common point between the +15 V supply and −15 V is called the *power supply common*. It is shown with a ground symbol for two reasons. First, *all* voltage measurements are made with respect to this point. Second, power supply common is usually wired to the third wire of the line cord that extends ground (usually from a water pipe in the basement) to the chassis containing the supply.

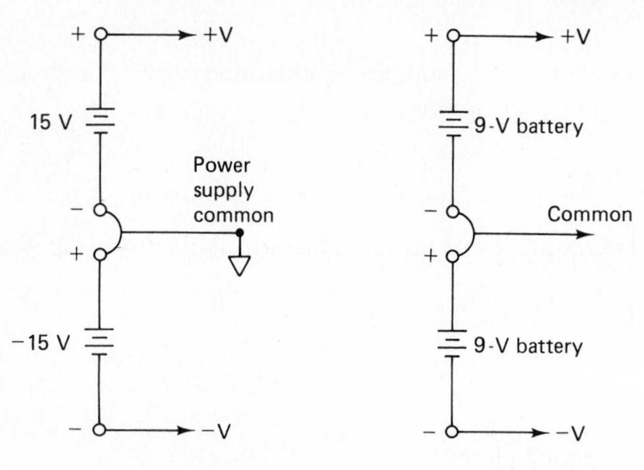

FIGURE 1-5 Power supplies for general-purpose op amps must be bipolar.

The schematic drawing of a portable supply is shown in Fig. 1-5(b). This is offered to reinforce the idea that a bipolar supply contains two separate power supplies connected in series aiding.

1-6.2 Breadboarding Suggestions

It is the author's intention that all circuits presented in this text will allow the reader to breadboard and test their performance. A few circuits require printed circuit board construction. Before we proceed to learn how to use an op amp, it is prudent to give some time-tested advice on breadboarding a circuit:

1. Do all wiring with power off.
2. Keep wiring and component leads as short as possible.

3. Wire the $+V$ and $-V$ supply leads *first* to the op amp. It is surprising how often this vital step is omitted.
4. Try to wire all ground leads to one tie point, the power supply common. This type of connection is called *star grounding*. Do not use a ground bus, because you may create a ground loop, thereby generating unwanted noise voltages.
5. Recheck wiring before applying power to the op amp.
6. Connect signal voltages to the circuit only after the op amp is powered.
7. Take all measurements with respect to ground. For example, if a resistor is connected between two terminals of an IC, do not connect either a meter or a CRO across the resistor; instead, measure the voltage on one side of the resistor and then on the other side and calculate the voltage across the resistor.
8. Avoid using ammeters, if possible. Measure the voltage as in step 7 and calculate current.
9. Disconnect the input signal before the dc power is removed. Otherwise, the IC may be destroyed.
10. These ICs will stand much abuse. But *never:*
    a. Reverse the polarity of the power supplies,
    b. Drive the op amp's input pins above or below the potentials at the $+V$ and $-V$ terminal, or
    c. Leave an input signal connected with no power on the IC.
11. If unwanted oscillations appear at the output and the circuit connections seem correct:
    a. Connect a 0.1-μF capacitor between the op amp's $+V$ pin and ground and another 0.1-μF capacitor between the op amp's $-V$ pin and ground.
    b. Shorten your leads, and
    c. Check the test instrument, signal generator, load, and power supply ground leads. They should come together at one point.
12. The same principles apply to all other linear ICs.

We now proceed to our first experience with an op amp.

PROBLEMS

1-1. In the term *operational amplifier,* what does the word *operational* stand for?

1-2. Is the LM324 op amp a single op amp housed in one package, a dual-op amp in one package, or a quad op amp in one package?

1-3. With respect to an op amp, what does the abbreviation PIN stand for?

1-4. Does the letter prefix of a PIN identify the manufacturer or the package style?

1-5. Does the letter suffix of a PIN identify the manufacturer or the package style?

1-6. Which manufacturer makes the AD741CN?

1-7. Does the tab on a metal can package identify pin 1 or pin 8?

1-8. Which pin is identified by the dot on an 8-pin minidip?

1-9. **(a)** How do you identify power supply common on a circuit schematic?
(b) Why do you need to do so?

1-10. When breadboarding an op amp circuit, should you use a ground bus or star grounding?

CHAPTER 2

First Experiences with an Op Amp

LEARNING OBJECTIVES

Upon completing this chapter on first experiences with an op amp, you will be able to:

- Briefly describe the task performed by the power supply and input and output terminals of an op amp.
- Show how the single-ended output voltage of an op amp depends on its open-loop gain and differential input voltage.
- Calculate differential input voltage E_d, and the resulting output voltage V_o.
- Draw the circuit schematic for an inverting or noninverting zero-crossing detector.
- Draw the output voltage waveshape of a zero-crossing detector if you are given the input voltage waveshape.
- Draw the output–input voltage characteristics of a zero-crossing detector.

- Sketch the schematic of a noninverting or inverting voltage-level detector.
- Describe at least two practical applications of voltage-level detectors.
- Analyze the action of a pulse-width modulator and tell how it can interface an analog signal with a microcomputer.

2-0 INTRODUCTION

The name *operational amplifier* was given to early high-gain vacuum-tube amplifiers designed to perform mathematical operations of addition, subtraction, multiplication, division, differentiation, and integration. They could also be interconnected to solve differential equations.

The modern successor of those amplifiers is the *linear integrated-circuit op amp*. It inherits the name, works at lower voltages, and is available in a variety of specialized forms. Today's op amp is so low in cost that millions are now used annually. Their low cost, versatility, and dependability have expanded their use far beyond applications envisioned by early designers. Some present-day uses for op amps are in the fields of process control, communications, computers, power and signal sources, displays, and testing or measuring systems. The op amp is still basically a very good high-gain dc amplifier.

One's first experience with a linear IC op amp should concentrate on its most important and fundamental properties. Accordingly, our objectives in this chapter will be to identify each terminal of the op amp and to learn its purpose, some of its electrical limitations, and how to apply it usefully.

2-1 OP AMP TERMINALS

Remember from Fig. 1-1 that the circuit symbol for an op amp is an arrowhead that symbolizes high gain and points from input to output in the direction of signal flow. Op amps have five basic terminals: two for supply power, two for input signals, and one for output. Internally they are complex, as was shown by the schematic diagram in Fig. 1-2. It is not necessary to know much about the internal operation of the op amp in order to use it. We will refer to certain internal circuitry, when appropriate. The people who design and build op amps have done such an outstanding job that external components connected to the op amp determine what the overall system will do.

The ideal op amp of Fig. 2-1 has infinite gain and infinite frequency response. The input terminals draw no signal or bias currents and exhibit infinite input resistance. Output impedance is zero ohms and the power supply voltages are without limit. We now examine the function of each op amp terminal to learn something about the limitations of a real op amp.

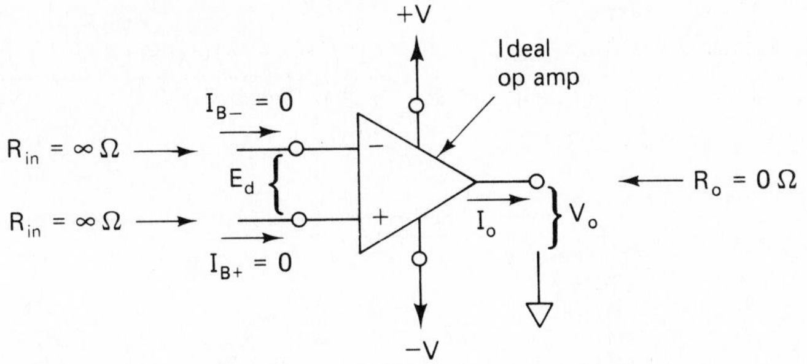

FIGURE 2-1 The ideal op amp has infinite gain and input resistances plus zero output resistance.

2-1.1 Power Supply Terminals

Op amp terminals labeled $+V$ and $-V$ identify those op amp terminals that must be connected to the power supply (see Fig. 2-2 and Appendices 1 and 2). Note that the power *supply* has *three* terminals: positive, negative, and power supply common. The power supply common terminal may or may not be wired to earth ground via the third wire of line cord. However, it has become standard practice to show power common as a ground symbol on a schematic diagram. Use of the term "ground" or the ground symbol is a convention which indicates that *all voltage measurements are made with respect to "ground."*

The power supply in Fig. 2-2 is called a bipolar or split supply and has typical values of ± 15 V. Special-purpose op amps may use a single polarity supply such as $+5$ or $+15$ V and ground. Note that the ground is *not* wired to the op amp in Fig. 2-2. Currents returning to the supply from the op amp must return through external circuit elements such as the load resistor R_L. The maximum supply voltage that can be applied between $+V$ and $-V$ is typically 36 V or ± 18 V.

2-1.2 Output Terminal

In Fig. 2-2 the op amp's output terminal is connected to one side of the load resistor R_L. The other side of R_L is wired to ground. Output voltage V_o is measured with respect to ground. Since there is only one output terminal in an op amp, it is called a *single-ended output*. There is a limit to the current that can be drawn from the output terminal of an op amp, usually on the order of 5 to 10 mA. There are also limits on the output terminal's voltage levels; these limits are set by the supply voltages and by output transistors Q_{14} and Q_{20} in Fig. 1-2 (see also Appendix 1, "Output Voltage Swing as a Function of Supply Voltage"). These transistors need about 1 to 2 V from collector to emitter to ensure that they are acting as amplifiers and not as switches. Thus the output terminal can rise approximately to within 1 V of $+V$ and

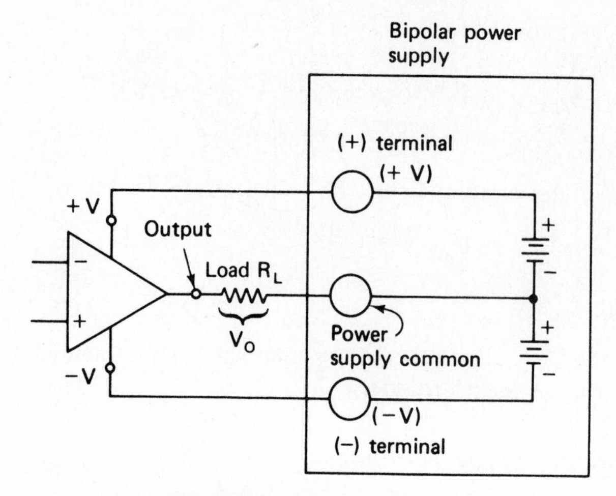

(a) Actual wiring from power supply to op amp

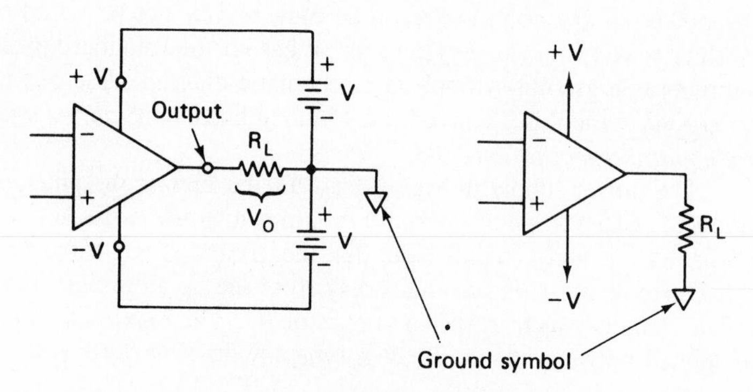

(b) Typical schematic representations of supplying power to an op amp

FIGURE 2-2 Wiring power and load to an op amp.

drop to within 2 V of $-V$. The upper limit of V_o is called the *positive saturation voltage*, $+V_{sat}$, and the lower limit is called the *negative saturation voltage*, $-V_{sat}$. For example, with a supply voltage of ± 15 V, $+V_{sat} = +14$ V and $-V_{sat} = -13$ V. Therefore, V_o is restricted to a symmetrical peak-to-peak swing of ± 13 V. Both current and voltage limits place a *minimum* value on the load resistance R_L of 2 kΩ. However, special-purpose op amps such as the CA3130 have MOS rather than bipolar output transistors. The output of a CA3130 can be brought to within millivolts of either $+V$ or $-V$.

Most op amps, like the 741, have internal circuitry that automatically limits current drawn from the output terminal. Even with a short circuit for R_L, output current is limited to about 25 mA, as noted in Appendix 1. This feature prevents destruction of the op amp in the event of a short circuit.

2-1.3 Input Terminals

In Fig. 2-3 there are two input terminals, labeled − and +. They are called *differential input terminals* because output voltage V_o depends on the *difference* in voltage between them, E_d, and the gain of the amplifier, A_{OL}. As shown in Fig. 2-3(a), the output terminal is positive with respect to ground when the (+) input is positive with

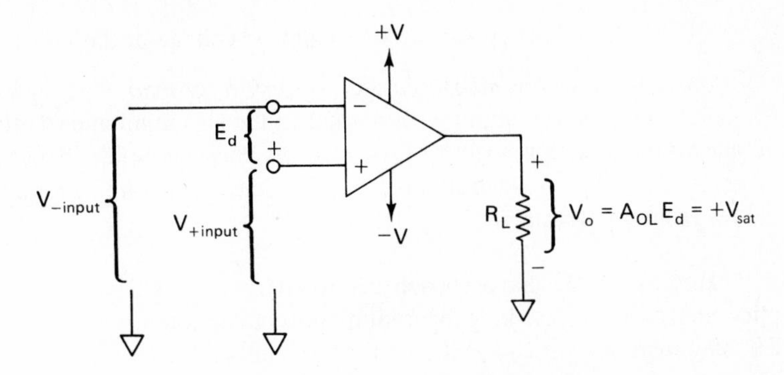

(a) V_o goes positive when the (+) input is more positive than (above) the (−) input, E_d = (+)

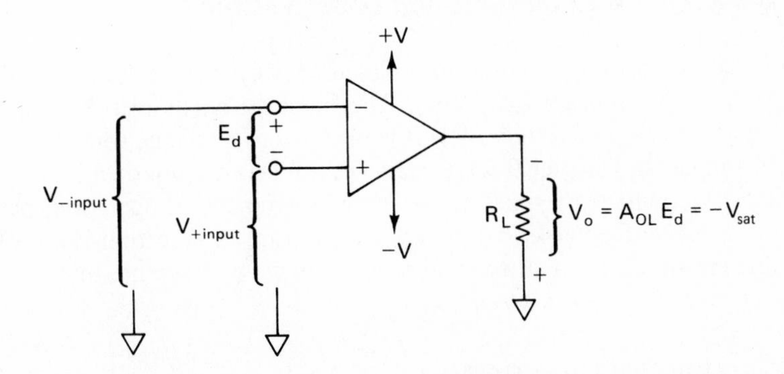

(b) V_o goes negative when the (+) input is less positive than (below) the (−) input, E_d = (−)

FIGURE 2-3 Polarity of *single-ended* output voltage V_o depends on the polarity of *differential* input voltage E_d. If the (+) input is *above* the (−) input, E_d is positive and V_o is *above* ground at $+V_{sat}$. If the (+) input is *below* the (−) input, E_d is negative and V_o is *below* ground at $-V_{sat}$.

respect to, or above, the (−) input. When E_d is reversed in Fig. 2-3(b) to make the (+) input negative with respect to, or below, the (−) input, V_o becomes negative with respect to ground.

We conclude from Fig. 2-3 that the polarity of the output terminal is the same as the polarity of (+) input terminal with respect to the (−) input terminal. Moreover, the polarity of the output terminal is opposite or inverted from the polarity of the (−) input terminal. For these reasons, the (−) input is designated the *inverting input* and the (+) input the *noninverting input* (see Appendix 1).

It is important to emphasize that the polarity of V_o depends only on the *difference* in voltage between inverting and noninverting inputs. This difference voltage can be found by

$$E_d = \text{voltage at the } (+) \text{ input} - \text{voltage at the } (-) \text{ input} \qquad (2\text{-}1)$$

Both input voltages are *measured with respect to ground*. The sign of E_d tells us (1) the polarity of the (+) input with respect to the (−) input and (2) the polarity of the output terminal with respect to ground. This equation holds if the inverting input is grounded, if the noninverting input is grounded, and even if both inputs are above or below ground potential.

Review. We have chosen the words in Fig. 2-3 very carefully. They simplify analysis of open-loop operation (no connection from output to either input). The key memory aid is—if the (+) input is *above* the (−) input, the output is *above* ground and at $+V_{sat}$. If the (+) input is *below* the (−) input, the output is *below* ground at $-V_{sat}$.

2-1.4 Input Bias Currents and Offset Voltage

The input terminals of real op amps draw tiny bias currents and signal currents to activate the internal transistors. The input terminals also have a small imbalance called *input offset voltage*, V_{io}. It is modeled as a voltage source V_{io} in series with the (+) input. In Chapter 9 we explain the effects of V_{io} in detail.

We must learn much more about op amp circuit operation, particularly involving negative feedback, before we can measure bias currents and offset voltage. For this reason, in these introductory chapters we will assume that both are negligible.

2-2 OPEN-LOOP VOLTAGE GAIN

2-2.1 Definition

Refer to Fig. 2-3. Output voltage V_o will be determined by both E_d and the *open-loop voltage gain*, A_{OL}. A_{OL} is called open-loop voltage gain because possible feedback connections from output terminal to input terminals are left open. Accordingly, V_o would be ideally expressed by the simple relationship output voltage = differential input voltage × open-loop gain (2-2)

$$V_o = E_d \times A_{OL}$$

2-2.2 Differential Input Voltage, E_d

The value of A_{OL} is extremely large, often 200,000 or more. Recall from Section 2-1.2 that V_o can never exceed positive or negative saturation voltages $+V_{sat}$ and $-V_{sat}$. For a ±15-V supply, saturation voltages would be about ±13 V. Thus, for the op amp to act as an amplifier, E_d must be limited to a maximum voltage of ±65 μV. This conclusion is reached from Eq. (2-2).

$$E_{d\,max} = \frac{+V_{sat}}{A_{OL}} = \frac{13\text{ V}}{200{,}000} = 65\ \mu\text{V}$$

$$-E_{d\,max} = \frac{-V_{sat}}{A_{OL}} = \frac{-13\text{ V}}{200{,}000} = -65\ \mu\text{V}$$

In the laboratory or shop it is difficult to measure 65 μV, because induced noise, 60-Hz hum, and leakage currents on the typical test setup can easily generate a millivolt (1000 μV). Furthermore, it is difficult and inconvenient to measure very high gains. The op amp also has tiny internal unbalances that *act* as a small voltage that may exceed E_d. As mentioned in Section 2-1.4, this small voltage is called an *offset voltage* and is discussed in Chapter 9.

2-2.3 Conclusions

There are three conclusions to be drawn from these brief comments. First, V_o in the circuit of Fig. 2-3 either will be at one of the limits $+V_{sat}$ or $-V_{sat}$ or will be oscillating between these limits. Don't be disturbed, because this behavior is what a high-gain amplifier usually does. Second, to maintain V_o between these limits we must go to a feedback type of circuit that forces V_o to depend on stable, precision elements such as resistors and capacitors rather than A_{OL} *and* E_d. Feedback circuits will be introduced in Chapter 3.

Without learning any more about the op amp, it is possible to understand basic comparator applications. In a comparator application, the op amp performs not as an amplifier but as a device that tells when an unknown voltage is below, above, or just equal to a known reference voltage. Before introducing the comparator in the next section, an example is given to illustrate ideas presented thus far.

Example 2-1

In Fig. 2-3, $+V = 15$ V, $-V = -15$ V, $+V_{sat} = +13$ V, $-V_{sat} = -13$ V, and gain $A_{OL} = 200{,}000$. Assuming ideal conditions, find the magnitude and polarity of V_o for each of the following input voltages. These input voltages are given with respect to ground.

	Voltage at (−) input	Voltage at (+) input
(a)	−10 μV	−15 μV
(b)	−10 μV	+15 μV
(c)	−10 μV	−5 μV
(d)	+1.000001 V	+1.000000 V
(e)	+5 mV	0 V
(f)	0 V	+5 mV

Solution The polarity of V_o is the same as the polarity of the (+) input with respect to the (−) input. The (+) input is more negative than the (−) input in (a), (d), and (e). This is shown by Eq. (2-1), and therefore V_o will go negative. From Eq. (2-2), the magnitude of V_o is A_{OL} times the difference, E_d, between voltages at the (+) and (−) inputs. But if $A_{OL} \times E_d$ exceeds $+V$ or $-V$, then V_o must stop at $+V_{sat}$ or $-V_{sat}$. Calculations are summarized as follows.

	E_d [using Eq. (2-1)]	Polarity of (+) input with respect to (−) input	Theoretical V_o [from Eq. (2-2)]	Actual V_o
(a)	−5 μV	−	−5 μV × 200,000 = −1.0 V	−13 V
(b)	25 μV	+	25 μV × 200,000 = 5.0 V	+13 V
(c)	5 μV	+	5 μV × 200,000 = 1.0 V	+13 V
(d)	−1 μV	−	−1 μV × 200,000 = −0.2 V	−13 V
(e)	−5 mV	−	−5 mV × 200,000 = −1000 V	−13 V
(f)	5 mV	+	5 mV × 200,000 = 1000 V	+13 V

Laboratory Exercise 2-1 shows how reality changes the theoretical predictions of V_o.

2-3 ZERO-CROSSING DETECTORS

2-3.1 Noninverting Zero-Crossing Detector

The op amp in Fig. 2.4(a) operates as a comparator. Its (+) input compares voltage E_i with a reference voltage of 0 V (V_{ref} = 0 V). When E_i is above V_{ref}, V_o equals $+V_{sat}$. This is because the voltage at the (+) input is more positive than the voltage at the (−) input. Therefore, the sign of E_d in Eq. (2-1) is positive. Consequently, V_o is positive, from Eq. (2-2).

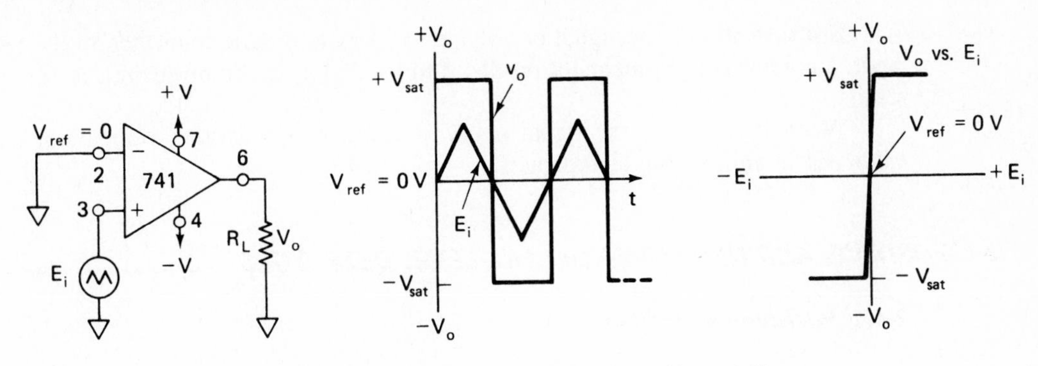

(a) Noninverting: when E_i is above V_{ref}, $V_o = +V_{sat}$

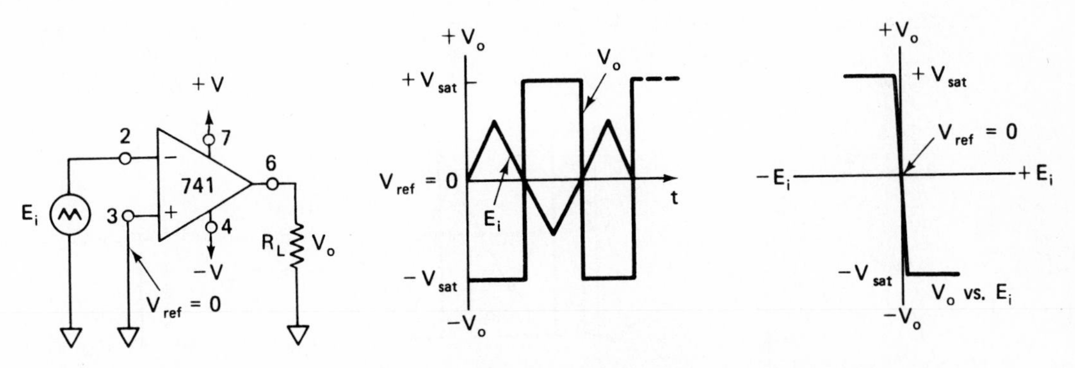

(b) Inverting: when E_i is above V_{ref}, $V_o = -V_{sat}$

FIGURE 2-4 Zero-crossing detectors, noninverting in (a) and inverting in (b). If the signal E_i is applied to the (+) input, the circuit action is noninverting.

The polarity of V_o tells if E_i is above or below V_{ref}. The *transition* of V_o tells *when* E_i crossed the reference and in what direction. For example, when V_o makes a positive-going transition from $-V_{sat}$ to $+V_{sat}$, it indicates that E_i just crossed 0 in the positive direction.

2-3.2 Inverting Zero-Crossing Detector

The op amp's (−) input in Fig. 2-4(b) compares E_i with a reference voltage of 0 V ($V_{ref} = 0$ V). This circuit is an *inverting zero-crossing detector*. The waveshapes of V_o versus time and V_o versus E_i can be explained by the following summary:

1. If E_i is above V_{ref}, V_o equals $-V_{sat}$.
2. Where E_i crosses the reference going positive, V_o makes a negative-going transition from $+V_{sat}$ to $-V_{sat}$.

Summary. If the signal or voltage to be monitored is connected to the (+) input, a noninverting comparator results; if to the (−) input, an inverting comparator results.

When $V_o = +V_{sat}$, the signal is *above* V_{ref} in a noninverting comparator and *below* V_{ref} in an inverting comparator.

2-4 POSITIVE- AND NEGATIVE-VOLTAGE-LEVEL DETECTORS

2-4.1 Positive-Level Detectors

In Fig. 2-5 a positive reference voltage V_{ref} is applied to one of the op amp's inputs. This means that the op amp is set up as a comparator to detect a positive voltage. If the voltage to be sensed, E_i, is applied to the op amp's (+) input, the result is a *non-*

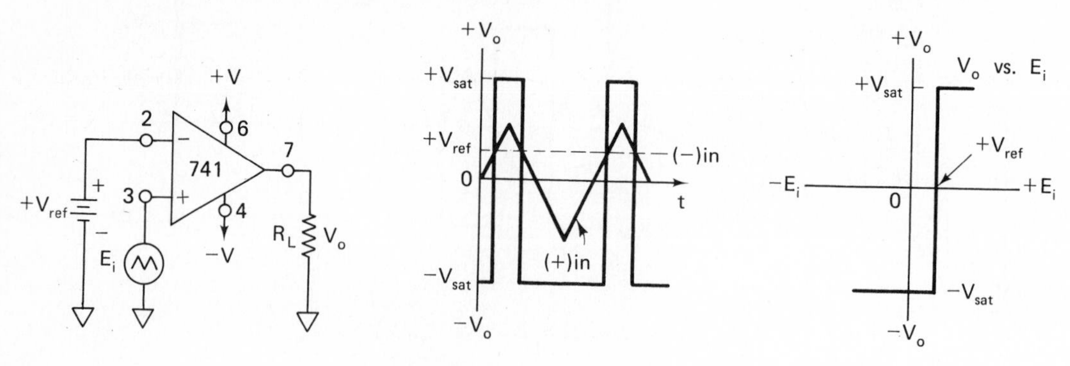

(a) Noninverting: when E_i is above V_{ref}, $V_o = +V_{sat}$

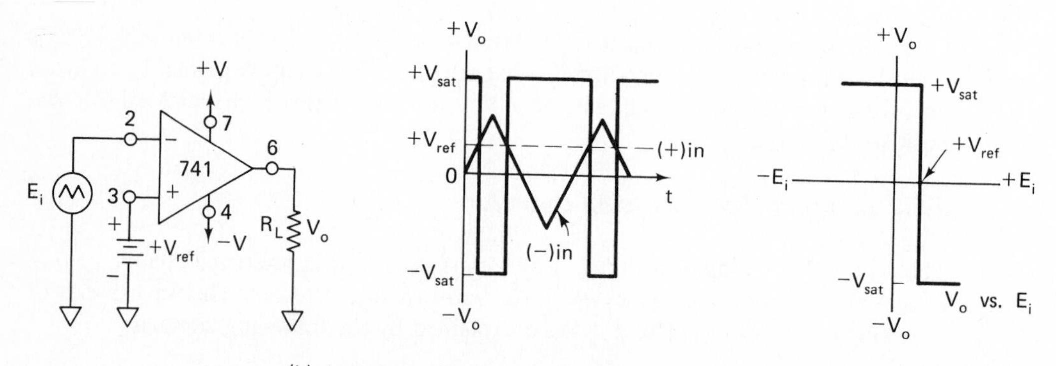

(b) Inverting: when E_i is above V_{ref}, $V_o = -V_{sat}$

FIGURE 2-5 Positive-voltage-level detector, noninverting in (a) and inverting in (b). If the signal E_i is applied to the (−) input, the circuit action is inverting.

inverting positive-level detector. Its operation is shown by the waveshapes in Fig. 2-5(a). When E_i is above V_{ref}, V_o equals $+V_{sat}$. When E_i is below V_{ref}, V_o equals $-V_{sat}$.

If E_i is applied to the inverting input as in Fig. 2-5(b), the circuit is an inverting positive-level detector. Its operation can be summarized by the statement: When E_i is above V_{ref}, V_o equals $-V_{sat}$. This circuit action can be seen more clearly by observing the plot of E_i and V_{ref} versus time in Fig. 2-5(b).

2-4.2 Negative-Level Detectors

Figure 2-6(a) is a *noninverting negative-level detector*. This circuit detects when input signal E_i crosses the negative voltage $-V_{ref}$. When E_i is above $-V_{ref}$, V_o equals $+V_{sat}$. When E_i is below $-V_{ref}$, $V_o = -V_{sat}$. The circuit of Fig. 2-6(b) is an *inverting*

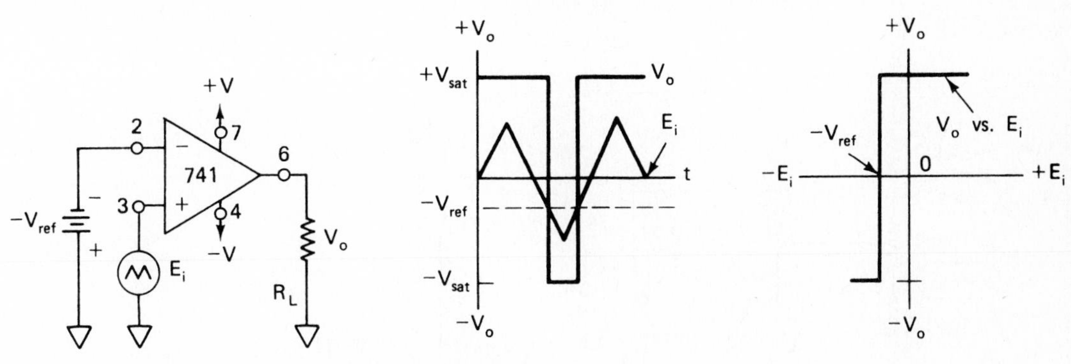

(a) Noninverting: when E_i is above V_{ref}, $V_o = +V_{sat}$

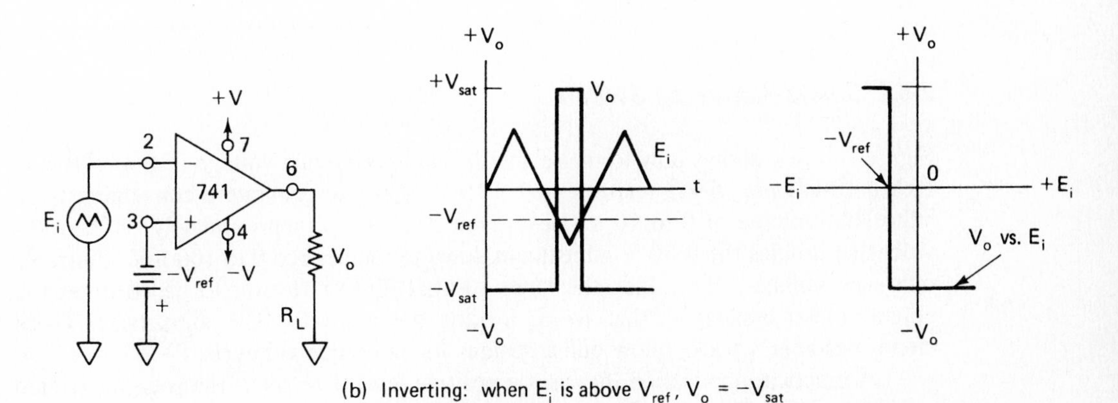

(b) Inverting: when E_i is above V_{ref}, $V_o = -V_{sat}$

FIGURE 2-6 Negative-voltage-level detector, noninverting in (a) and inverting in (b).

negative-level detector. When E_i is above $-V_{ref}$, V_o equals $-V_{sat}$, and when E_i is below $-V_{ref}$, V_o equals $+V_{sat}$.

2-5 TYPICAL APPLICATIONS OF VOLTAGE-LEVEL DETECTORS

2-5.1 Adjustable Reference Voltage

Figure 2-7 shows how to make an adjustable reference voltage. Two 10-kΩ resistors and a 10-kΩ potentiometer are connected in series to make a 1-mA voltage divider. Each kilohm of resistance corresponds to a voltage drop of 1 V. V_{ref} can be set to any value between −5 and +5 V. Remove the $-V$ connection to the bottom 10-kΩ resistor and substitute a ground. You now have a 0.5 mA divider, and V_{ref} can be adjusted from 5 to 10 V.

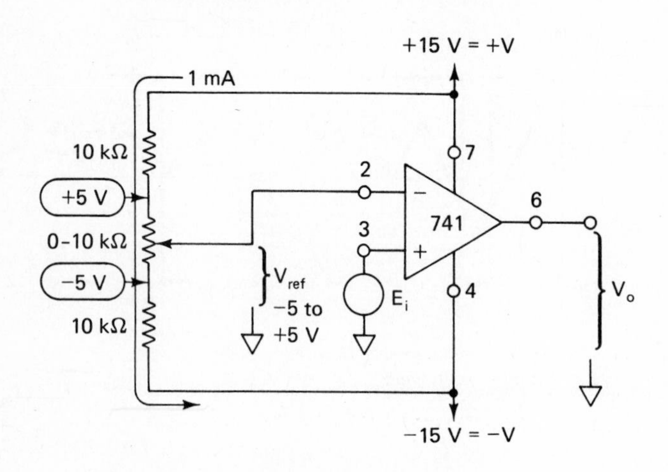

FIGURE 2-7 A variable reference voltage can be obtained by using the op amp's bipolar supply along with a voltage-divider network.

2-5.2 Sound-Activated Switch

Figure 2-8 first shows how to make an adjustable reference voltage of 0 to 100 mV. Pick a 10-kΩ pot, 5-kΩ resistor, and +15-V supply to generate a convenient large adjustable voltage of 0 to 10 V. Next connect a 100 : 1 (approximately) voltage divider that divides the 0-10 V adjustment down to the desired 0 to 100 mV adjustable reference voltage. *Note:* Pick the large 100-kΩ divider resistor to be 10 times the potentiometer resistance; this avoids loading down the 0-10 V adjustment. These circuit designer's tricks allow quick designs for reference voltages.

A practical application that uses a positive level detector is the sound-activated switch shown in Fig. 2-8. Signal source E_i is a microphone and an alarm circuit is connected to the output. The procedure to arm the sound switch is as follows:

1. Open the reset switch to turn off both SCR and alarm.
2. In a quiet environment, adjust the sensitivity control until V_o just swings to $-V_{sat}$.
3. Close the reset switch. The alarm should remain off.

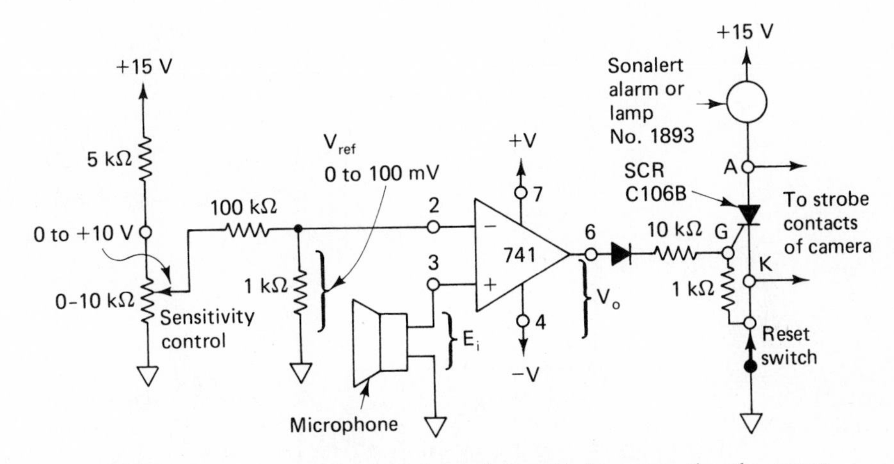

FIGURE 2-8 A sound-activated switch is made by connecting the output of a noninverting voltage-level detector to an alarm circuit.

Any noise signal will now generate an ac voltage and be picked up by the microphone as an input. The first positive swing of E_i above V_{ref} drives V_o to $+V_{sat}$. The diode now conducts a current pulse of about 1 mA into the gate (G) of the silicon-controlled rectifier (SCR). Normally, the SCR's anode, A, and cathode, K, terminals act like an open switch. However, the gate current pulse makes the SCR turn on, and now the anode and cathode terminals act like a closed switch. The audible or visual alarm is now activated. Furthermore, the alarm stays on because once an SCR has been turned on, it stays on until its anode–cathode circuit is opened.

The circuit of Fig. 2-8 can be modified to photograph high-speed events such as a bullet penetrating a glass bulb. Some cameras have mechanical switch contacts that close to activate a stroboscopic flash. To build this sound-activated flash circuit, remove the alarm and connect anode and cathode terminals to the strobe input in place of the camera switch. Turn off the room lights. Open the camera shutter and fire the rifle at the glass bulb. The rifle's sound activates the switch. The strobe does the work of apparently stopping the bullet in midair. Close the shutter. The position of the bullet in relation to the bulb in the picture can be adjusted experimentally by moving the microphone closer or farther away from the rifle.

2-5.3 Light Column Voltmeter

A light column voltmeter displays a column of light whose height is proportional to voltage. Manufacturers of audio and medical equipment may replace analog meter panels with light column voltmeters because they are easier to read at a distance.

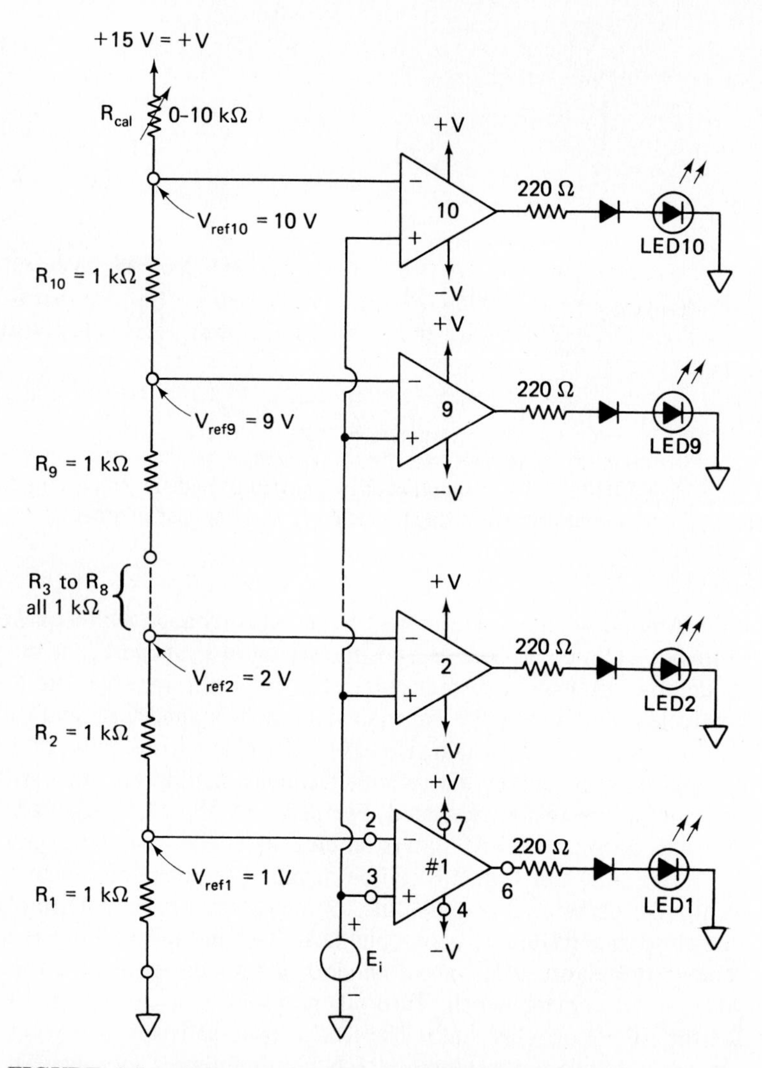

FIGURE 2-9 Light-column voltmeter. Reference voltages to each op amp are in steps of 1 V. As E_i is increased from 1 V to 10 V, LED1 through LED10 light in sequence. R_1 to R_{10} are 1% resistors. Op amps are 741 8-pin mini-DIPS.

A light column voltmeter is constructed in the circuit of Fig. 2-9. R_{cal} is adjusted so that 1 mA flows through the equal resistor divider network R_1 to R_{10}. Ten separate reference voltages are established in 1 V steps from 1 V to 10 V.

When $E_i = 0$ V or is less than 1 V, the outputs of all op amps are at $-V_{sat}$. The silicon diodes protect the light-emitting diodes against excessive reverse bias voltage. When E_i is increased to a value between 1 and 2 V, only the output of op amp 1 goes positive to light LED1. Note that the op amp's output current is automatically limited by the op amp to its short-circuit value of about 20 mA. The 220-Ω output resistors divert heat away from the op amp.

As E_i is increased, the LEDs light in numerical order. This circuit can also be built using two and one-half LM324 quad op amps.

2-5.4 Smoke Detector

Another practical application of a voltage-level detector is a smoke detector, as shown in Fig. 2-10. The lamp and photoconductive cell are mounted in an enclosed chamber that admits smoke but not external light. The photoconductor is a light-sensitive resistor. In the absence of smoke, very little light strikes the photoconductor and its resistance stays at some high value, typically several hundred kilohms. The 10-kΩ sensitivity control is adjusted until the alarm turns off.

Any smoke entering the chamber causes light to reflect off the smoke particles and strike the photoconductor. This, in turn, causes the photoconductor's resistance

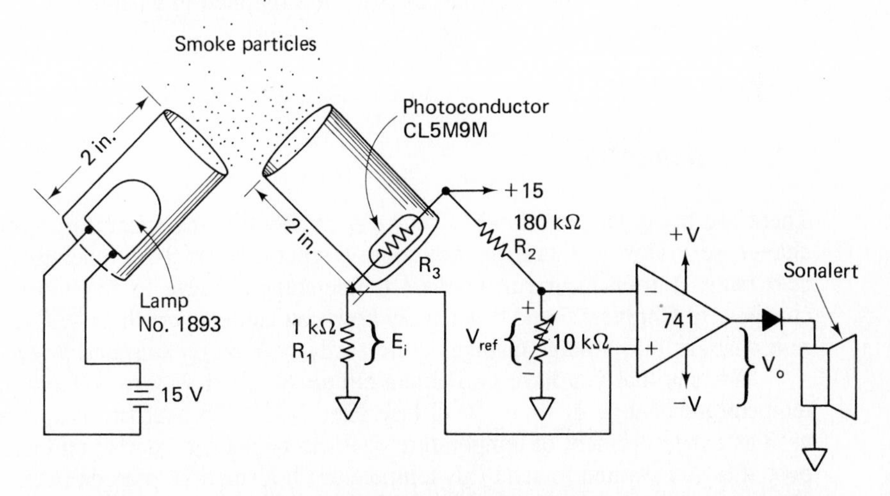

FIGURE 2-10 With no smoke present the 10-kΩ sensitivity control is adjusted until the alarm stops. Light reflected off any smoke particles causes the alarm to sound.

to decrease and the voltage across R_1 to increase. As E_i increases above V_{ref}, V_o switches from $-V_{sat}$ to $+V_{sat}$, causing the alarm to sound. The alarm circuit of Fig. 2-10 does not include an SCR. Therefore, when the smoke particles leave the chamber, the photoconductor's resistance increases and the alarm turns off. If you want the alarm to stay on, use the SCR alarm circuit shown in Fig. 2-8. Lamp and photoresistor must be mounted in a flat black, lightproof box that admits smoke. Ambient (room) light prevents proper operation.

2-6 SIGNAL PROCESSING WITH VOLTAGE-LEVEL DETECTORS

2-6.1 Introduction

Armed with only the knowledge gained thus far, we will make a sine-to-square wave converter, analog-to-digital converter, and pulse-width modulator out of the versatile op amp. These open-loop comparator (or voltage-level detector) applications are offered to show how easy it is to use op amps.

2-6.2 Sine-to-Square Wave Converter

The zero-crossing detector of Fig. 2-4 will convert the output of a sine-wave generator such as a variable-frequency audio oscillator into a variable-frequency square wave. If E_i is a sine, triangular, or wave of any other shape that is symmetrical around zero, the zero-crossing detector's output will be square. The frequency of E_i should be below 100 Hz, for reasons that are explained in Chapter 10.

2-7 COMPUTER INTERFACING WITH VOLTAGE-LEVEL DETECTORS

2-7.1 Introduction

There are many characteristics of our environment or manufacturing processes that change very slowly. Examples are room temperature or the temperature of a large acid bath. A transducer can convert temperature changes to resistance or current changes. In Chapters 5 and 8 we show how you can convert these resistance or current changes into voltage changes quite easily with an op amp and a few parts.

Assume that you have available a circuit which gives 0 to 5 V out for a room-temperature change of 0° to 50°C (see Fig. 2-11). The output, V_{temp}, can now be used as a measurement of temperature or it can be used to control temperature. Suppose that you wanted to send this temperature information to a computer so that the computer could monitor, control, or change room temperature. A voltage-level detector can accomplish this task. To understand how this can be done, we present the *pulse-width modulator* in Section 2-7.3. Before we do that, we will look at a special-

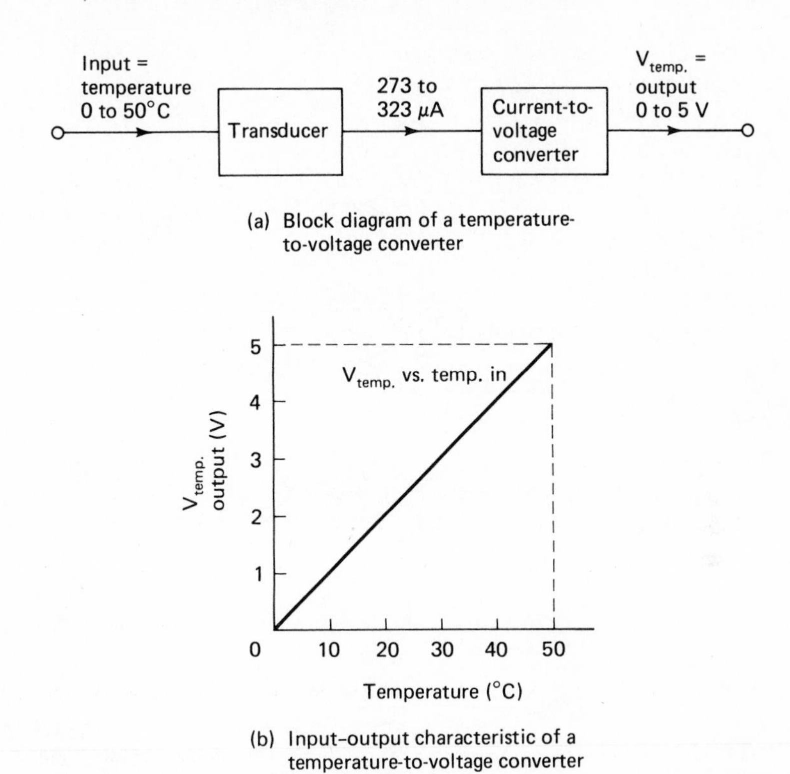

FIGURE 2-11 An example of how room or a process temperature is measured electronically.

ized op amp that makes it easy to interface a ±15-V analog system with a 0 to 5-V digital system.

2-7.2 Quad Voltage Comparator, LM339

Pinouts and operation of a specialized op amp, the LM339, are shown in Figs. 2-12 and 2-13. The LM339 houses four independent op amps that have been specially designed to be flexible voltage comparators. We examine its operation by analyzing the role played by each terminal.

Power supply terminals. Pins 3 and 12 are positive and negative supply voltage terminals, respectively, for all four comparators. Maximum supply voltage between pins 3 and 12 is ±18 *V*. In most applications, the negative supply terminal, pin 12, is grounded. Then pin 3 can be any voltage from 2 to 36 V_{dc}. The LM339 is used primarily for single-supply operation.

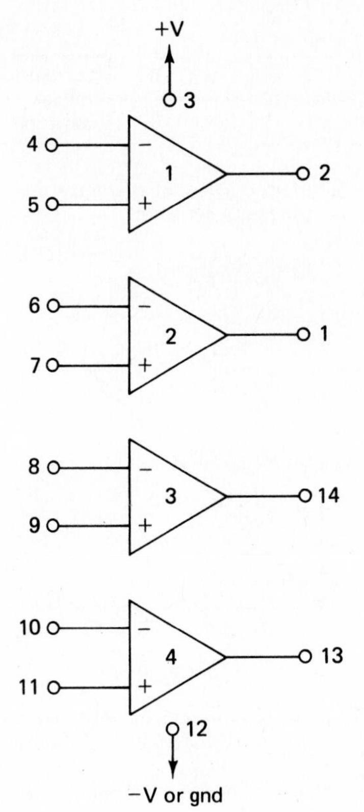

FIGURE 2-12 Connection diagram for the LM339 quad comparator. Four voltage comparators are containted in one 14-pin dual-in-line package.

Output terminals. The output terminal of each op amp is an open-collector *npn* transistor. Each transistor collector is connected to the respective output terminals 2, 1, 14, and 13. All emitters are connected together and then to pin 12. If pin 12 is grounded, the output terminal acts like a switch. A closed switch extends the ground from pin 12 to the output terminal [See Fig. 2-13(b)].

If you want the output to go high when the switch is open, you must install a pull-up resistor and an external voltage source. As shown in Fig. 2-13(a), this feature allows easy interfacing between a ±15-V analog system and a 5-V digital system. The output terminal should not sink more than 16 mA.

Input terminals. The input terminals are differential. Use Eq. (2-1) to determine the sign for E_d. If E_d is *positive,* the output switch is *open,* as in Fig. 2-13(a). If E_d is *negative,* the output switch is *closed* as in Fig. 2-13(b). Unlike many other op amps, the input terminals can be brought down to ground potential when pin 12 is grounded.

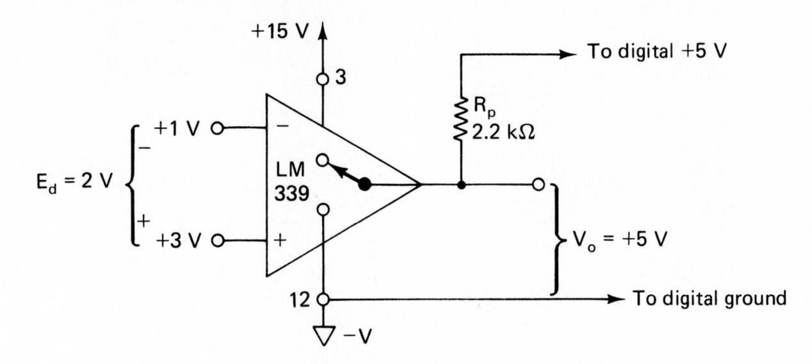

(a) If (+) input is above (−) input, the output switch is open; V_o is set by the digital system

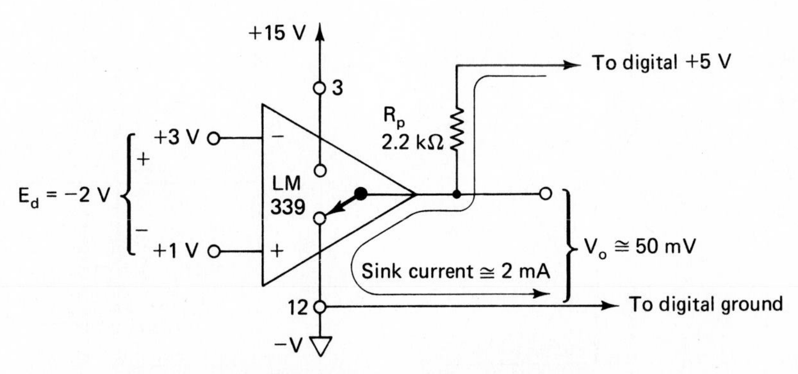

(b) If (+) input is below (−) input, the output switch is closed and sinks 2 mA of current from pull-up resistor R_p and the 5 V supply

FIGURE 2-13 Operation of an LM339 (open collector output) comparator. When E_d is positive in (a), V_o goes high. V_o is determined by an external positive supply, pull-up resistor R_p and any external load resistor. If E_d is negative as in (b), the output goes low to essentially ground potential.

Summary. If the (+) input of an LM339 is *above* its (−) input, the output is pulled *high* by the pull-up resistor. If the (+) input is *below* the (−) input, the output is pulled *down* to the ground potential at pin 12. We now have information to analyze the pulse-width modulator.

2-7.3 Pulse-Width Modulator, Noninverting

The LM339 comparator in Fig. 2-14(a) compares two input voltages, E_c and V_{temp}. [Figure 2-14(b) is similar to Fig. 2-5(b).] A sawtooth wave, E_c, with constant frequency is connected to the (−) input. It is called a *carrier wave*. V_{temp} is a temperature-controlled voltage. Its rate of change must be much less than that of E_c.

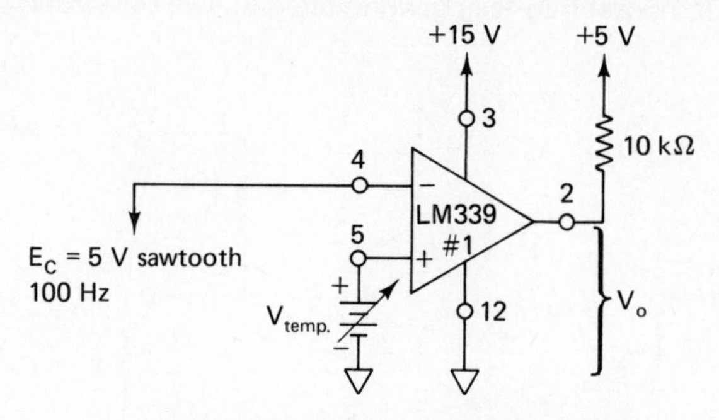

(a) Noninverting pulse-width-modulator circuit

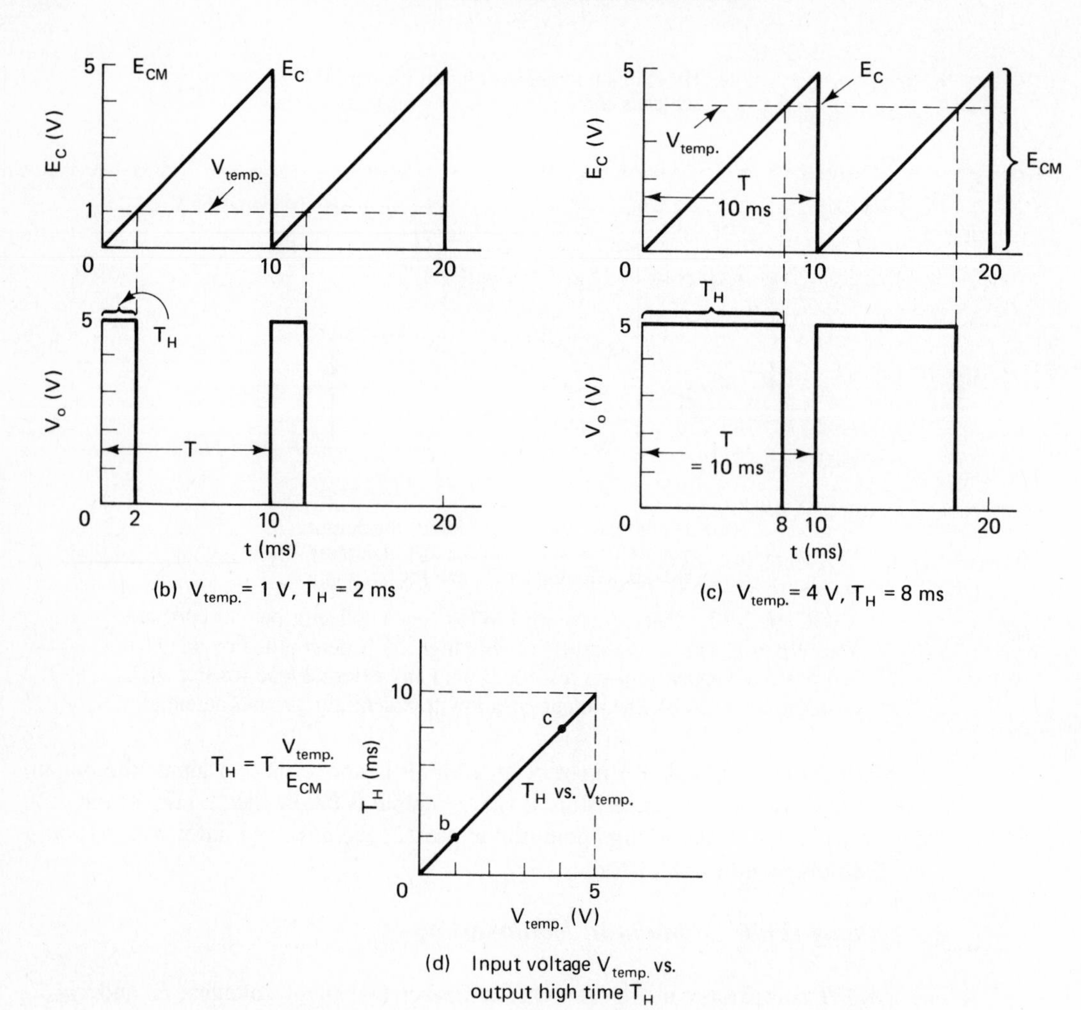

(b) $V_{temp.}$ = 1 V, T_H = 2 ms

(c) $V_{temp.}$ = 4 V, T_H = 8 ms

(d) Input voltage $V_{temp.}$ vs. output high time T_H

FIGURE 2-14 V_{temp} is defined as the input signal in (a). As V_{temp} *increases* from 0 to 5 V, the high time of output voltage V_o *increases* from 0 to 10 ms. The circuit is called a noninverting, *pulse-width modulator*.

In this circuit *the input signal is defined* as V_{temp}. *The output is defined as the high time,* T_H of V_o. In Fig. 2-14(b), the output stays high for 2 ms when $V_{temp} = 1$ V. If V_{temp} increases to 4 V, high time T_H increases to 8 ms as in Fig. 2-14(c).

Operation of the circuit is summarized by the input–output characteristics in Fig. 2-14(d). The width of output pulse T_H is changed (modulated) by V_{temp}. The constant period of the output wave is set by E_c. *Thus* E_c *carries the information contained in* V_{temp}. V_o is then said to be a pulse-width-modulated wave. The input–output equation is

$$\text{output } T_H = (V_{temp})\frac{T}{E_{CM}} \tag{2-3}$$

where T = period of sawtooth carrier wave
E_{CM} = maximum peak voltage of a sawtooth carrier

Example 2-2

A 10-V 50-Hz sawtooth wave is pulse-width-modulated by a 4-V signal. Find the output's (a) high time; (b) duty cycle.

Solution Period T is found from the reciprocal of the frequency:

$$T = \frac{1}{f} = \frac{1}{50 \text{ Hz}} = 20 \text{ ms}$$

(a) From Eq. (2-3),

$$T_H = (4 \text{ V})\frac{20 \text{ ms}}{10 \text{ V}} = 8 \text{ ms}$$

(b) Duty cycle is defined as the ratio of high time to the period and is expressed in percent:

$$\begin{aligned} \text{duty cycle} &= \frac{T_H}{T} \times 100 \\ &= \frac{8 \text{ ms}}{20 \text{ ms}} \times 100 \end{aligned} \tag{2-4}$$

Thus the output stays high for 40% of each signal.

Example 2-2 shows that the pulse-width modulator can also be called a *duty-cycle controller*.

2-7.4 Inverting and Noninverting Pulse-Width Modulators

Figure 2-15 shows the difference between noninverting and inverting pulse-width modulators. If signal V_{temp} is applied to the (+) input, the circuit is defined as noninverting [see Fig. 2-15(a), (b), and (c)]. The *slope* of T_H versus V_{temp} rises to the right and is *positive* or *noninverting*.

V_{temp} is applied to the (−) input in Fig. 2-15(d). As V_{temp} increases, T_H decreases. The *slope* of T_H versus V_{temp} is shown in Fig. 2-14(f) and is *negative*. The inverting performance equation is

$$T_H = T\left(1 - \frac{V_{temp}}{E_{CM}}\right) \tag{2-5}$$

Example 2-3

Calculate the output high time if $V_{temp} = 4$ V Fig. 2-15(d).

Solution From Eq. (2-5),

$$T_H = 10 \text{ ms}\left(1 - \frac{4 \text{ V}}{5 \text{ V}}\right) = 2 \text{ ms}$$

2-8 ANALOG-TO-DIGITAL CONVERSION WITH A MICROCOMPUTER AND A PULSE-WIDTH MODULATOR

The pulse-width modulator can interface an analog signal with an input port of a microprocessor or microcomputer (see Fig. 2-16). An analog temperature is converted first to a voltage. A noninverting pulse-width modulator converts this analog input voltage to an output that is digital in nature. That is, its output is either high or low. High time is directly proportional to temperature.

The computer programmer actually performs the analog-to-digital conversion of high time to a digital code. This can be done by using a 1-ms timing loop and counting the number of times that the timing loop is executed.

LABORATORY EXERCISES

2-1. Every circuit in this chapter can be breadboarded, tested, and used as a laboratory experiment. However, it is revealing to first perform the experiment shown in Fig. LE2-1. Clearly, differential input voltage, E_d, equals zero. Mathematically, Eq. (2-2) predicts that

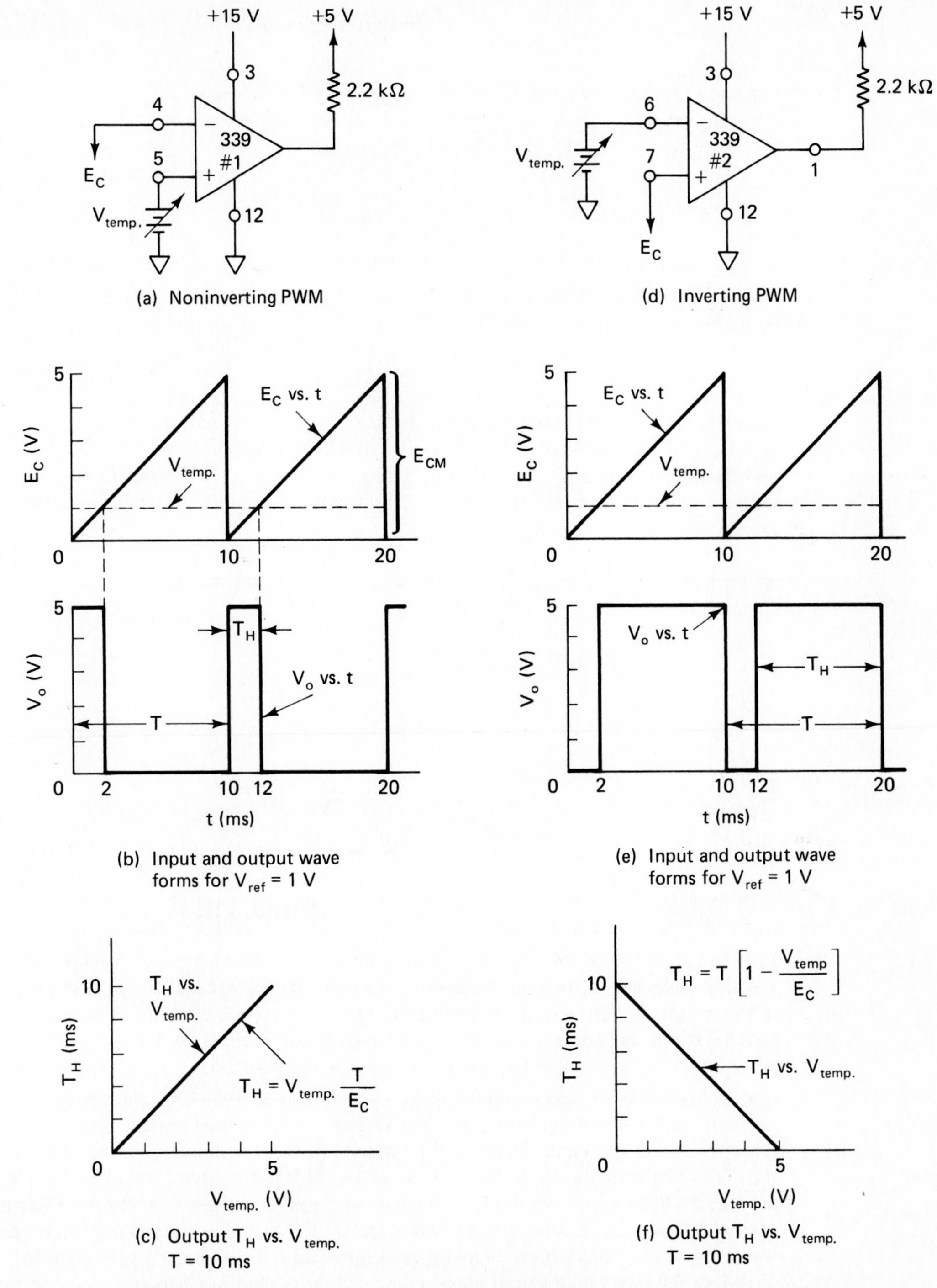

FIGURE 2-15 Output high time *increases* as input V_{temp} *increases* in a noninverting pulse-width modulator [see (a), (b), and (c)]. Output high time *decreases* as V_{temp} *increases* in an *inverting* pulse-width modulator.

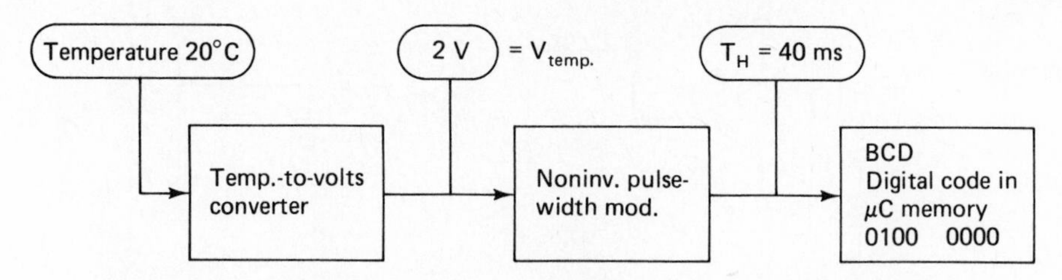

FIGURE 2-16 Block diagram of a computerized temperature measurement.

$$V_o = A_{OL} E_d = (200{,}000)(0 \text{ V}) = 0 \text{ V}$$

However, if twelve 741 op amps are wired as shown, roughly half the outputs will lock at $+V_{sat}$ and the others at $-V_{sat}$. This gives you a good opportunity to measure $\pm V_{sat}$ and the supply voltages $\pm V$. Your data will show you that a 741's output can rise to within about 0.8 V of $+V$. However, it can drop to only about 2 V above $-V$. This is typical of most general-purpose op amps.

This apparent violation of the laws of mathematics is quite normal. It is caused by an op amp characteristic called "input offset voltage" and will be faced squarely in Chapter 9.

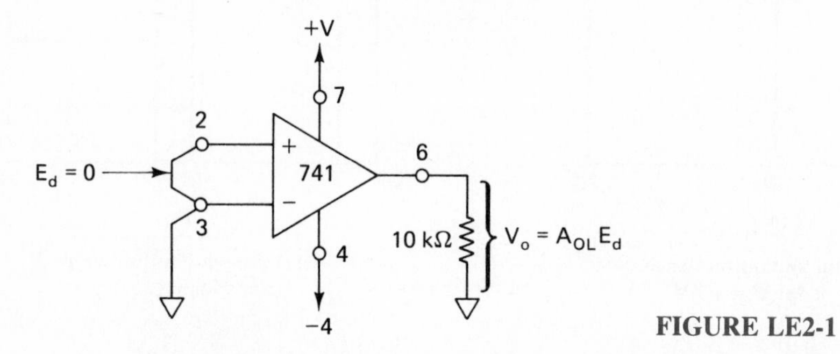

FIGURE LE2-1

2-2. *Lab measurement of time-varying waveshapes*. Use a dual-trace oscilloscope and a function generator to recreate the plots of input and output voltage versus time for any of the circuits in this chapter. The input voltage is best connected to the A or number 1 channel of the cathode-ray oscilloscope. V_o goes to the B channel. *Always use dc coupling;* otherwise, you won't know if the output is in saturation.

2-3. *Lab measurement of input–output voltage characteristics*. Your CRO must have an *x-y* position on the time-base knob, in order to plot output voltage versus input voltage. Always connect the input (independent variable) to the *x* or horizontal axis. Connect the dependent variable, V_o, to the vertical *y*-axis. This is consistent, sound engineering practice. When you see a published characteristic you know how to connect a CRO to verify it. Again, use dc coupling for both *x* and *y* CRO amplifiers or you may view performance of the CRO's input coupling capacitors and not the circuit's performance.

More advanced comparator applications will be studied in Chapter 4, but first we will learn in Chapter 3 how easy it is to make precise and versatile voltage amplifiers with one op amp and a few resistors.

PROBLEMS

2-1. Name the five basic terminals of an op amp.

2-2. Name the manufacturer of an AD741 op amp.

2-3. A 741 op amp is manufactured in an 8-pin dual-in-line package. What are the terminal numbers for the **(a)** inverting input; **(b)** noninverting input; **(c)** output?

2-4. A 741 op amp is connected to a ±15-V supply. What are the output terminal's operating limits under normal conditions with respect to **(a)** output voltage; **(b)** output current?

2-5. When the load resistor of an op amp is short-circuited, what is the op amp's **(a)** output voltage; **(b)** approximate output current?

2-6. Both op amps of Fig. P2-6 are in 14-pin dual-in-line packages. **(a)** Number the terminals. **(b)** Calculate E_d. **(c)** Find V_o.

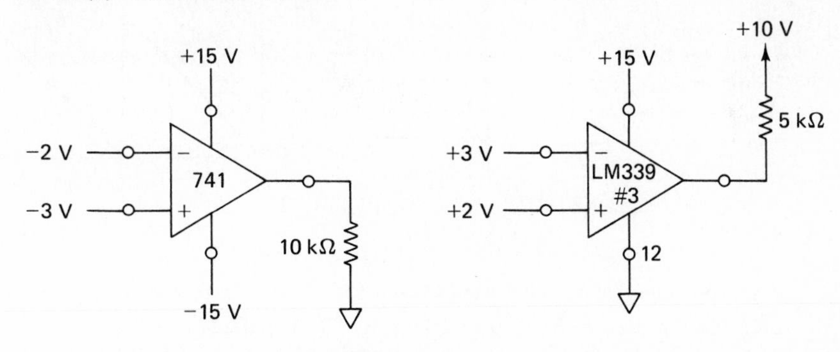

FIGURE P2-6

2-7. E_i is applied to the (−) input and ground to the (+) input of a 741 in Fig. P2-7. Sketch accurately **(a)** V_o vs. t and **(b)** V_o vs. E_i.

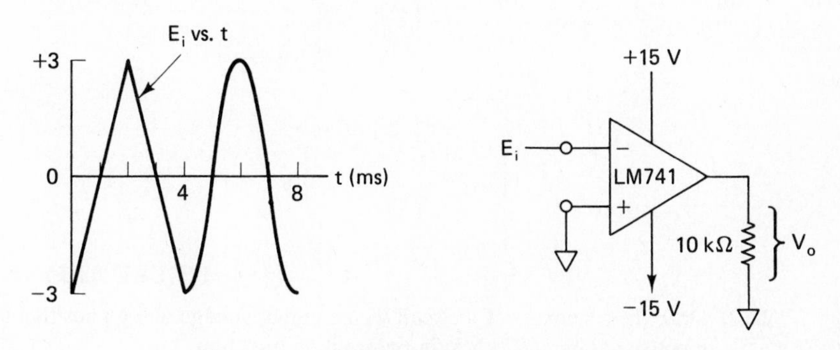

FIGURE P2-7

2-8. Swap the input connections to E_i and ground in Fig. P2-7. Sketch **(a)** V_o vs. t and **(b)** V_o vs. E_i.

2-9. Which circuit in Problems 2-7 and 2-8 is the noninverting and inverting zero-crossing detector?

2-10. To which input would you connect a reference voltage to make an inverting level detector?

2-11. You need a 741 noninverting voltage-level detector. **(a)** Will the output be at $+V_{sat}$ or $-V_{sat}$ when the signal voltage is above the reference voltage? **(b)** To which input do you connect the signal?

2-12. Design a reference voltage that can be varied from 0 to −5 V. Assume that the negative supply voltage is −15 V.

2-13. Design a 0 to +50 mV adjustable reference voltage. Derive it from the +15-V supply.

2-14. The frequency of carrier wave E_c is constant at 50 Hz in Fig. P2-14. If $V_{temp} = 5$ V, **(a)** calculate high time T_H; **(b)** plot V_o vs. time.

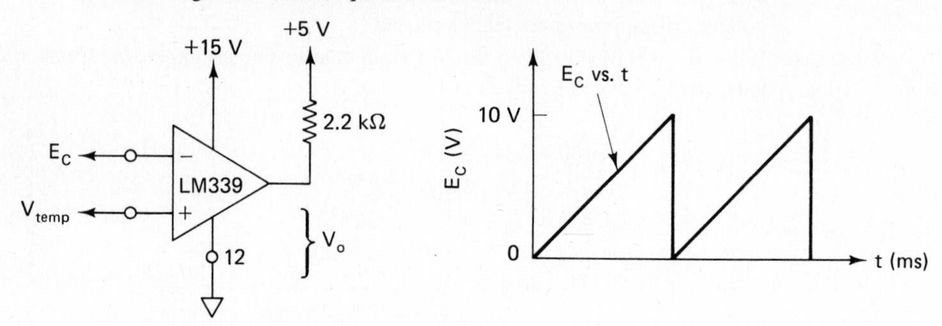

FIGURE P2-14

2-15. Assume that V_{temp} is varied from 0 V to +10 V in Problem 2-14. Plot T_H vs. V_{ref}.

2-16. In Fig. P2-16, E_{in} is a triangle wave. The amplitude is −5 to +5 V and the frequency is 100 Hz. Sketch accurately the graphs of **(a)** V_o vs. E_{in}; **(b)** V_o vs. t.

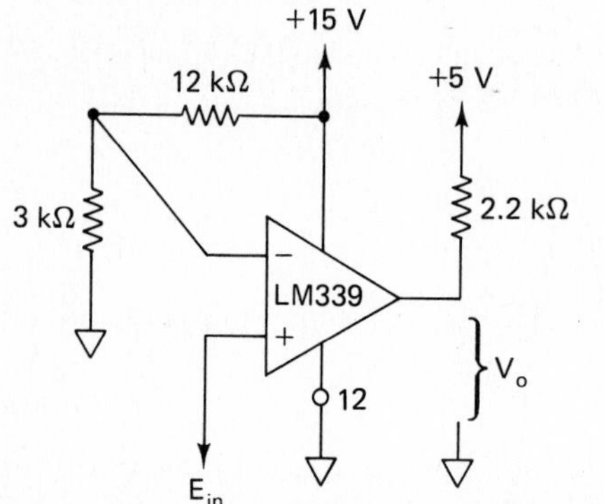

FIGURE P2-16

2-17. Draw the schematic of a circuit whose output voltage will go positive to $+V_{sat}$ when the input signal crosses +5 V in the positive direction.

2-18. Is the solution of Problem 2-17 classified as an inverting or noninverting comparator?

2-19. Draw a circuit whose output goes to $+V_{sat}$ when the input signal is below −4 V. The output should be at $-V_{sat}$ when the input is above −4 V.

2-20. Does the solution circuit for Problem 2-19 represent an **(a)** inverting or noninverting, **(b)** positive- or negative-voltage-level detector?

CHAPTER 3

Inverting and Noninverting Amplifiers

LEARNING OBJECTIVES

Upon completing this chapter on inverting and noninverting amplifiers, you will be able to:

- Draw the circuit for an inverting amplifier and calculate all voltages and currents for a given input signal.
- Draw the circuit for a noninverting amplifier and calculate all voltages and currents.
- Plot the output voltage waveshape and output–input characteristics of either an inverting or a noninverting amplifier for any input voltage waveshape.
- Design an amplifier to meet a gain and input resistance specification.

- Build an inverting or noninverting adder and audio mixer.
- Use a voltage follower to make an ideal voltage source.
- Add a dc offset voltage to an ac signal voltage.
- Measure the average value of several signals.
- Build a subtractor.
- Describe how a servoamplifier gives a delayed response to a signal and how to calculate that delay.

3-0 INTRODUCTION

This chapter uses the op amp in one of its most important applications—making an amplifier. An *amplifier* is a circuit that receives a signal at its input and delivers an undistorted larger version of the signal at its output. All circuits in this chapter have one feature in common: An external feedback resistor is connected between the output terminal and (−) input terminal. This type of circuit is called a *negative feedback circuit*.

There are many advantages obtained with negative feedback, all based on the fact that circuit performance no longer depends on the open-loop gain of the op amp, A_{OL}. By adding the feedback resistor, we form a loop from output to (−) input. The resulting circuit now has a *closed-loop gain* or *amplifier gain,* A_{CL}, which is independent of A_{OL} (provided that A_{OL} is much larger than A_{CL}).

As will be shown, closed-loop gain, A_{CL}, depends only on external resistors. For best results 1% resistors should be used, and A_{CL} will be known within 1%. Note that adding external resistors does not change open-loop gain A_{OL}. A_{OL} still varies from op amp to op amp. So adding negative feedback will allow us to ignore changes in A_{OL} as long as A_{OL} is large. We begin with the inverting amplifier to show that A_{CL} depends simply on the ratio of two resistors.

3-1 THE INVERTING AMPLIFIER

3-1.1 Introduction

The circuit of Fig. 3-1 is one of the most widely used op amp circuits. It is an amplifier whose closed-loop gain from E_i to V_o is set by R_f and R_i. It can amplify ac or dc signals. To understand how this circuit operates, we make two realistic simplifying assumptions that were introduced in Chapter 2.

1. The voltage E_d between (+) and (−) inputs is essentially 0, if V_o is not in saturation.
2. The current drawn by either the (+) or the (−) input terminal is negligible.

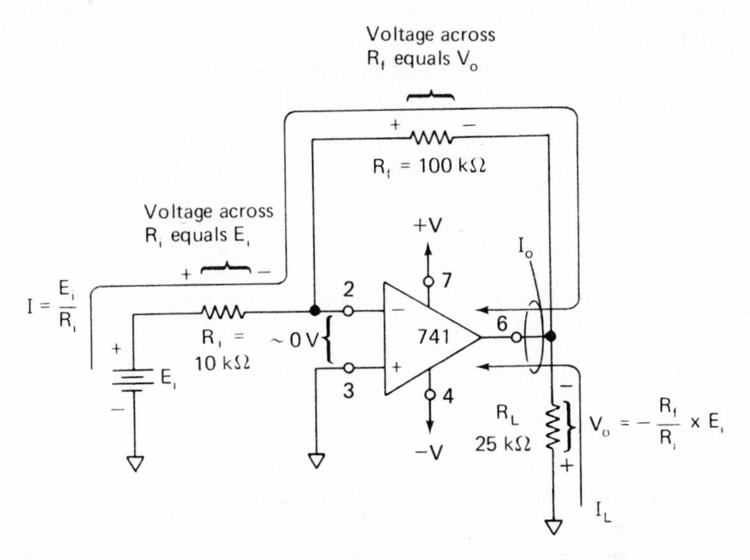

FIGURE 3-1 A positive input voltage is applied to the (−) input of an inverting amplifier. R_i converts this voltage to a current, I; R_f converts I back into an amplified version of E_i.

3-1.2 Positive Voltage Applied to the Inverting Input

In Fig. 3-1, positive voltage E_i is applied through input resistor R_i to the op amp's (−) input. Negative feedback is provided by feedback resistor R_f. The voltage between (+) and (−) inputs is essentially equal to 0 V. Therefore, the (−) input terminal is also at 0 V, so ground potential is at the (−) input. For this reason, the (−) input is said to be at *virtual* ground.

Since one side of R_i is at E_i and the other is at 0 V, the voltage drop across R_i is E_i. The current I through R_i is found from Ohm's Law:

$$I = \frac{E_i}{R_i} \tag{3-1a}$$

R_i includes the resistance of the signal generator. This point is discussed further in Section 3-5.2.

All of the input current I flows through R_f, since a negligible amount is drawn by the (−) input terminal. Note that the current through R_f is set by R_i and E_i; *not* by R_f, V_o, or the op amp.

The voltage drop across R_f is simply $I(R_f)$, or

$$V_{Rf} = I \times R_f = \frac{E_i}{R_i} R_f \tag{3-1b}$$

As shown in Fig. 3-1, one side of R_f and one side of load R_L are connected. The voltage from this connection to ground is V_o. The other sides of R_f and of R_L are at ground potential. Therefore, V_o equals V_{Rf} (the voltage across R_f). To obtain the polarity of V_o, note that the left side of R_f is at ground potential. The current direction established by E_i forces the right side of R_f to go negative. Therefore, V_o is negative when E_i is positive. Equating V_o with V_{Rf} and adding a minus sign to signify that V_o goes negative when E_i goes positive, we have

$$V_o = -E_i \frac{R_f}{R_i} \tag{3-2a}$$

Now, introducing the definition that the closed-loop gain of the amplifier is A_{CL}, we rewrite Eq. (3-2a) as

$$A_{CL} = \frac{V_o}{E_i} = \frac{-R_f}{R_i} \tag{3-2b}$$

The minus sign in Eq. (3-2b) shows that the polarity of the output V_o is inverted with respect to E_i. For this reason, the circuit of Fig. 3-1 is called an *inverting amplifier*.

3-1.3 Load and Output Currents

The load current I_L that flows through R_L is determined only by R_L and V_o and is furnished from the op amp's output terminal. Thus $I_L = V_O/R_L$. The current I through R_f must also be furnished by the output terminal. Therefore, the op amp output current I_o is

$$I_o = I + I_L \tag{3-3}$$

The maximum value of I_o is set by the op amp; it is usually between 5 and 10 mA.

Example 3-1

For Fig. 3-1, let $R_f = 100\text{ k}\Omega$, $R_i = 10\text{ k}\Omega$, and $E_i = 1\text{ V}$. Calculate (a) I; (b) V_o; (c) A_{CL}.

Solution (a) From Eq. (3-1a),

$$I = \frac{E_i}{R_i} = \frac{1\text{ V}}{10\text{ k}\Omega} = 0.1\text{ mA}$$

(b) From Eq. (3-2a),

$$V_o = -\frac{R_f}{R_i} \times E_i = -\frac{100\text{ k}\Omega}{10\text{ k}\Omega}(1\text{ V}) = -10\text{ V}$$

(c) Using Eq. (3-2b), we obtain

$$A_{CL} = -\frac{R_f}{R_i} = -\frac{100\text{ k}\Omega}{10\text{ k}\Omega} = -10$$

This answer may be checked by taking the ratio of V_o to E_i:

$$A_{CL} = \frac{V_o}{E_i} = \frac{-10\text{ V}}{1\text{ V}} = -10$$

Example 3-2

Using the values given in Example 3-1 and $R_L = 25\text{ k}\Omega$, determine (a) I_L; (b) the total current into the output pin of the op amp.

Solution (a) Using the value of V_o calculated in Example 3-1, we obtain

$$I_L = \frac{V_o}{R_L} = \frac{10\text{ V}}{25\text{ k}\Omega} = 0.4\text{ mA}$$

The direction of current is shown in Fig. 3-1.
(b) Using Eq. (3-3) and the value of I from Example 3-1, we obtain

$$I_o = I + I_L = 0.1\text{ mA} + 0.4\text{ mA} = 0.5\text{ mA}$$

The input resistance seen by E_i is R_i. In order to keep input resistance of the *circuit* high, R_i should be equal to or greater than 10 kΩ.

3-1.4 Negative Voltage Applied to the Inverting Input

Figure 3-2 shows a negative voltage, E_i, applied via R_i to the inverting input. All the principles and equations of Sections 3-1.1 to 3-1.3 still apply. The only difference between Figs. 3-1 and 3-2 is the direction of the currents. Reversing the polarity of the input voltage, E_i, reverses the direction of all currents and the voltage polarities. Now the output of the amplifier will go positive when E_i goes negative.

Example 3-3

For Fig. 3-2, let $R_f = 250\text{ k}\Omega$, $R_i = 10\text{ k}\Omega$, and $E_i = -0.5$ V. Calculate (a) I; (b) the voltage across R_f; (c) V_o.

Solution (a) From Eq. (3-1a),

$$I = \frac{E_i}{R_i} = \frac{0.5\text{ V}}{10\text{ k}\Omega} = 50\ \mu\text{A} = 0.05\text{ mA}$$

(b) From Eq. (3-1b),

$$V_{R_f} = I \times R_f$$
$$= (50\ \mu\text{A})(250\ \text{k}\Omega) = 12.5\ \text{V}$$

(c) From Eq. (3-2a),

$$V_o = -\frac{R_f}{R_i} \times E_i = -\frac{250\ \text{k}\Omega}{10\ \text{k}\Omega}(-0.5\ \text{V}) = +12.5\ \text{V}$$

Thus the magnitude of the output voltage does equal the voltage across R_f, and $A_{CL} = -25$.

Example 3-4

Using the values in Example 3-3, determine (a) R_L for a load current of 2 mA; (b) I_o; (c) the circuit's input resistance.

Solution (a) Using Ohm's Law and V_o from Example 3-3,

$$R_L = \frac{V_o}{I_L} = \frac{12.5\ \text{V}}{2\ \text{mA}} = 6.25\ \text{k}\Omega$$

(b) From Eq. (3-3) and Example 3-3,

$$I_o = I + I_L = 0.05\ \text{mA} + 2\ \text{mA} = 2.05\ \text{mA}$$

(c) The circuit's input resistance, or the resistance seen by E_i, is $R_i = 10\ \text{k}\Omega$.

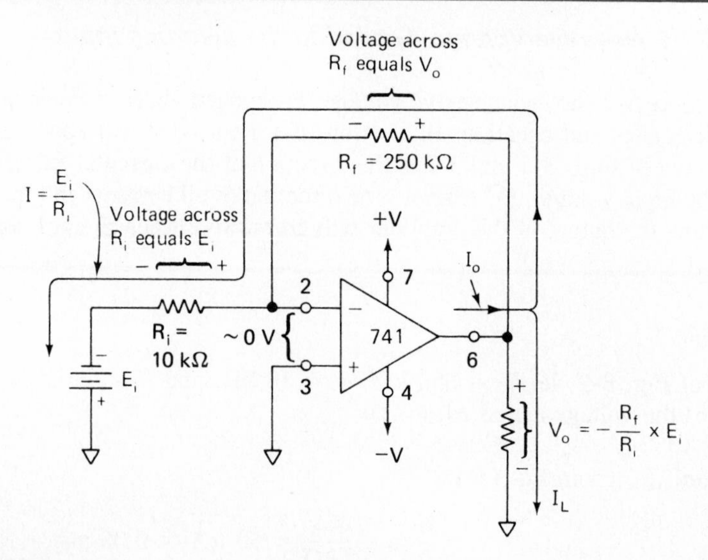

FIGURE 3-2 Negative voltage applied to the (−) input of an inverting amplifier.

3-1.5 AC Voltage Applied to the Inverting Input

Figure 3-3(a) shows an ac signal voltage E_i applied via R_i to the inverting input. For the positive half-cycle, the voltage polarities and the direction of currents are the same as in Fig. 3-1. For the negative half-cycle voltage, the polarities and direction of currents are the same as in Fig. 3-2. The output waveform is the negative (or 180° out of phase) of the input wave as shown in Fig. 3-3(b). That is, when E_i is positive,

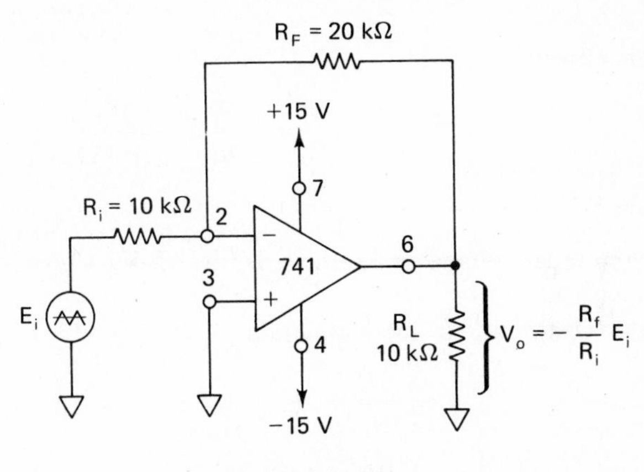

(a) Ac input voltage E_i is amplified by −2

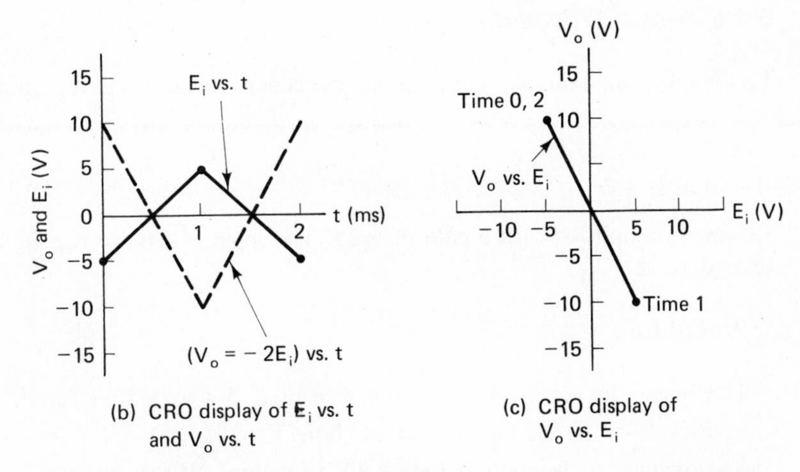

(b) CRO display of E_i vs. t and V_o vs. t

(c) CRO display of V_o vs. E_i

FIGURE 3-3 The inverting amplifier circuit in (a) has an ac input signal and a gain of −2. Time plots are shown in (b) and the output–input characteristic in (c). Note that the slope of V_o versus E_i in (c) is the closed loop gain A_{CL} (rise/run $= V_o/E_i$).

V_o is negative; and vice versa. The equations developed in Section 3-1.2 are applicable to Fig. 3-3 for ac voltages.

Example 3-5

For the circuit of Fig. 3-3, $R_f = 20\ \text{k}\Omega$, and $R_i = 10\ \text{k}\Omega$, calculate the voltage gain A_{CL}.

Solution From Eq. (3-2b),

$$A_{CL} = -\frac{R_f}{R_i} = \frac{-20\ \text{k}\Omega}{10\ \text{k}\Omega} = -2$$

Example 3-6

If the input voltage in Example 3-5 is -5 V, determine the output voltage.

Solution Using Eq. (3-2b), we obtain

$$V_o = \frac{-R_f}{R_i} \times E_i = A_{CL} E_i = (-2)(-5\ \text{V}) = 10\ \text{V}$$

See time 0 in Fig. 3-3(b) and (c). The frequency of the output and input signals is the same.

3-1.6 Design Procedure

Following is an example of the design procedure for an inverting amplifier.

Design Example 3-7

Design an amplifier with a gain of -25. The input resistance R_{in} should equal or exceed 10 kΩ.

Design Procedure

1. Choose the circuit type illustrated in Figs. 3-1 to 3-3.
2. Pick $R_i = 10\ \text{k}\Omega$ (safe, prudent choice).
3. Calculate R_f from $R_f = (\text{gain})(R_i)$. (For this calculation, use the magnitude of gain.)

3-1.7 Analysis Procedure

You are interviewing for a job in the electronics field. The technical interviewer asks you to analyze the circuit. Assume that you recognize the circuit as that of an inverting amplifier. Then

1. Look at R_i. State that the input resistance of the circuit equals the resistance of R_i.
2. Divide the value of R_f by the value of R_i. State that the magnitude of gain equals R_f/R_i. Also, the output voltage will be negative when the input voltage is positive.

3-2 INVERTING ADDER AND AUDIO MIXER

3-2.1 Inverting Adder

In the circuit of Fig. 3-4, V_o equals the sum of the input voltages with polarity reversed. Expressed mathematically,

$$V_o = -(E_1 + E_2 + E_3) \tag{3-4}$$

Circuit operation is explained by noting that the summing point S and the (−) input are at ground potential. Current I_1 is set by E_1 and R, I_2 by E_2 and R, and I_3 by E_3 and R. Expressed mathematically,

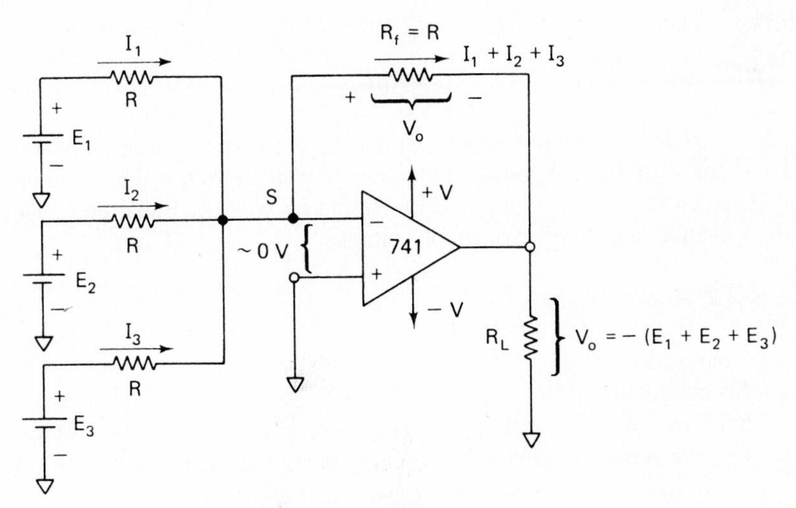

FIGURE 3-4 Inverting adder, $R = 10\text{k}\Omega$.

$$I_1 = \frac{E_1}{R}, \qquad I_2 = \frac{E_2}{R}, \qquad I_3 = \frac{E_3}{R} \tag{3-5}$$

Since the (−) input draws negligible current, I_1, I_2, and I_3 all flow through R_f. That is, the sum of the input currents flows through R_f and sets up a voltage drop across R_f equal to V_o, or

$$V_o = -(I_1 + I_2 + I_3)R_f$$

Substituting for the currents from Eq. (3-5) and substituting R for R_f, we obtain Eq. (3-4):

$$V_o = -\left(\frac{E_1}{R} + \frac{E_2}{R} + \frac{E_3}{R}\right)R = -(E_1 + E_2 + E_3)$$

Example 3-8

In Fig. 3-4, $E_1 = 2$ V, $E_2 = 3$ V, $E_3 = 1$ V, and all resistors are 10 kΩ. Evaluate V_o.

Solution From Eq. (3-4), $V_o = -(2\text{ V} + 3\text{ V} + 1\text{ V}) = -6$ V.

Example 3-9

If the polarity of E_3 is reversed in Fig. 3-4 but the values are the same as in Example 3-8, find V_o.

Solution From Eq. (3-4), $V_o = -(2\text{ V} + 3\text{ V} - 1\text{ V}) = -4$ V.

If only two input signals E_1 and E_2 are needed, simply replace E_3 with a short circuit to ground. If four signals must be added, simply add another equal resistor R between the fourth signal and the summing point S. Equation (3-4) can be changed to include any number of input voltages.

3-2.2 Audio Mixer

In the adder of Fig. 3-4, all the input currents flow through feedback resistor R_f. This means that I_1 does not affect I_2 or I_3. More generally, the input currents do *not* affect one another because each sees ground potential at the summing node. Therefore, the input currents—and consequently the input voltages E_1, E_2, and E_3—do *not* interact.

This feature is especially desirable in an audio mixer. For example, let E_1, E_2, and E_3 be replaced by microphones. The ac voltages from each microphone will be added or mixed at every instant. Then if one microphone is carrying guitar music, it will not come out of a second microphone facing the singer. If a 100-kΩ volume control is installed between each microphone and associated input resistor, their relative volumes can be adjusted and added. A weak singer can then be heard above a very loud guitar.

3-2.3 DC Offsetting an AC Signal

Some applications require that you add a dc offset voltage or current to an ac signal. Suppose that you must transmit an audio signal via an infrared emitting diode (IRED) or light-emitting diode. It is first necessary to bias the IRED on with a dc current. Then the audio signal can be superimposed as an ac current that rides on or modulates the dc current. The result is a light or infrared beam whose intensity changes directly with the audio signal. We shall illustrate this principle by an example.

Example 3-10

Design a circuit that allows you to add a dc voltage to a triangle wave.

Solution Select a two-channel adder circuit as in Fig. 3-5(a). A variable dc offset voltage E_{dc}, is connected to one channel. The ac signal, E_{ac}, is connected to the other.

Circuit analysis. If E_{dc} is 0 V, E_{ac} appears inverted at V_o (gain is -1) [see Fig. 3-5(b) and (c)]. If E_{dc} is -5 V, it appears at the output as a $+5$-V dc offset voltage upon which rides the inverted E_{ac}. If E_{dc} is $+7$ V, then E_{ac} is shifted *down* by 7 V. Most function generators contain this type of circuit.

3-3 MULTICHANNEL AMPLIFIER

3-3.1 The Need for a Multichannel Amplifier

Suppose you had a low-, medium-, and high-level signal source. You need to combine them and make their relative amplitudes reasonably equal. You can use a three-input adder circuit to combine the signals. The versatile adder circuit will also allow you to equalize the signal amplitudes at its output. Simply design the required gain

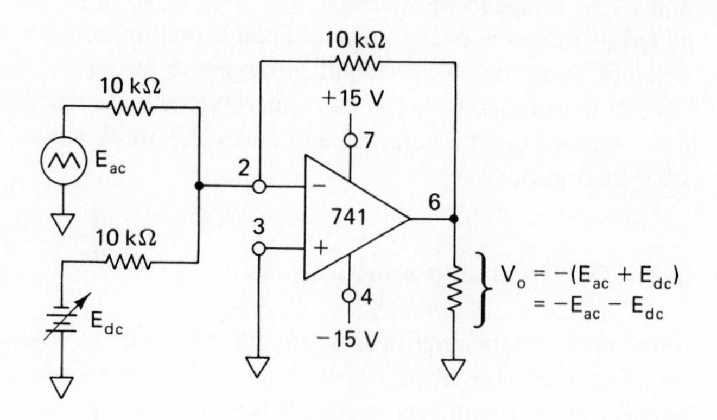

(a) Circuit to add a dc offset voltage E_{dc} to on ac signal voltage E_{ac}

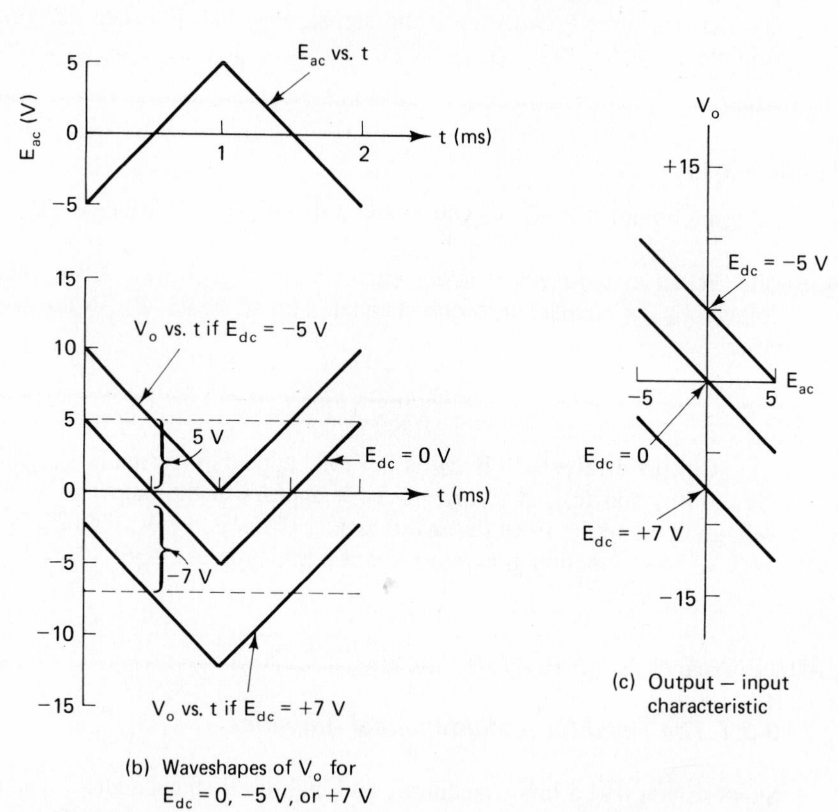

(b) Waveshapes of V_o for E_{dc} = 0, −5 V, or +7 V

(c) Output — input characteristic

FIGURE 3-5 E_{ac} is transmitted with a gain of −1. If E_{dc} is *positive,* the average (dc) value of V_o is shifted *negative* by the same value.

for each input channel by the selection of R_f and input resistors R_1, R_2, and R_3 as shown in Fig. 3-6.

3-3.2 Circuit Analysis

As shown in Fig. 3-6, each channel input signal sees its associated input resistor connected to a virtual ground at the op amp's (−) input. Therefore, the input resistance of each channel is equal to the corresponding value selected for R_1, R_2, or R_3.

Input currents I_1, I_2, and I_3 are added in feedback resistor R_f and then converted back to a voltage V_{Rf}.

$$V_{Rf} = (I_1 + I_2 + I_3)R_f \tag{3-6a}$$

where

$$I_1 = \frac{E_1}{R_1}, \qquad I_2 = \frac{E_2}{R_2}, \qquad I_3 = \frac{E_3}{R_3} \tag{3-6b}$$

As was shown in Section 3-2.1, output voltage V_o equals minus V_{Rf}. Therefore,

$$V_o = -\left(E_1\frac{R_f}{R_1} + E_2\frac{R_f}{R_2} + E_3\frac{R_f}{R_3}\right) \tag{3-7a}$$

Equation (3-7a) shows that the gain of each channel can be changed independently of the others by simply changing its input resistor.

$$A_{CL1} = -\frac{R_f}{R_1}, \qquad A_{CL2} = -\frac{R_f}{R_2}, \qquad A_{CL3} = -\frac{R_f}{R_3} \tag{3-7b}$$

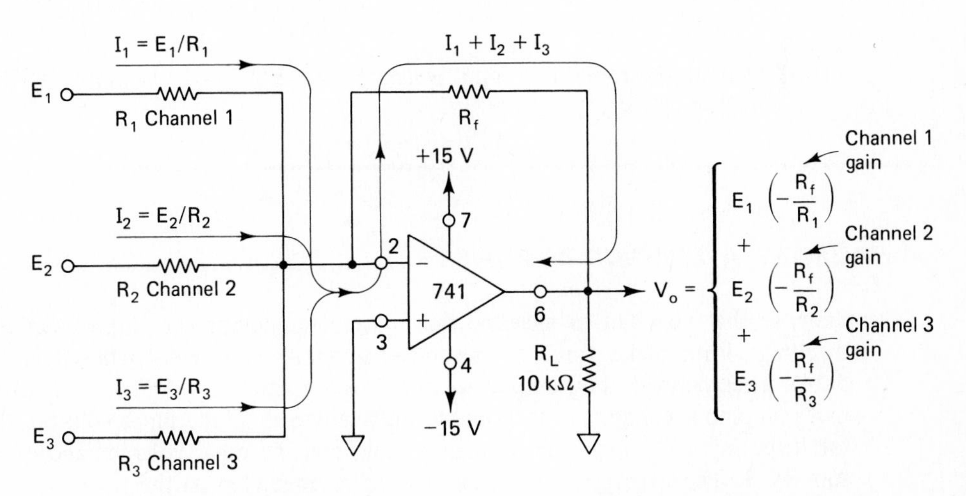

FIGURE 3-6 Multichannel amplifier. The inverting voltage gain of each channel depends on the values of its input resistor and R_F.

or

$$V_o = E_1 A_{CL1} + E_2 A_{CL2} + E_3 A_{CL3}$$

3-3.3 Design Procedure

Following is an example of the design procedure for a multichannel amplifier.

Design Example 3-11

Design a three-channel inverting amplifier. The gains for each channel will be:

Channel number	Voltage gain
1	−10
2	−5
3	−2

Design Procedure

1. Select a 10-kΩ resistor for the input resistance of the channel with the highest gain. Choose $R_1 = 10\ \text{k}\Omega$ since A_{CL1} is the largest.
2. Calculate feedback resistor R_f from Eq. (3-7b):

$$A_{CL1} = -\frac{R_f}{R_1}, \qquad -10 = -\frac{R_f}{10\ \text{k}\Omega}, \qquad R_f = 100\ \text{k}\Omega$$

3. Calculate the remaining input resistors from Eq. (3-7b) to get $R_2 = 20\ \text{k}\Omega$ and $R_3 = 50\ \text{k}\Omega$.

3-4 INVERTING AVERAGING AMPLIFIER

Suppose that you had to measure the average temperature at three locations in a dwelling. First make three temperature-to-voltage converters (shown in Section 5-14). Then connect their outputs to an *averaging* amplifier. An *averaging* amplifier gives an output voltage proportional to the average of all the input voltages. If there are three input voltages, the averager should add the input voltages and divide the sum by 3. The averager is the same circuit arrangement as the inverting adder in Fig. 3-4 or the inverting adder with gain in Fig. 3-6. The difference is that the input resistors are made equal to some convenient value R and the feedback resistor is made equal to R divided by the number of inputs. Let n equal the number of inputs. Then for a three-input averager, $n = 3$ and $R_f = R/3$. Proof is found by substituting into Eq. (3-7a), for $R_f = R/3$ and $R_1 = R_2 = R_3 = R$ to show that

$$V_o = -\left(\frac{E_1 + E_2 + E_3}{n}\right) \tag{3-8}$$

Example 3-12

In Fig. 3-4, $R_1 = R_2 = R_3 = R = 100\ \text{k}\Omega$ and $R_f = 100\ \text{k}\Omega/3 = 33\ \text{k}\Omega$. If $E_1 = +5$ V, $E_2 = +5$ V, and $E_3 = -1$ V, find V_o.

Solution Since $R_f = R/3$, the amplifier is an averager, and from Eq. (3-8) with $n = 3$, we have

$$V_o = -\left[\frac{5\text{ V} + 5\text{ V} + (-1\text{ V})}{3}\right] = -\frac{9\text{ V}}{3} = -3\text{ V}$$

Up to now we have dealt with amplifiers whose input signals were applied via R_i to the op amp's inverting input. We turn our attention next to amplifiers where E_i is applied directly to the op amp's noninverting input.

3-5 VOLTAGE FOLLOWER

3-5.1 Introduction

The circuit of Fig. 3-7 is called a *voltage follower*, but it is also referred to as a *source follower*, *unity-gain amplifier*, *buffer amplifier*, or *isolation amplifier*. The input voltage, E_i, is applied directly to the (+) input. Since the voltage between (+) and (−) pins of the op amp can be considered 0,

$$V_o = E_i \tag{3-9a}$$

Note that the output voltage equals the input voltage in both magnitude and sign. Therefore, as the name of the circuit implies, the output voltage *follows* the input or source voltage. The voltage gain is 1 (or unity), as shown by

$$A_{CL} = \frac{V_o}{E_i} = 1 \tag{3-9b}$$

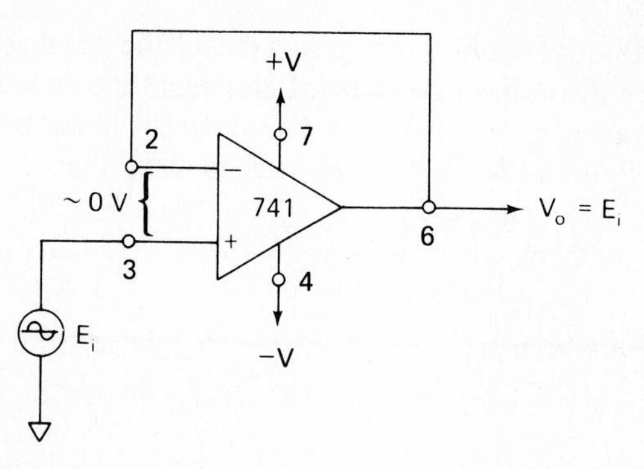

FIGURE 3-7 Voltage follower.

Example 3-13

For Fig. 3-8(a), determine (a) V_o; (b) I_L; (c) I_o.

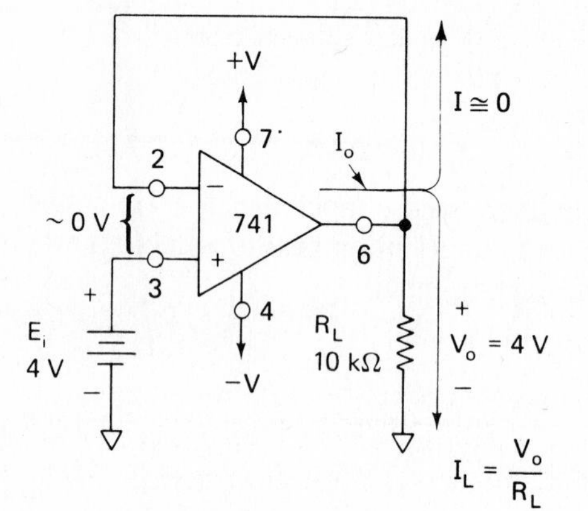

(a) Voltage follower for a positive input voltage

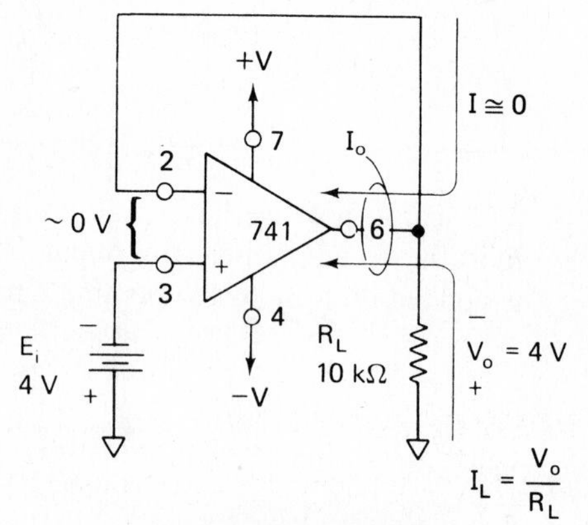

(b) Voltage follower for a negative input voltage

FIGURE 3-8 Circuits for Example 3-13.

Solution (a) From Eq. (3-9a),

$$V_o = E_i = 4 \text{ V}$$

(b) From Ohm's Law,

$$I_L = \frac{V_o}{R_L} = \frac{4 \text{ V}}{10 \text{ k}\Omega} = 0.4 \text{ mA}$$

(c) From Eq. (3-3),

$$I_o = I + I_L$$

This circuit is still a negative-feedback amplifier because there is a connection between output and (−) input. Remember that it is negative feedback that forces E_d to be 0 V. Also $I \approx 0$, since input terminals of op amps draw negligible current; therefore,

$$I_o = 0 + 0.4 \text{ mA} = 0.4 \text{ mA}$$

If E_i were reversed, the polarity of V_o and the direction of currents would be reversed, as shown in Fig. 3-8(b).

3-5.2 Using the Voltage Follower

A question that arises quite often is: Why bother to use an amplifier with a gain of 1? The answer is best seen if we compare a voltage follower with an inverting amplifier. In this example, we are not primarily concerned with the polarity of voltage gain but rather with the input loading effect.

The voltage follower is used because its input resistance is high (many megohms). Therefore, it draws negligible current from a signal source. For example, in Fig. 3-9(a) the signal source has an open circuit or generator voltage, E_{gen}, of 1.0 V. The generator's internal resistance is 90 kΩ. Since the input terminal of the op amp draws negligible current, the voltage drop across R_{int} is 0 V. The terminal voltage E_i of the signal source becomes the input voltage to the amplifier and equals E_{gen}. Thus

$$V_o = E_i = E_{gen}$$

Now let us consider the same signal source connected to an inverting amplifier whose gain is −1 [see Fig. 3-9(b)]. As stated in Section 3-1.3, the input resistance to an inverting amplifier is R_i. This causes the generator voltage E_{gen} to divide between R_{int} and R_i. Using the voltage division law to find the generator terminal voltage E_i yields

$$E_i = \frac{R_i}{R_{int} + R_i} \times E_{gen} = \frac{10 \text{ k}\Omega}{10 \text{ k}\Omega + 90 \text{ k}\Omega} \times (1.0 \text{ V}) = 0.1 \text{ V}$$

Thus it is this 0.1 V that becomes the input voltage to the inverting amplifier. If the inverting amplifier has a gain of only −1, the output voltage V_o is −0.1 V.

In conclusion, if a high-impedance source is connected to an inverting amplifier, the voltage gain from V_o to E_{gen} is not set by R_f and R_i as given in Eq. (3-2b). The actual gain must include R_{int}, as

$$\frac{V_o}{E_{gen}} = -\frac{R_f}{R_i + R_{int}} = -\frac{10 \text{ k}\Omega}{100 \text{ k}\Omega} = -0.1$$

If you must amplify and invert a signal from a high-impedance source and wish to draw no signal current, first *buffer* the source with a voltage follower. Then

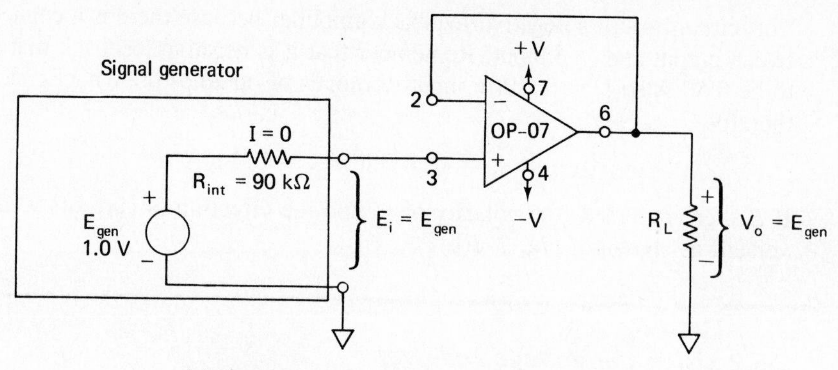

(a) Essentially no current is drawn from E_{gen}. The output terminal of the op amp can supply up to 5 mA with a voltage held constant at E_{gen}

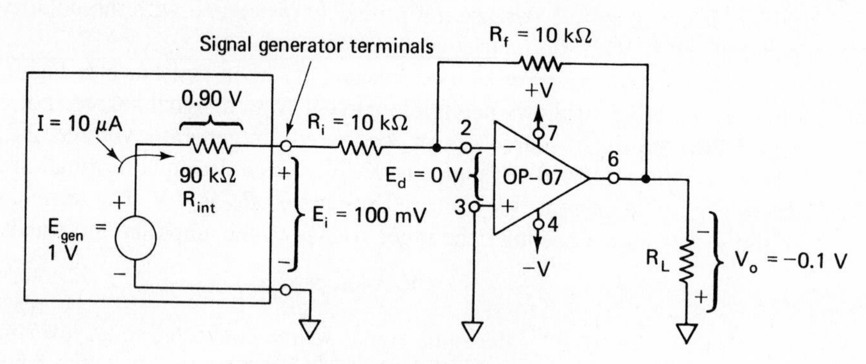

(b) E_{gen} divides between its own internal resistance and amplifier input resistance

FIGURE 3-9 Comparison of loading effect between inverting and noninverting amplifiers on a high-resistance source.

feed the follower's output into an inverter. We now turn to a circuit that will amplify and buffer but *not* invert a signal source, the noninverting amplifier.

3-6 NONINVERTING AMPLIFIER

3-6.1 Circuit Analysis

Figure 3-10 is a noninverting amplifier; that is, the output voltage, V_o, is the same polarity as the input voltage, E_i. The input resistance of the inverting amplifier (section 3-1) is R_i, but the input resistance of the noninverting amplifier is extremely

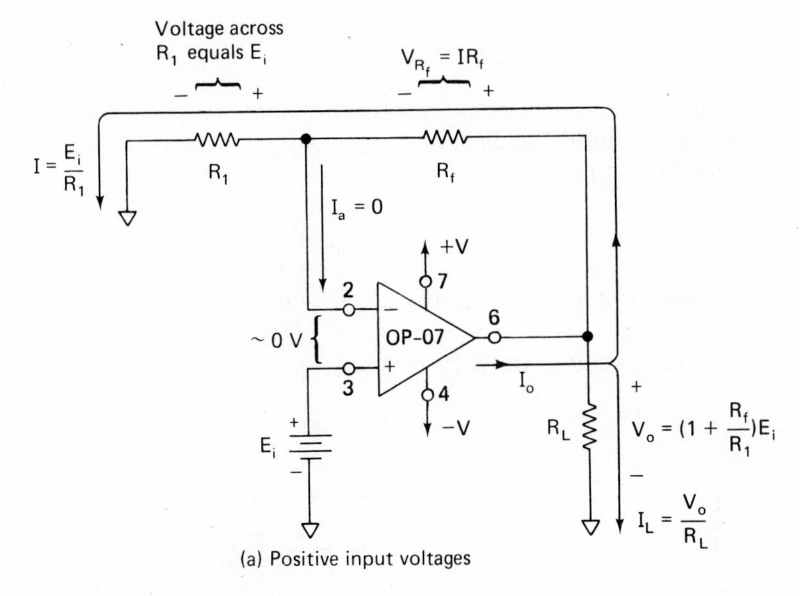

(a) Positive input voltages

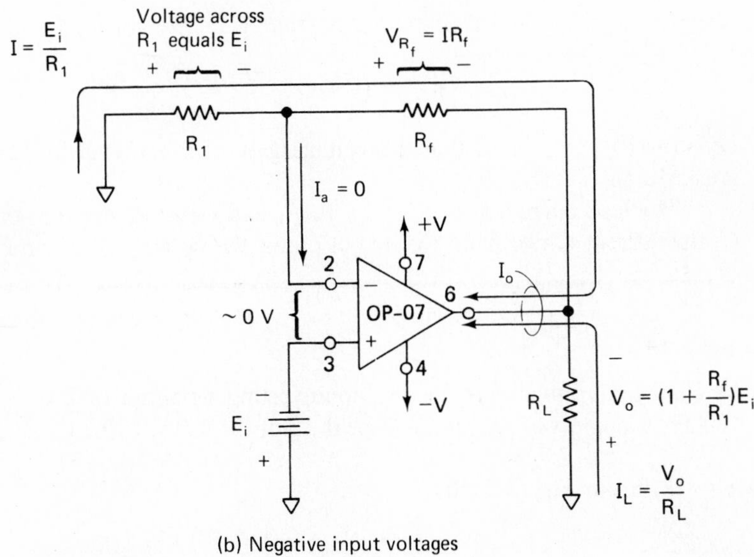

(b) Negative input voltages

FIGURE 3-10 Voltage polarities and direction of currents for noninverting amplifiers.

large, typically exceeding 100 MΩ. Since there is practically 0 voltage between the (+) and (−) pins of the op amp, both pins are at the same potential E_i. Therefore, E_i appears across R_1. E_i causes current I to flow as given by

$$I = \frac{E_i}{R_1} \tag{3-10a}$$

The direction of I depends on the polarity of E_i. Compare Fig. 3-10(a) and (b). The input current to the op amp's (−) terminal is negligible. Therefore, I flows through R_f and the voltage drop across R_f is represented by V_{Rf} and expressed as

$$V_{Rf} = I(R_f) = \frac{R_f}{R_1}E_i \tag{3-10b}$$

Equations (3-10a) and (3-10b) are similar to Eqs. (3-1a) and (3-1b).

The output voltage V_o is found by adding the voltage drop across R_1, which is E_i, to the voltage across R_f, which is V_{Rf}:

$$V_o = E_i + \frac{R_f}{R_1}E_i$$

or

$$V_o = \left(1 + \frac{R_f}{R_1}\right)E_i \tag{3-11a}$$

Rearranging Eq. (3-11a) to express voltage gain, we get

$$A_{CL} = \frac{V_o}{E_i} = 1 + \frac{R_f}{R_1} = \frac{R_f + R_i}{R_i} \tag{3-11b}$$

Equation (3-11b) shows that the voltage gain of a noninverting amplifier is always greater than 1.

The load current I_L is given by V_o/R_L and therefore depends only on V_o and R_L. I_o, the current drawn from the output pin of the op amp, is given by Eq. (3-3).

Example 3-14

(a) Find the voltage gain for the noninverting amplifier of Fig. 3-11. If E_i is a 100-Hz triangle wave with a 2-V peak, plot (b) V_o vs. t; (c) V_o vs. E_i.

Solution (a) From Eq. (3-11b),

$$A_{CL} = \frac{R_f + R_i}{R_i} = \frac{(40 + 10)\text{ k}\Omega}{10\text{ k}\Omega} = 5$$

(b) See Fig. 3-11(b). These are the waveshapes that would be seen on a dc-coupled, dual-trace CRO.

(c) See Fig. 3-11(c). Set a CRO for x-y display, verticle 5 V/div, horizontal 1 V/div. Note the slope rises to the right and is positive. Rise over run gives you the gain magnitude of +5.

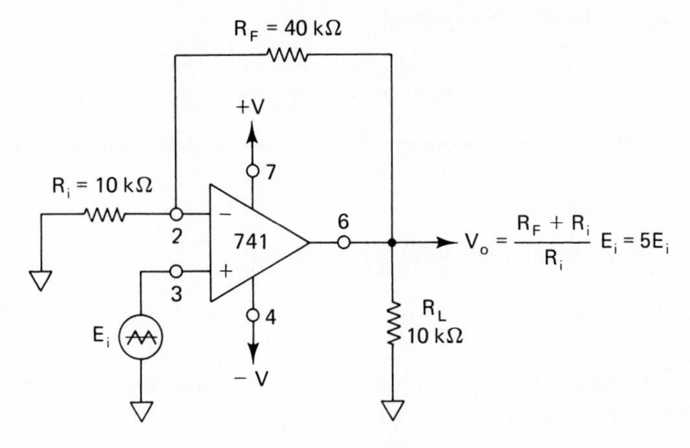

(a) Noninverting amplifier circuit with gain of +5

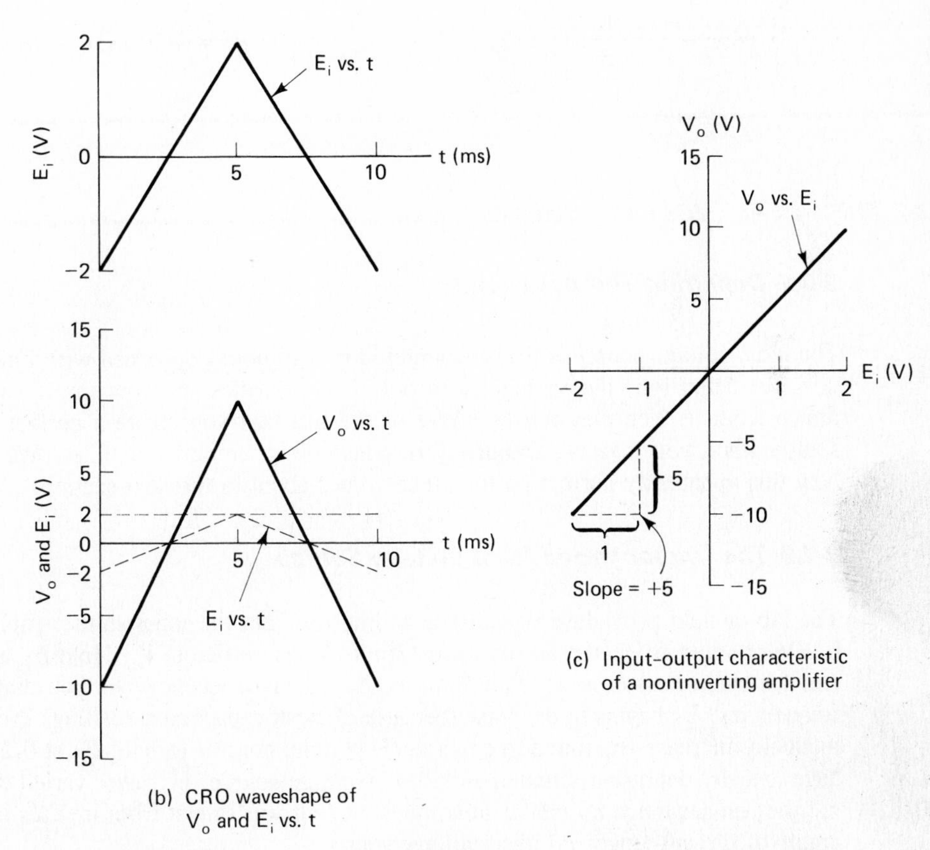

(b) CRO waveshape of V_o and E_i vs. t

(c) Input-output characteristic of a noninverting amplifier

FIGURE 3-11 Noninverting amplifier circuit analysis for Example 3-14.

3-6.2 Design Procedure

Following is an example of the design procedure for a noninverting amplifier.

Design Example 3-15

Design an amplifier with a gain of +10.

Design Procedure

1. Since the gain is positive, select a noninverting amplifier. That is, we apply E_i to the op amp's (+) input.
2. Choose $R_i = 10\ \text{k}\Omega$.
3. Calculate R_f from Eq. (3-11b).

$$A_{CL} = 1 + \frac{R_f}{R_i}, \qquad 10 = 1 + \frac{R_f}{10\ \text{k}\Omega}, \qquad R_f = 9(10\ \text{k}\Omega) = 90\ \text{k}\Omega$$

3-7 THE "IDEAL" VOLTAGE SOURCE

3-7.1 Definition and Awareness

The ideal voltage source is first encountered in textbooks concerned with fundamentals. By definition, the voltage does not vary regardless of how much current is drawn from it. You may not be aware of the fact that you create a perfect voltage source when you measure frequency response of an amplifier or filter. We explain how this apparently perfect performance comes about in the next section.

3-7.2 The Unrecognized Ideal Voltage Source

The lab or field procedure typically goes like this: Set the input signal amplitude at 0.2 V rms and frequency at the lowest limit. Measure output V_o. Hold E_{in} at 0.2 V rms for each measurement. Plot V_o or V_o/E_{in} versus frequency. As you dial higher frequencies, E_{in} begins to decrease (because of input capacitance loading). You automatically increase the function generator's volume control to hold E_{in} at 0.2 V. *You* have just, by definition, created an "ideal" voltage source. E_{in} never varied throughout the test sequence no matter how much current was drawn from it. This is an example of the unrecognized ideal voltage source.

3-7.3 The Practical Ideal Voltage Source

A circuit schematic shows a battery symbol labeled −7.5 V. Your job is to make one. The convenient +15-V supply voltage is available and a simple voltage divider network gives 7.5 V as shown in Fig. 3-12(a). This 7.5-V source is fine as long as you never use it by connecting a load.

As shown in Fig. 3-12(b), R_i of the inverter appears in parallel with R_2 to form

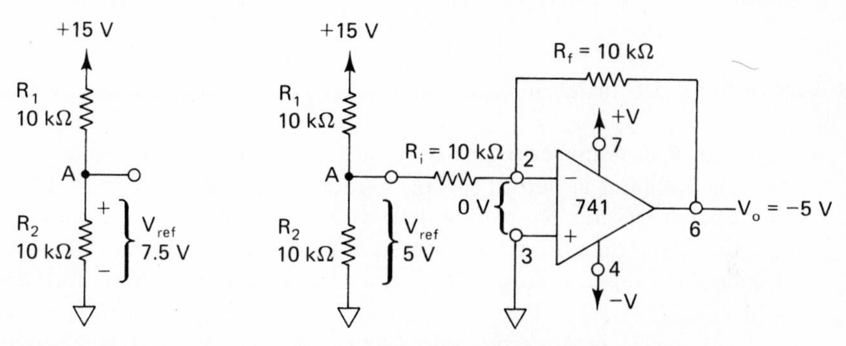

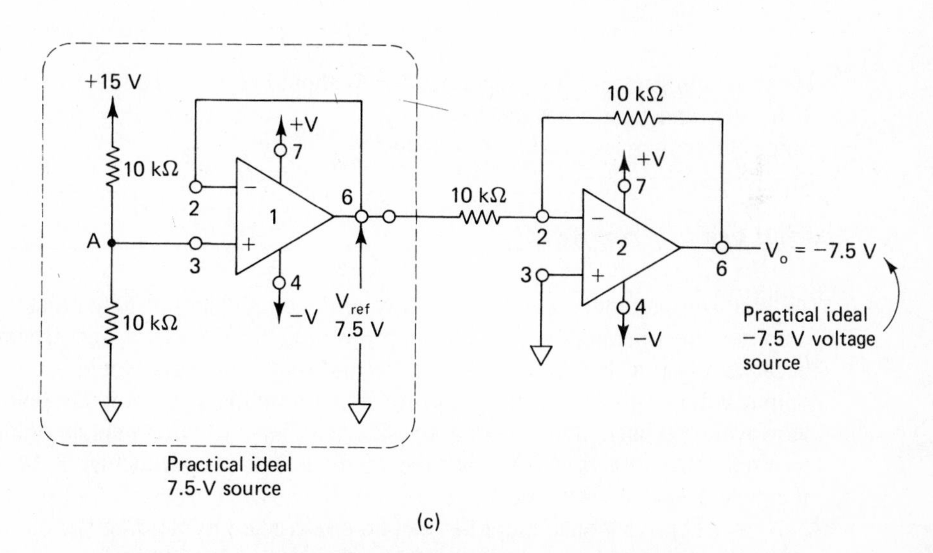

FIGURE 3-12 A voltage divider and (+) supply voltage gives a 7.5-V test or reference voltage in (a). V_{ref} drops to 5 V in (b) when connected to an inverter. A voltage follower converts the voltage divider into an ideal voltage source in (c).

an equivalent resistance of 10 kΩ ∥ 10 kΩ = 5 kΩ. The 15-V supply divides between R_1 = 10 kΩ and 5 kΩ and V_{ref} drops to 5 V.

To preserve the value of any reference voltage, simply *buffer* it with a voltage follower. The 7.5-V reference voltage is connected to a voltage follower in Fig. 3-12(c). The output of the follower equals V_{ref}. You can extract more than 5 mA from the follower's output with no change in V_{ref}.

The buffer makes an excellent clandestine bug. You can monitor what is going on at any circuit point. Since a follower has a high input impedance, it draws no current from the circuit. Therefore, it is nearly impossible to detect.

3-8 NONINVERTING ADDER

A three-input noninverting adder is constructed with a passive averager and noninverting amplifier as shown in Fig. 3-13(a). The passive averager circuit consists of three equal resistors R_A and the three voltages to be added. Output of the passive averager is E_{in}, where E_{in} is the average of E_1, E_2, and E_3 or $E_{in} = (E_1 + E_2 + E_3)/3$. Connect a voltage follower to E_{in} if you need a noninverting averager (in contrast with Sec. 3-4).

Output V_o results from amplifying E_{in} by a gain equal to the number of inputs n. In Fig. 3-13, $n = 3$. Design the amplifier by choosing a convenient value for resistor R. Then find R_f from

$$R_f = R(n - 1) \tag{3-12}$$

As shown in Fig. 3-13(a), the value for R_f should be R_f = 10 kΩ(3 − 1) = 20 kΩ. If E_1, E_2, and E_3 are not ideal voltage sources, such as a battery or output of an op amp, buffer them with followers as in Fig. 3-13(b).

3-9 SINGLE-SUPPLY OPERATION

AC voltages cannot be amplified by any of the amplifiers presented thus far if the op amp must be operated by a single supply voltage (e.g., +15 V and ground). This is because negative half-cycles or positive half-cycles signals would try to drive the output voltage of noninverting and inverting amplifiers respectively below ground. However, the basic noninverting amplifier can be used with a single polarity supply to amplify ac signals if we make the modifications shown in Fig. 3-14. (This amplifier will *not* amplify a dc signal.)

A single-supply ac amplifier can be constructed by holding the op amp's input and output terminals at some convenient dc voltage that is usually half of the single supply voltage. For example, in Fig. 3-14 the equal 220-kΩ resistors R_B divide the 30-V supply voltage in half to set point B at +15 V with respect to ground. Point C must go to +15 V because the op amp's differential input voltage E_d equals 0 V. No

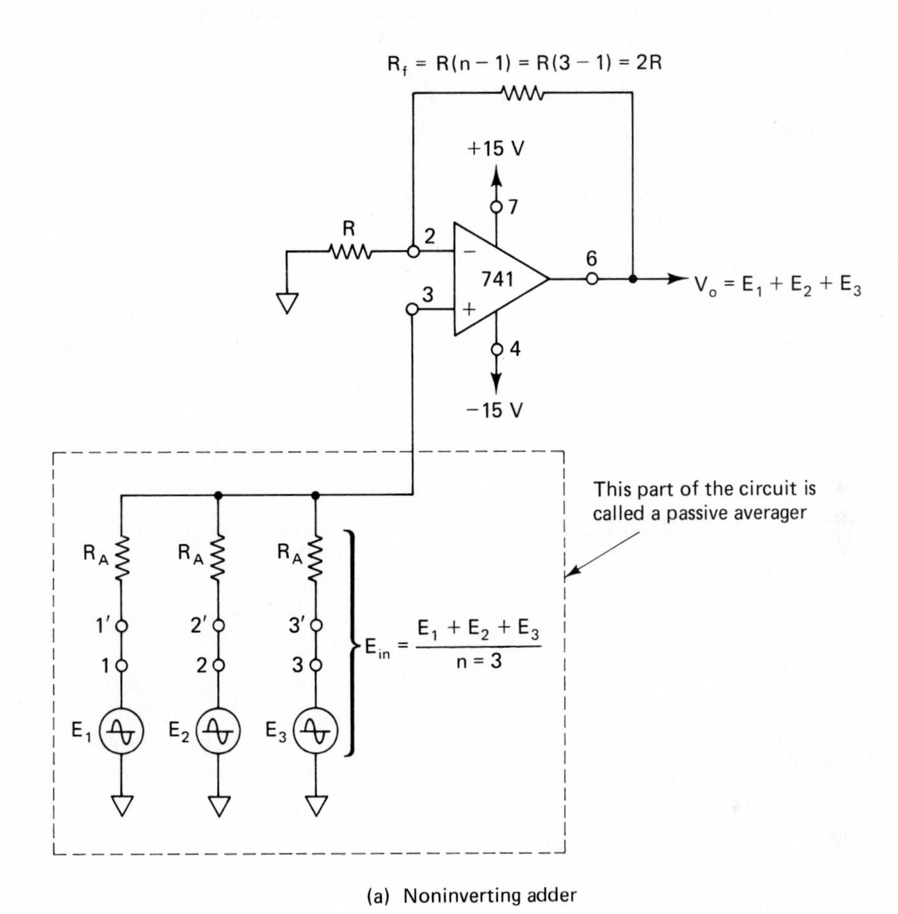

(a) Noninverting adder

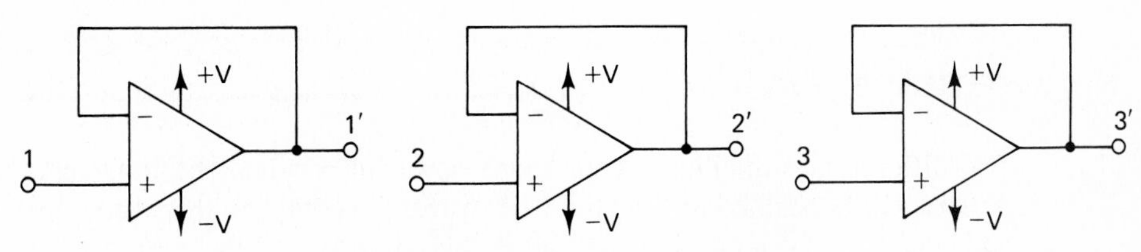

(b) If E_1, E_2, and E_3 are not ideal voltage sources, simply buffer each one with a voltage follower

FIGURE 3-13 All resistors of an n-input noninverting adder are equal except the feedback resistor; choose $R = 10\ \text{k}\Omega$ and $R_A = 10\ \text{k}\Omega$. Then R_f equals R times the number of inputs minus one: $R_f = R(n - 1)$.

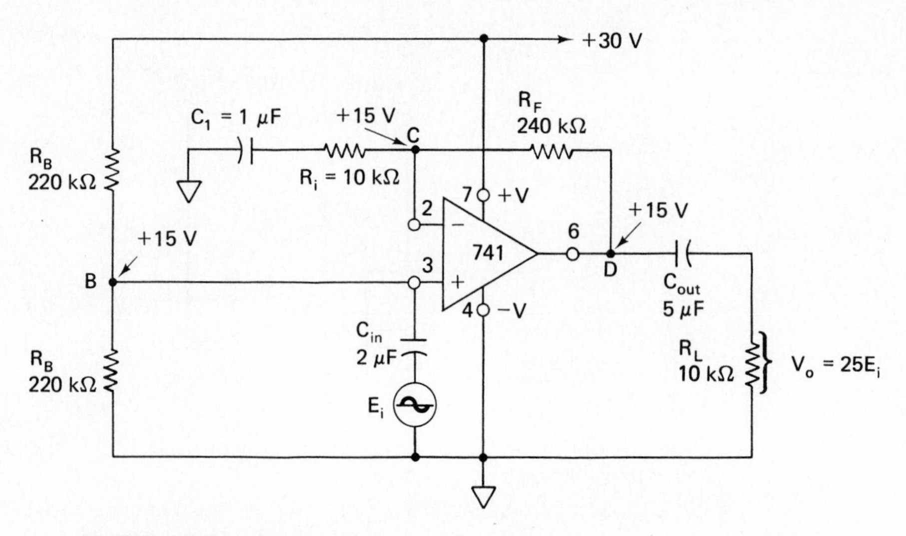

FIGURE 3-14 Construction of an ac noninverting amplifier with an op amp and a single-polarity power supply.

dc current flows through R_i, and consequently R_f, because of capacitor C_1. Therefore, point D is at +15 V.

Only the ac component of signal source E_i is coupled through C_{in} to the op amp's (+) input. E_i sees an input resistance equal to the parallel combination of the 220-kΩ resistors, or 110 kΩ. R_i and R_f form a noninverting amplifier for ac signals with a gain of $(R_f + R_i)/R_i = 25$. C_o blocks the 15-V bias voltage at point D and transmits only the amplified ac signals to load R_L.

To make an inverting ac amplifier with a gain of 24:

1. Replace E_i with a short circuit.
2. Connect E_i in series with C_1 and ground.

3-10 DIFFERENCE AMPLIFIERS

The differential amplifier and its more powerful relative, the instrumentation amplifier, will be studied in Chapter 8. However, to complete this chapter on inverting and noninverting amplifiers, we offer two examples of the *difference* amplifier in this section and end with a *servoamplifier* in the next.

3-10.1 The Subtractor

A circuit that takes the difference between two signals is called a *subtractor* [see Fig. 3-15(a)]. It is made by connecting an inverting amplifier to a two-input inverting averager. To analyze this circuit, note that E_1 is transmitted through op amp A

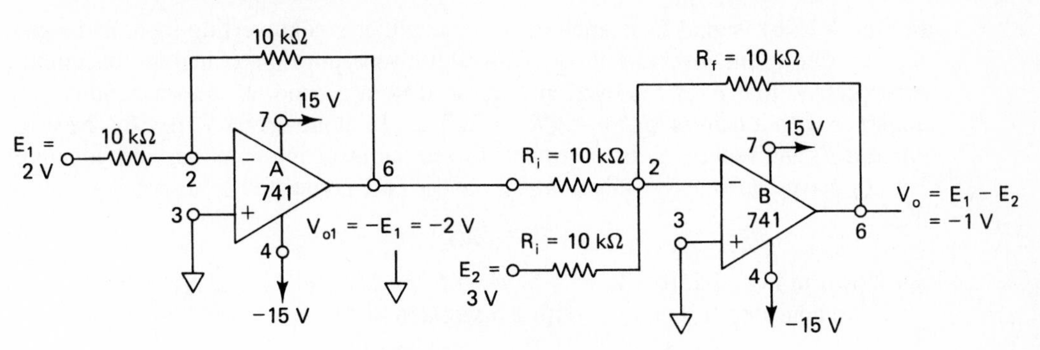

(a) An inverting amplifier and a two-input inverting adder makes a subtractor. $V_o = E_1 - E_2$

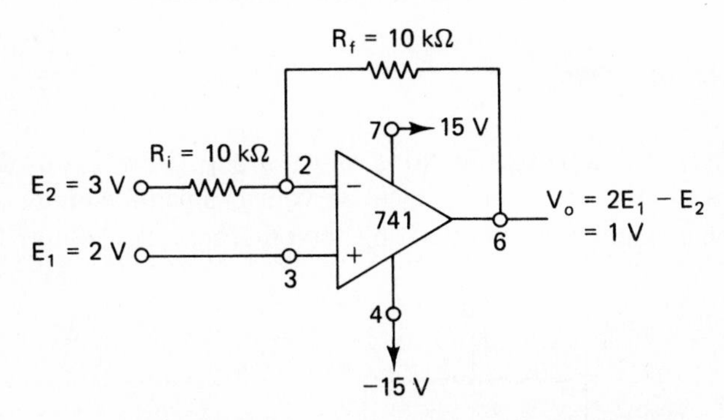

(b) Both amplifier inputs are used to make an amplifier that calculates the difference between $2E_1$ and E_2

FIGURE 3-15 Two examples of different amplifiers are the subtractor in (a) and using the op amp as both an inverting and a noninverting amplifier in (b).

with a gain of -1 and appears as $V_{o1} = -E_1$. V_{o1} is then inverted (times -1) by the top channel of the inverting amplifier B. Thus E_1 is inverted once by op amp A and again by op amp B to appear at V_o as E_1.

E_2 is inverted by the bottom channel of op amp B and drives V_o to $-E_2$. Thus V_o responds to the difference between E_1 and E_2, or

$$V_o = E_1 - E_2 \tag{3-13a}$$

As shown in Fig. 3-15, for $E_1 = 2$ V and $E_2 = 3$ V, $V_o = 2 - 3 = -1$ V. If the value of R_f is made larger than R_i, the subtractor will have gain

$$V_o = \frac{R_f}{R_i}(E_1 - E_2) \tag{3-13b}$$

3-10.2 Inverting–Noninverting Amplifier

In Fig. 3-15(b), signal E_1 is applied to the amplifier's noninverting input and signal E_2 is applied to the inverting input. We will use superposition to analyze this circuit. First pretend that E_2 is removed and replaced by a ground. E_1 sees a noninverting amplifier with a gain of $(R_f + R_i)/R_i$ or 2. Thus E_1 alone drives V_o to $2E_1$. Next reconnect E_2 and replace E_1 by a ground. E_1 sees an inverting amplifier with a gain of -1. E_1 drives V_o to $-E_2$. When *both* E_1 and E_2 are connected V_o responds to

$$V_o = 2E_1 - E_2 \tag{3-14}$$

As shown in Fig. 3-15(b), $V_o = 1$ V when $E_1 = 2$ V and $E_2 = 3$ V.

We conclude this chapter with a discussion of the servoamplifier.

3-11 SERVOAMPLIFIER

3-11.1 Introduction

A simplified servoamplifier circuit is shown in Fig. 3-16. Let us draw an analogy between this circuit and a mechanical servomechanism. Aiming a tank cannon is a very graphic example of an operating servomechanism. Assume that the gunner has

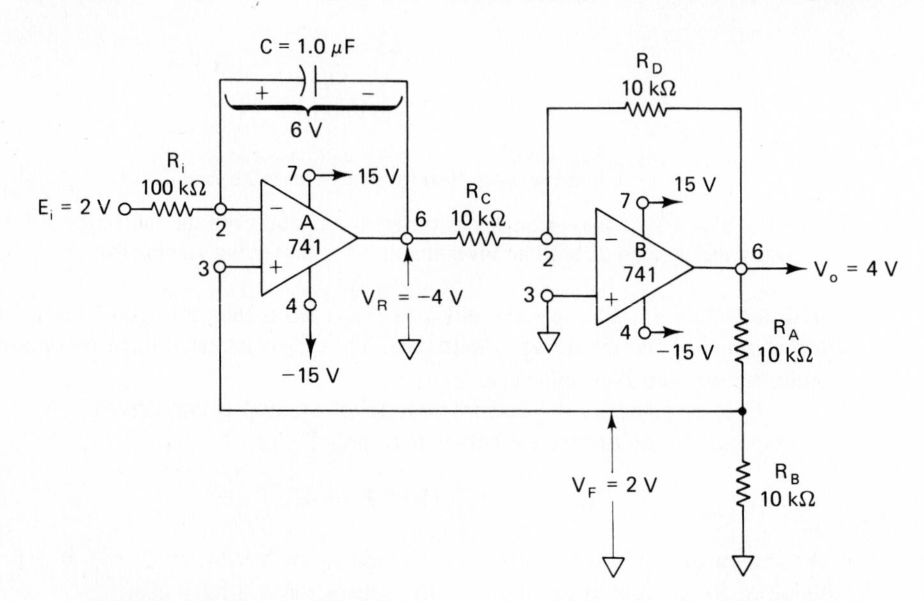

FIGURE 3-16 V_o exhibits a delayed response to a change of E_i in this servoamplifier circuit.

the aiming device pointed straight ahead and also that the cannon is pointing straight ahead. Call this an equilibrium. The aiming device controls the cannon via two servomechanisms, one for azimuth (side to side) and one for elevation (up and down).

Suddenly, the gunner changes aim by 90° right. The input ($E_i = 2$ V in Fig. 3-16) to the servo system has just changed. The cannon now must swing from its straight-ahead equilibrium position ($V_o = 4$ V in Fig. 3-16) to a new equilibrium position 90° right. We conclude that the output of a servo system follows the input, but always with a delay. So we will look for the answer to two questions about circuit operation in Fig. 3-16.

1. If E_i is in equilibrium, what is V_o in equilibrium?
2. How long will it take for V_o to change from one equilibrium to another?

3-11.2 Servoamplifier Circuit Analysis

We analyze the circuit behavior of Fig. 3-16 at *equilibrium* as follows:

1. Assume that $E_i = 2$ V, the capacitor C is charged, no current flows through R_i, and its voltage drop is zero.
2. The voltage at pin 2 equals E_i (since R_i current equals 0) and negative feedback makes voltages equal at pins 2 and 3 of op amp A.
3. Therefore, $V_F = E_i$.
4. V_F causes a current through R_B of $I = V_F/R_B$. This current flows through R_A. V_o is set by I flowing through *both* R_A and R_B.

$$V_o = I(R_A + R_B) = \frac{V_F}{R_B}(R_A + R_B)$$

For $R_A = R_B = 10\ \text{k}\Omega$, $V_o = 2V_F$.

5. Since op amp B has a gain of -1, $V_o = -V_R$, or rather, $V_R = -V_o$.
6. Capacitor voltage V_{cap} is an equilibrium at $E_i - V_R$.

Summary

$$V_o = 2V_F = 2E_i = -V_R \tag{3-15a}$$

$$V_{cap} = E_i - V_R = 3E_i \tag{3-15b}$$

We have answered question 1 in Section 3-11. Next we use two examples to prepare an answer for question 2.

Example 3-16

Calculate the equilibrium voltages for the servoamplifier in Fig. 3-16.

Solution From Eqs. (3-15a) and (3-15b):

1. $E_i = 2$ V, forcing V_F to 2 V.
2. V_F forces V_o to $2V_F = 4$ V.
3. V_o forces V_R to -4 V.
4. V_{cap} stabilizes at $3E_i = 6$ V.

Example 3-17

If E_i is abruptly stepped to 4 V, find the new equilibrium voltages.

Solution

1. $E_i = 4$ V forces V_F to 4 V, forcing $V_o = 8$ V.
2. V_R decreases toward -8 V.
3. V_{cap} must charge to 12 V.

3-11.3 Delay Action

Examples 3-16 and 3-17 shows that V_o must servo from 4 to 8 V when E_i is stepped from 2 to 4 V. A delay will occur (as V_o servos toward 8 V) because the capacitor must charge from 6 to 12 V. The time constant for the capacitor charge is

$$T = 3R_iC \tag{3-16a}$$

Assume that we need 5 time constants for the capacitor to fully charge. Thus equilibrium will be achieved in

$$\text{equilibrium time} = 5T \tag{3-16b}$$

Example 3-18

How long does it take for V_o to reach equilibrium in the servoamplifier of Fig. 3-16?

Solution From Eqs. (3-16a) and (3-16b):

$$T = 3R_iC = 3(1 \times 10^5\ \Omega)(1 \times 10^{-6}\ \text{F}) = 0.3\ \text{s}$$

$$\text{equilibrium time} = 5T = 5 \times 0.3\ \text{s} = 1.5\ \text{s}$$

LABORATORY EXERCISES

All of the circuits in this chapter can be used or modified for laboratory experiments. Use a test frequency of 100 Hz. The authors suggest the following order:

3-1. *Inverting amplifiers*

(a) Use Fig. 3-3 to view the 180° phase shift of an inverting amplifier.

(b) Learn how to calculate gain from an *x-y* CRO display.

(c) Set the scope for a plot of V_o vs. time; increase E_i until V_o clips. You have driven the op amp's output into saturation. Observe the *x-y* plot of V_o vs. E_i and measure the saturation voltages.

3-2. *Noninverting amplifiers*

(a) Design a noninverting amplifier with a gain of 2.

(b) Compare waveshapes and an *x-y* plot of V_o vs E_i with those of Fig. 3-11. What happens when you overdrive the amplifier?

3-3. *Measuring input resistance of a voltage follower*

(a) Refer to Fig. LE3-3 to measure R_{in}. Adjust E_i for 5.00 V rms at 100 Hz (sine wave). Pinouts are for an 8-pin minidip package. Do *not* measure V_{in} because any meter will load the circuit down.

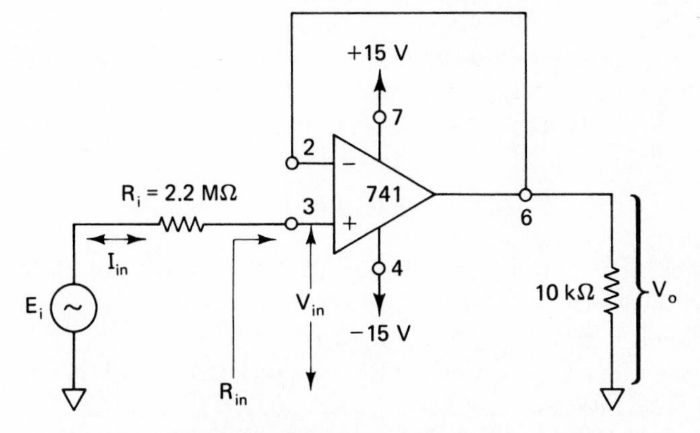

FIGURE LE3-3

(b) Measure V_o. V_o will equal V_{in}. Note that V_o will be very close to E_i.

(c) Calculate I_{in} from

$$I_{in} = \frac{E_i - V_o}{R_i}$$

(d) Calculate R_{in} from

$$R_{in} = \frac{E_i}{I_{in}}$$

The conclusion is that the input resistance is very high and difficult to measure. In Chapter 9 we explain why E_i cannot be a dc voltage. Chapter 10 will explain why the frequency should be less than 1 kHz.

One of the most important lessons to be learned from a laboratory experience with negative feedback circuits is this. With a CRO set for dc coupling, measure the voltage at the (+) input with respect to ground. Then measure voltage at the (−) input with respect to ground. If they are equal, the circuit is probably working.

PROBLEMS

3-1. What type of feedback is applied to an op amp when an external component is connected between the output terminal and the inverting input?

3-2. If the open-loop gain is very large, does the closed-loop gain depend on the external components or the op amp?

3-3. What two assumptions have been used to analyze the circuits of this chapter?

3-4. Identify the circuit in Fig. P3-4.

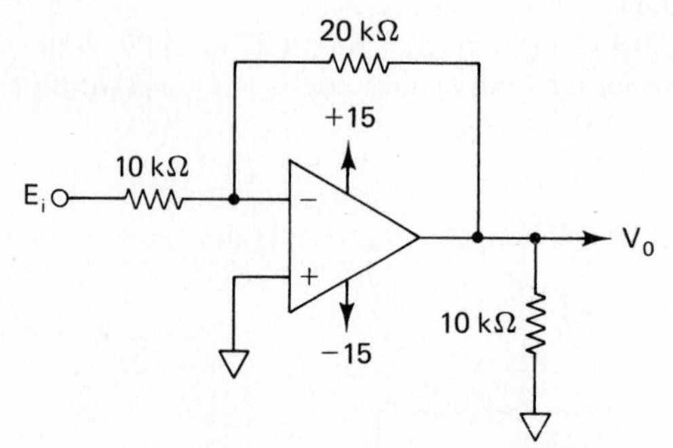

FIGURE P3-4

3-5. Calculate V_o and the op amp's output current in Fig. P3-4 if E_i equals **(a)** +5 V; **(b)** −2 V. For each situation, state if the op amp sources or sinks current.

3-6. Calculate E_i in Fig. P3-4 if V_o equals **(a)** +5 V; **(b)** −2 V.

3-7. Let E_i be a triangle wave with a frequency of 100 Hz and a peak value of 5 V in Fig. P3-4. **(a)** Plot E_i and V_o vs. time; **(b)** V_o vs. E_i.

3-8. Repeat Problem 3-7 but let E_i be increased in amplitude to 8 V. (Assume that $\pm V_{sat} = \pm 15$ V for ease of plotting.)

3-9. Identify the circuit in Fig. P3-9 and calculate V_o if E_i equals **(a)** +5 V; **(b)** −2 V. Compare your results with Problem 3-5.

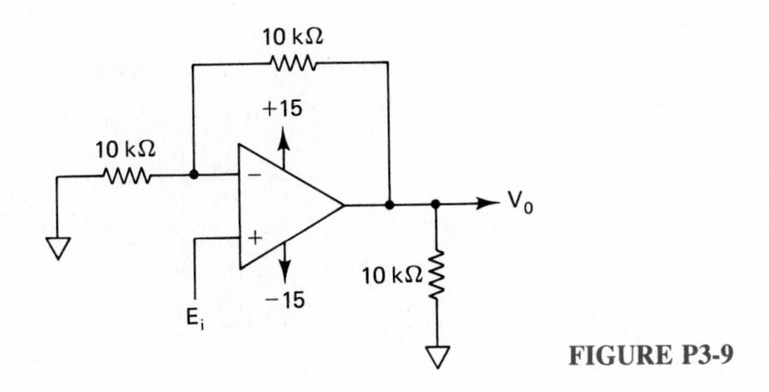

FIGURE P3-9

3-10. Repeat Problem 3-7 except apply it to Fig. P3-9. Compare solutions of both problems to distinguish between inverting and noninverting operation.

3-11. Design an inverting amplifier with a gain of −5 and an input resistance of 10 kΩ.

3-12. Design a noninverting amplifier with a gain of 5.

3-13. Input–output characteristics are shown for three different circuits in Fig. P3-13. Design circuits to recreate plots A, B, and C.

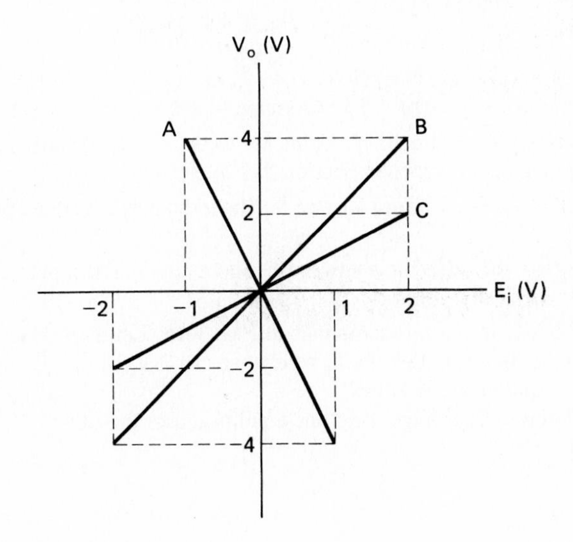

FIGURE P3-13

3-14. The circuit of Fig. P3-14 is called a "subtractor." Is E_1 subtracted from E_2, or vice versa?

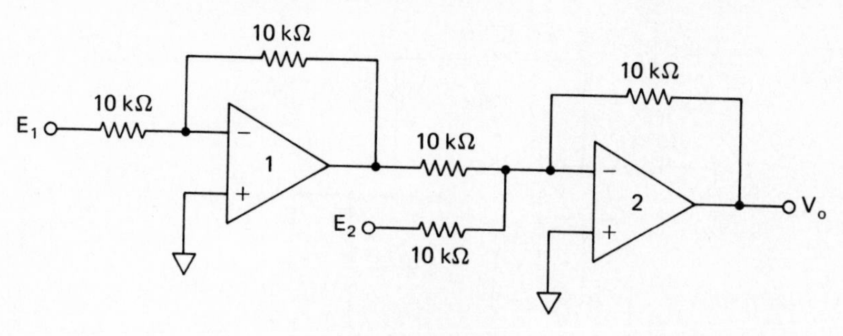

FIGURE P3-14

3-15. A 5-V peak-to-peak sine wave, E_i, is applied to (−) In of Fig. P3-15. Plot V_o vs. E_i if voltage at (+) In is **(a)** +5 V; **(b)** −5V.

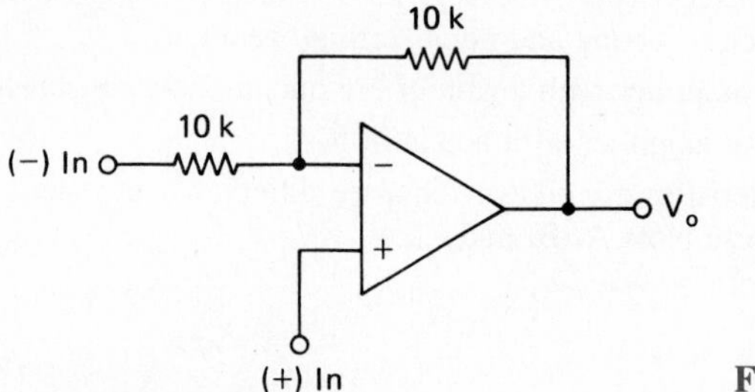

FIGURE P3-15

3-16. A 5-V peak-to-peak sine wave, E_i, is applied to (+) In of Fig. P3-15. Plot V_o vs. E_i if the voltage of (−) In is **(a)** +5 V; **(b)** −5 V. (Assume that $\pm V_{sat} = \pm 15$ V.)

3-17. Design a three-channel inverting amplifier. Gains are to be −1 for channel 1, −3 for channel 2, and −5 for channel 3 (refer to Section 3-3.2).

3-18. Draw a circuit to subtract $V_B = 1$ V from $V_C = 3$ V. Show the output voltage present at each op amp.

3-19. Design a circuit to amplify the difference between E_1 and E_2 by 5. The inputs E_1 and E_2 should be buffered.

3-20. Let R_i be changed to 1 MΩ in the servoamplifier of Fig. 3-16. Let $E_i = 1$ V.
(a) Find the equilibrium voltages. Now let E_i be changed to 3 V.
(b) What are the new equilibrium voltages?
(c) How long will it take V_o to change from one equilibrium to the other?

CHAPTER 4

Comparators and Controls

LEARNING OBJECTIVES

Upon completion of this chapter on comparators and controls, you will be able to:

- Draw the circuit for a zero-crossing detector and plot its output–input characteristic.
- Identify on an output–input characteristic both the upper and lower threshold voltages.
- Calculate hysteresis voltage if you know the threshold voltages.
- Explain how hysteresis gives a measure of noise immunity to comparator circuits.
- Explain why hysteresis must be present in all on–off control circuitry, using the familiar wall thermostat as an example.
- Make a battery-charger control circuit.
- Build and calibrate an independently adjustable setpoint controller.

- Describe the operation of an LM311 precision comparator.
- Connect two LM311 comparators to make a window detector.
- Give the definition of propagation delay and know how to measure it.

4-0 INTRODUCTION

A comparator compares a signal voltage on one input with a reference voltage on the other input. Voltage-level-detector circuits are also comparators and were introduced in Chapter 2 to show how easy it is to use op amps to solve some types of signal comparison applications without the need to know much about the op amp itself. The general-purpose op amp was used as a substitute for ICs designed only for comparator applications.

Unfortunately, the general-purpose op amp's output voltage does not change very rapidly. Also, its output changes between limits fixed by the saturation voltages, $+V_{sat}$ and $-V_{sat}$ that are typically about ± 13 V. Therefore, the output cannot drive devices, such as TTL digital logic ICs, that require voltage levels between 0 and +5 V. These disadvantages are eliminated by ICs that have been specifically designed to act as a comparator. One such device is the 311 comparator. It will be introduced at the end of this chapter.

Neither the general-purpose op amp nor the comparator can operate properly if noise is present at either input. To solve this problem, we will learn how the addition of *positive feedback* overcomes the noise problem. Note that positive feedback does not eliminate the noise but makes the op amp less responsive to it. These circuits will show how to make better voltage-level detectors and also build a foundation to understand square-wave generators (multivibrators) and single-pulse generators (one-shots) that are covered in Chapter 6.

4-1 EFFECT OF NOISE ON COMPARATOR CIRCUITS

Input signal E_i is applied to the (−) input of a 301 op amp in Fig. 4-1 (the 301 is a general-purpose op amp). If no noise is present, the circuit operates as an inverting zero-crossing detector because $V_{ref} = 0$.

Noise voltage is shown, for simplicity, as a square wave in series with E_i. To show the effect of noise voltage, the op amp's input signal voltage is drawn both with and without noise in Fig. 4-2. The waveshape of V_o vs. time shows clearly how the addition of noise causes false output signals. V_o should indicate only the crossings of E_i, *not* the crossings of E_i plus noise voltage.

If E_i approaches V_{ref} very slowly or actually hovers close to V_{ref}, V_o can either follow all the noise voltage oscillations or burst into high-frequency oscillation. These false crossings can be eliminated by *positive feedback*.

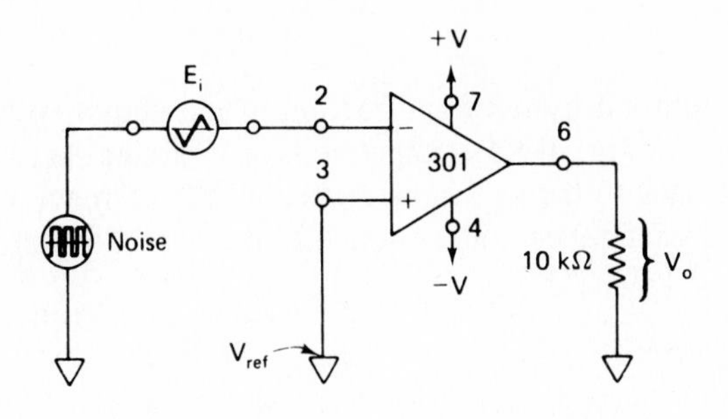

FIGURE 4-1 Inverting zero-crossing detector.

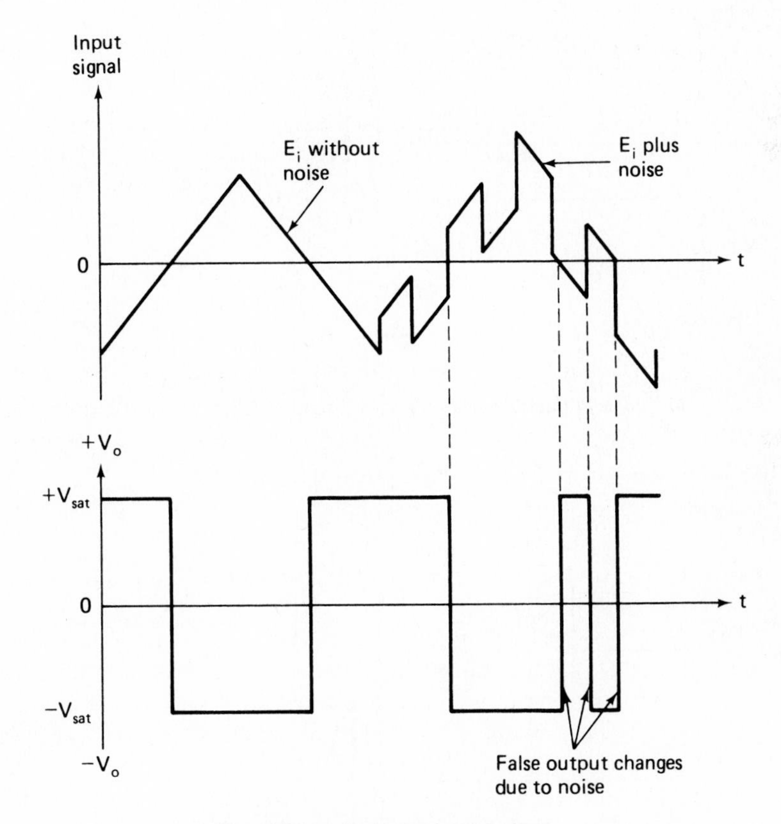

The addition of noise voltage at the input causes false zero crossings.

FIGURE 4-2 Effect of noise on a zero-crossing detector.

4-2 POSITIVE FEEDBACK

4-2.1 Introduction

Positive feedback is accomplished by taking a fraction of the output voltage V_o and applying it to the (+) input. In Fig. 4-3(a), output voltage V_o divides between R_1 and R_2. A fraction of V_o is fed back to the (+) input and creates a reference voltage that depends on V_o. The idea of a reference voltage was introduced in Chapter 2. We will

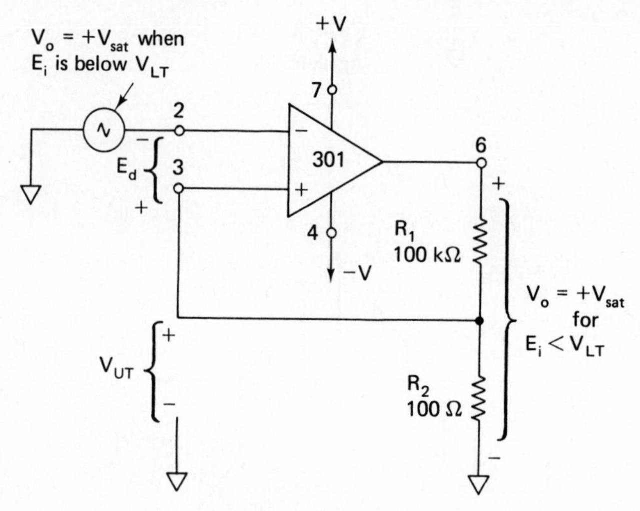

(a) Upper-threshold voltage, V_{UT}

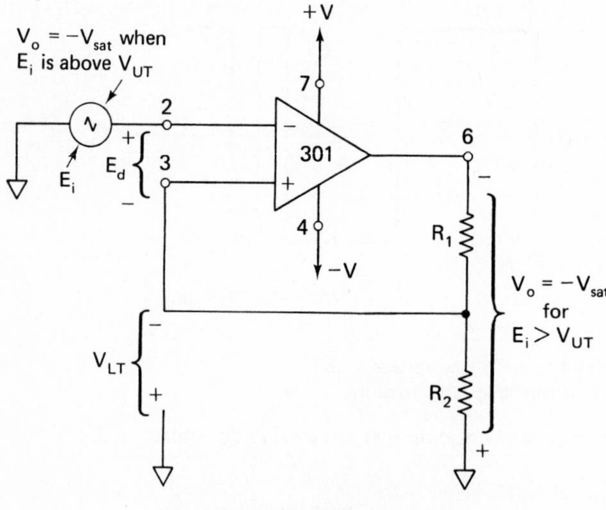

(b) Lower-threshold voltage, V_{LT}

FIGURE 4-3 R_1 and R_2 feed back a reference voltage from the output to the (+) input terminal.

now study positive feedback and how it can be used to eliminate false output changes due to noise.

4-2.2 Upper-Threshold Voltage

In Fig. 4-3(a), output voltage V_o divides between R_1 and R_2. A fraction of V_o is fed back to the (+) input. When $V_o = +V_{sat}$, the fed-back voltage is called the *upper-threshold voltage* V_{UT}. V_{UT} is expressed from the voltage divider as

$$V_{UT} = \frac{R_2}{R_1 + R_2}(+V_{sat}) \tag{4-1}$$

For E_i values below V_{UT}, the voltage at the (+) input is above the voltage at the (−) input. Therefore, V_o is locked at $+V_{sat}$.

If E_i is made slightly more positive than V_{UT}, the polarity of E_d, as shown, reverses and V_o begins to drop in value. Now the fraction of V_o fed back to the positive input is smaller, so E_d becomes larger. V_o then drops even faster and is driven quickly to $-V_{sat}$. The circuit is then stable at the condition shown in Fig. 4-3(b).

4-2.3 Lower-Threshold Voltage

When V_o is at $-V_{sat}$, the voltage fed back to the (+) input is called *lower-threshold voltage* V_{LT} and is given by

$$V_{LT} = \frac{R_2}{R_1 + R_2}(-V_{sat}) \tag{4-2}$$

Note that V_{LT} is negative with respect to ground. Therefore, V_o will stay at $-V_{sat}$ as long as E_i is above, or positive with respect to, V_{LT}. V_o will switch back to $+V_{sat}$ if E_i goes more negative than, or below, V_{LT}.

We conclude that positive feedback induces a snap action to switch V_o faster from one limit to the other. Once V_o begins to change, it causes a regenerative action that makes V_o change even faster. If the threshold voltages are larger than the peak noise voltages, positive feedback will eliminate false output transitions. This principle is investigated in the following examples.

Example 4-1

If $+V_{sat} = 14$ V in Fig. 4-3(a), find V_{UT}.

Solution By Eq. (4-1),

$$V_{UT} = \frac{100\ \Omega}{100{,}100\ \Omega}(14\text{ V}) \approx 14\text{ mV}$$

Example 4-2

If $-V_{sat} = -13$ V in Fig. 4-3(b), find V_{LT}.

Solution By Eq. (4-2),

$$V_{LT} = \frac{100\ \Omega}{100{,}100\ \Omega}(-13\text{ V}) \approx -13\text{ mV}$$

Example 4-3

In Fig. 4-4, E_i is a triangular wave applied to the (−) input in Fig. 4-3(a). Find the resultant output voltage.

Solution The dashed lines drawn on E_i in Fig. 4-4 locate V_{UT} and V_{LT}. At time $t = 0$, E_i is below V_{LT}, so V_o is at $+V_{sat}$ (as in Fig. 4-4). When E_i goes above V_{UT}, at times (a) and (c), V_o switches quickly to $-V_{sat}$. When E_i again goes below V_{LT}, at times (b) and (d), V_o switches quickly to $+V_{sat}$. Observe how positive feedback has eliminated the false crossings.

4-3 ZERO-CROSSING DETECTOR WITH HYSTERESIS

4-3.1 Defining Hysteresis

There is a standard technique of showing comparator performance on one graph instead of two graphs, as in Fig. 4-4. By plotting E_i on the horizontal axis and V_o on the vertical axis, we obtain the output-input voltage characteristic, as in Fig. 4-5. For E_i less than V_{LT}, $V_o = +V_{sat}$. The vertical line (a) shows V_o going from $+V_{sat}$ to $-V_{sat}$ as E_i becomes greater than V_{UT}. Vertical line (b) shows V_o changing from $-V_{sat}$ to $+V_{sat}$ when E_i becomes less than V_{LT}. The difference in voltage between V_{UT} and V_{LT} is called the *hysteresis voltage,* V_H.

Whenever any circuit changes from one state to a second state at some input signal and then reverts from the second to the first state at a *different* input signal, the circuit is said to exhibit *hysteresis*. For the positive-feedback comparator, the difference in input signals is

$$V_H = V_{UT} - V_{LT} \tag{4-3}$$

For Examples 4-1 and 4-2, the hysteresis voltage is 14 mV − (−13 mV) = 27 mV.

If the hysteresis voltage is designed to be greater than the *peak-to-peak* noise voltage, there will be no false output crossings. Thus V_H tells us how much peak-to-peak noise the circuit can withstand.

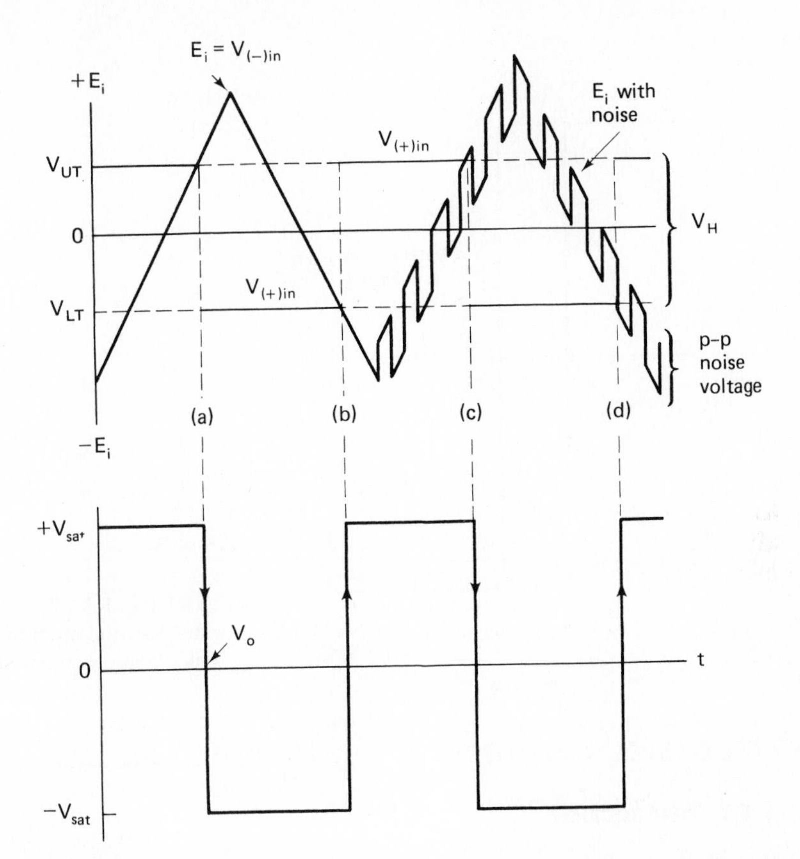

FIGURE 4-4 Solution to Example 4-3. When E_i goes above V_{UT} at time (c), V_o goes to $-V_{sat}$. The peak-to-peak noise voltage would have to equal or exceed V_H to pull E_i below V_{LT} and generate a false crossing. Thus V_H tells us the margin against peak-to-peak noise voltage.

4-3.2 Zero-Crossing Detector with Hysteresis as a Memory Element

If E_i has a value that lies between V_{LT} and V_{UT}, it is impossible to predict the value of V_o unless *you already know* the value of V_o. For example, suppose that you substitute ground for $E_i (E_i = 0\text{ V})$ in Fig. 4-3 and turn on the power. The op amp will go to *either* $+V_{sat}$ or $-V_{sat}$, depending on the inevitable presence of noise. If the op amp goes to $+V_{sat}$, E_i must then go above V_{UT} in order to change the output. If V_o had gone to $-V_{sat}$, then E_i would have to go below V_{LT} to change V_o.

Thus the comparator with hysteresis exhibits the property of *memory*. That is, if E_i lies between V_{UT} and V_{LT} (within the hysteresis voltage), the op amp remembers whether the last switching value of E_i was above V_{UT} or below V_{LT}.

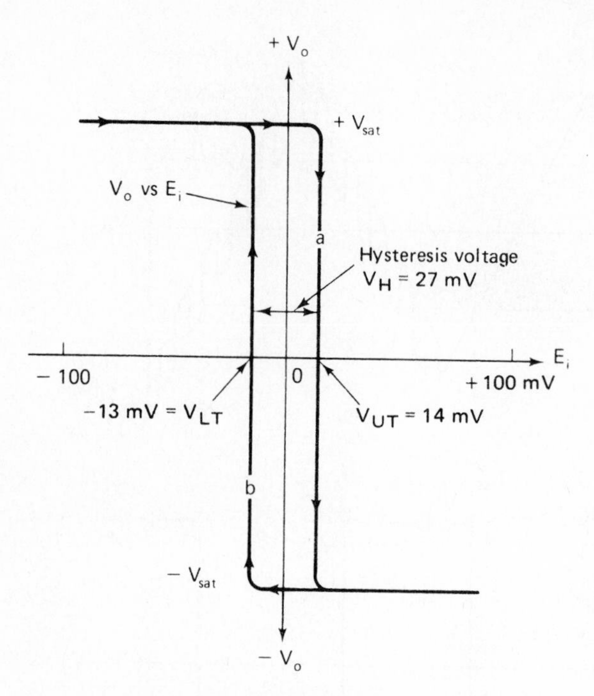

FIGURE 4-5 Plot of V_o vs. E_i illustrates the amount of hysteresis voltage in a comparator circuit.

4-4 VOLTAGE-LEVEL DETECTORS WITH HYSTERESIS

4-4.1 Introduction

In the zero-crossing detectors of Sections 4-2 and 4-3, the hysteresis voltage V_H is centered on the zero reference voltage V_{ref}. It is also desirable to have a collection of circuits that exhibit hysteresis about a center voltage that is either positive or negative. For example, an application may require a positive output, V_o, when an input E_i goes above an upper threshold voltage of $V_{UT} = 12$ V. Also, we may wish V_o to go negative when E_i goes below a lower threshold voltage of, for example, $V_{LT} = 8$ V. These requirements are summarized on the plot of V_o vs. E_i in Fig. 4-6. V_H is evaluated from Eq. (4-3) as

$$V_H = V_{UT} - V_{LT} = 12\text{ V} - 8\text{ V} = 4\text{ V}$$

The hysteresis voltage V_H should be centered on the average of V_{UT} and V_{LT}. This average is called center voltage V_{ctr}, where

$$V_{ctr} = \frac{V_{UT} + V_{LT}}{2} = \frac{12\text{ V} + 8\text{ V}}{2} = 10\text{ V}$$

When we try to build this type of voltage-level detector, it is desirable to have four features: (1) an adjustable resistor to set and refine the value of V_H; (2) a sepa-

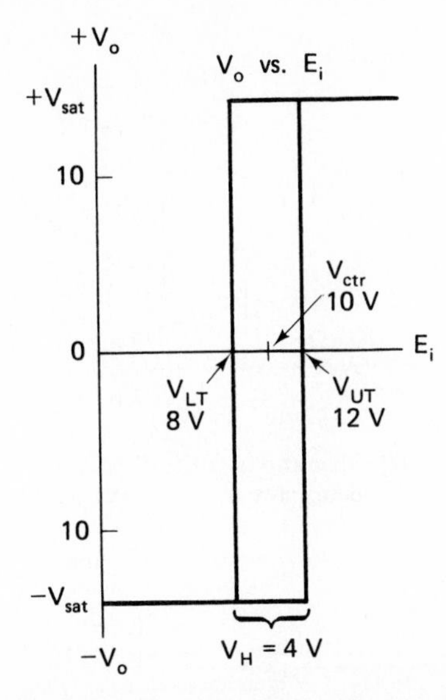

FIGURE 4-6 Positive-voltage-level detector. Hysteresis voltage V_H is symmetrical about the center voltage V_{ctr}. This voltage-level detector is a noninverting type because V_o goes positive when E_i goes above V_{UT}.

rate adjustable resistor to set the value of V_{ctr}; (3) the setting of V_H and V_{ctr} should *not* interact; and (4) the center voltage V_{ctr} should equal or be simply related to an external reference voltage V_{ref}. For the lowest possible parts count, the op amp's regulated supply voltage and a resistor network can be used for selecting V_{ref}.

Sections 4-4.2 and 4-4.3 deal with circuits that do not have all these features but are low in parts count and consequently cost. In Section 4-5 we present a circuit that has all four features but at the cost of a higher parts count.

4-4.2 Noninverting Voltage-Level Detector with Hysteresis

The positive feedback resistor from output to (+) input indicates the presence of hysteresis in the circuit of Fig. 4-7. E_i is applied via R to the (+) input, so the circuit is noninverting. (Note that E_i must be a low-impedance source or the output of either a voltage follower or op amp amplifier.) The reference voltage V_{ref} is applied to the op amp's (−) input.

The upper- and lower-threshold voltages can be found from the following equations:

$$V_{UT} = V_{ref}\left(1 + \frac{1}{n}\right) - \frac{-V_{sat}}{n} \tag{4-4a}$$

$$V_{LT} = V_{ref}\left(1 + \frac{1}{n}\right) - \frac{+V_{sat}}{n} \tag{4-4b}$$

Hysteresis voltage V_H is expressed by

$$V_H = V_{UT} - V_{LT} = \frac{(+V_{sat}) - (-V_{sat})}{n} \tag{4-5}$$

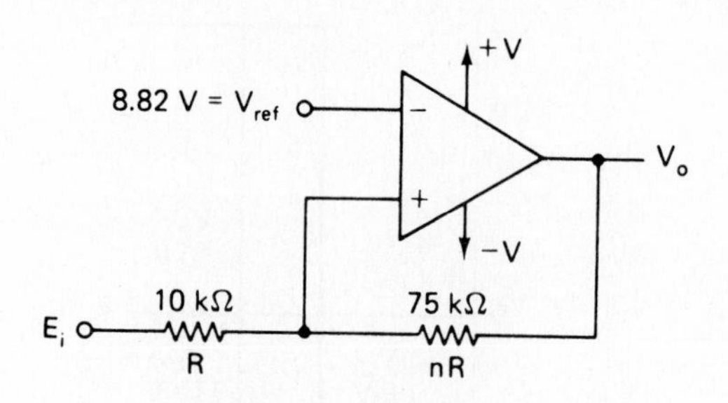

(a) The ratio of nR to R or n and V_{ref} determine V_{UT}, V_{LT}, V_H, and V_{ctr}

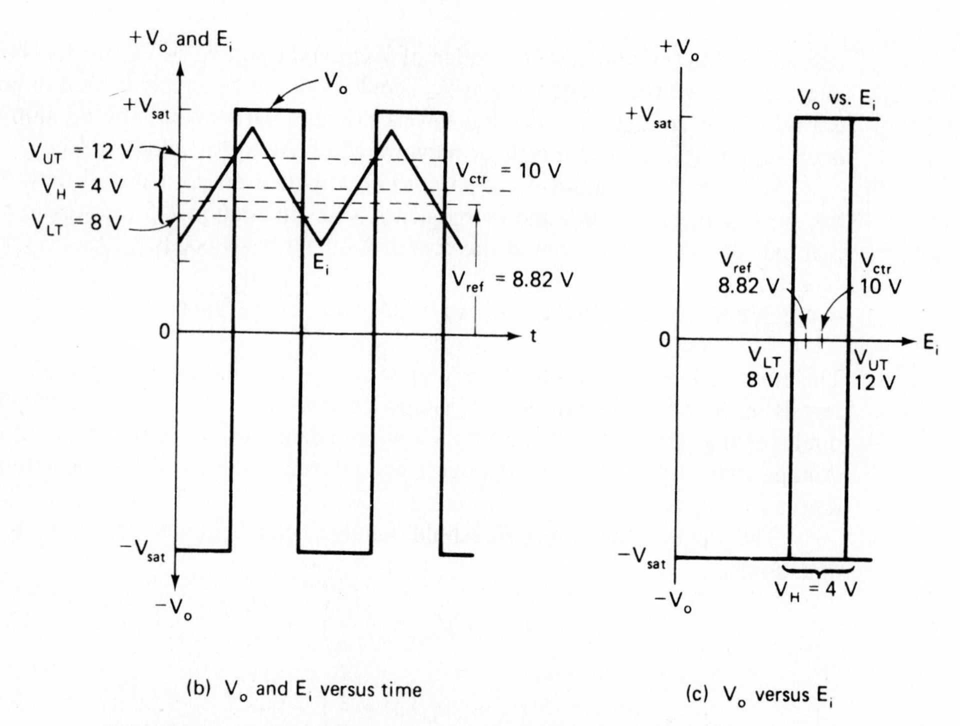

(b) V_o and E_i versus time

(c) V_o versus E_i

FIGURE 4-7 Noninverting voltage-level detector with hysteresis. Center voltage V_{ctr} and hysteresis voltage V_H cannot be adjusted independently since both depend on the ratio n.

In zero-crossing detectors, V_H is centered on the zero-volts reference. For the circuit of Fig. 4-7, V_H is *not* centered on V_{ref} but is symmetrical about the *average* value of V_{UT} and V_{LT}. This average value is called *center voltage* V_{ctr} and is found from

$$V_{ctr} = \frac{V_{UT} + V_{LT}}{2} = V_{ref}\left(1 + \frac{1}{n}\right) \tag{4-6}$$

Compare the locations of V_{ctr} and V_{ref} in Figs. 4-6 and 4-7(c). Also compare Eqs. (4-5) and (4-6) to see that n appears in *both* equations. This means that any adjustment in resistor nR affects *both* V_{ctr} and V_H.

Design Example 4-4

Design the circuit of Fig. 4-7 to have $V_{UT} = 12$ V and $V_{LT} = 8$ V. Assume that $\pm V_{sat} = \pm 15$ V.

Design Procedure

1. From Eqs. (4-5) and (4-6), calculate V_H and V_{ctr}:

$$V_H = 12 \text{ V} - 8 \text{ V} = 4 \text{ V}, \qquad V_{ctr} = \frac{12 \text{ V} + 8 \text{ V}}{2} = 10 \text{ V}$$

2. Find n from Eq. (4-5):

$$n = \frac{+V_{sat} - (-V_{sat})}{V_H} = \frac{+15 \text{ V} - (-15 \text{ V})}{4} = 7.5$$

3. Find V_{ref} from Eq. (4-6):

$$V_{ref} = \frac{V_{ctr}}{1 + 1/n} = \frac{10 \text{ V}}{1 + 1/7.5} = 8.82 \text{ V}$$

4. Select $R = 10$ kΩ and $nR = 7.5 \times 10$ kΩ $= 75$ kΩ. The relationships between E_i and V_o are shown in Fig. 4-7(b) and (c).

4-4.3 Inverting Voltage-Level Detector with Hysteresis

If E_i and V_{ref} are interchanged in Fig. 4-7(a), the result is the inverting voltage-level detector with hysteresis (see Fig. 4-8). The expressions for V_{UT} and V_{LT} are

$$V_{UT} = \frac{n}{n+1}(V_{ref}) + \frac{+V_{sat}}{n+1} \tag{4-7a}$$

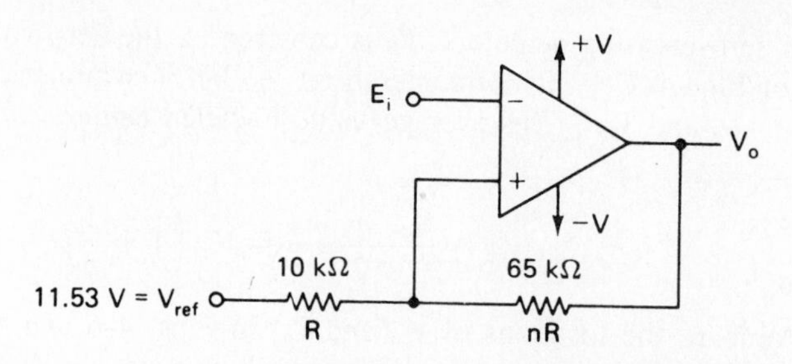

(a) The ratio of nR to R or n and V_{ref} determine V_{UT}, V_{LT}, V_H, and V_{ctr}

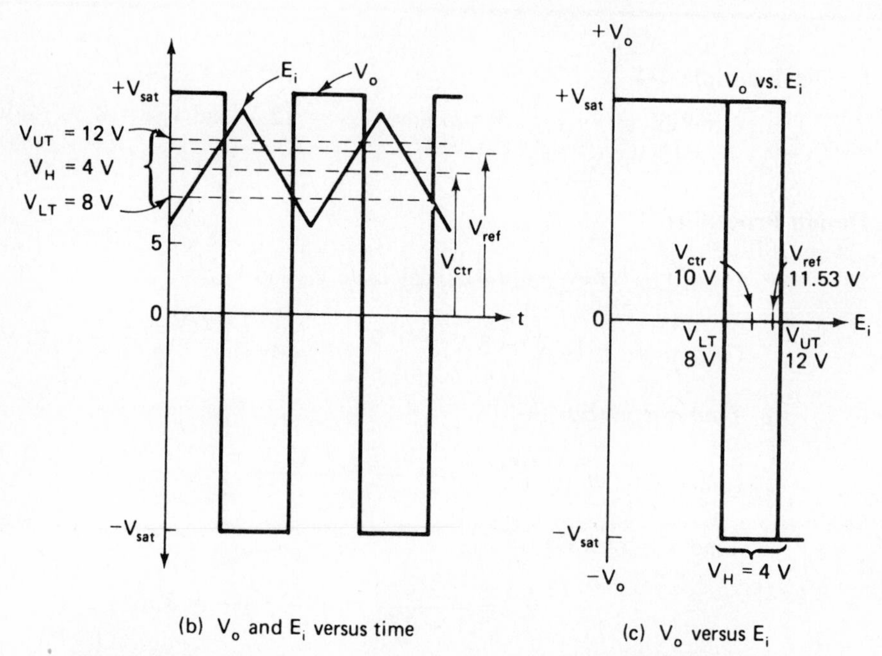

(b) V_o and E_i versus time

(c) V_o versus E_i

FIGURE 4-8 Inverting voltage-level detector with hysteresis. Center voltage V_{ctr} and V_H cannot be adjusted independently since both depend on n.

$$V_{LT} = \frac{n}{n+1}(V_{ref}) + \frac{-V_{sat}}{n+1} \tag{4-7b}$$

V_{ctr} and V_H are then found to be

$$V_{ctr} = \frac{V_{UT} + V_{LT}}{2} = \left(\frac{n}{n+1}\right)V_{ref} \tag{4-8}$$

$$V_H = V_{UT} - V_{LT} = \frac{(+V_{sat}) - (-V_{sat})}{n+1} \tag{4-9}$$

Note that V_{ctr} and V_H both depend on n and therefore are *not* independently adjustable.

Design Example 4-5

Complete a design for Fig. 4-8 that has $V_{UT} = 12$ V and $V_{LT} = 8$ V. To make this example comparable with Example 4-5, assume that $\pm V_{sat} = \pm 15$ V. Therefore, $V_{ctr} = 10$ V and $V_H = 4$ V.

Design Procedure

1. Find n from Eq. (4-9):

$$n = \frac{(+V_{sat}) - (-V_{sat})}{V_H} - 1 = \frac{15\text{ V} - (-15\text{ V})}{4\text{ V}} - 1 = 6.5$$

2. Find V_{ref} from Eq. (4-8):

$$V_{ref} = \frac{n+1}{n}(V_{ctr}) = \frac{6.5 + 1}{6.5}(10) = 11.53\text{ V}$$

3. Choose $R = 10$ kΩ; therefore, resistor nR will be 6.5×10 kΩ $= 65$ kΩ. These circuit values and waveshapes are shown in Fig. 4-8.

4-5 VOLTAGE-LEVEL DETECTOR WITH INDEPENDENT ADJUSTMENT OF HYSTERESIS AND CENTER VOLTAGE

4-5.1 Introduction

The circuit of Fig. 4-9 is a noninverting voltage-level detector with *independent adjustment of hysteresis and center voltage*. In this circuit, the center voltage V_{ctr} is determined by both resistor mR and the reference voltage V_{ref}. V_{ref} can be either supply voltage $+V$ or $-V$. Remember that the op amp's supply voltage is being used for a lower parts count. Hysteresis voltage V_H is determined by resistor nR. If resistor nR is adjustable, then V_H can be adjusted *independently of* V_{ctr}. Adjusting resistor mR adjusts V_{ctr} without affecting V_H. Note that the signal source, E_i, must be a low-impedance source.* The key voltages are shown in Fig. 4-9 and are designed or evaluated from the following equations:

$$V_{UT} = -\frac{-V_{sat}}{n} - \frac{V_{ref}}{m} \tag{4-10a}$$

* If not, buffer E_i with a voltage follower as shown in Section 3-5.

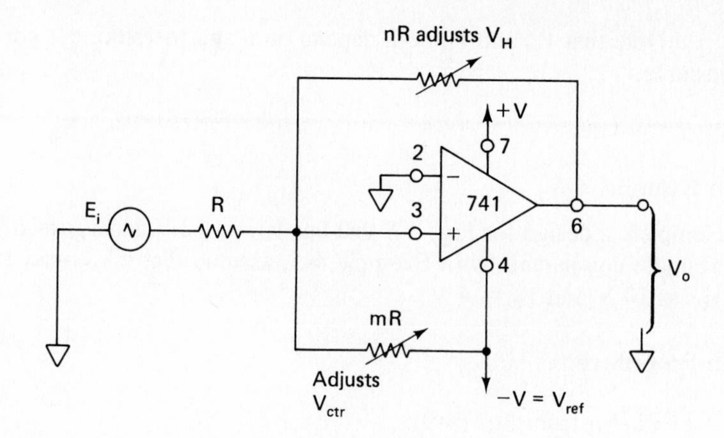

(a) Comparator with independent adjustments for hysteresis and reference voltage

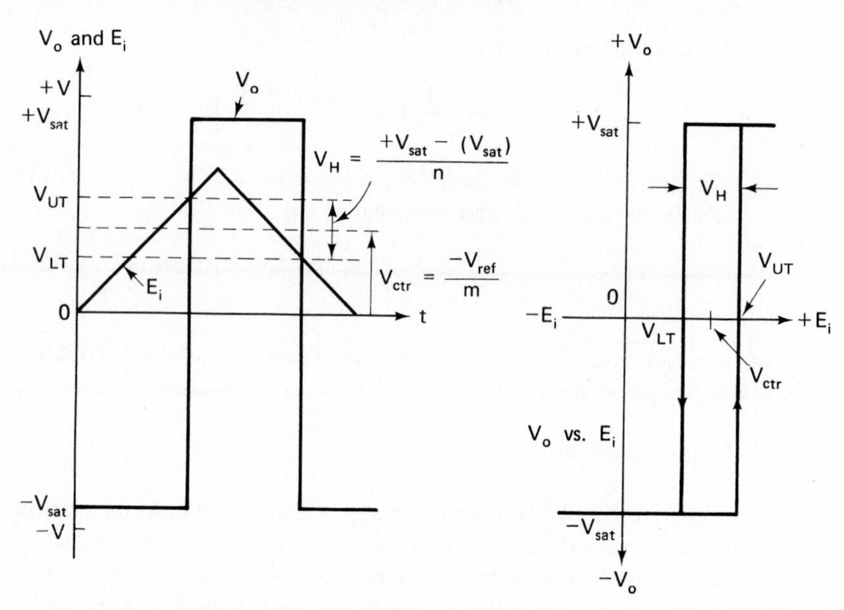

(b) Waveshapes for V_o and E_i

FIGURE 4-9 Resistor mR and supply voltage $-V$ establish the center voltage V_{ctr}. Resistor nR allows independent adjustment of the hysteresis voltage V_H, symmetrically around V_{ctr}.

$$V_{LT} = \frac{-V_{ref}}{m} - \frac{+V_{sat}}{n} \tag{4-10b}$$

$$V_H = V_{UT} - V_{LT} = \frac{(+V_{sat}) - (-V_{sat})}{n} \tag{4-11}$$

$$V_{ctr} = \frac{V_{UT} + V_{LT}}{2} = -\frac{V_{ref}}{m} - \frac{+V_{sat} + (-V_{sat})}{2n} \tag{4-12a}$$

The general equation for V_{ctr} seems complex. However, if the magnitudes of $+V_{sat}$ and $-V_{sat}$ are nearly equal, then V_{ctr} is expressed simply by

$$V_{ctr} = -\frac{V_{ref}}{m} \tag{4-12b}$$

So V_{ctr} depends only on m, and V_H depends only on n.

The following example shows how easy it is to design a battery-charger control circuit using the principles studied in this section.

4-5.2 Battery-Charger Control Circuit

Following is an example of the design procedure for a battery-charger control circuit.

Design Example 4-6

Assume that you want to monitor a 12-V battery. When the battery's voltage drops below 10.5 V, you want to connect it to a charger. When the battery voltage reaches 13.5 V, you want the charger to be disconnected. Therefore, $V_{LT} = 10.5$ V and $V_{UT} = 13.5$ V. Let us use the $-V$ supply voltage for V_{ref} and assume that it equals -15.0 V. Further, let us assume that $\pm V_{sat} = \pm 13.0$ V. Find (a) V_H and V_{ctr}; (b) resistor mR; (c) resistor nR.

Design Procedure

1. From Eqs. (4-11) and (4-12), find V_H and V_{ctr}.

$$V_H = V_{UT} - V_{LT} = 13.5\text{ V} - 10.5\text{ V} = 3.0\text{ V}$$

$$V_{ctr} = \frac{V_{UT} + V_{LT}}{2} = \frac{13.5\text{ V} + 10.5\text{ V}}{2} = 12.0\text{ V}$$

Note that the center voltage is the battery's nominal voltage.

2. Arbitrarily choose resistor R to be a readily available value of 10 kΩ. From Eq. (4-12b), choose V_{ref} as -15 V to make the sign of m be positive:

$$m = -\left(\frac{V_{ref}}{V_{ctr}}\right) = -\left(\frac{-15\text{ V}}{12\text{ V}}\right) = 1.25$$

Therefore, $mR = 1.25 \times 10\text{ k}\Omega = 12.5\text{ k}\Omega$.

3. From Eq. (4-11), find n.

$$n = \frac{+V_{sat} - (-V_{sat})}{V_H} = \frac{13\text{ V} - (-13\text{ V})}{3} = 8.66$$

Therefore, $nR = 86.6\text{ k}\Omega$.

The final circuit is shown in Fig. 4-10. When E_i drops below 10.5 V, V_o goes negative, releasing the relay to its normally closed position. The relay's normally closed contacts (NC) connect the charger to battery E_i. Diode D_1 protects the transistor against excessive reverse bias when $V_o = -V_{sat}$. When the battery charges to 13.5 V, V_o switched to $+V_{sat}$, which turns on the transistor and operates the relay. Its NC contacts open to disconnect the charger. Diode D_2 protects both op amp and transistor against transients developed by the relay's collapsing magnetic field.

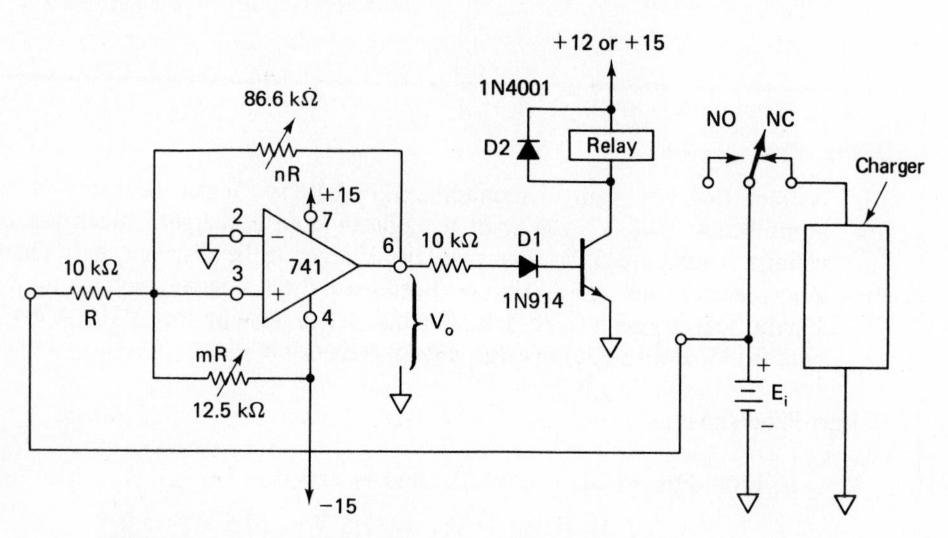

FIGURE 4-10 Battery-charger control for solution to Example 4-6. Adjust mR for $V_{ctr} = 12$ V in the test circuit of Fig. 4-9 and adjust nR for $V_H = 3$ V centered on V_{ctr}.

One final note. Suppose that the application requires an inverting voltage-level detector with hysteresis. That is, V_o must go low when E_i goes above V_{UT} and V_o must go high when E_i drops below V_{LT}. For this application, do not change the circuit or design procedure for the noninverting voltage-level detectors, simply add an inverting amplifier, or inverting comparator, to the output V_o.

4-6 ON–OFF CONTROL PRINCIPLES

4-6.1 Comparators in Process Control

Design Example 4-6 illustrates one of the most important applications of positive feedback circuits with hysteresis. They make excellent low-cost on–off controls. The circuit of Fig. 4-10 turns a charger on when battery voltage is below 10.5 V. It also turns the charger off when battery voltage exceeds 13.5 V. Note that the 10.5- to 13.5-V area is the *memory* or hysteresis range. If the battery voltage is 12.0 V, it can be either in the process of charging or discharging, depending on the last command.

4-6.2 The Room Thermostat as a Comparator

The temperature control in a room is a familiar example of on–off control. Let's draw an analogy to our comparator circuits. You adjust the temperature pointer to 65°F. This corresponds to V_{ctr}. The manufacturer builds in the hysteresis. Turn heat *on* if temperature is below 63°F. Turn heat *off* if temperature is above 67°F. If temperature is in the memory range (63 to 67°F), remember the last command. The memory range corresponds to V_H.

4-6.3 Selection/Design Guideline

Controls that are to be operated by the general public all share common characteristics with the comparators of Sections 4-4 and 4-5. The customer can only adjust V_{ctr}. V_H is present or, in some applications, the V_H control is brought out for the customer. These types of control circuits are fail safe. If temperature is set to 50°F, the room simply gets colder. They do not contain the possibility of catastrophic failure. That is, they won't mix up V_{UT} and V_{LT}.

The control circuit of the next section gives high-precision adjustments for the upper and lower setpoints of process control. It contains the possibility of runaway operation. Controls for V_{UT} and V_{LT} must be available only to a knowledgeable person and never to the general public.

4-7 AN INDEPENDENTLY ADJUSTABLE SETPOINT CONTROLLER

4-7.1 Principle of Operation

The circuit to be presented will allow both upper setpoint voltage V_{UT} and lower setpoint voltage V_{LT} to be adjusted independently and with precision. The principle of operation is straightforward. Set up a voltage with one fixed resistor and two adjustable resistors as in Fig. 4-11.

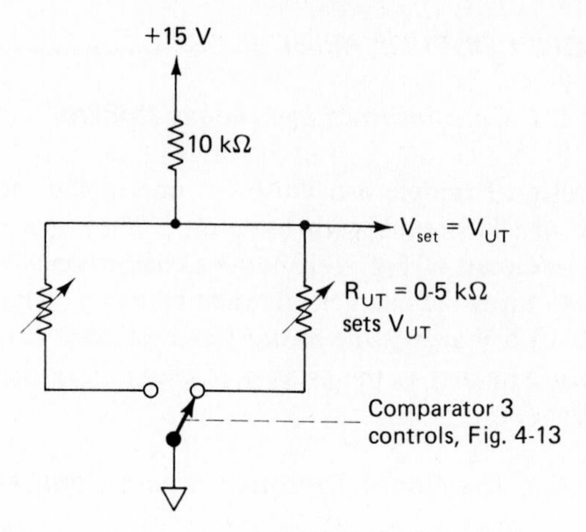

(a) If $E_i > V_{UT}$ switch is thrown to the right and grounds the upper setpoint adjust pot. If $V_{LT} < E_i < V_{UT}$, stay there (memory)

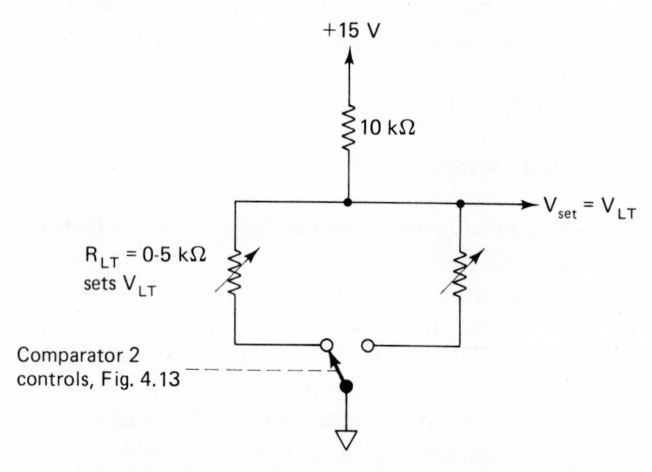

(b) If $E_i < V_{LT}$, switch is thrown to the left and grounds the lower setpoint adjust pot. If $V_{LT} < E_i < V_{UT}$, stay there (remember last command)

FIGURE 4-11 Separately adjustable setpoint or threshold voltages are designed with an electronically controlled S.P.D.T. switch, two pots, one resistor, and a power supply. For the resistor values shown, setpoint voltages are independently adjusted for any value between 0 and 5 V.

When a single-pole double-throw (S.P.D.T.) switch is thrown to one position, it grounds the upper setpoint adjust pot. Adjust R_{UT} so that the upper setpoint voltage V_{UT} appears at the setpoint voltage output, V_{set} [see Fig. 4-11(a)]. Then throw the switch to its remaining position as in Fig. 4-11(b). Adjust R_{LT} so that V_{LT} appears on the V_{set} line.

Before we design and analyze the hardware to make an electronically controlled S.P.D.T. switch, let us define the required output–input characteristics.

4-7.2 Output–Input Characteristics of an Independently Adjustable Setpoint Controller

Two outputs are required from our basic controller. The first is shown in Fig. 4-12(a). The setpoint voltage output V_{set} is needed for two reasons. First, it will be applied to one input of the in-out comparator (number 1 in Fig. 4-13). Second, it must be available for system test or calibration to trained test persons (not the public).

The second output voltage required is shown in Fig. 4-12(b). It is the input–output characteristic of the control system (number 1 in Fig. 4-13). Both output characteristics exhibit hysteresis.

4-7.3 Choice of Setpoint Voltages

In Fig. 4-11, the resistor components were chosen to give a range of 0 to 5 V for the setpoint voltage adjustments. In Fig. 4-12, V_{UT} was chosen to be 2.0 V and V_{LT} was chosen to be 0.5 V.

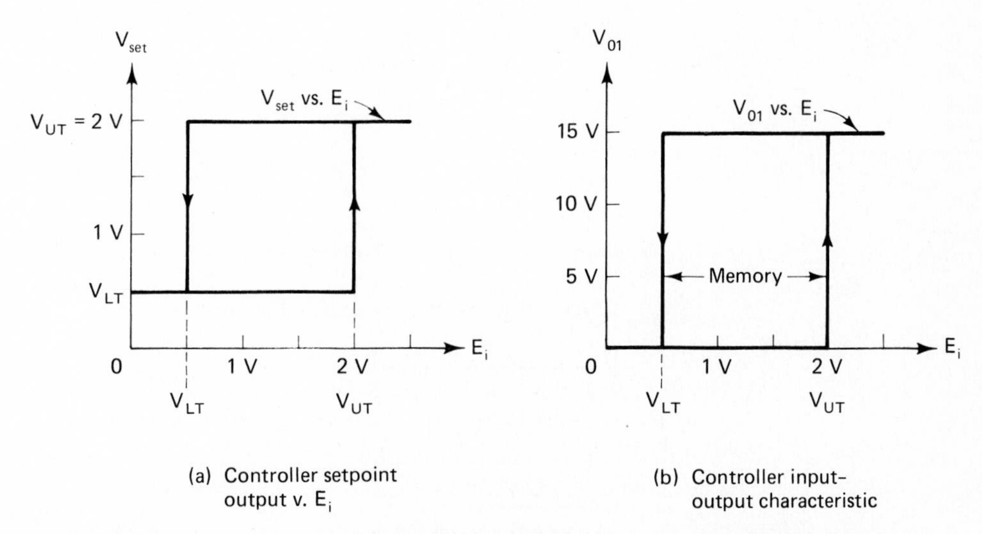

FIGURE 4-12 An on–off controller with independent setpoint voltages, V_{UT} and V_{LT}, has two distinct characteristics. The setpoint voltage depends on the input voltage as in (a). The input-output characteristics exhibits hysteresis as in (b).

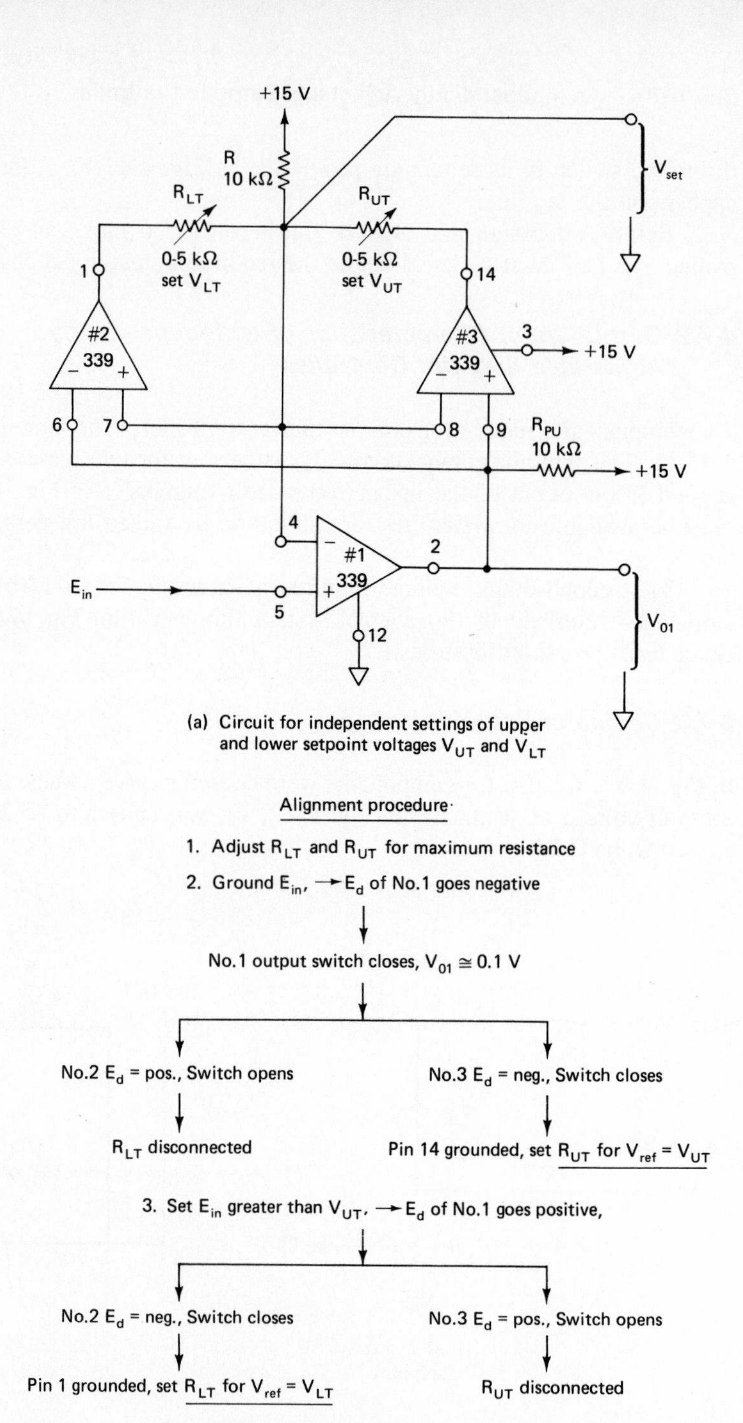

(a) Circuit for independent settings of upper and lower setpoint voltages V_{UT} and V_{LT}

(b) Alignment procedure and operating sequence

FIGURE 4-13 Circuit, alignment procedure, and operating sequence for a control module that allows independent adjustment of upper and lower setpoint voltages.

4-7.4 Circuit for Independently Adjustable Setpoint Voltage

We only have to add two more parts to the basic voltage divider of Fig. 4-11. One is a 10-kΩ resistor and the other is three-fourths of an LM339 open-collector comparator.

Recall from Chapter 2 that the LM339 usually works single supply with a ground at pin 12 on its negative rail. If differential input voltage E_d is negative for any one of its four comparators, the corresponding output switch is closed. This grounds the comparator's output terminal (and incidentally a set point pot). If E_d is positive [(+)in above (−)in] the output switch is open (and incidentally disconnects an adjustment pot).

The control circuit is finally presented in Fig. 4-13(a). The alignment procedure is presented as a flowchart showing the cause-and-effect sequence in Fig. 4-13(b). Study the alignment procedure carefully. It explains exactly how the circuit works.

You will probably draw the following conclusions from your study.

1. Comparators 2 and 3 form a single-pole double-throw switch to ground the bottom terminal and activate either R_{UT} or R_{LT} (see Fig. 4-11).
2. Comparator 1 is the output–input control. V_{set} is either at V_{UT} or at V_{LT} and is applied to 1's (−) input. Since E_i is applied to the (+) input, the circuit is inherently noninverting [see Fig. 4-12(b)].
3. This circuit makes an excellent laboratory experiment since it has a parts count of only five.

In practice, V_{o1} will usually drive a relay or optocoupler with triac output. Therefore, V_{o1} should be buffered or use the remaining LM339 comparator to (a) invert the output–input characteristic, or (b) avoid loading R_{PU} below either setpoint voltages.

4-7.5 Precautions

One last time, suppose that V_{UT} is adjusted to a value below V_{LT}. Upon power UP, one of two events will occur: (1) the system will not start up, or (2) the system will turn on and remain on until destruction occurs.

4-8 IC PRECISION COMPARATOR, 111/311

4-8.1 Introduction

The 111 (military) or 311 (commercial) comparator is an IC that has been designed and optimized for superior performance in voltage-level-detector applications. A comparator should be fast. That is, its output should respond quickly to changes at its inputs. The 311 is much faster than the 741 or 301 but not as fast as the 710 and NE522 high-speed comparators. The subject of speed is discussed in Section 4-10, "Propagation Delay."

The 311 is an excellent choice for a comparator because of its versatility. Its output is designed *not* to bounce between $\pm V_{sat}$ but can be changed quite easily. As a matter of fact, if you are interfacing to a system with a different supply voltage, you simply connect the output to the new supply voltage via an appropriate resistor. We begin by examining operation of the output terminal.

4-8.2 Output Terminal Operation

A simplified model of the 311 in Fig. 4-14(a) shows that its output behaves like a switch, Sw, connected between output pin 7 and pin 1. Pin 7 can be wired to any voltage V^{++} with magnitudes up to 40 V more positive than the $-V$ supply terminal (pin 4). When (+) input pin 2 is more positive than (−) input pin 3, the 311's equivalent output switch is open. V_o is then determined by V^{++} and is +5 *V*.

When the (+) input is less positive than (below) the (−) input, the 311's equivalent output switch closes and extends the ground at pin 1 to output pin 7. Here is one important difference between the 311 and the 339. The 339 has no equivalent to pin 1. There is no separate switch return terminal on the 339 as there is on the 311.

R_f and R_i add about 50 mV of hysteresis to minimize noise effects so that pin 2 is essentially at 0 V. Waveshapes for V_o and E_i are shown in Fig. 4-14(b). V_o is 0 V (switch closed) for positive half-cycles of E_i. V_o is +5 V (switch open) for negative half-cycles of E_i. This is a typical interface circuit; that is, voltages may vary between levels of +15 V and −15 V, but V_o is restrained between +5 V and 0 V, which are typical digital signal levels. So the 311 can be used for converting analog voltage levels to digital voltage levels (interfacing).

4-8.3 Strobe Terminal Operation

The strobe terminal of the 311 is pin 6 (see also Appendix 3). This strobe feature allows the comparator output either to respond to input signals or to be independent of input signals. Fig. 4-15 uses the 311 comparator as a zero-crossing detector. A 10-kΩ resistor is connected to the strobe terminal. The other side of the resistor is connected to a switch. With the strobe switch open, the 311 operates normally. That is, the output voltage is at V^{++} for negative values of E_i and at 0 for positive values of E_i. When the strobe switch is closed (connecting the 10 kΩ to ground), the output voltage goes to V^{++} regardless of the input signal. V_o will stay at V^{++} as long as the strobe switch is closed [see Fig. 4-15(b)]. The output is then independent of the inputs until the strobe switch is again opened.

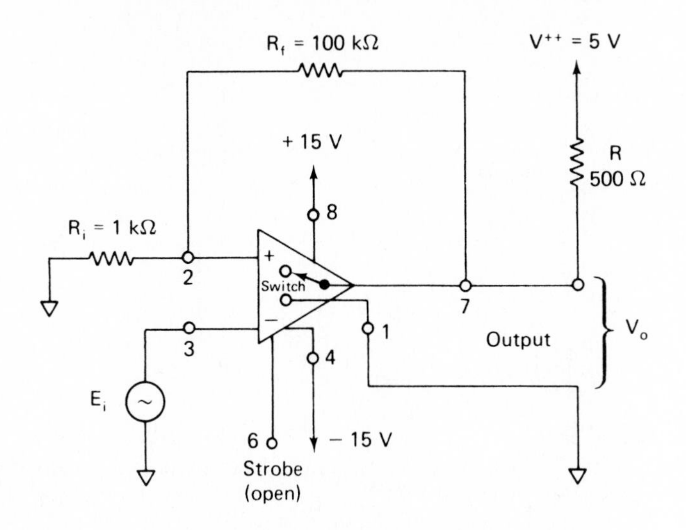

(a) 311 0-crossing detector with hysteresis

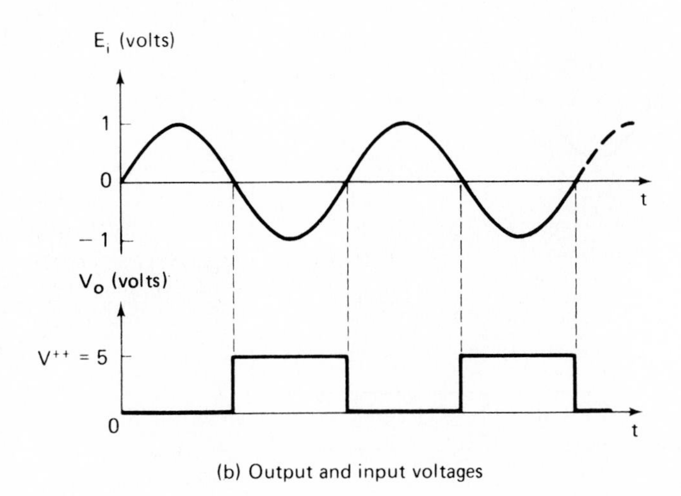

(b) Output and input voltages

FIGURE 4-14 Simplified model of the 311 comparator with input and output voltage waveforms.

The strobe feature is useful when a comparator is used to determine what type of signal is to be read out of a computer memory. The strobe switch is closed to ignore extraneous input signals that may occur up until the readout is due. Then during the readout time the switch is opened, and the 311 performs as a regular comparator. Current from the strobe terminal should be limited to about 3 mA. If the strobe feature is not to be used, the strobe terminal is left open or wired to $+V$ (see Appendix 3).

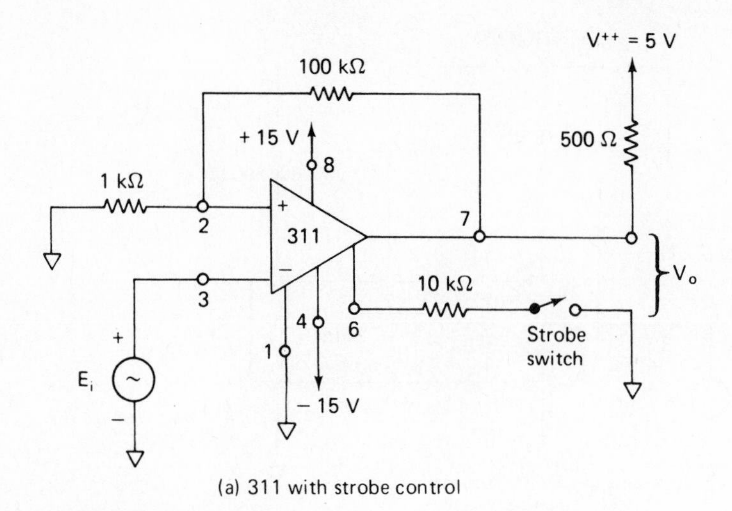

(a) 311 with strobe control

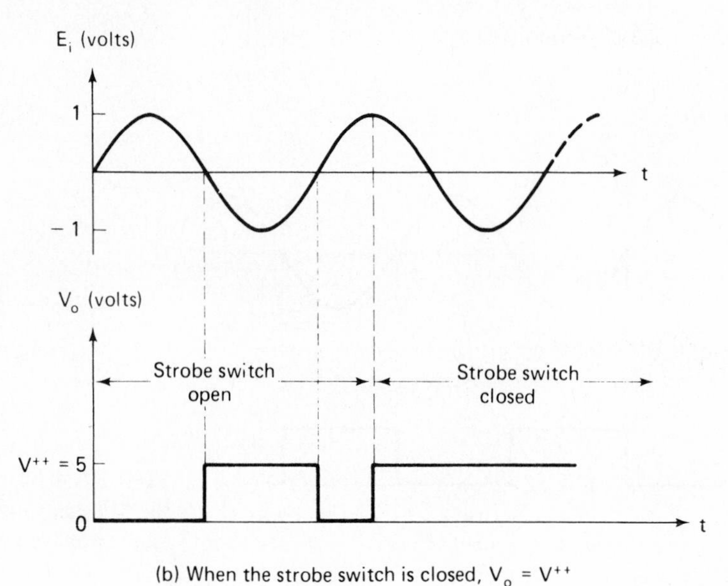

(b) When the strobe switch is closed, $V_o = V^{++}$

FIGURE 4-15 Operation of the strobe terminal.

4-9 WINDOW DETECTOR

4-9.1 Introduction

The circuit of Fig. 4-16 is designed to monitor an input voltage and indicate when this voltage goes either above or below prescribed limits. For example, IC logic power supplies for TTL must be regulated to 5.0 V. If the supply voltage should ex-

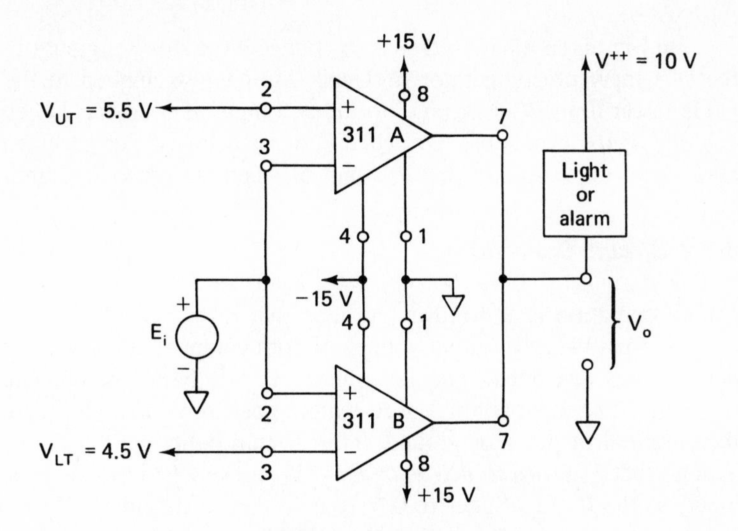

(a) Window detector circuit

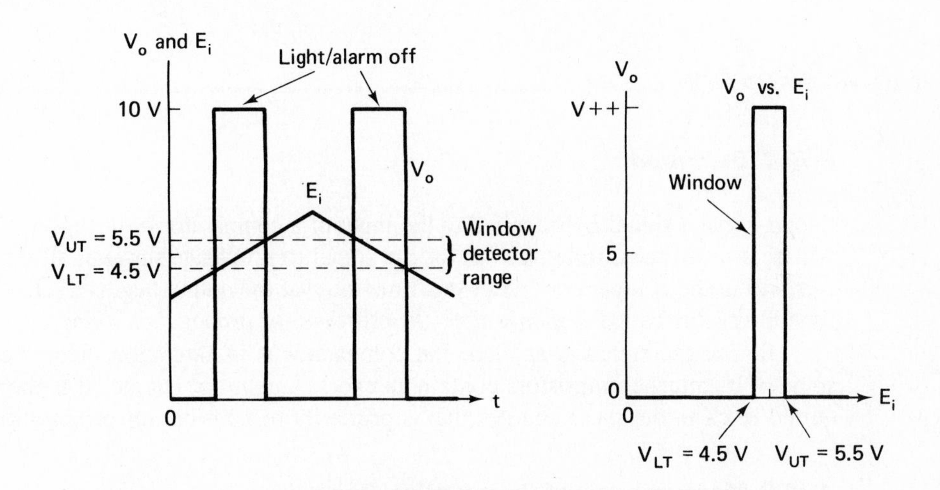

(b) Waveshapes for window detector

FIGURE 4-16 Upper- and lower-threshold voltages are independently adjustable in the window detector circuit.

ceed 5.5 V, the logic may be damaged, and if the supply voltage should drop below 4.5 V, the logic may exhibit marginal operation. Therefore, the limits for TTL power supplies are 4.5 V and 5.5 V. The power supply should be looking through a window whose limits are 4.5 V and 5.5 V, hence the name *window detector*. This circuit is sometimes called a *double-ended limit detector*.

In Fig. 4-16 input voltage E_i is connected to the (−) input of comparator *A* and the (+) input of comparator *B*. Upper limit V_{UT} is applied to the (+) input of *A*, while lower limit V_{LT} is applied to the (−) input of *B*. When E_i lies between V_{LT} and V_{UT}, the light/alarm is off. But when E_i drops below V_{LT} or goes above V_{UT}, the light/alarm goes on to signify that E_i is not between the prescribed limits.

4-9.2 Circuit Operation

Circuit operation is as follows. Assume that $E_i = 5$ V. Since E_i is greater than V_{LT} and less than V_{UT}, the output voltage of both comparators is at V^{++} because both output switches are open. The lamp/alarm is off. Next, assume that $E_i = 6.0$ V or $E_i > V_{UT}$. The input at pin 3 of *A* is more positive than at pin 2, so the *A* output is at the potential of pin 1 or ground. This ground lights the lamp, and $V_o = 0$ V. Now assume that E_i drops to 4.0 V *or* $E_i < V_{LT}$. The (+) input of *B* is less than its (−) input, so the *B* output goes to 0 V (the voltage at its pin 1). Once again this ground causes the lamp/alarm to light. Note that this application shows that output pins of 311 can be connected together and the output is at V^{++} only when the output of each comparator is at V^{++}.

4-10 PROPAGATION DELAY

4-10.1 Definition

Suppose that a signal E_i is applied to the input of a comparator as in Fig. 4-17. There will be a measurable time interval for the signal to propagate through all the transistors within the comparator. After this time interval the output begins to change. This time interval is called *response time, transit time,* or *propagation delay*.

Before the signal is applied, the comparator is in saturation. This means that some of its internal transistors contain an excess amount of charge. It is the time required to clean out these charges that is primarily responsible for propagation delay.

4-10.2 Measurement of Propagation Delay

The comparator's depth of saturation depends directly on the amount of differential input voltage. The conventional method of comparing performance of one comparator with another is to first connect a +100-mV reference voltage to one input. In Fig. 4-17(a) the reference voltage is connected to the (−) input. The other (+) input is connected to 0 V. This forces all comparators, under test for propagation delay, into the same initial state of saturation. In Fig. 4-17(b) the outputs are shown at about 0.4 V before time 0.

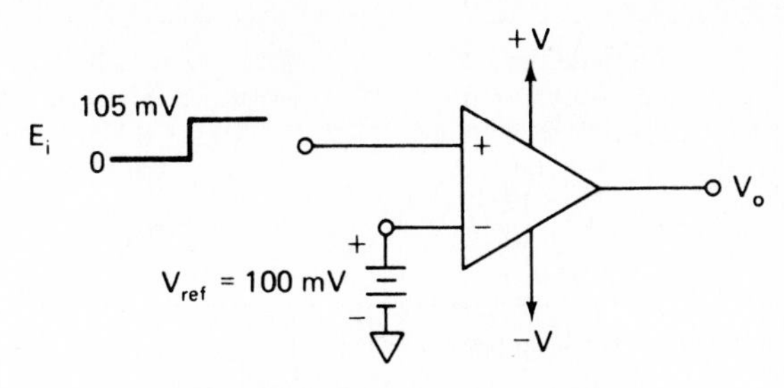

(a) Test circuit for propagation delay

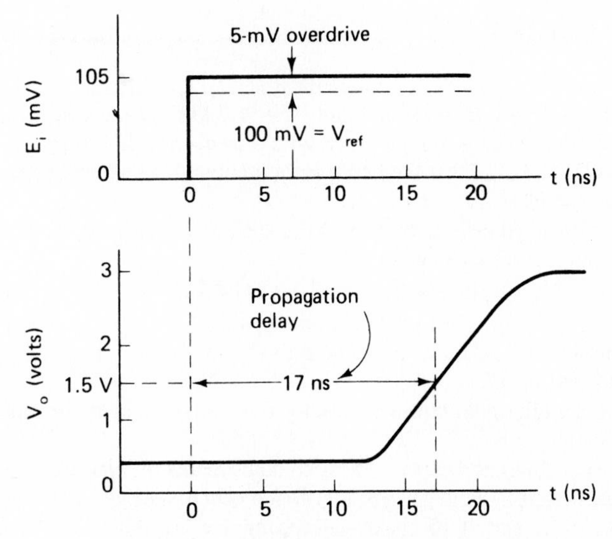

(b) Propagation delay is the time interval between start of an input step voltage and the output rise to 1.5 V (NE522 comparator)

FIGURE 4-17 Propagation delay is measured by the test circuit in (a) and defined by the waveshapes in (b).

A fast-rising signal voltage E_i is then applied to the (+) input at time $t = 0$ in Fig. 4-17. If E_i is brought up to 100 mV, the comparator will be on the verge of switching but will *not* switch. However, if E_i is brought up quickly to 100 mV plus a small amount of *overdrive,* the overdrive signal will propagate through the comparator. After a propagation delay the output comes out of saturation and rises to a specified voltage. This voltage is typically 1.5 V.

As shown in Fig. 4-17(b), a 5-mV overdrive results in a propagation delay of 17 ns for a NE522 comparator. Increasing the overdrive to 100 mV will reduce propagation delay to 10 ns. Typical response times for the 311, 522, and 710 comparators, and the 301 op amp are:

Comparator	Response time for 5-mV overdrive (ns)	Response time for 20-mV overdrive (ns)
311[a]	170	100
522	17	15
710	40	20
301	>10,000	>10,000

[a] $V^{++} = 5$ V with a 500-Ω pull-up resistor.

LABORATORY EXERCISES

4-1. Build the circuit of Fig. 4.3 with $R_1 = 10$ kΩ and $R_2 = 4.7$ kΩ. This gives threshold voltages of about ±5 V. Let E_i be a triangle wave with a frequency of 100 Hz.

(a) Set E_i peak for 1.0 V and note that the comparator's output does *not* change. This shows operation in the memory range.

(b) Increase E_i to 10 V peak. Sketch E_i vs. t and V_o vs. t from a dual-trace CRO. Then sketch V_o vs. E_i from a CRO *x-y* plot.

(c) Measure from, and label on, the sketches: V_{UT}, V_{LT}, and V_H.

4-2. Use Design Example 4-6 and Fig. 4-9 to design a circuit whose output is at $+V_{sat}$ when E_i exceeds 2.0 V, and at $-V_{sat}$ when E_i is below 0.5 V. Choose $R = 10$ kΩ. Install convenient pots for nR and/or mR.

(a) Use a laboratory-type adjustable power supply for E_i or a low-impedance-function generator.

(b) Observe V_o vs. E_i on a CRO and experiment with adjustments of mR and nR. Note that nR adjusts the hysteresis voltage; mR adjusts the center voltage.

The circuits of Figs. 4-13 and 4-16 make interesting lab experiments. Look at waveshapes of E_i vs. t and V_o vs. t on a dual-trace CRO. It is good experience to learn how to measure V_{UT} and V_{LT} from these plots. Identify the value of E_i, where the V_o transitions occur.

PROBLEMS

4-1. How do you recognize when positive feedback is present in the schematic of an op amp circuit?

4-2. In Fig. P4-2, $R_1 = 25$ kΩ and $R_2 = 5$ kΩ. Assume for simplicity that $\pm V_{sat} = \pm 15$ V. Calculate **(a)** V_{UT}; **(b)** V_{LT}; **(c)** V_H.

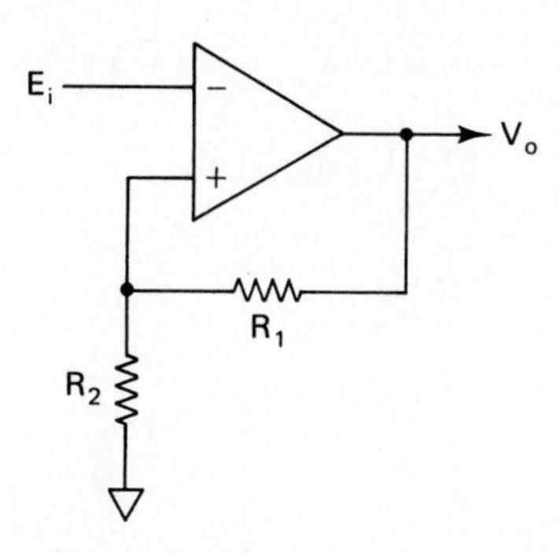

FIGURE P4-2

4-3. For the values given in Problem 4-2, plot **(a)** E_i vs. t; **(b)** V_o vs. t; **(c)** V_o vs. E_i. Let E_i be a 100-Hz triangular wave with peak values of ± 10 V.

4-4. Label V_{UT}, V_{LT}, and V_H on your sketches from Problem 4-3.

4-5. Given the waveshapes of E_i vs. t and V_o vs. t in Fig. P4-5, identify **(a)** the frequency of E_i; **(b)** the peak amplitude of E_i; **(c)** the value of V_{UT}; **(d)** the value of V_{LT}; **(e)** V_H.

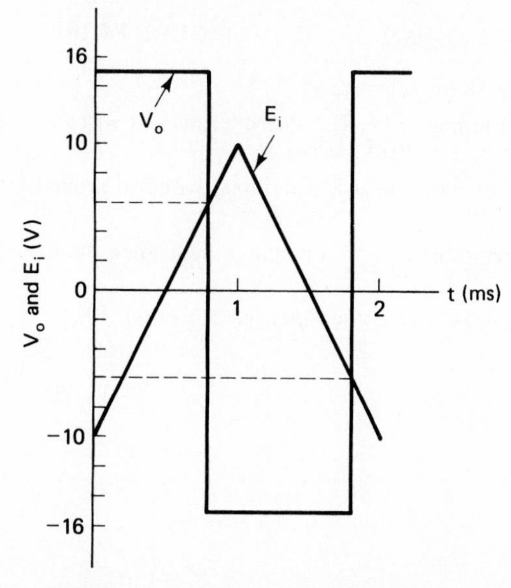

FIGURE P4-5

4-6. Use Fig. 4-7 and Design Example 4-4 for guidance. Design a noninverting voltage-level detector with $V_{UT} = 2.0$ V and $V_{LT} = 0.5$ V.

4-7. To see how negative threshold voltages are handled, redesign the voltage-level detector of Problem 4-6 for $V_{UT} = -0.5$ V and $V_{LT} = -2.0$ V. (Note that $V_H = +1.5$ V in both problems.)

4-8. Refer to Fig. 4-8 and Design Example 4-5 for guidance. Design a circuit **(a)** whose output is at $(+)V_{sat}$ when its input is *below* $V_{LT} = 0.5$ V; **(b)** whose output is at $(-)V_{sat}$ when its input is above $V_{UT} = 2.0$ V.

4-9. Redesign the circuit of Fig. 4-9 for $V_{UT} = 2.0$ V and $V_{LT} = 0.5$ V (see Design Example 4-6).

4-10. For the circuit of Fig. P4-10 calculate **(a)** V_{ctr}; **(b)** V_H; **(c)** V_{UT}; **(d)** V_{LT}. Assume that $\pm V_{sat} = \pm 15$ V.

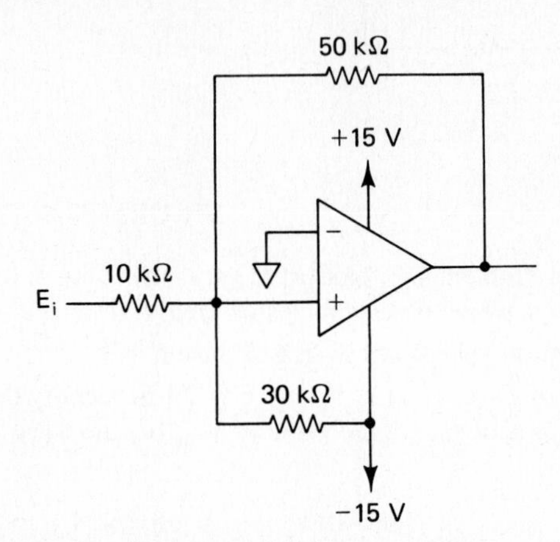

FIGURE P4-10

4-11. If E_i is grounded in Fig. 4-14, calculate V_o.

4-12. Refer to the 311 circuit in Fig. 4-14. The strobe terminal is wired to +15 V. Find the value of V_o when **(a)** $E_i = 1$ V; **(b)** $E_i = -1$ V.

4-13. Repeat Problem 4-12 but with the strobe terminal wired to ground via a 10 kΩ resistor.

4-14. Design a window detector circuit whose output is high when the input voltage is between +2 and +0.5 V.

4-15. Which comparator has a faster response time, the 311 or the 301?

CHAPTER 5

Selected Applications of Op Amps

LEARNING OBJECTIVES

Upon completing this chapter on selected applications of op amps, you will be able to:

- Appreciate how one or two op amps plus a few components can give an inexpensive solution to a number of practical applications.
- Make a universal high-resistance voltmeter.
- Test diodes, LEDs, IREDs, and low-voltage zeners with a constant current.
- Draw a circuit that interfaces a teleprinter to a microcomputer.
- Measure the power received by a solar cell, photodiode, or photoresistor.
- Explain how to measure solar energy.
- Shift the phase angle of a fixed-frequency sine wave by a precise amount and independent of its amplitude.
- Explain the problems encountered during the magnetic recording process and how they are solved by an op amp equalizer circuit.

- Sketch a tone-control circuit.
- Show how to make an electronic Celsius or Fahrenheit thermometer with an AD590 temperature transducer and a current-to-voltage converter.

5-0 INTRODUCTION

Why is the op amp such a popular device? This chapter attempts to answer that question by presenting a wide selection of applications. They were selected to show that often the op amp can perform as a very nearly ideal device. Moreover, the diversity of operations that the op amp can perform is almost without limit. In fact, applications that are normally very difficult, such as measuring shortcircuit current, are rendered easy by the op amp. Together with a few resistors and a power supply, the op amp can, for example, measure the output from photodetectors, give audio tone control, equalize tones of different amplitudes, control high currents, and allow matching of semiconductor device characteristics. We begin with selecting an op amp circuit to make a high-resistance dc and ac voltmeter.

5-1 HIGH-RESISTANCE DC VOLTMETER

5-1.1 Basic Voltage-Measuring Circuit

Figure 5-1 shows a simple but very effective high-input-resistance dc voltmeter. The voltage to be measured, E_i, is applied to the (+) input terminal. Since the differential input voltage is 0 V, E_i is developed across R_i. The meter current I_m is set by E_i and R_i just as in the noninverting amplifier.

$$I_m = \frac{E_i}{R_i} \tag{5-1}$$

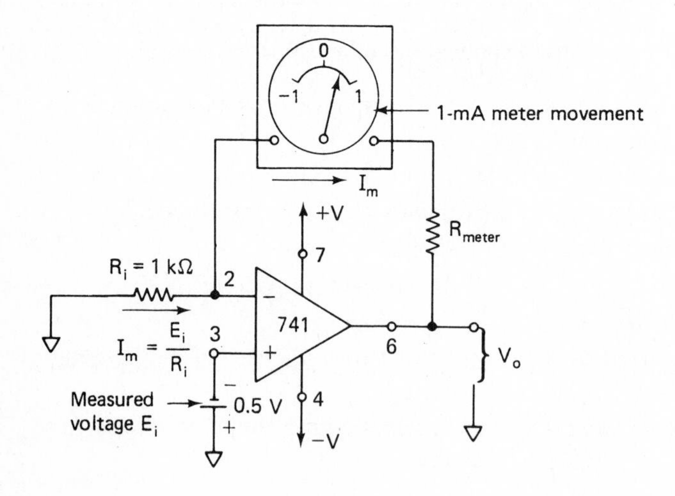

FIGURE 5-1 High-input-resistance dc voltmeter.

If R_i is 1 kΩ, then 1 mA of meter current will flow for $E_i = 1$ V dc. Therefore, the milliammeter can be calibrated directly in volts. As shown, this circuit can measure any dc voltage from −1 V to +1 V.

Example 5-1

Find I_m in Fig. 5-1.

Solution From Eq. (5-1), $I_m = 0.5 \text{ V}/1 \text{ k}\Omega = 0.5$ mA. The needle is deflected halfway between 0 and +1 mA.

One advantage of Fig. 5-1 is that E_i sees the very high input impedance of the (+) input. Since the (+) input draws negligible current, it will not load down or change the voltage being measured. Another advantage of placing the meter in the feedback loop is that if the meter resistance should vary, it will have no effect on meter current. Even if we added a resistor in series with the meter, within the feedback loop, it would not affect I_m. The reason is that I_m is set only by E_i and R_i. The output voltage will change if meter resistance changes, but in this circuit we are not concerned with V_o. This circuit is sometimes called a *voltage-to-current* converter.

5-1.2 Voltmeter Scale Changing

Since the input voltage in Fig. 5-1 must be less than the power supply voltages (±15 V), a convenient maximum limit to impose on E_i is ±10 V. The simplest way to convert Fig. 5-1 from a ±1-V voltmeter to a ±10-V voltmeter is to change R_i to 10 kΩ. In other words, pick R_i so that the full-scale input voltage E_{FS} equals R_i times the full-scale meter current I_{FS} or

$$R_i = \frac{E_{FS}}{I_{FS}} \tag{5-2}$$

Example 5-2

A microammeter with $50\ \mu\text{A} = I_{FS}$ is to be used in Fig. 5-1. Calculate R_i for $E_{FS} = 5$ V.

Solution By Eq. (5-2), $R_i = 5 \text{ V}/50\ \mu\text{A} = 100$ kΩ.

To measure higher input voltages, use a voltage-divider circuit. The output of the divider is applied to the (+) input.

5-2 UNIVERSAL HIGH-RESISTANCE VOLTMETER

5-2.1 Circuit Operation

The voltage-to-current converter of Fig. 5-2 can be used as a universal voltmeter. That is, it can be used to measure positive or negative dc voltage or the rms, peak, or peak-to-peak (p-p) value of a *sine wave*. To change from one type of voltmeter to another, it is necessary to change only a single resistor. The voltage to be measured, E_i, is applied to the op amp's (+) input. Therefore, the meter circuit has a high input resistance.

When E_i is positive, current flows through the meter movement and diodes D_3 and D_4. When E_i is negative, current flows in the *same* direction through the meter and diodes D_1 and D_2. Thus meter current direction is the same whether E_i is positive or negative.

A dc meter movement measures the *average* value of current. Suppose that a basic meter movement is rated to give full-scale deflection when conducting a current of 50 μA. A voltmeter circuit containing the basic meter movement is to indicate at full scale when E_i is a sine wave with a peak voltage of 5 V. The meter *face* should be calibrated linearly from 0 V to +5 V instead of 0 to 50 μA. The *circuit* and meter movement would then be called a *peak reading voltmeter* (for sine waves only) with a full-scale deflection for E_{ip} = 5 V. The following section shows how easy it is to design a universal voltmeter.

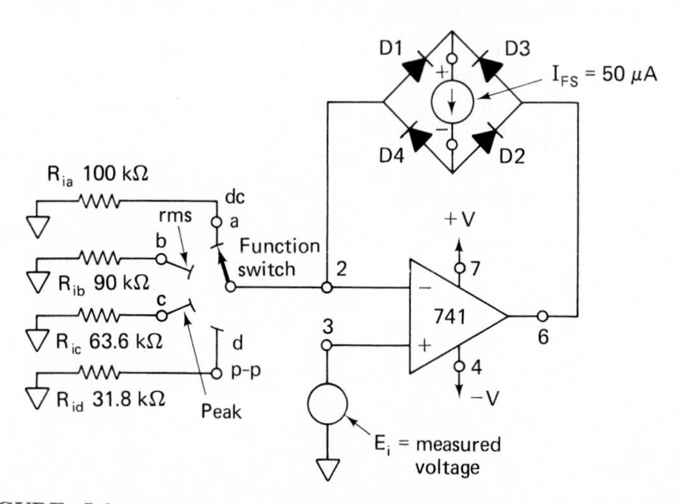

FIGURE 5-2 Basic high-resistance universal voltmeter circuit. The meaning of a full-scale meter deflection depends on the function switch position as follows: 5 V dc on position *a*, 5 V ac rms on position *b*, 5 V peak ac on position *c*, and 5 V ac p-p on position *d*.

5-2.2 Design Procedure

The design procedure is as follows: Calculate R_i according to the application from one of the following equations:

1. Dc voltmeter:

$$R_i = \frac{\text{full-scale } E_{dc}}{I_{FS}} \tag{5-3a}$$

2. Rms ac voltmeter (sine wave only):

$$R_i = 0.90\frac{\text{full-scale } E_{rms}}{I_{FS}} \tag{5-3b}$$

3. Peak reading voltmeter (sine wave only)

$$R_i = 0.636\frac{\text{full-scale } E_{peak}}{I_{FS}} \tag{5-3c}$$

4. Peak-to-peak ac voltmeter (sine wave only)

$$R_i = 0.318\frac{\text{full-scale } E_{p/p}}{I_{FS}} \tag{5-3d}$$

where I_{FS} is the meter's full-scale current rating in amperes. The design procedure is illustrated by an example.

Design Example 5-3

A basic meter movement (such as the Simpson 260) is rated at 50 μA for full-scale deflection (with a meter resistance of 5 kΩ). Design a simple switching arrangement and select resistors to indicate full-scale deflection when the voltage to be measured is (a) 5 V dc; (b) 5 V rms; (c) 5 V peak; (d) 5 V p-p.

Design Procedure From Eqs. (5-3a) to (5-3d):

(a) $R_{ia} = \dfrac{5\text{ V}}{50\ \mu\text{A}} = 100\text{ k}\Omega$ (b) $R_{ib} = 0.9\dfrac{5\text{ V}}{50\ \mu\text{A}} = 90\text{ k}\Omega$

(c) $R_{ic} = 0.636\dfrac{5\text{ V}}{50\ \mu\text{A}} = 63.6\text{ k}\Omega$ (d) $R_{id} = 0.318\dfrac{5\text{ V}}{50\ \mu\text{A}} = 31.8\text{ k}\Omega$

The resulting circuit is shown in Fig. 5-2.

It must be emphasized that neither meter resistance nor diode voltage drops affect meter current. Only R_i and E_i determine average or dc meter current.

5-3 VOLTAGE-TO-CURRENT CONVERTERS: FLOATING LOADS

5-3.1 Voltage Control of Load Current

From Sections 5-1 and 5-2 we learned not just how to make a voltmeter but that current in the feedback loop depends on the input voltage and R_i. There are applications where we need to pass a constant current through a load and hold it constant despite any changes in load resistance or load voltage. If the load does not have to be grounded, we simply place the load in the feedback loop and control both input and load current by the principle developed in Section 5-1.

5-3.2 Zener Diode Tester

Suppose that we have to test the breakdown voltage of a number of zener diodes at a current of precisely 5 mA. If we connect the zener in the feedback loop as in Fig. 5-3(a), our voltmeter circuit of Fig. 5-1 becomes a zener diode tester. That is, E_i and R_i set the load or zener current at a constant value. E_i forces V_o to go negative until the zener breaks down and clamps the zener voltage at V_z. R_i converts E_i to a current, and as long as R_i and E_i are constant, the load current will be constant regardless of the value of the zener voltage. Zener breakdown voltage can be calculated from V_o and E_i as $V_z = V_o - E_i$.

Example 5-4

In the circuit of Fig. 5-3(a), $V_o = 10.3$ V, $E_i = 5$ V, and $R_i = 1$ kΩ. Find (a) the zener current; (b) the zener voltage.

Solution (a) From Eq. (3-1), $I = E_i/R_i$ or $I = 5\text{ V}/1\text{ k}\Omega = 5$ mA. (b) From Fig. 5-3(a), rewrite the equation for V_o.

$$V_z = V_o - E_i = 10.3\text{ V} - 5\text{V} = 5.3\text{ V}$$

5-3.3 Diode Tester

Suppose that we needed to select diodes from a production batch and find pairs with matching voltage drops at a particular value of diode current. Place the diode in the feedback loop as shown in Fig. 5-3(b). E_i and R_i will set the value of I. The (−) input draws negligible current, so I passes through the diode. As long as E_i and R_i are constant, current through any diode I will be constant at $I = E_i/R_i$. V_o will equal the diode voltage for the same reasons that V_o was equal to V_{R_f} in the inverting amplifier (see Section 3-1).

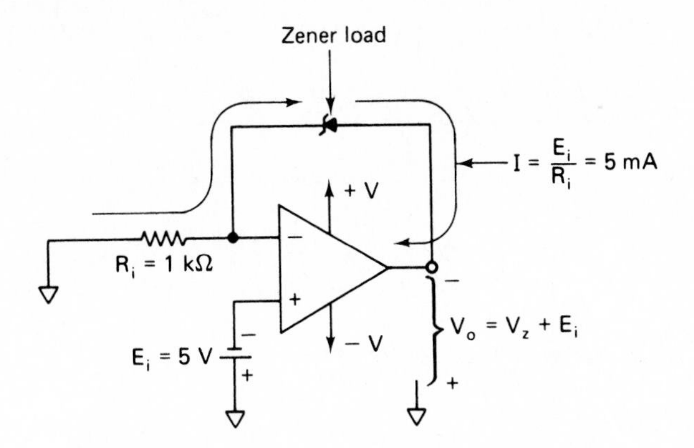

(a) Negligible current drawn from E_i, load current furnished by op amp

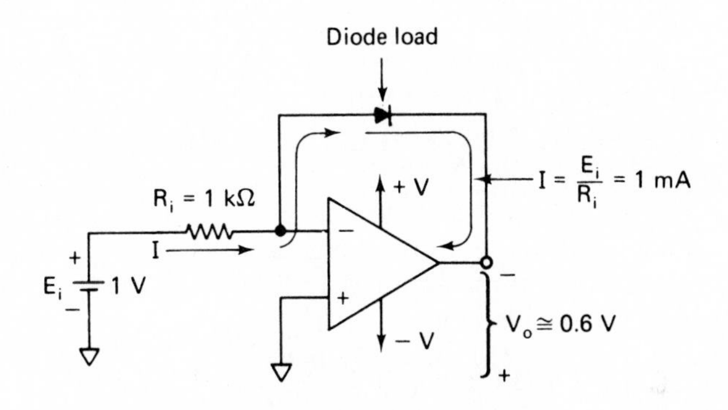

(b) Load current equals input current

FIGURE 5-3 Voltage-controlled load currents with loads in the feedback loop.

Example 5-5

$E_i = 1$ V, $R_i = 1$ kΩ, and $V_o = 0.6$ V in Fig. 5-3(b). Find (a) the diode current; (b) the voltage drop across the diode.

Solution (a) $I = E_i/R_i = 1\text{ V}/1\text{ k}\Omega = 1$ mA. (b) $V_{\text{diode}} = V_o = 0.6$ V.

There is one disadvantage with the circuit of Fig. 5-3(b): E_i must be able to furnish the current. Both circuits in Fig. 5-3 can only furnish currents up to 10 mA because of the op amp's output current limitation. Higher-load currents can be fur-

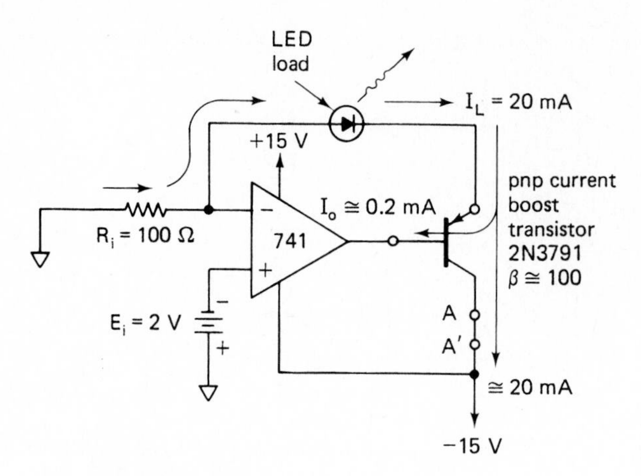

FIGURE 5-4 Voltage-to-high current converter.

nished from the power supply terminal and a current boost transistor as shown in Fig. 5-4.

5-4 LIGHT-EMITTING-DIODE TESTER

The circuit of Fig. 5-4 converts E_i to a 20-mA load current based on the same principles discussed in Sections 5-1 to 5-3. Since the 741's output terminal can only supply about 5 to 10 mA, we cannot use the circuits of Figs. 5-1 to 5-3 for higher load currents. But if we add a transistor as in Fig. 5-4, load current is furnished from the negative supply voltage. The op amp's output terminal is required to furnish only base current, which is typically $\frac{1}{100}$ of the load current. The factor $\frac{1}{100}$ comes from assuming that the transistor's beta equals 100. Since the op amp can furnish an output current of up to 5 mA into the transistor's base, this circuit can supply a maximum load current of 5 mA $\times$ 100 = 0.5 A.

A light-emitting diode such as the MLED50 is specified to have a typical brightness of 750 fL provided that the forward diode current is 20 mA. E_i and R_i will set the diode current I_L equal to $E_i/R_i = 2\text{ V}/100\ \Omega = 20\text{ mA}$. Now brightness of LEDs can be measured easily one after another for test or matching purposes, because the current through each diode will be exactly 20 mA regardless of the LED's forward voltage.

It is worthwhile to note that a load of two LEDs can be connected in series with the feedback loop and both would conduct 20 mA. The load could also be connected in Fig. 5-4 between points AA' (which is in series with the transistor's collector) and still conduct about 20 mA. This is because the collector and emitter currents of a transistor are essentially equal. A load in the feedback loop is called a *floating load*. If one side of the load is grounded, it is a *grounded load*. To supply a constant current to a grounded load, another type of circuit must be selected, as shown in Section 5-5.

5-5 FURNISHING A CONSTANT CURRENT TO A GROUNDED LOAD

5-5.1 Differential Voltage-to-Current Converter

The circuit of Fig. 5-5 can be called a differential voltage-to-current converter because the load current I_L depends on the *difference* between input voltage E_1 and E_2 and resistor R. I_L does *not* depend on load resistor R_L. Therefore, if E_1 and E_2 are constant, the grounded load is driven by a constant current. Load current can flow in either direction, so this circuit can either source or sink current.

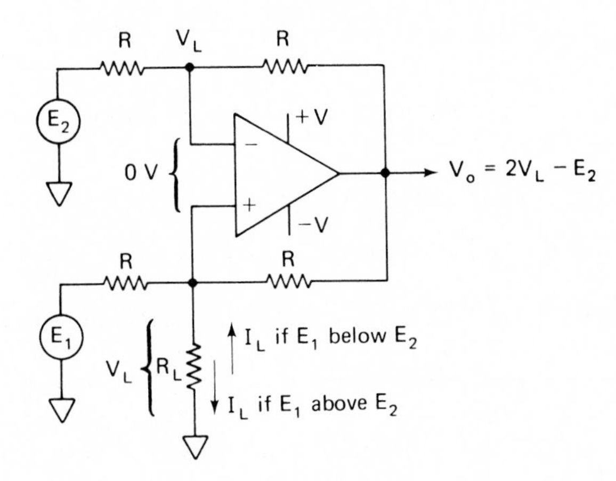

FIGURE 5-5 Differential voltage-to-current converter or constant-current source with grounded load.

Load current I_L is determined by

$$I_L = \frac{E_1 - E_2}{R} \tag{5-4}$$

A positive value for I_L signifies that it flows downward in Fig. 5-5 and V_L is positive with respect to ground. A negative value of I_L means that V_L is negative with respect to ground and current flows upward.

Load voltage V_L (not I_L) depends on load resistor R_L from

$$V_L = I_L R_L \tag{5-5}$$

To ensure that the op amp does not saturate, V_o must be known and can be calculated from

$$V_o = 2V_L - E_2 \tag{5-6}$$

Circuit operation is illustrated by the following examples.

Example 5-6

In Fig. 5-5, $R = 10\ \text{k}\Omega$, $E_2 = 0$, $R_L = 5\ \text{k}\Omega$, and $E_1 = 5$ V. Find (a) I_L; (b) V_L; (c) V_o.

Solution (a) From Eq. (5-4),

$$I_L = \frac{5\ \text{V} - 0}{10\ \text{k}\Omega} = 0.5\ \text{mA}$$

(b) From Eq. (5-5),

$$V_L = 0.5\ \text{mA} \times 5\ \text{k}\Omega = 2.5\ \text{V}$$

(c) From Eq. (5-6),

$$V_o = 2 \times 2.5\ \text{V} = 5\ \text{V}$$

Reversing the polarity of E_1 reverses I_L and the polarity of V_o and V_L.

Example 5-7

In Fig. 5-5, $R = 10\ \text{k}\Omega$, $E_2 = 5$ V, $R_L = 5\ \text{k}\Omega$, and $E_1 = 0$. Find (a) I_L; (b) V_L; (c) V_o. Compare this example with Example 5-6.

Solution (a) From Eq. (5-4),

$$I_L = \frac{0 - 5\ \text{V}}{10\ \text{k}\Omega} = -0.5\ \text{mA}$$

(b) From Eq. (5-5),

$$V_L = -0.5\ \text{mA} \times 5\ \text{k}\Omega = -2.5\ \text{V}$$

(c) From Eq. (5-6),

$$V_o = 2(-2.5\ \text{V}) - 5\ \text{V} = -10\ \text{V}$$

Note: V_L and I_L are reversed in polarity and direction, respectively, from Example 5-6. If the polarity of E_2 is reversed, I_L and V_L change sign but *not* magnitude.

5-5.2 Constant-High-Current Source, Grounded Load

In certain applications, such as electroplating, it is desirable to furnish a high current, of constant value, to a grounded load. The circuit of Fig. 5-6 will furnish constant currents above 500 mA provided that the transistor is heat-sinked properly (above 5 W) and has a high beta ($\beta > 100$). The circuit operates as follows. The

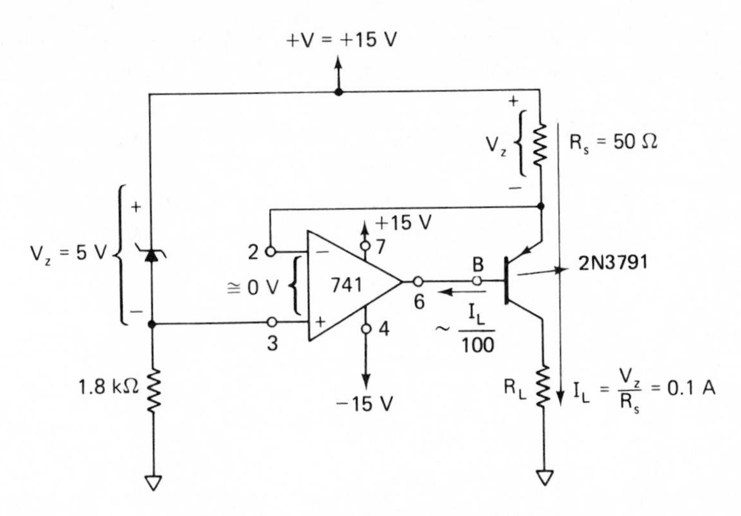

FIGURE 5-6 Constant-high-current source.

zener diode voltage is applied to one end of current sense resistor R_s and the op amp's positive input. Since the differential input voltage is 0 V, the zener voltage is developed across R_s. R_s and V_z set the emitter current, I_E, constant at V_z/R_s. The emitter and collector currents of a bipolar junction transistor are essentially equal. Since the collector current is load current I_L and $I_L \approx I_E$, the load current I_L is set by V_z and R_s.

If the op amp can furnish a base current drive of over 5 mA and if the beta of the transistor is greater than 100, then I_L can exceed 5 mA × 100 = 500 mA. The voltage across the load must not exceed the difference between the supply and the zener voltage; otherwise, the transistor and the op amp will go into saturation. (If oscillations occur, install a 100-Ω resistor between terminals 6 and B.)

5-5.3 Interfacing a Microcomputer to a Teleprinter

A TTL digital circuit, microprocessor, or microcomputer communicates to the outside world in a binary language that has only two symbols, 0 and 1. Their corresponding electrical voltages are low (<0.8 V) and high (>2.4 V). Teleprinters need a serial pulse train of either 20- or 4-mA current pulses coded to form numerical and alphabetical characters that can be understood by human beings. The 20-mA current pulse energizes a selector magnet and the 4-mA current pulse releases the magnet to control a rotating mechanical selector mechanism.

Five current pulses in sequence give $2^5 = 32$ possible character combinations. Actually, 30 combinations are used for the alphabet plus space, period, and so on. The thirty-first combination downshifts the selector print head to print numeral and other codes such as the ampersand (comma, period, colon, etc.). The thirty-second combination upshifts the printhead to print alphabetical characters. The printhead has an upper row of alphabet characters and a lower row of numerical characters.

The principles of operation for a circuit that will interface between a microcomputer and teleprinter is shown in Fig. 5-7 and analyzed in Section 5-5.4.

5-5.4 Digitally Controlled 4- to 20-mA Current Source

In the circuit of Fig. 5-7(a), resistors R_1 and R_2 form an unloaded voltage divider. Since E_d of the op amp is zero volts, the 2-V drop always appears between the posi-

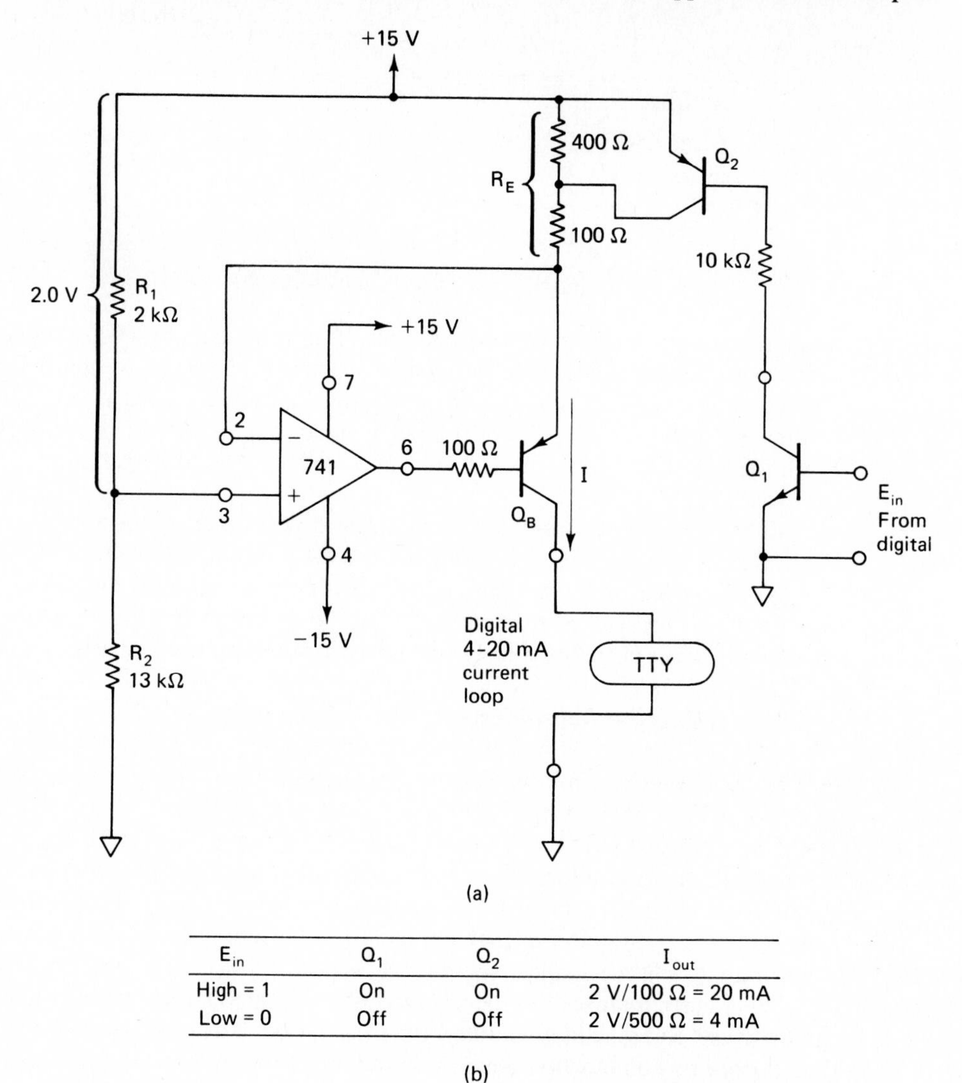

E_{in}	Q_1	Q_2	I_{out}
High = 1	On	On	2 V/100 Ω = 20 mA
Low = 0	Off	Off	2 V/500 Ω = 4 mA

(b)

FIGURE 5-7 Digitally controlled 4-to-20-mA current source.

tive rail and emitter of the current boost transistor Q_B. Operation of the circuit is summarized in Fig. 5-7(b).

If the 400-Ω resistor is *not* shorted out, the current through emitter resistor R_E (and collector or loop current I) equals 2 V/500 Ω = 4 mA. If the 400-Ω resistor *is* shorted out by Q_2, the loop current I equals 2 V/100 Ω = 20 mA.

The selection of either 4 or 20 mA is determined by E_{in}. E_{in} can be (1) a TTL or (other logic family) open-collector gate circuit ouput, (2) an output port from a single-chip microprocessor or microcomputer, or (3) a discrete (Q_1) bipolar junction transistor. (For a stand-alone circuit add a 2.2-kΩ resistor in series with the base).

When E_{in} is high, Q_1 saturates and I_{CE1} equals about 1.4 mA. The collector current of Q_1 is the base current of Q_2 and Q_2 saturates. When Q_2 saturates, it effectively shorts out the 400-Ω resistor and fixes Q_B's emitter current and thus the loop current to 20 mA.

When E_{in} is low, Q_1 is cut off, which in turn cuts off Q_2. Q_2 then appears as an open circuit to the 400-Ω resistor, and the loop current, I, is clamped at 2 V/500 Ω = 4 mA. The compliance voltage of this circuit is 12 V. Allow 12 V for V_{R1} plus 1 V to keep Q_B out of saturation [15 − (2 + 1) = 12 V].

5-6 SHORT-CIRCUIT CURRENT MEASUREMENT AND CURRENT-TO-VOLTAGE CONVERSION

5-6.1 Introduction

Transducers such as phonograph pickups and solar cells convert some physical quantity into electrical signals. For convenience, the transducers may be modeled by a signal generator as in Fig. 5-8(a). It is often desirable to measure their maximum output current under short-circuit conditions; that is, we should place a short circuit across the output terminals and measure current through the short circuit. This technique is particularly suited to signal sources with very high internal resistance. For example, in Fig. 5-8(a), the short-circuit current I_{SC} should be 2.5 V/50 kΩ = 50 μA. However, if we place a microammeter across the output terminals of the generator, we no longer have a short circuit but a 5000-Ω resistance. The meter indication is

$$\frac{2.5\text{ V}}{50\text{ k}\Omega + 5\text{ k}\Omega} \cong 45\ \mu\text{A}$$

High-resistance sources are better modeled by an equivalent Norton circuit. This model is simply the ideal short-circuit current, I_{SC}, in parallel with its own internal resistance as in Fig. 5-8(b). This figure shows how I_{SC} splits between its internal resistance and the meter resistance. To eliminate this current split, we will use the op amp.

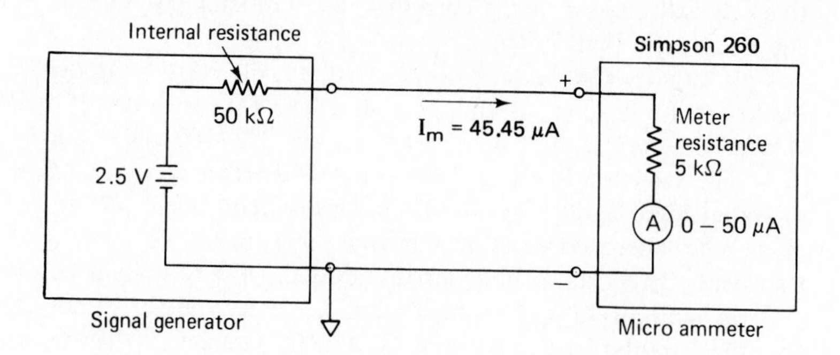

(a) Ammeter resistance reduces short-circuit current from the signal generator

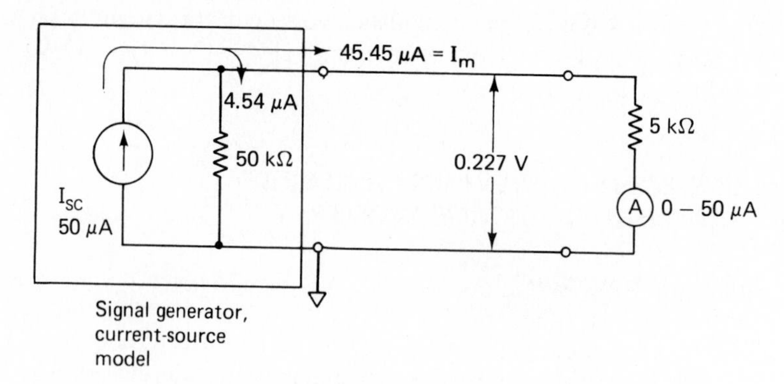

(b) Current-source model of signal generator in (a)

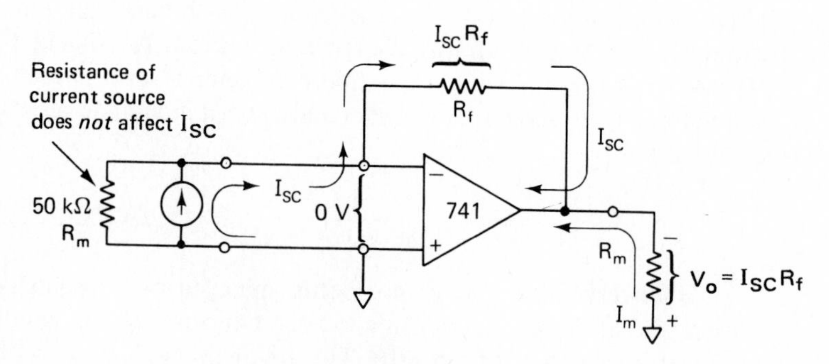

(c) Current-to-voltage converter

FIGURE 5-8 Current-measuring circuits.

5-6.2 Using the Op Amp to Measure Short-Circuit Current

The op amp circuit of Fig. 5-8(c) effectively places a short circuit around the current source. The (−) input is at virtual ground because the differential input voltage is almost 0 V. The current source "sees" ground potential at both of its terminals, or the equivalent of a short circuit. *All* of I_{SC} flows toward the (−) input and on through R_f. R_f converts I_{SC} to an output voltage, revealing the basic nature of this circuit to be a *current-to-voltage converter*.

Example 5-8

V_o measures 5 V in Fig. 5-8(c), and $R_f = 100\ \text{k}\Omega$. Find the short-circuit current I_{SC}.

Solution From Fig. 5-8(c),

$$I_{SC} = \frac{V_o}{R_f} = \frac{5\ \text{V}}{100\ \text{k}\Omega} = 50\ \mu\text{A}$$

The resistance R_m is the resistance of either the voltmeter or the CRO. The current I_m needed to drive R_m comes from the op amp and not from I_{SC}.

5-7 MEASURING CURRENT FROM PHOTODETECTORS

5-7.1 Photoconductive Cell

With the switch at position 1 in Fig. 5-9, a photoconductive cell, sometimes called a light-sensitive resistor (LSR), is connected in series with the (−) input and E_i. The

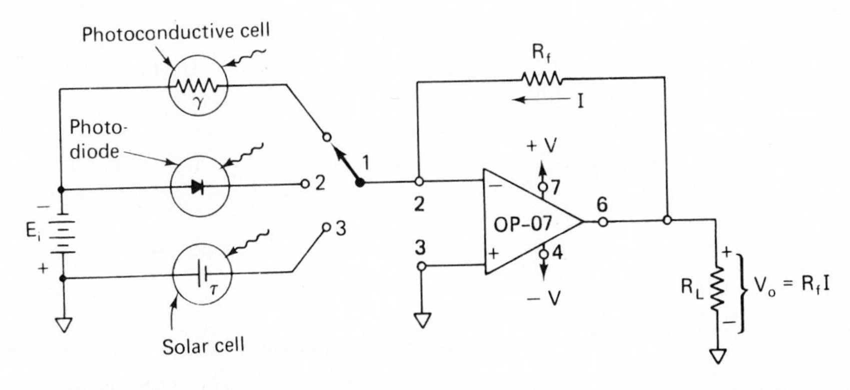

FIGURE 5-9 Using the op amp to measure output current from photodetectors.

resistance of a photoconductive cell is very high in darkness and much lower when illuminated. Typically, its dark resistance is greater than 500 kΩ and its light resistance in bright sun is approximately 5 kΩ. If $E_i = 5$ V, then current through the photoconductive cell, I, would be 5 V/500 kΩ = 10 μA in darkness and 5 V/5 kΩ = 1 mA in sunlight.

Example 5-9

In Fig. 5-9 the switch is in position 1 and $R_f = 10$ kΩ. If the current through the photoconductive cell is 10 μA in darkness and 1 mA in sunlight, find V_o for (a) the dark condition; (b) the light condition.

Solution From Fig. 5-9 $V_o = R_f I$. (a) $V_o = 10\ \text{k}\Omega \times 10\ \mu\text{A} = 0.1$ V; (b) $V_o = 10\ \text{k}\Omega \times 1\ \text{mA} = 10$ V. Thus the circuit of Fig. 5-9 converts the output current from the photoconductive cell into an output voltage (a current-to-voltage converter).

5-7.2 Photodiode

When the switch is in position 2 in Fig. 5-9, E_i is on one side of the photodiode and virtual ground on the other. The photodiode is reverse-biased, as it must be for normal operation. In darkness the photodiode conducts a small leakage current on the order of nanoamperes. But depending on the radiant energy striking the diode, it will conduct 50 μA or more. Therefore, current I depends only on the energy striking the photodiode and not on E_i. This current is converted to a voltage by R_f.

Example 5-10

With the switch in position 2 in Fig. 5-9 and $R_f = 100$ kΩ, find V_o as the light changes photodiode current from (a) 1 μA to (b) 50 μA.

Solution From $V_o = R_f I_L$, (a) $V_o = 100\ \text{k}\Omega \times 1\ \mu\text{A} = 0.1$ V; (b) $V_o = 100\ \text{k}\Omega \times 50\ \mu\text{A} = 5.0$ V.

5-8 CURRENT AMPLIFIER

Characteristics of high-resistance signal sources were introduced in Section 5-6.1. There is no point in converting a current into an equal current, but a circuit that converts a small current into a large current can be very useful. The circuit of Fig. 5-10 is a current multiplier or current amplifier (technically, a current-to-current converter). The signal current source I_{SC} is effectively short-circuited by the input terminals of the op amps. All of I_{SC} flows through resistor mR, and the voltage across it is mRI_{SC}. (Resistor mR is known as a multiplying resistor and m the multiplier.) Since R and mR are in parallel, the voltage across R is also mRI_{SC}. Therefore, the current through R must be mI_{SC}. Both currents add to form load current I_L. I_L is an amplified version of I_{SC} and is found simply from

$$I_L = (1 + m)I_{SC} \tag{5-7}$$

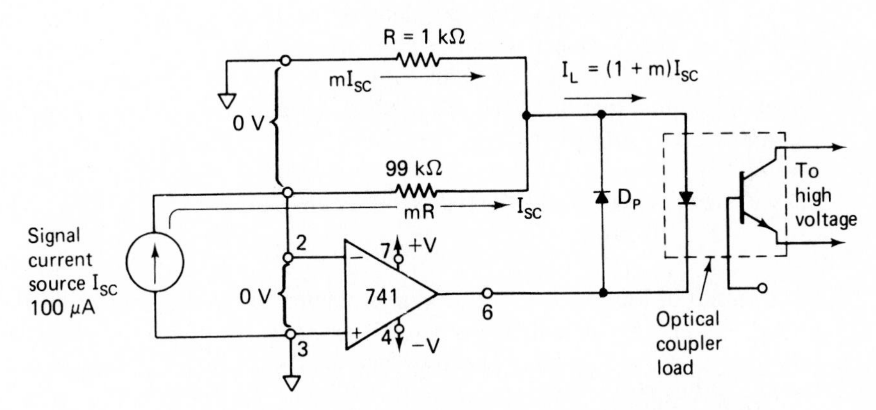

FIGURE 5-10 Current amplifier with optical coupler load.

Analysis Example 5-11

In Fig. 5-10, $R = 1\ \text{k}\Omega$ and $mR = 99\ \text{k}\Omega$. Therefore, $m = 99\ \text{k}\Omega/1\ \text{k}\Omega = 99$. Find the current I_L through the emitting diode of the optical coupler.

Solution By Eq. (5-7), $I_L = (1 + 99)(10\ \mu\text{A}) = 1.0$ mA.

It is important to note that the load does not determine load current. Only the multiplier m and I_{SC} determine load current. For variable current gain, mR and R can be replaced by a single 100-kΩ potentiometer. The wiper goes to the emitting diode, one end to ground and the other end to the (−) input. The optical coupler isolates the op amp circuit from any high-voltage load. D_P is an ordinary silicon diode that protects the emitting diode against reverse bias.

5-9 SOLAR CELL ENERGY MEASUREMENTS

5-9.1 Introduction to the Problems

A solar cell (also called a photovoltaic cell) is a device that converts light energy directly into electrical energy. The best way to record the amount of power received by the solar cell is to measure its short-circuit current. For example, one type of solar cell furnishes a short-circuit current I_{SC} that ranges from 0 to 0.5 A as sunlight varies from complete darkness to maximum brightness.

One problem facing users of these devices is to convert the 0 to 0.5-A solar cell output current to 0 to 10 V so that its performance can be monitored with a strip-chart recorder. Another problem is to measure $\frac{1}{2}$ A of current with a low-current meter movement (0 to 0.1 mA). To solve this problem, I_{SC} must be divided so that it can be measured on site with an inexpensive basic meter movement. The final problem is that the value of I_{SC} is too large to be used with the op amp circuits studied thus far in this chapter.

5-9.2 Converting Solar Cell Short-Circuit Current to a Voltage

The circuit of Fig. 5-11 solves several problems. First, the solar cell sees the (−) input of the op amp as a virtual ground. Therefore, it can deliver its short-circuit current I_{SC}. A second problem is solved when I_{SC} is converted by R_f to a voltage V_o. To obtain a 0 to 10-V output for a 0 to 0.5-A input, R_f should have a value of

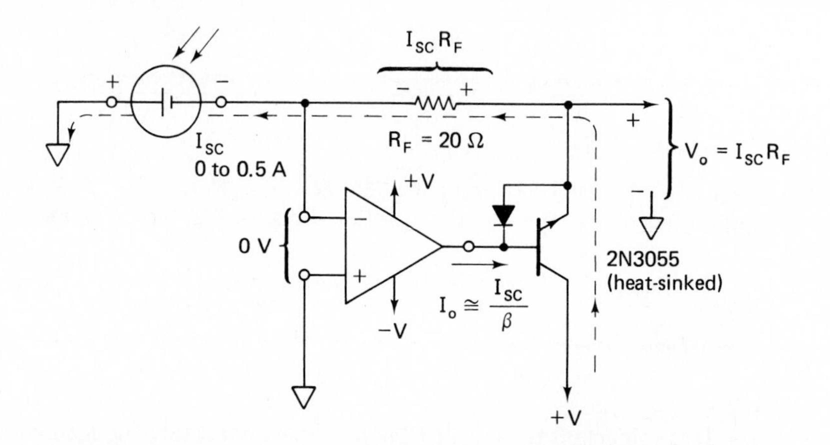

FIGURE 5-11 This circuit forces the solar cell to deliver a short-circuit current I_{SC}. I_{SC} is converted to a voltage by R_f. Current boost is furnished by the *npn* transistor. The diode protects the base-emitter junction of the 2N3055 against accidental excessive reverse bias.

$$R_f = \frac{V_o(\text{full scale})}{I_{SC}(\text{max})} = \frac{10 \text{ V}}{0.5 \text{ A}} = 20.0 \ \Omega$$

V_o should be buffered by a voltage follower. The solar cell current of 0.5 A is too large to be handled by the op amp. This problem is solved by adding an *npn* current boost transistor.

The solar cell current flows through the emitter and collector of the boost transistor to $+V$. Current gain of the transistor should exceed $\beta = 100$ to ensure that the op amp has to furnish no more than 0.5 A/100 = 5 mA, when $I_{SC} = 0.5$ A.

5-9.3 Current-Divider Circuit (Current-to-Current Converter)

Only a slight addition to the circuit of Fig. 5-11 allows us to measure I_{SC} with a low-current milliammeter or microammeter. The current-divider resistance dR_f is shown in Fig. 5-12. Resistance dR_f is made up of the meter resistance R_m plus the scale resistor R_{scale}.

The short-circuit current develops a voltage drop across R_f equal to V_o. V_o is also equal to the voltage across resistance dR_f. Thus the current divider d can be found by equating the voltage drops across dR_f and R_f.

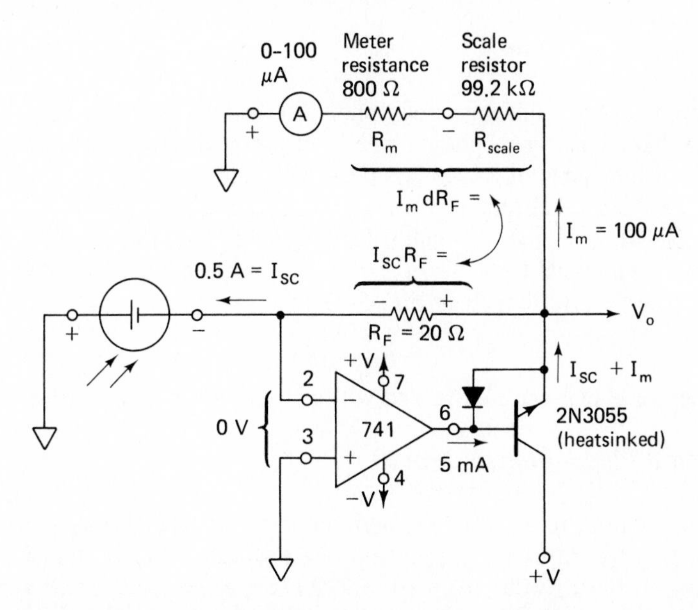

FIGURE 5-12 Current-to-current converter. Divider resistance dR_f equals the sum of meter resistance R_m and scale resistance R_{scale}. Short-circuit current $I_{SC} = 0.5$ A is converted by d down to 100 μA for measurement by a low-current meter.

$$V_o = I_{SC} R_f = I_m dR_f \tag{5-8a}$$

so

$$d = \frac{I_{SC}}{I_m} \tag{5-8b}$$

Design Example 5-12

If the meter in Fig. 5-12 is to indicate full scale at $I_m = 100\ \mu A$ when $I_{SC} = 0.5$ A, find resistance dR_f and R_{scale}.

Design Procedure From Eq. (5-8b), $d = 0.5\ A/100\ \mu A = 5000$ and $dR_f = 5000 \times 20\ \Omega = 100\ k\Omega$. Then $R_{scale} = dR_f - R_m = 100\ k\Omega - 0.8\ k\Omega = 99.2\ k\Omega$.

5-10 PHASE SHIFTER

5-10.1 Introduction

An ideal phase-shifting circuit should transmit a wave without changing its amplitude but changing its phase angle by a preset amount. For example, a sine wave E_i with a frequency of 1 kHz and peak value of 1 V is the input of the phase shifter in Fig. 5-13(a). The output V_o has the same frequency and amplitude but lags E_i by 90°. That is, V_o goes through 0 V 90° *after* E_i goes through 0 V. Mathematically, V_o can be expressed by $V_o = E_i \angle -90°$. A general expression for the output voltage of the phase-shifter circuit in Fig. 5-13(b) is given by

$$V_o = E_i \angle -\theta \tag{5-9}$$

where θ is the phase angle and will be found from Eq. (5-10a).

5-10.2 Phase-Shifter Circuit

One op amp, three resistors, and one capacitor are all that is required as shown in Fig. 5-13(b) to make an excellent phase shifter. The resistors R must be equal, and any convenient value from 10 to 220 kΩ may be used. Phase angle θ depends only on R_i, C_i, and the frequency f of E_i. The relationship is

$$\theta = 2 \arctan 2\pi f R_i C_i \tag{5-10a}$$

where θ is in degrees, f in hertz, R_i in ohms, and C_i in farads. Equation (5-10a) is

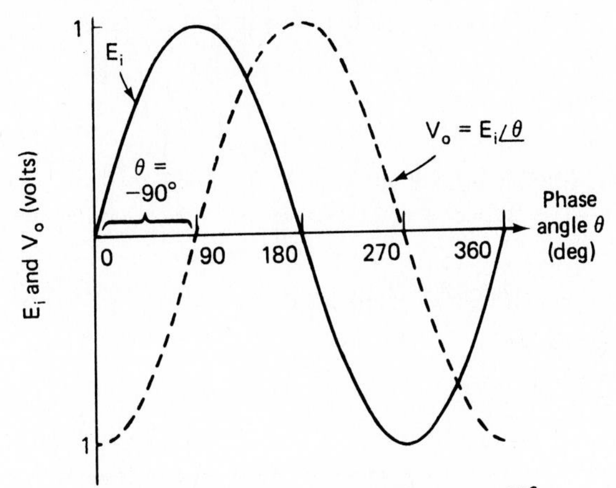

(a) Input and output voltages for $\theta = -90°$

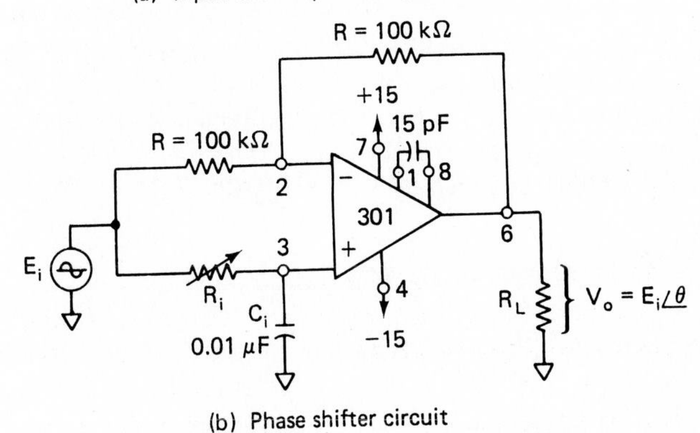

(b) Phase shifter circuit

FIGURE 5-13 Phase shifter.

useful to find the phase angle if f, R_i, and C_i are known. If the desired phase angle is known, choose a value for C_i and solve for R_i:

$$R_i = \frac{\tan(\theta/2)}{2\pi f C_i} \tag{5-10b}$$

Design Example 5-13

Find R_i in Fig. 5-13(b) so that V_o will lag E_i by 90°. The frequency of E_i is 1 kHz.

Design Procedure Since $\theta = 90°$, $\tan(90°/2) = \tan(45°) = 1$; from Eq. (5-10b),

$$R_i = \frac{1}{2\pi \times 1000 \times 0.01 \times 10^{-6}} = 15.9\ \text{k}\Omega$$

With $R_i = 15.9 \text{ k}\Omega$, V_o will have the phase angle shown in Fig. 5-13(a). This waveform is a negative cosine wave.

Analysis Example 5-14

If $R_i = 100 \text{ k}\Omega$ in Fig. 5-13(b), find the phase angle θ.

Solution From Eq. (5-10a),

$$\theta = 2 \arctan (2\pi)(1 \times 10^3)(100 \times 10^3)(0.01 \times 10^{-6})$$

$$= 2 \arctan 6.28$$

$$= 2 \times 81° = 162° \qquad \text{and} \qquad V_o = E_i \underline{/-162°}$$

It can be shown from Eq. (5-10a) that $\theta = -90°$ when R_i equals the reactance of C_i, or $1/(2\pi f C_i)$. As R_i is varied from 1 kΩ to 100 kΩ, θ varies from approximately −12 to −168°. Thus the phase shifter can shift phase angles over a range approaching 180°. If R_i and C_i are interchanged in Fig. 5-12(b), the phase angle is positive, and the circuit becomes a leading phase-angle shifter. The magnitude of θ is found from Eq. (5-10a), but the output is given by $V_o = E_i \underline{/180° - \theta}$.

5-11 THE CONSTANT-VELOCITY RECORDING PROCESS

5-11.1 Introduction to Record-Cutting Problems

The process of recording data or music on a record is accomplished by a heated chisel-shaped cutting stylus that vibrates from side to side (laterally) in the record groove. Each groove is about 1 mil (0.001 in.) wide. The cutting stylus is vibrated by electromechanical transducers that are activated by the magnetic fields from a driving coil and a feedback coil. The coil-transducer-stylus combination is called the cutting head.

If the *amplitude* of the input signal current is held *constant*, the stylus cuts laterally at a constant velocity. When the frequency of the input signal is varied, the cutting rate or lateral *velocity* of the stylus remains *constant, provided that amplitude of the input signal is held constant*. This type of a recording process is called a *constant-velocity recording*.

5-11.2 Groove Modulation with Constant-Velocity Recording

Groove modulation is defined as the peak-to-peak lateral cutting distance and should depend only on the amplitude of the input signal, *not* on its frequency. Unfortunately, this is *not* the case for constant-velocity recording, as shown in Fig. 5-14.

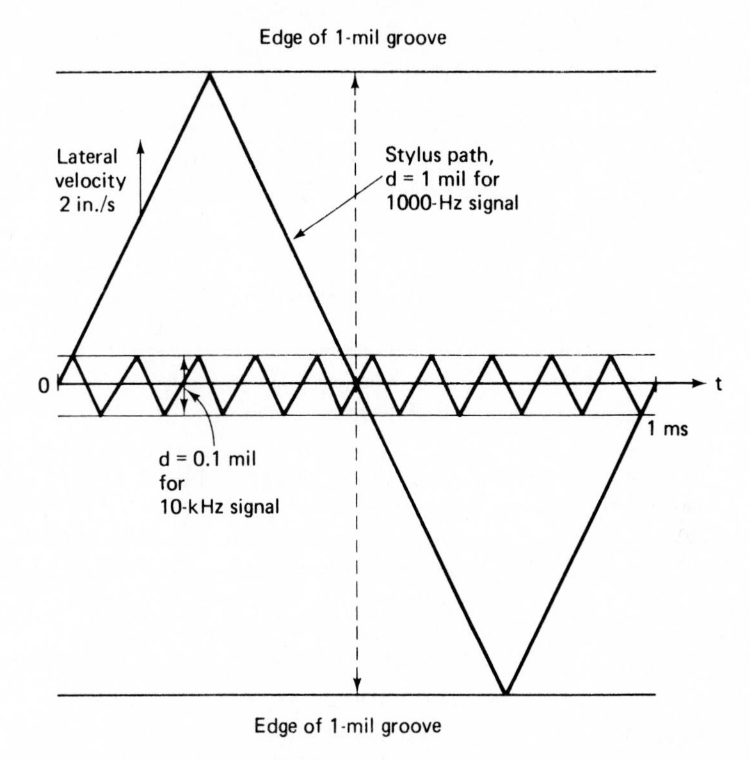

FIGURE 5-14 If signal amplitude is held constant across a cutting head, the resulting constant-velocity recording process overcuts at low frequencies. The smaller cuts at high frequencies become indistinguishable from record noise due to surface imperfections.

If amplitude at the input signal is held constant at 10 mV, the cutter's lateral velocity V will be typically 2 in./s. The p-p lateral distance d can then be found from

$$d = \frac{V}{2f} \tag{5-11}$$

where f is the signal frequency in hertz. The problem that results, because groove modulation depends on signal frequency, is brought out by an example.

Example 5-15

Find the peak-to-peak lateral cutting distance for signals of (a) 1000 Hz reference; (b) 10,000 Hz. Assume that the input signal is 10 mV and the stylus velocity is 2 in./s.

Solution (a) From Eq. (5-11),

$$d = \frac{2 \text{ in./s}}{2f} \times \frac{1 \text{ (s)}}{(2)1000 \text{ (cycles)}} = 1 \text{ mil}$$

(b) From Eq. (5-11),

$$d = \frac{2 \text{ in./s}}{2f} \times \frac{1 \text{ (s)}}{(2)10{,}000 \text{ (cycles)}} = 0.1 \text{ mil}$$

5-11.3 Record Cutover and Noise

Cutting paths for the stylus are shown in Fig. 5-14, for both 1- and 10-kHz signals over a time period of 1 ms for a constant input voltage of 10 mV. There are two conclusions to be drawn from this figure. First, all frequencies below 1 kHz will cause cutover into the adjacent grooves. Second, the lateral cutting distance decreases as frequency increases, so that eventually the groove modulation will become indistinguishable from surface imperfections.

Suppose that the amplitude of the recording signal was increased to 100 mV, signifying a louder tone. This would cause a cutting velocity of 20 in./s. As shown in Fig. 5-15, this would result in cutting 10 grooves at 1 kHz. The problem with constant-velocity recording is summarized in Fig. 5-15. It shows the lack of dynamic range in fitting loud and soft tones of different frequencies into a 1-mil groove.

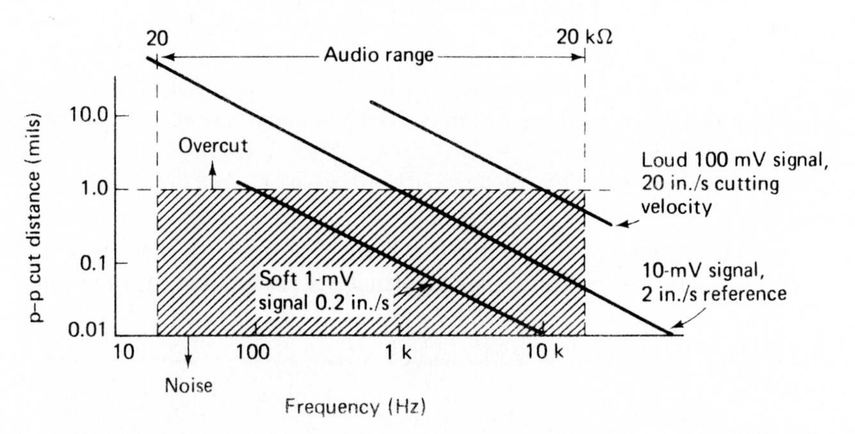

FIGURE 5-15 Acceptable constant-velocity recording process levels and frequencies lie within the crosshatched area.

5-11.4 Solution to Record Cutover and Noise Problems

The solution to frequency cutover and noise problems is to attenuate the low-frequency signals and boost the high-frequency signals applied to the record cutter. The circuit that does this is called a *recording preequalizer*. For a given input signal, its output will cause the cutting stylus to cut a constant peak-to-peak distance independent of frequency. The process is shown conceptually in Fig. 5-16(b). With a preequalizer, the dynamic volume range can approach 40 dB or a distinction between loud and soft of 100:1.

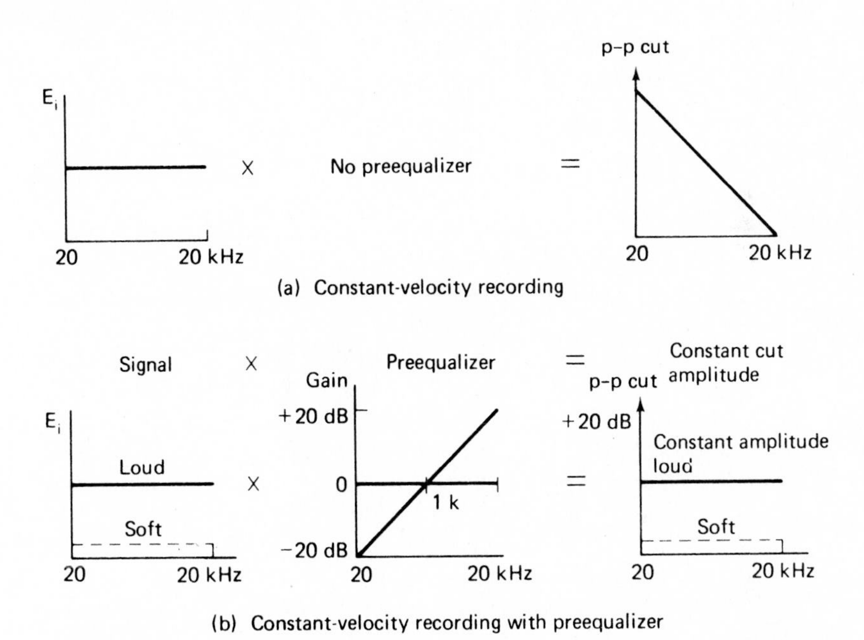

FIGURE 5-16 When the signal to be recorded is transmitted through a preequalizer, the stylus gives a peak-to-peak cut that is independent of the signal frequency and depends only on the signal's amplitude.

5-12 RECORD PLAYBACK

5-12.1 Need for Playback Equalization

It was shown in Section 5-11 that constant-velocity cutting heads would make lateral cuts whose amplitude will depend on signal amplitude and not on signal frequency, *if a preequalizer* circuit was installed. The stylus of a magnetic pickup cartridge moves a magnet within a coil. The output of the coil is proportional to the magnet's movement. The stylus will move faster (laterally) to follow groove modulation as frequency is increased. Therefore, output voltage of the pickup will increase directly with increasing frequency for the same peak-to-peak lateral cut (see Fig. 5-17).

Thus, if the output signal voltage at 1 kHz is just right, the output voltage will get progressively lower at lower frequencies and progressively higher at higher frequencies. This means that the output of the magnetic cartridge must be equalized in amplitude by a *playback equalizer*. The circuit that does the amplitude equalization is called a *preamplifier* and its approximate relative and absolute gains are shown in Fig. 5-17. Low frequencies are amplified and high frequencies are attenuated. Thus the output of the equalizer gives a voltage that is proportional to amplitude of the lateral groove cut and not its frequency.

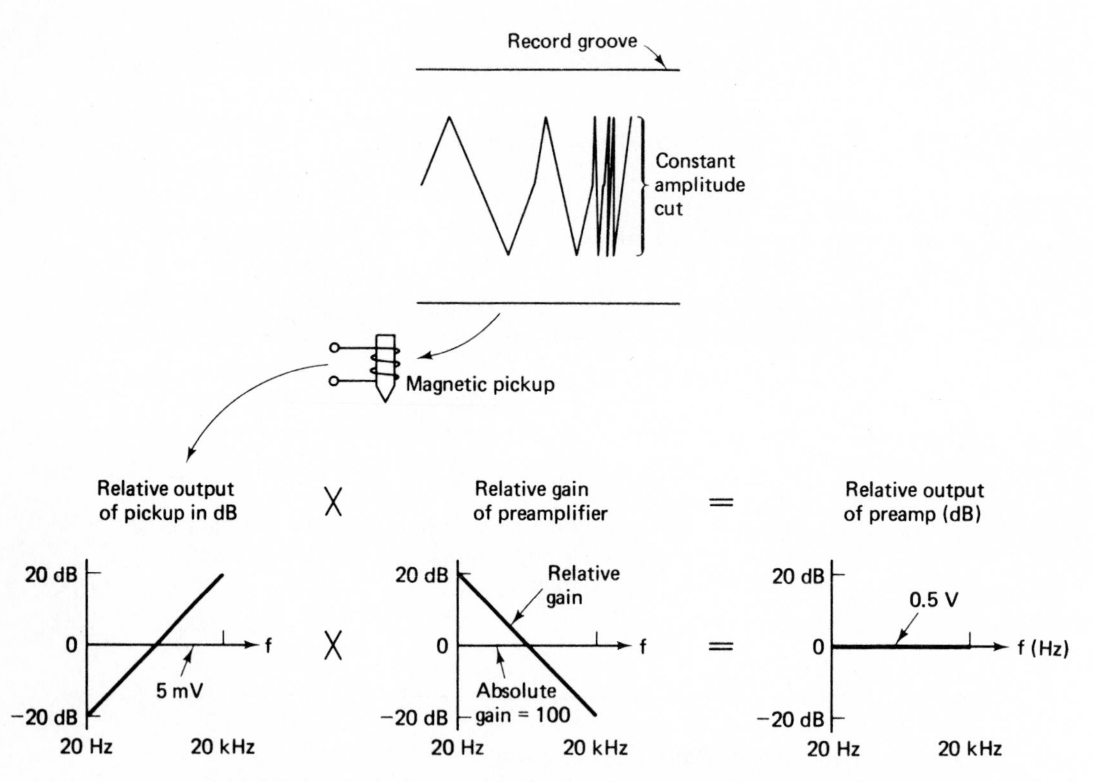

FIGURE 5-17 When a variable-frequency constant-amplitude groove modulation is applied to a magnetic pickup, its output voltage increases with increasing frequency. Gain of the playback equalizer decreases with increasing frequency so that its output is flat.

5-12.2 Preamplifier Gain and Signal Voltage Levels

The typical magnetic cartridge produces 5 mV of output for a needle velocity of 5 cm/s. The phonograph preamplifier must also provide different gains at different frequencies to bring this 5-mV output up to about 0.2 to 0.5 V at all frequencies to drive an audio amplifier. Typical *approximate* gains and signal levels are shown below for a velocity of 5 cm/s from a constant-amplitude cut at different frequencies.

Output of pickup (mV)	Gain of preamplifier		Output of preamplifier (V)	Frequency (Hz)
	Absolute	Relative		
0.05	1000	+20 dB	0.5	20
5	100	0	0.5	1,000
50	10	−20 dB	0.5	20,000

5-12.3 Playback Preamplifier Circuit Operation

The ideal RIAA (Record Industry Association of America) playback equalization curve is shown in Fig. 5-18(a) as a dashed line. An inexpensive circuit that accomplishes playback equalization is shown in Fig. 5-18(b). The RC4739 has two low-noise op amps on one chip. They are internally compensated (see Chapter 10). (The μA739 or MC1303 may also be used as pin-for-pin replacements provided that external compensation is installed.) One RC4739 can equalize both channels of a stereo system. The first number on each terminal identifies the A-channel op amp and the second the B-channel op amp.

Operation of the circuit is analyzed by looking at the role of each capacitor:

1. At zero frequency (dc), all capacitors are open circuits and the gain $V_o/E_i = +1$.
2. As frequency is increased, the reactance of C_1 begins to decrease at 0.03 Hz and becomes negligible at about 26 Hz. In this low-frequency range the gain increases from +1 to a value set by

$$A_{CL} = \frac{R_{F1} + R_i}{R_i} = 834$$

3. At 54 Hz the reactance of capacitor C_2 begins to decrease until at 580 Hz its reactance becomes negligible. (C_1 = short, C_3 = open in this frequency range.) R_{F1} is now connected in parallel with R_{F2} to reduce gain at 580 Hz to about 77.

4. When the frequency of E_i increases above 2.3 kHz, C_3 begins to bypass R_{F2} and R_{F1}, reducing gain at 20 dB/decade until the gain settles to unity at about 178 kHz.

The resultant practical playback equalization curve is shown in Fig. 5-18(a).

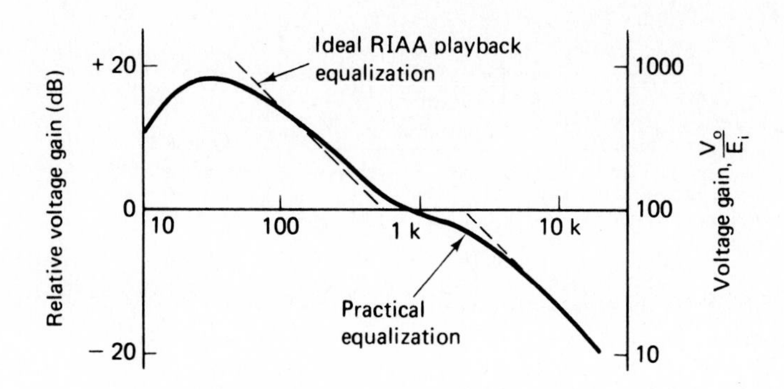

(a) RIAA Playback equalization curve

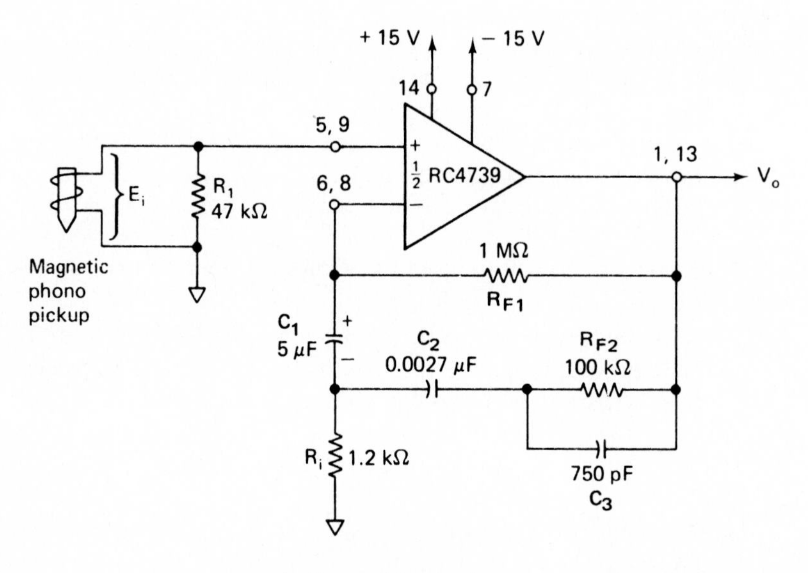

(b) Preamplifier circuit

FIGURE 5-18 RIAA playback equalization curve and preamplifier.

5-13 TONE CONTROL

5-13.1 Introduction

The preamplifier of Section 5-12 will deliver a flat frequency response at its output. In most high-fidelity systems, the owner wants to have a tone-control feature that allows boosting or cutting the volume of bass or treble frequencies. A frequency-controlling network, made of resistors and capacitors, could be installed in series with the output of the preamplifier. However, this network would attenuate some of the frequencies by as much as $1/100$ or -40 dB. Much of the gain so carefully built into the preamplifier would be lost.

5-13.2 Tone-Control Circuit

The practical tone-control circuit shown in Fig. 5-19(a) (1) features boost or cut of bass frequencies below 500 Hz and of treble frequencies above 2 kHz, and (2) eliminates attenuation. The top 50-kΩ audio taper potentiometer is the bass frequency control. With the wiper adjusted to full boost position, the voltage gain at 10 Hz is about $10R/R$ or 10. With the wiper at full bass cut, the voltage gain at 10 Hz is

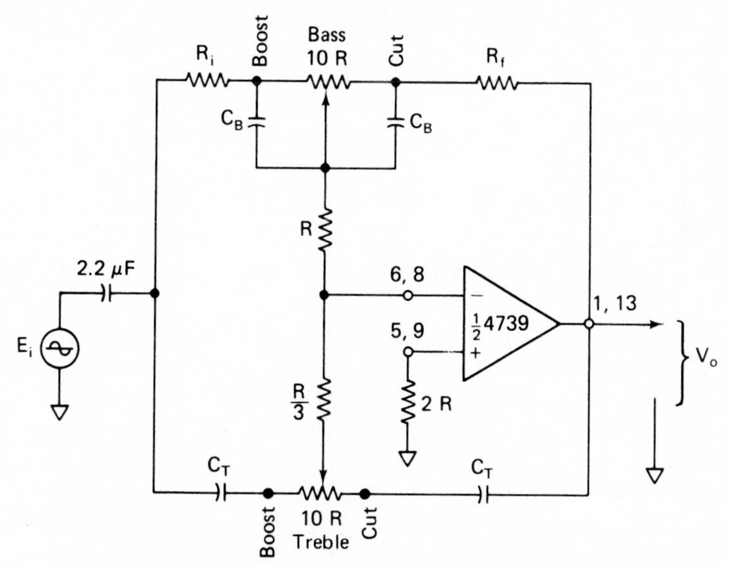

(a) Tone-control circuit; $C_B = C_T = 0.068\ \mu F$, $R_i = R_f = R = 5\ k\Omega$. Connections for $+V$ and $-V$ shown in Fig. 5-13(b)

FIGURE 5-19 The tone-control circuit in (a) has the frequency response curves shown in (b).

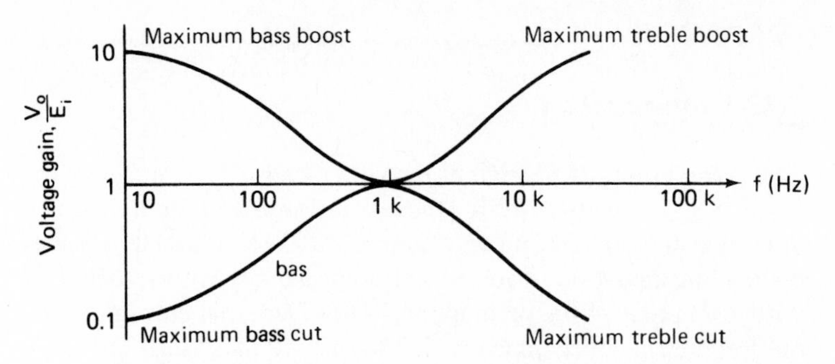

(b) Tone-control circuit frequency response curves

FIGURE 5-19 *(cont.)*

about $R/10R = 0.1$. In effect, the $10R$ pot is adjusted to be in series with R_i for cut or R_f for boost. The boost capacitors C_B begin to bypass the pot at frequencies between 50 Hz and 500 Hz, as shown in Fig. 5-19(b).

When adjusted to full boost, the treble control, $R/3$, and the capacitors C_T set the gain at 20 kHz to 10. At full cut, the gain is 0.1, as shown in Fig. 5-19(b). With both bass and treble control pots adjusted to the center of their rotation, the frequency response of the tone-control circuit will be flat. The input signal E_i delivered from the preamplifier should be about 0.2 V rms at 1 kHz. Therefore, the output of the tone control should be at about the same level.

Our final op amp application will be a temperature-to-voltage converter (electronic thermometer).

5-14 TEMPERATURE-TO-VOLTAGE CONVERTERS

5-14.1 AD590 Temperature Transducer

An electronic thermometer can be made from a temperature transducer, an op amp, and resistors. We shall select the AD590, manufactured by Analog Devices, as the temperature transducer. The AD590 converts its ambient temperature in degrees Kelvin into an output current, I_T, that is 1 μA for every degree Kelvin. In terms of Celsius temperatures, $I_T = 273\ \mu$A at 0°C (273°K) and 373 μA at 100°C (373°K). In terms of Fahrenheit temperatures, $I_T = 255\ \mu$A at 0°F and 310 μA at 100°F. Thus the AD590 acts as a current source that depends on temperature. If, however, we need a voltage reading to indicate temperature, such as 10 mV/°C or 10 mV/°F, a current-to-voltage converter circuit is required.

The circuit symbol for an AD590 is the same as a current source, as shown in Fig. 5-20(a) and (b). Also, the AD590 requires a supply voltage exceeding 4V to bias its internal transistor circuitry. Let's use this device to build a Celsius or Fahrenheit thermometer.

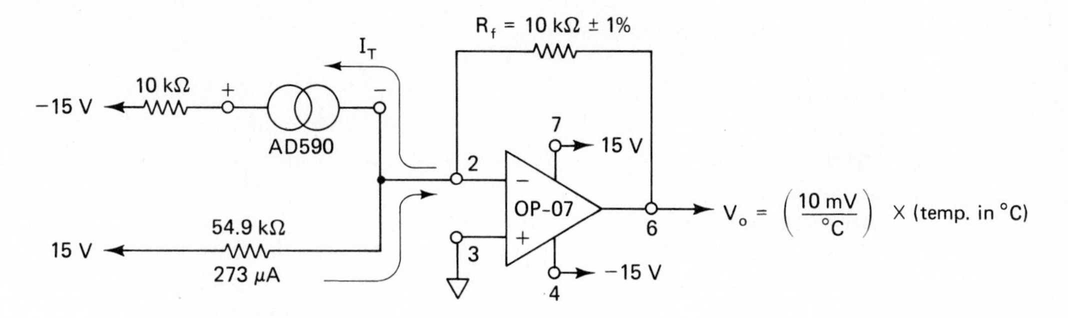

(a) V_o = 0 V at 0°C and 1000 mV at 100°C

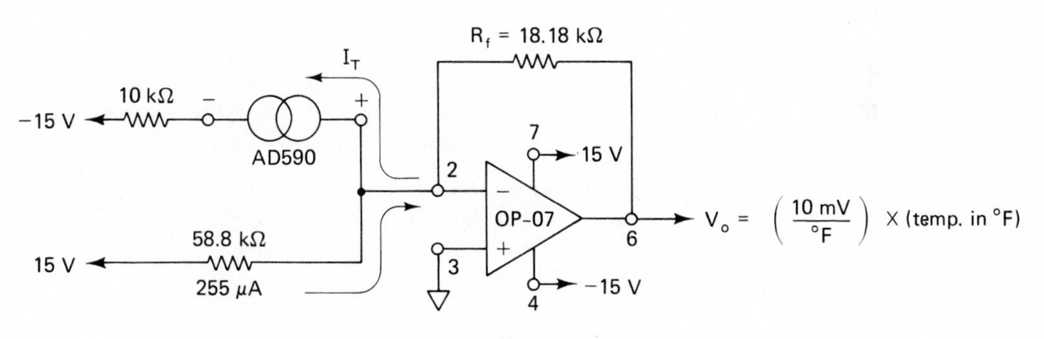

(b) V_o = 0 V at 0°F and 1000 mV at 100°F

FIGURE 5-20 Temperature-to-voltage converters for Celsius degrees in (a) to Fahrenheit degrees in (b).

5-14.2 Celsius Thermometer

In the Celsius thermometer of Fig. 5-20(a), all of the AD590's current is steered into the virtual ground at pin 2 and flows through the 10-kΩ feedback resistor, producing a voltage drop equal to V_o. Each microampere of current thus causes V_o to go positive by (1 μA × 10 kΩ) = 10 mV. A change of 1°C causes I_T to change by 1 μA and consequently produces a change in V_o of 10 mV. The temperature-to-voltage converter thus has a conversion gain of 10 mV/°C.

At 0°C, I_T = 273 μA. But we want V_o to equal zero volts. For this reason, an equal and opposite current of 273 μA through the 15-V supply and 54.9-kΩ resistor is required. This results in the net current through R_f to be zero and thus V_o to be zero volts. For every increase of 1 μA/°C above 0°C, the net current through R_f increases by 1 μA and V_o increases by 10 mV.

5-14.3 Fahrenheit Thermometer

A circuit for a Fahrenheit thermometer is shown in Fig. 5-20(b). At 0°F we want V_o to equal 0 V. Since $I_T = 255\ \mu A$ at 0°F, it must be nulled out by an equal but opposite current through R_f. This current is generated by the 15-V supply and the 58.8 kΩ resistor.

An increase of 1°F corresponds to an increase of $\frac{5}{9}$°C or 0.555°C. Thus the AD590 increases its output current by 0.555 μA/°F. This increase is converted by R_f into a voltage of $0.555\ \mu A/°F \times 18.18\ k\Omega = 10\ mV/°F$. In conclusion, for every temperature increase of 1°F above 0°F, V_o will increase by 10 mV above 0 V.

LABORATORY EXERCISES

All of the circuits in this chapter are suited for lab experiments. If oscillations are encountered with circuits containing a transistor, shorten up the wiring and connect a 30pF capacitor between collector and base. The following circuits are basic and educational.

5-1. Build the high-resistance voltmeter of Fig. 5-2. Use 1N914 diodes. The 50-μA meter movement may be obtained from a Simpson 260 (or similar VOM) by selecting the 0–50 μA scale. Use a jumper wire for the function switch and a resistor decade box to simulate required values for R_i. For best accuracy, hold the frequency of E_i between 100 and 1000 Hz.

5-2. Use a 5.1-V zener and 1N914 (or equivalent diode) to gain experience with the voltage-to-current converters of Fig. 5-3.

5-3. Build the constant-current source of Fig. 5-6. Take data to plot I_L versus R_L and V_L versus R_L for $R_L = 0$ to 150Ω. If a 5-V zener is not available, replace it with a 5-kΩ resistor and replace the 1.8-kΩ resistor with a 10-kΩ resistor.

5-4. Design, build, and test a phase-shifter circuit (see Fig. 5-13) to have a phase shift of −90° at a frequency of 1 kHz. Display V_o vs. E_i on a CRO. Refine R_i (decade box) or E_i (frequency) to observe a perfect circle.

5-5. Redesign your circuit to get a phase shift of −90° at 1590 Hz. Take data to plot θ vs. frequency on semilog paper from 15 Hz to 15 kHz. $E_i = 1$ V peak.

PROBLEMS

5-1. Refer to Example 5-1 and Fig. 5-1. Assume that $I_{FS} = 1$ mA and meter winding resistance $R_m = 1$ kΩ. If $E_i = -1.0$ V and $R_i = 1$ kΩ, find **(a)** I_m; **(b)** V_o.

5-2. A 1-mA movement, with $R_m = 1$ kΩ, is to be substituted in the circuit of Fig. 5-2. Redesign the R_i resistors for full-scale meter deflection when **(a)** $E_i = \pm 6$ V dc; **(b)** $E_i = 6$ V rms; **(c)** $E_i = 6$ V p-p; **(d)** $E_i = 6$ V peak.

5-3. In Fig. P5-3 complete the schematic wiring between op amp, diodes, and milliammeter. The current through the meter must be steered from right to left.

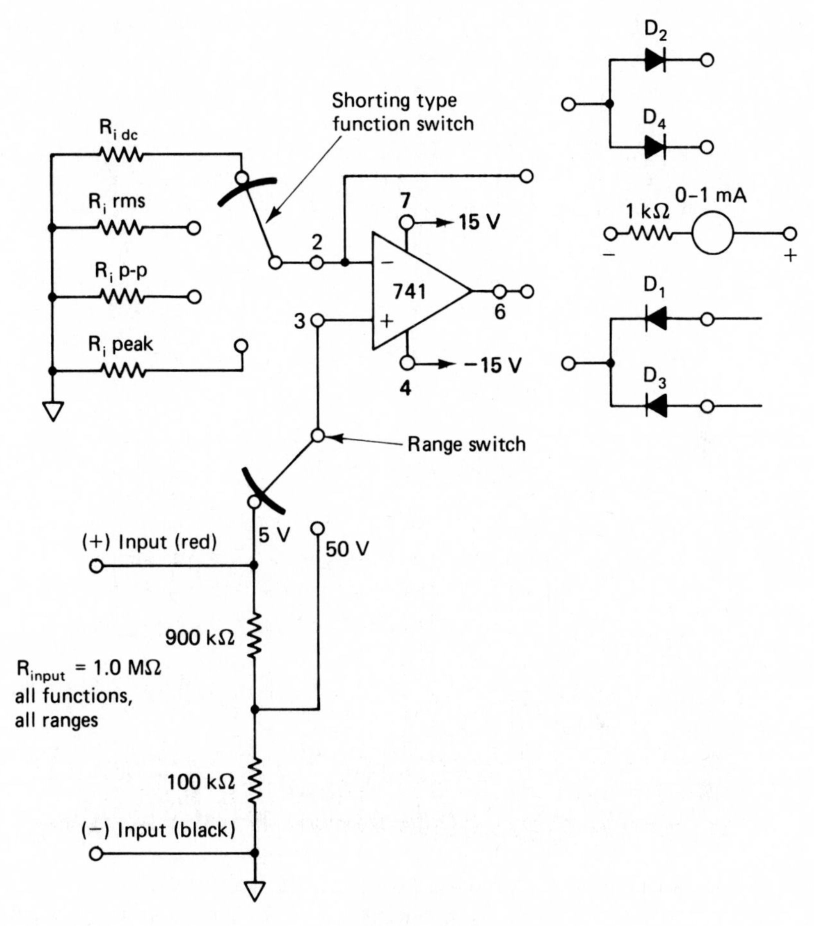

FIGURE P5-3

5-4. Calculate a value for R_{idc} in Fig. P5-3 so that the meter reads full scale when $E_i = 5$ V and the range switch is on the 5-V position.

5-5. Consider that the range switch is in the 5-V position in Fig. P5-3. Calculate values for the following resistors to give a full-scale meter deflection of 5 V: **(a)** R_{irms} for $E_i =$ 5 V rms; **(b)** R_i p-p for $E_i = 5$ V p-p; **(c)** R_i peak for $E_i = 5$ V peak.

5-6. With the circuit conditions shown in Problem 5-4, **(a)** which diodes are conducting? **(b)** Find V_o. Assume that diode drops are 0.6 V.

5-7. For the constant-current source shown in Fig. P5-7: **(a)** draw the emitter arrow and state if the transistor is *npn* or *pnp*; **(b)** find I; **(c)** find V_L.

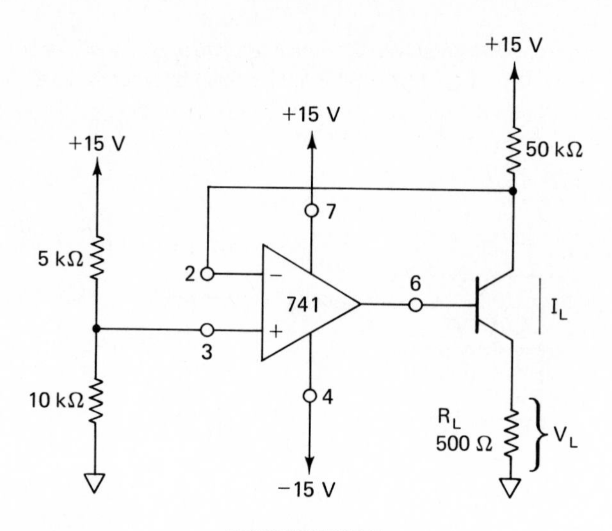

FIGURE P5-7

5-8. If V_o = 11 V and E_i = 5 V in Fig. 5-3, find V_z.

5-9. I_1 must equal 20 mA in Fig. 5-4 when $E_i = -10$ V. Find R_i.

5-10. Define a floating load.

5-11. In Fig. 5-5, E_2 = 0 V, R = 10 kΩ, and R_L = 5 kΩ. Find I_L, V_L, and V_o for **(a)** $E_1 = -2$ V; **(b)** $E_1 = +2$ V.

5-12. In Fig. 5-5, E_1 = 0 V, R = 10 kΩ, and R_L = 1 kΩ. Find I_L, V_L, and V_o for **(a)** $E_2 = -2$ V; **(b)** $E_2 = +2$ V.

5-13. In Fig. 5-5, $E_1 = E_2 = -5$ V, and $R = R_L$ = 5 kΩ. Find I_L, V_L, and V_o.

5-14. Replace V_z in Fig. 5-6 with a 900-Ω resistor. Find I_L.

5-15. Sketch an op amp circuit that will draw short-circuit current from a signal source and convert the short-circuit current to a voltage.

5-16. A CL5M9M photocell has a resistance of about 10 kΩ under an illumination of 2 fc. If $E_i = -10$ V in Fig. 5-9, calculate R_f for an output V_o of 0.2 V when the photoconductive cell is illuminated by 2 fc.

5-17. Change multiplier resistor mR in Fig. 5-10 to 49 kΩ. Find I_L.

5-18. A solar cell is installed in the circuit of Fig. 5-12 that has a maximum short-circuit current of 0.1 A = I_{SC}. **(a)** Select R_F to give V_o = 10 V when I_{SC} = 0.1 A. **(b)** A 50-μA meter movement is to indicate full scale when I_{SC} = 0.1 A. Find R_{scale} if R_M = 5 kΩ.

5-19. Resistor R_i is changed to 10 kΩ in Example 5-14. Find the phase angle θ.

5-20. Find the peak-to-peak lateral cutting distance in Example 5-15 at a signal frequency of 20 kHz.

5-21. Does a record playback preamplifier circuit provide greater amplification for the lower frequencies or the higher frequencies?

5-22. In the tone-control circuit of Fig. 5-19, what is the **(a)** gain at 10 Hz when the base control is at full boost; **(b)** gain at 20 kHz when the treble control is at full cut?

5-23. Design a phase shifter to give a $-90°$ shift at 1 Hz. Choose C_i from 0.001, 0.01, 0.1, or 1.0 μF. R_i must lie between 2 and 100 kΩ (see Laboratory Exercise 5-4).

5-24. Design a $-90°$ phase shift at 1590 Hz. Then for your design, calculate **(a)** θ at 15 Hz; **(b)** θ at 15 kHz (see Laboratory Exercise 5-5).

5-25. Calculate the net current through R_f in Fig. 5-20a if the AD590 temperature is 100°C. Find V_o.

5-26. Calculate the net current through R_f in Fig. 5-20b when the temperature is 100°F. Find V_o.

CHAPTER 6

Signal Generators

LEARNING OBJECTIVES

Upon completion of this chapter on signal generators, you will be able to:

- Explain the operation of a multivibrator circuit, sketch its output voltage waveshapes, and calculate its frequency of oscillation.
- Make a one-shot multivibrator and explain the purpose of this circuit.
- Show how two op amps, three resistors, and one capacitor can be connected to form an inexpensive triangle/square-wave generator.
- Predict the frequency of oscillation and amplitude of the voltages in a bipolar or unipolar triangle-wave generator and identify its disadvantages.
- Build a sawtooth wave generator and tell how it can be used as a voltage-to-frequency converter, frequency modulator, or frequency shift key circuit.
- Connect an AD630 balanced modulator/demodulator to operate as a switched gain amplifier.

- Connect the AD630 to an op amp circuit to make a *precision* triangle-wave generator whose output voltage amplitude can be adjusted independently of the oscillating frequency, and vice versa.
- Build, test, measure, and explain the operation of an AD639-universal trigonometric function generator when it is wired to generate sine functions.
- Connect the AD639 to the triangle-wave generator to make a superb *precision* sine-wave generator. Its oscillating frequency can be adjusted over a wide frequency range by a *single* resistor, without changing amplitude.

6-0 INTRODUCTION

Up to now our main concern has been to use the op amp in circuits that process signals. In this chapter we concentrate on op amp circuits that *generate* signals. Four of the most common and useful signals are described by their shape when viewed on a cathode ray oscilloscope. They are the square wave, triangular wave, sawtooth wave, and sine wave. Accordingly, the signal generator is classified by the shape of the wave it generates. Some circuits are so widely used that they have been assigned a special name. For example, the first circuit presented in Section 6-1 is a multivibrator that generates primarily square waves and exponential waves.

6-1 FREE-RUNNING MULTIVIBRATOR

6-1.1 Multivibrator Action

A *free-running* or *astable multivibrator* is a square-wave generator. The circuit of Fig. 6-1 is a multivibrator circuit and looks something like a comparator with hysteresis (Chapter 4), except that the input voltage is replaced by a capacitor. Resistors R_1 and R_2 form a voltage divider to feed back a fraction of the output to the (+) input. When V_o is at $+V_{sat}$, as shown in Fig. 6-1(a), the feedback voltage is called the upper-threshold voltage V_{UT}. V_{UT} is given in Eq. (4-1) and repeated here for convenience:

$$V_{UT} = \frac{R_2}{R_1 + R_2}(+V_{sat}) \tag{6-1}$$

Resistor R_f provides a feedback path to the (−) input. When V_o is at $+V_{sat}$, current I^+ flows through R_f to charge capacitor C toward V_{UT}. As long as the capacitor voltage V_C is less than V_{UT}, the output voltage remains at $+V_{sat}$.

When V_C charges to a value slightly greater than V_{UT}, the (−) input goes positive with respect to the (+) input. This switches the output from $+V_{sat}$ to $-V_{sat}$. The (+) input is now held negative with respect to ground because the feedback voltage is negative and given by

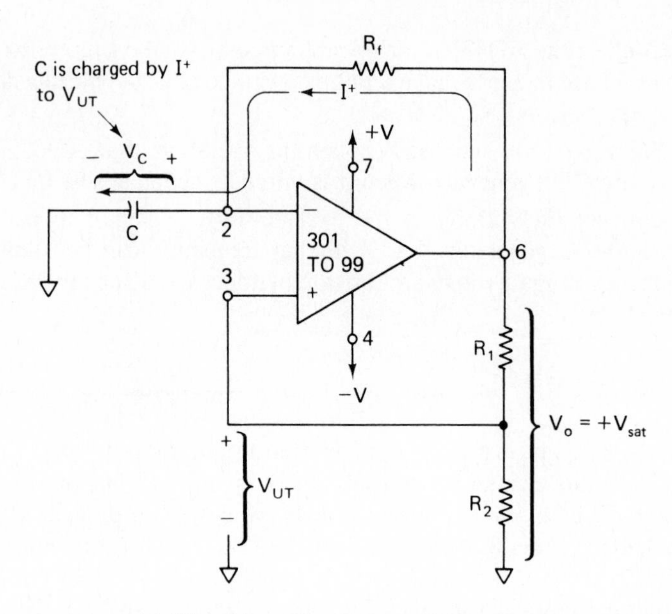

(a) When $V_o = +V_{sat}$, V_C charges toward V_{UT}

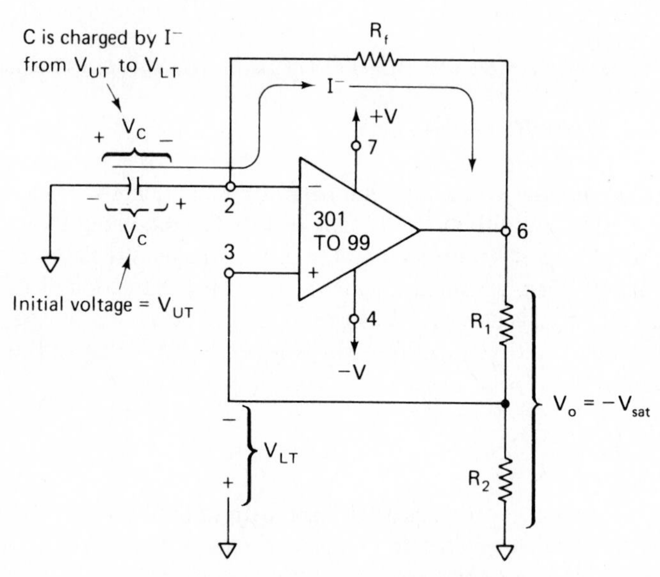

(b) When $V_o = -V_{sat}$, V_C charges toward V_{LT}

FIGURE 6-1 Free-running multivibrator ($R_1 = 100\text{ k}\Omega$, $R_2 = 86\text{ k}\Omega$). Output-voltage waveforms shown in Fig. 6-2.

$$V_{LT} = \frac{R_2}{R_1 + R_2}(-V_{sat}) \tag{6-2}$$

Equation (6-2) is the same as Eq. (4-2). Just after V_o switches to $-V_{sat}$, the capacitor has an initial voltage equal to V_{UT} [see Fig. 6-1(b)]. Now current I^- discharges C to 0 V and recharges C to V_{LT}. When V_C becomes slightly more negative than the feedback voltage V_{LT}, output voltage V_o switches back to $+V_{sat}$. The condition in Fig. 6-1(a) is reestablished except that C now has an initial charge equal to V_{LT}. The capacitor will discharge from V_{LT} to 0 V and then recharge to V_{UT}, and the process is repeating. Free-running multivibrator action is summarized as follows:

1. When $V_o = -V_{sat}$, C discharges from V_{UT} to V_{LT} and switches V_o to $+V_{sat}$.
2. When $V_o = +V_{sat}$, C charges from V_{LT} to V_{UT} and switches V_o to $-V_{sat}$.

The time needed for C to charge and discharge determines the frequency of the multivibrator.

6-1.2 Frequency of Oscillation

The capacitor and output voltage waveforms for the free-running multivibrator are shown in Fig. 6-2. Resistor R_2 is chosen to equal $0.86R_1$ to simplify calculation of capacitor charge time. Time intervals t_1 and t_2 show how V_C and V_o change with time for Fig. 6-1(a) and (b), respectively. Time intervals t_1 and t_2 are equal to the product of R_f and C.

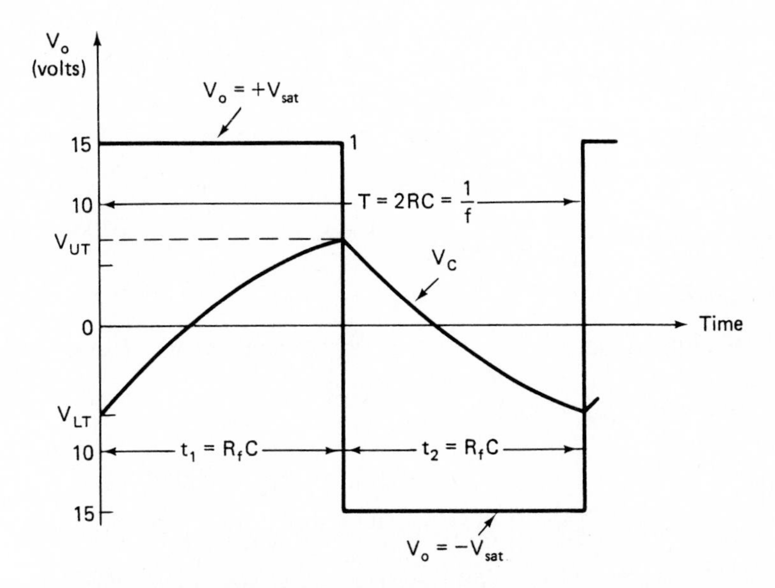

FIGURE 6-2 Voltage waveshapes for the multivibrator of Fig. 6-1.

The period of oscillation, T, is the time needed for one complete cycle. Since T is the sum of t_1 and t_2,

$$T = 2R_f C \qquad \text{for } R_2 = 0.86R_1 \tag{6-3a}$$

The frequency of oscillation f is the reciprocal of the period T and is expressed by

$$f = \frac{1}{T} = \frac{1}{2R_f C} \tag{6-3b}$$

where T is in seconds, f in hertz, R_f in ohms, and C in farads.

Example 6-1

In Fig. 6-1, if $R_1 = 100\ \text{k}\Omega$, $R_2 = 86\ \text{k}\Omega$, $+V_{sat} = +15$ V, and $-V_{sat} = -15$ V, find (a) V_{UT}; (b) V_{LT}.

Solution (a) By Eq. (6-1),

$$V_{UT} = \frac{86\ \text{k}\Omega}{186\ \text{k}\Omega} \times 15\ \text{V} \approx 7\ \text{V}$$

(b) By Eq. (6-2),

$$V_{LT} = \frac{86\ \text{k}\Omega}{186\ \text{k}\Omega}(-15\ \text{V}) = -7\ \text{V}$$

Example 6-2

Find the period of the multivibrator in Example 6-1 if $R_f = 100\ \text{k}\Omega$ and $C = 0.1\ \mu\text{F}$.

Solution Using Eq. (6-3a), $T = (2)(100\ \text{k}\Omega)(0.1\ \mu\text{F}) = 0.020\ \text{s} = 20\ \text{ms}$.

Example 6-3

Find the frequency of oscillation for the multivibrator of Example 6-2.

Solution From Eq. (6-3b),

$$f = \frac{1}{20 \times 10^{-3}\ \text{s}} = 50\ \text{Hz}$$

Example 6-4

Show why $T = 2R_fC$ when $R_2 = 0.86R$, as stated in Eq. (6-3a).

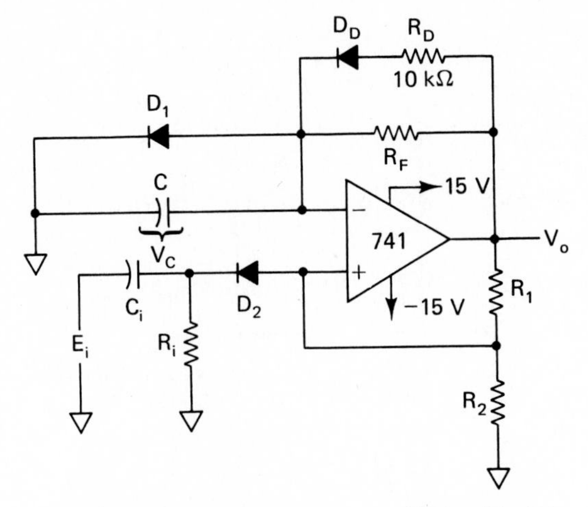

(a) Discharge diode D_D and R_D are added to Fig. 6-4

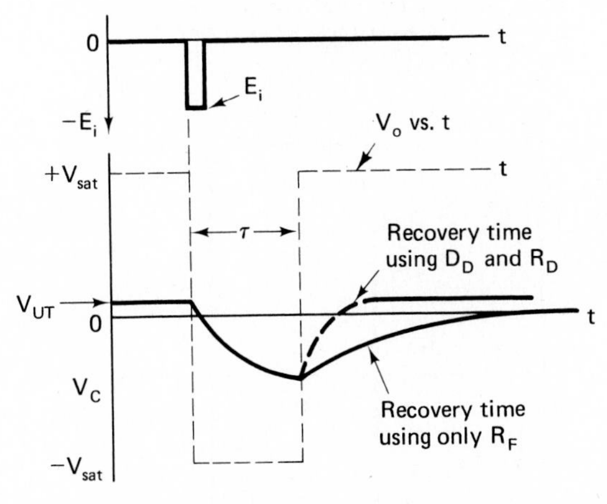

(b) Recovery time is reduced by D_D and R_D

FIGURE 6-5 The recovery time of a one-shot multivibrator is reduced by adding discharge diode D_D and R_D. R_D should be about one-tenth of R_f to reduce recovery time by one-tenth.

time is reduced. Typically, if $R_D = 0.1R_f$, recovery time is reduced by one-tenth. Diode D_D prevents R_D from affecting the timing-cycle interval τ.

6-3 TRIANGLE-WAVE GENERATORS

6-3.1 Theory of Operation

A basic bipolar triangle-wave generator circuit is presented in Fig. 6-6. The triangle wave, V_A, is available at the output of the 741 integrator circuit. An additional square-wave signal, V_B, is available at the output of the 301 comparator.

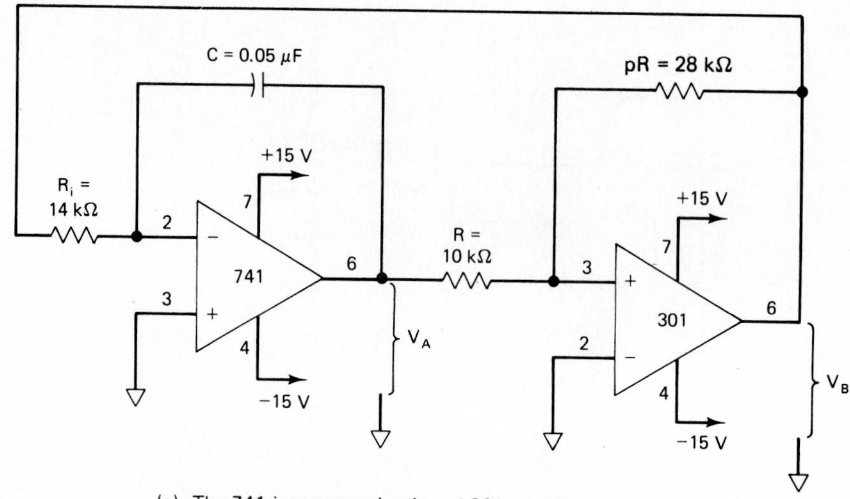

(a) The 741 integrator circuit and 301 comparator circuit are wired to make a triangle-wave generator

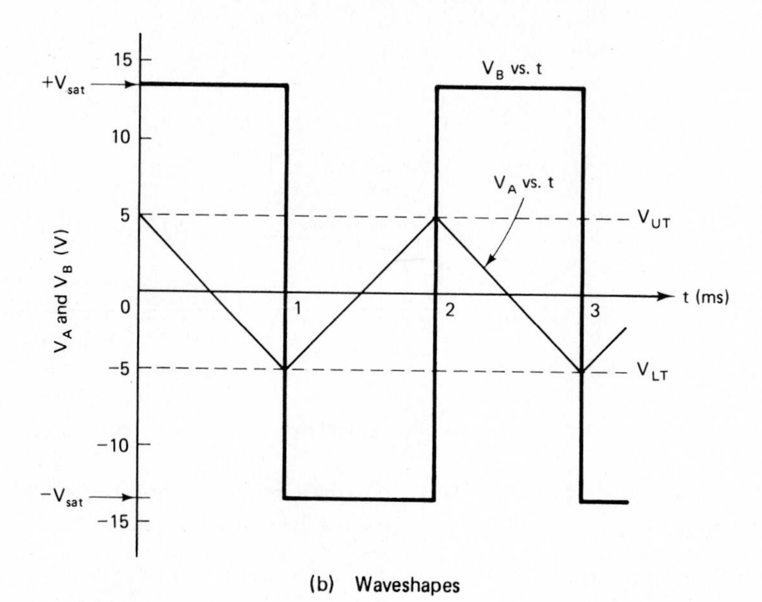

(b) Waveshapes

FIGURE 6-6 The bipolar triangle-wave generator circuit in (a) generates triangle-wave and square-wave oscillator signals as in (b). (a) Basic bipolar triangle-wave generator oscillator frequency for 1000 Hz; (b) output voltage waveshapes.

To understand circuit operation, refer to time interval 0 to 1 ms in Fig. 6-6. Assume that V_B is high at $+V_{sat}$. This forces a constant current (V_{sat}/R_i) through C (left to right) to drive V_A negative from V_{UT} to V_{LT}. When V_A reaches V_{LT}, pin 3 of the 301 goes negative and V_B snaps to $-V_{sat}$ and $t = 1$ ms.

When V_B is at $-V_{sat}$, it forces a constant current (right to left) through C to drive V_A positive from V_{LT} toward V_{UT} (see the time interval 1 to 2 ms). When V_A reaches V_{UT} at $t = 2$ ms, pin 3 of the 301 goes positive and V_B snaps to $+V_{sat}$. This initiates the next cycle of oscillation.

6-3.2 Frequency of Operation

The peak values of the triangular wave are established by the *ratio* of resistor pR to R and the saturation voltages. They are given by

$$V_{UT} = -\frac{-V_{sat}}{p} \tag{6-5a}$$

$$V_{LT} = -\frac{+V_{sat}}{p} \tag{6-5b}$$

where

$$p = \frac{pR}{R} \tag{6-5c}$$

If the saturation voltages are reasonably equal, the frequency of oscillation, f, is given by

$$f = \frac{p}{4R_i C} \tag{6-6}$$

Example 6-6

A triangle-wave generator oscillates at a frequency of 1000 Hz with peak values of approximately +5 V. Calculate the required values for pR, R_i, and C in Fig. 6-6.

Solution *First* we work on the calculation for the comparator resistor ratio p that controls peak triangle-wave output voltages, V_{UT} and V_{LT}. $+V_{sat}$ is practically + 14.2 V and $-V_{sat}$ is typically -13.8 V for a ±15-V supply. This observation points out one deficiency in our low-cost triangle-wave generator. It does *not* have *precisely* equal positive and negative peak outputs. (We will remedy this problem, at a higher cost, in Section 6-6.) From Eq. (6-5a), solve for p:

$$p = -\frac{-V_{sat}}{V_{UT}} = -\frac{-13.8 \text{ V}}{5 \text{ V}} = +2.76 \approx 2.8$$

Choose $R = 10\ \text{k}\Omega$. Then from Eq. (6-5c) we solve for pR as

$$pR = (p)R = 2.8(10\ \text{k}\Omega) = 28\ \text{k}\Omega$$

Next we select R_i and C. Begin by making a *trial* choice for $C = 0.05\ \mu\text{F}$. Then calculate a value for R_i to see if R_i comes up greater than 10.0 kΩ. From Eq. (6-6),

$$R_i = \frac{p}{4fC} = \frac{2.8}{4(1000\ \text{Hz})(0.05\ \mu\text{F})} = 14\ \text{k}\Omega$$

In practice it would be prudent to make R_i up from a 12-kΩ resistor in series with a 0 to 5-kΩ pot. The 5-kΩ pot would then be adjusted for an oscillation frequency of precisely 1.00 kHz.

6-3.3 Unipolar Triangle-Wave Generator

The bipolar triangle-wave generator circuit of Fig. 6-6 can be changed to output a unipolar triangle wave. Simply add a diode in series with pR as shown in Fig. 6-7. Circuit operation is studied by reference to the waveshapes in Fig. 6-7.

When V_B is at $+V_{sat}$, the diode stops current flow through pR and sets V_{LT} at 0 V. When V_B is at $-V_{sat}$, the diode allows current flow through pR and sets V_{UT} at a value of

$$V_{UT} = -\frac{-V_{sat} + 0.6\ \text{V}}{p} \tag{6-7a}$$

Frequency of oscillation is then given approximately by

$$f \simeq \frac{p}{2R_iC} \tag{6-7b}$$

Example 6-7

Find the approximate peak voltage and frequency for the unipolar triangle-wave generator in Fig. 6-7.

Solution Calculate

$$p = \frac{pR}{R} = \frac{28\ \text{k}\Omega}{10\ \text{k}\Omega} = 2.8$$

Find the peak value of V_A from Eq. (6-7a):

$$V_{UT} = -\left(\frac{-V_{sat} + 0.6\ \text{V}}{p}\right) = -\left(\frac{-13.8\ \text{V} + 0.6\ \text{V}}{2.8}\right) \simeq 4.7\ \text{V}$$

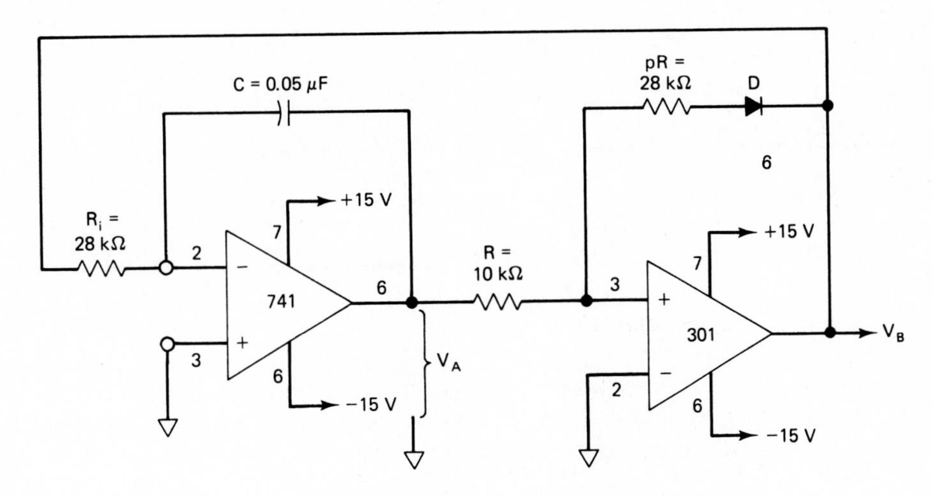

(a) Unipolar triange-wave generator

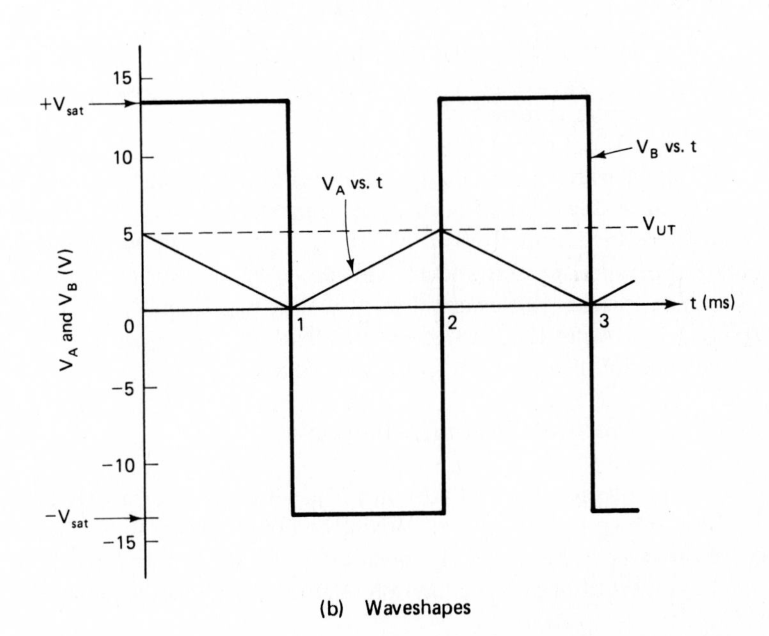

(b) Waveshapes

FIGURE 6-7 Diode D in (a) converts the bipolar triangle wave-generator into a unipolar triangle-wave generator. Waveshapes are shown in (b). (a) Basic unipolar triangle-wave generator, oscillating frequency is 1000 Hz; (b) output voltage waveshapes.

From Eq. (6-7b),

$$f = \frac{p}{2R_iC} = \frac{2.8}{2(28\ \text{k}\Omega)(0.05\ \mu\text{F})} = 1000\ \text{Hz}$$

6-4 SAWTOOTH-WAVE GENERATOR

6-4.1 Circuit Operation

A low-parts-count sawtooth-wave generator circuit is shown in Fig. 6-8(a). Op amp A is a ramp generator. Since E_i is negative, $V_{o\,\text{ramp}}$ can only ramp up. The rate of rise of the ramp voltage is constant at

$$\frac{V_{o\,\text{ramp}}}{t} = \frac{E_i}{R_iC} \tag{6-8}$$

The ramp voltage is monitored by the (+) input of comparator 301B. If $V_{o\,\text{ramp}}$ is below V_{ref}, the comparator's output is negative. Diodes protect the transistors against excessive reverse bias.

When $V_{o\,\text{ramp}}$ rises to just exceed V_{ref}, the output $V_{o\,\text{comp}}$ goes to positive saturation. This forward biases "dump" transistor Q_D into saturation. The saturated transistor acts as a short circuit across the integrating capacitor C. C discharges quickly through Q_D to essentially 0 V. When $V_{o\,\text{comp}}$ goes positive, it turns on Q_1 to short-circuit the 10-kΩ potentiometer. This drops V_{ref} to almost zero volts.

As C discharges toward 0 V, it drives $V_{o\,\text{ramp}}$ rapidly toward 0 V. $V_{o\,\text{ramp}}$ drops below V_{ref}, causing $V_{o\,\text{comp}}$ to go negative and turn off Q_D. C begins charging linearly and generation of a new sawtooth wave begins.

6-4.2 Sawtooth Waveshape Analysis

The ramp voltage rises at a rate of 1 V per millisecond in Fig. 6-8(b). Meanwhile, $V_{o\,\text{comp}}$ is shown to be negative. When the ramp crosses V_{ref}, $V_{o\,\text{comp}}$ snaps positive to drive the ramp voltage quickly toward 0 V. As $V_{o\,\text{ramp}}$ snaps to 0 V, the comparator's output is reset to negative saturation. Ramp operation is summarized in Fig. 6-8(c).

6-4.3 Design Procedure

The time for one sawtooth-wave period can be derived most efficiently by analogy with a familiar experience.

$$\text{time (of rise)} = \frac{\text{distance (of rise)}}{\text{speed (of rise)}} \tag{6-9a}$$

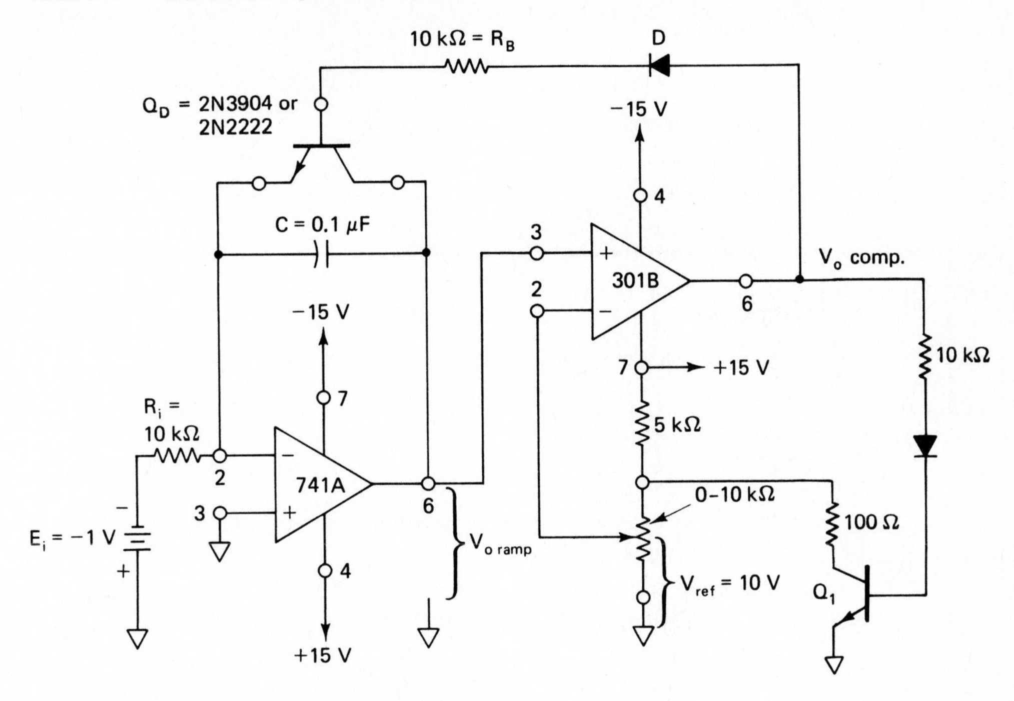

(a) Sawtooth-wave generator circuit

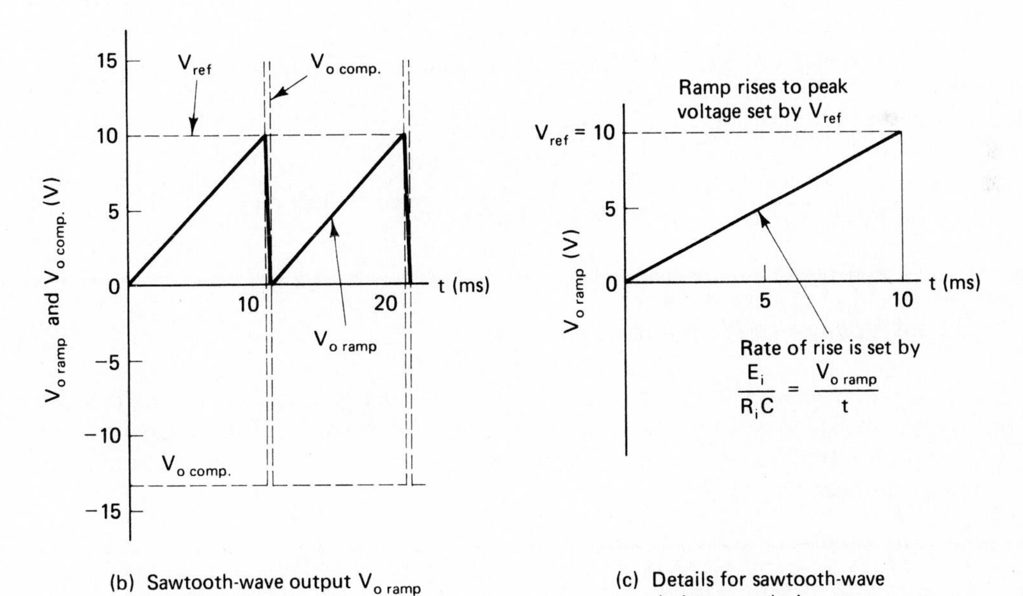

(b) Sawtooth-wave output $V_{o\,ramp}$ and comparator output

(c) Details for sawtooth-wave design or analysis

FIGURE 6-8 The sawtooth-wave generator circuit in (a) has the waveshapes shown in (b) and (c). Oscillating frequency is 100 Hz or $f = (1/R_i C)(E_i/V_{ref})$.

$$\text{period } T = \frac{V_{ref}}{E_i/R_i C} \tag{6-9b}$$

Since frequency is the reciprocal of the period

$$f = \left(\frac{1}{R_i C}\right)\frac{E_i}{V_{ref}} \tag{6-9c}$$

Design Example 6-8

Design a sawtooth-wave generator to have a 10-V peak output and a frequency of 100 Hz.

Design Procedure

1. Design a voltage divider to give a reference voltage $V_{ref} = +10$ V for the 301 op amp in Fig. 6-8.
2. Let's select a ramp rate rise of 1 V/ms. Pick any $R_i C$ combination to give 1.0 ms. Therefore, let's select $R_i = 10$ kΩ and $C = 0.1$ μF [see Eq. (6-9a)].
3. E_i should be made from a voltage divider and voltage follower to make an ideal voltage source.
4. The resulting circuit is shown in Fig. 6-7.
5. Alternatively, you could pick a trial value for $R_i C$ and solve for E_i in Eq. 6-9b.
6. Check the design values in Eq. 6-9c.

$$f = \frac{1}{(10\text{ k}\Omega)(0.1\ \mu\text{F})}\left(\frac{1\text{ V}}{10\text{ V}}\right) = 100\text{ Hz}$$

6-4.4 Voltage-to-Frequency Converter

There are two ways to change or modulate the oscillating frequency of Fig. 6-8. We see from Eq. (6-9c) that the frequency is directly proportional to E_i and inversely proportional to V_{ref}. The advantages and disadvantages of each method are examined with an example.

Example 6-9

If E_i is doubled to -2 V in Fig. 6-8, find the new frequency of oscillation.

Solution In Eq.(6-9c) use $|E_i|$

$$f = \left(\frac{1}{R_i C}\right)\frac{E_i}{V_{ref}} = \frac{1}{(10 \times 10^3\ \Omega)(0.1 \times 10^{-6}\ \text{F})}\frac{E_i}{10\ \text{V}}$$

$$= \frac{1}{1.0 \times 10^{-3}\ \text{s}}\frac{E_i}{10\ \text{V}} = \left(\frac{100\ \text{Hz}}{\text{volt}}\right)E_i$$

For $E_i = -2$ V, $f = (2\ \text{V})(100\ \text{Hz/V}) = 200$ Hz. Thus as E_i changes from 0 V to -10 V, frequency changes from 0 Hz to 1 kHz. The peak amplitude of the sawtooth wave remains equal to V_{ref} (10 V) for all frequencies.

Example 6-10

Keep E_i, R_i, and C at their value shown in Fig. 6-7(a). Reduce V_{ref} by one-half to 5 V. Is the frequency doubled or halved?

Solution From Example 6-8 and Eq. 6-9c,

$$f = \frac{1\ \text{V}}{(\text{ms})V_{ref}} = \frac{(1000\ \text{Hz})/\text{V}}{V_{ref}}$$

For $V_{ref} = 10$ V, $f = 100$ Hz. For $V_{ref} = 5$ V the frequency is doubled to 200 Hz. As V_{ref} is *reduced* from 10 V to 0 V, frequency increased from 100 Hz to a very high value.

This type of frequency modulation by V_{ref} has two disadvantages with respect to control frequency by E_i. First, the relationship between input voltage V_{ref} and output frequency is *not* linear. Second, the sawtooth's peak output voltage is not constant since it varies directly with V_{ref}.

6-4.5 Frequency Modulation and Frequency Shift Keying

Examples 6-9 and 6-10 indicate one way of achieving *frequency modulation* (FM). Thus, if the amplitude of E_i varies, the frequency of the sawtooth oscillator will be changed or modulated. If E_i is keyed between two voltage levels, the sawtooth oscillator changes frequencies. This type of application is called *frequency shift keying* (FSK) and is used for data transmission. These two preset frequencies correspond to "0" and "1" states (commonly called *space* and *mark*) in binary.

6-4.6 Disadvantages

The triangle-wave generators of Section 6-3 are inexpensive and reliable. However, they have two disadvantages. The rates of rise and fall of the triangle wave are unequal. Also, the peak values of both triangle-wave and square-wave outputs are unequal. This is because the magnitudes of $+V_{sat}$ and $-V_{sat}$ are unequal.

In the next section we substitute an AD630 for the comparator. This will give the equivalent of precisely equal square-wave ± voltages that will *also* be equal to the ± peak values of triangle-wave voltage. Once we have made a *precision* triangle-wave generator, we will use it to drive a new state-of-the-art trigonometric function generator to make a precision sine-wave generator.

6-5 BALANCED MODULATOR/DEMODULATOR, THE AD630

6-5.1 Introduction

The AD630 is an advanced integrated circuit. It has 20 pins, which allow this versatile switched voltage gain IC to act as a modulator, demodulator, phase detector and multiplexer, as well as performing other signal conditioning tasks. We connect the AD630, as in Fig. 6-9(a), as a controlled switched gain (+1 or −1) amplifier. This particular application will be examined by discussing the role performed by the dominant terminals.

6-5.2 Input and Output Terminals

The input signal V_{ref} is connected to *modulation* pins 16 and 17 in Fig. 6-9, and thus to the inputs of two amplifiers *A* and *B*. The gain of *A* is programmed for −1 and *B* for +1 by shorting terminals (1) 13 to 14, (2) 15 to 19 to 20, (3) 16 to 17, and (4) grounding pin 1.

The carrier input terminal, pin 9 (in this application), determines which amplifier *A* or *B* is connected to the output terminal. If pin 9 is *above* the voltage at pin 10 (ground), amplifier *B* is selected. Voltage at output pin 13 then equals V_{ref} times (+1).

If pin 9 voltage is *below* ground (negative), amplifier *A* is selected and output pin 13 equals V_{ref} times (−1). (Note that in communication circuits, V_{ref} is called the *analog data* or *signal voltage*. V_C is called a chopper or *carrier voltage*. V_o is the *modulated* output. That is, the *amplitude* of the low-frequency signal voltage is impressed upon the higher-frequency carrier wave—hence the names selected for the AD630's input and output terminals.)

6-5.3 Input–Output Waveforms

V_{ref} is a dc voltage of 5.0 V in Fig. 6-9(b). V_C is a 100-Hz square wave with peak amplitudes that must exceed ±1 mV. Output voltage V_o is shown to switch synchronously with V_C from $+V_{ref}$ to $-V_{ref}$, and vice versa, in Fig. 6-9(b). We are go-

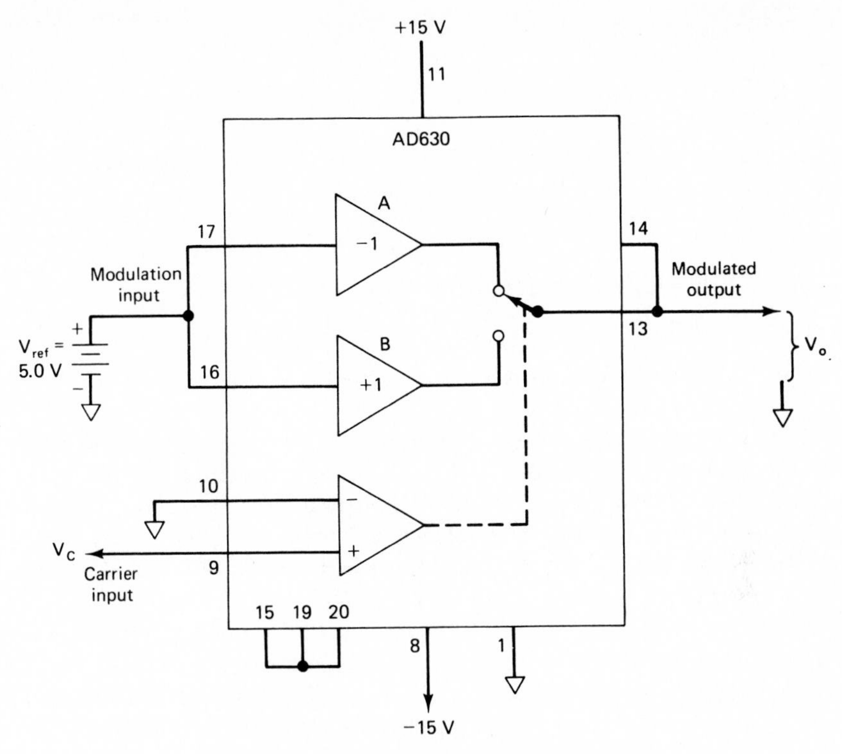

(a) The AD630 is wired as a switch gain amplifier

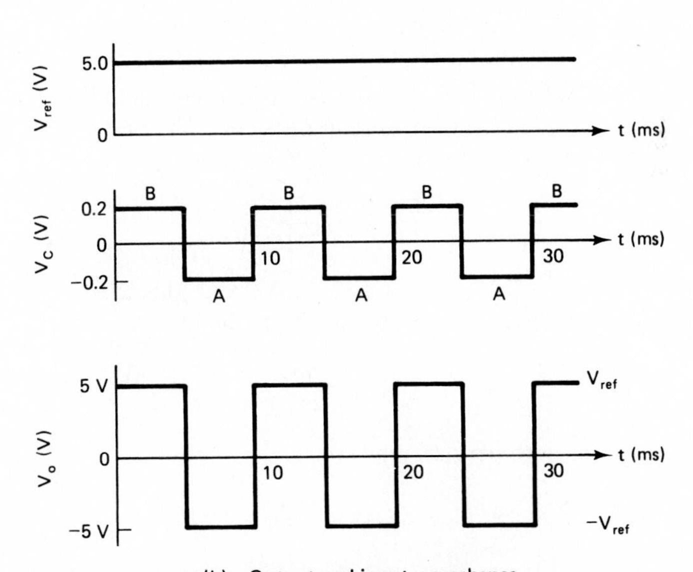

(b) Output and input waveshapes

FIGURE 6-9 Operation of the AD630 balanced modulator/demodulator as a switched gain amplifier. (a) Wiring for switched gains of +1 or −1; (b) carrier V_C selects gains of +1 or −1 for input V_{ref}. Output V_o is equal to precisely V_{ref} or $-V_{ref}$.

ing to replace the unpredictable $\pm V_{sat}$ of the 301 comparator in Figs. 6-6 or 6-7 with precisely + or $-V_{ref}$. Moreover, V_{ref} can be adjusted easily to any required value. As shown in the next section, V_{ref} will set the positive and negative peak values of both triangle-wave and square-wave generators.

6-6 PRECISION TRIANGLE/SQUARE-WAVE GENERATOR

6-6.1 Circuit Operation

Only six parts plus voltage source, V_{ref}, make up a versatile precision triangle-and square-wave generator of Fig. 6-10(a). Circuit operation is explained by reference to the waveshapes in Fig. 6-10(b). We begin at time zero. Square-wave output V_{os} begins at $-V_{ref}$ or -5 V. This forces the triangle-wave V_{oT} to go positive from a starting point of $-V_{ref} = -5$ V. During this time, pin 9 is below ground to select an AD630 gain of -1 and holds V_{os} at -5 V.

At time $T/2 = 0.5$ ms, V_{oT} reaches $+5$ V, where pin 9 is driven slightly positive to select an AD630 gain of $+1$. This snaps V_{os} to $V_{ref} = +5$ V. V_{os} then drives V_{oT} negative. When V_{oT} reaches -5 V, pin 9 goes negative at $T = 1.0$ ms and snaps V_{os} negative to -5 V. This completes one cycle of oscillation and begins another.

6-6.2 Frequency of Oscillation

The easiest way to find the frequency of oscillation is to begin with the rate of rise of the triangle wave, V_{oT}/t, in volts per second. The rate of rise of the triangle wave, from 0 to 0.5 ms in Fig. 6-10(b), is found from

$$\frac{V_{oT}}{t} = \frac{V_{ref}}{R_i C} \tag{6-10}$$

The time t for a half-cycle is $T/2$, and during this time, V_{oT} changes by $2V_{ref}$. We substitute for t and V_{oT} into Eq. (6-10) to obtain

$$\frac{2V_{ref}}{T/2} = \frac{V_{ref}}{R_i C} \tag{6-11}$$

and solve for both period T and frequency of oscillation f:

$$T = 4R_i C \qquad \text{and} \qquad f = \frac{1}{T} = \frac{1}{4R_i C} \tag{6-12}$$

Note that V_{ref} cancels out in Eqs. (6-11) and (6-12). This is a very important advantage. The peak output voltages of both square- and triangle-wave signals are set by $+V_{ref}$. As V_{ref} is adjusted, the frequency of oscillation does *not* change.

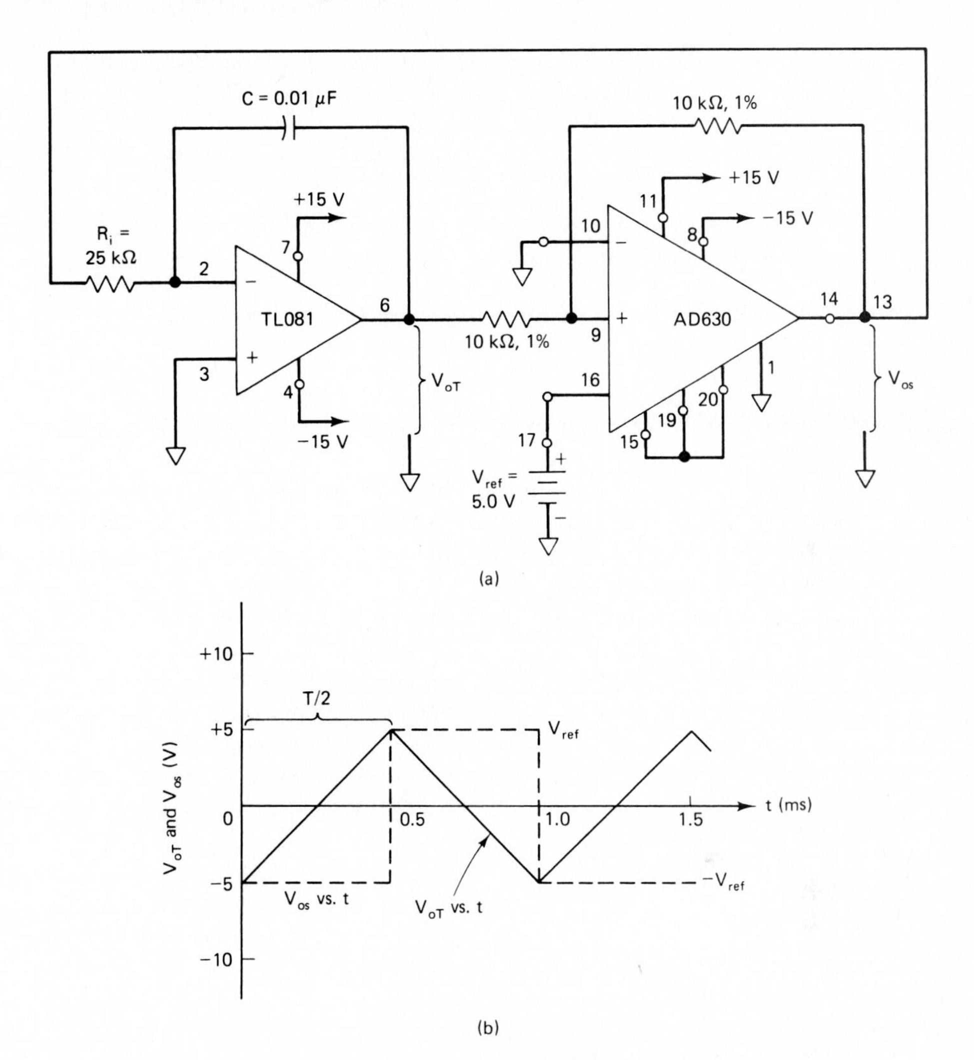

FIGURE 6-10 The precision triangle/square-wave oscillator in (a) has the output waveshapes in (b). V_{ref} should be buffered for a low-impedance source voltage. (a) Precision triangle/square-wave oscillator (compare with Fig. 6-6). V_{ref} must be a low-impedance source. V_{ref} sets the ± peak values and R_i adjusts the frequency. (b) Square- and triangular-wave output waveshapes.

Example 6-11

Make a triangle/square-wave generator that has peak voltages of ±5 V and oscillates at a frequency of 1.0 kHz.

Solution Choose $V_{ref} = 5.0$ V. For low impedance, V_{ref} should be the output of an op amp. Arbitrarily choose $C = 0.01\ \mu$F. From Eq. (6-12),

$$R_i = \frac{1}{4fC} = \frac{1}{4(1000)(0.01\ \mu\text{F})} = 25.0\ \text{k}\Omega$$

For a fine adjustment of the output frequency, make R_i from a 22-kΩ resistor in series with a 5- or 10-kΩ variable resistor.

6-7 SINE-WAVE GENERATION SURVEY

Commercial function generators produce triangular, square, and sinusoidal signals whose frequency and amplitude can be changed by the user. To obtain a sine-wave output, the triangle wave is passed through a shaping network made of carefully selected resistors and diodes. The sine waves thus produced are reasonably good. However, there is inevitably some distortion, particularly at the peaks of the sine wave.

When an application requires a single-frequency sine wave, conventional oscillators use phase-shifting techniques that usually employ (1) two *RC* tuning networks, and (2) complex amplitude limiting circuitry. To minimize distortion, the limit circuit must usually be custom-adjusted for each oscillator. The frequency of this oscillator is difficult to vary because two *RC* networks must be varied and their values must track within ±1%.

The recent invention of two state-of-the-art ICs has eliminated the disadvantages of difficult frequency adjustment and difficult amplitude control. They are the AD630 and AD639. The AD630 has already been used to generate a precision triangle wave whose frequency and amplitudes are precise and easy to adjust. We will connect the triangle-wave output V_{oT} of Fig. 6-10(a) to an AD639 universal trigonometric function generator. The resulting circuit will have the best qualities of a *precision sine-wave generator whose frequency will be easily adjustable*.

6-8 UNIVERSAL TRIGONOMETRIC FUNCTION GENERATOR, THE AD639

6-8.1 Introduction

The AD639 is a state-of-the-art trigonometric function generator. It will perform all trigonometric functions *in real time,* including sin, cos, tan, cosec, sec, and cotan. When a calculator performs a trig function, the operator punches in a number corresponding to the number of angular degrees and punches SIN. The calculator pauses, then displays a number indicating the sine of the angle. That is, a number for angle θ is entered and the calculator produces a number for sin θ.

The AD639 accepts an input voltage that represents the angle. It is called the *angle voltage*, V_{ang}. For the AD639, the angle voltage is found from

$$V_{ang} = \left(\frac{20 \text{ mV}}{1°}\right)\theta = \left(\frac{1 \text{ V}}{50°}\right)\theta \tag{6-13}$$

Four input terminals are available. However, we shall look at only the single active input that generates sin functions. The output voltage will equal sin θ or 10 sin θ, depending how the internal gain control is pin programmed.

6-8.2 Sine Function Operation

The AD639 is wired to output $V_o = 1 \sin \theta$ in Fig. 6-11. There are four input terminals: 1, 2, 7, and 8. Wired as shown, the chip performs a sine function. Pins 3, 4, and 10 control gain. Normally, 3 and 4 are grounded so that pin 10 can activate the internal gain control. A gain of 1 results when pin 10 is wired to $-V_s$ or pin 9. Wire pin 10 to $+V_s$ or pin 16 and obtain a gain of +10. *Then* $V_o = 10 \sin \theta$. Pin 6 is a precision 1.80-V reference voltage that corresponds to an angle voltage of 90° [see Eq. (6-13)]. We analyze sin function operation by an example.

Example 6-12

Calculate the required input angle voltage and resultant output voltage for angles of (a) ±45°; (b) ±90°; (c) ±225°; (d) ±405°.

Solution From Eq. (6-13) and Fig. 6-11:

(a) $V_{ang} = \left(\frac{20 \text{ mV}}{1°}\right)(\pm 45°) = \pm 0.90 \text{ V}, V_o = 1 \sin (\pm 45°) = \pm 0.707 \text{ V}.$

(b) $V_{ang} = \left(\frac{20 \text{ mV}}{1°}\right)(\pm 90°) = \pm 1.80 \text{ V}, V_o = 1 \sin (\pm 90°) = \pm 1.0 \text{ V}.$

(c) $V_{ang} = \left(\frac{20 \text{ mV}}{1°}\right)(\pm 225°) = \pm 4.50 \text{ V}, V_o = 1 \sin (\pm 225°) = \pm 0.707 \text{ V}.$

(d) $V_{ang} = \left(\frac{20 \text{ mV}}{1°}\right)(\pm 405°) = \pm 8.10 \text{ V}, V_o = 1 \sin (\pm 405°) = \pm 0.707 \text{ V}.$

Example 6-12 clearly illustrates that the AD639, remarkable as it is, cannot output the sine of, for example, 36,000°. This would require an angle voltage of 720 V. The normal ±15-V supply limits the guaranteed usable input angle to ±500°

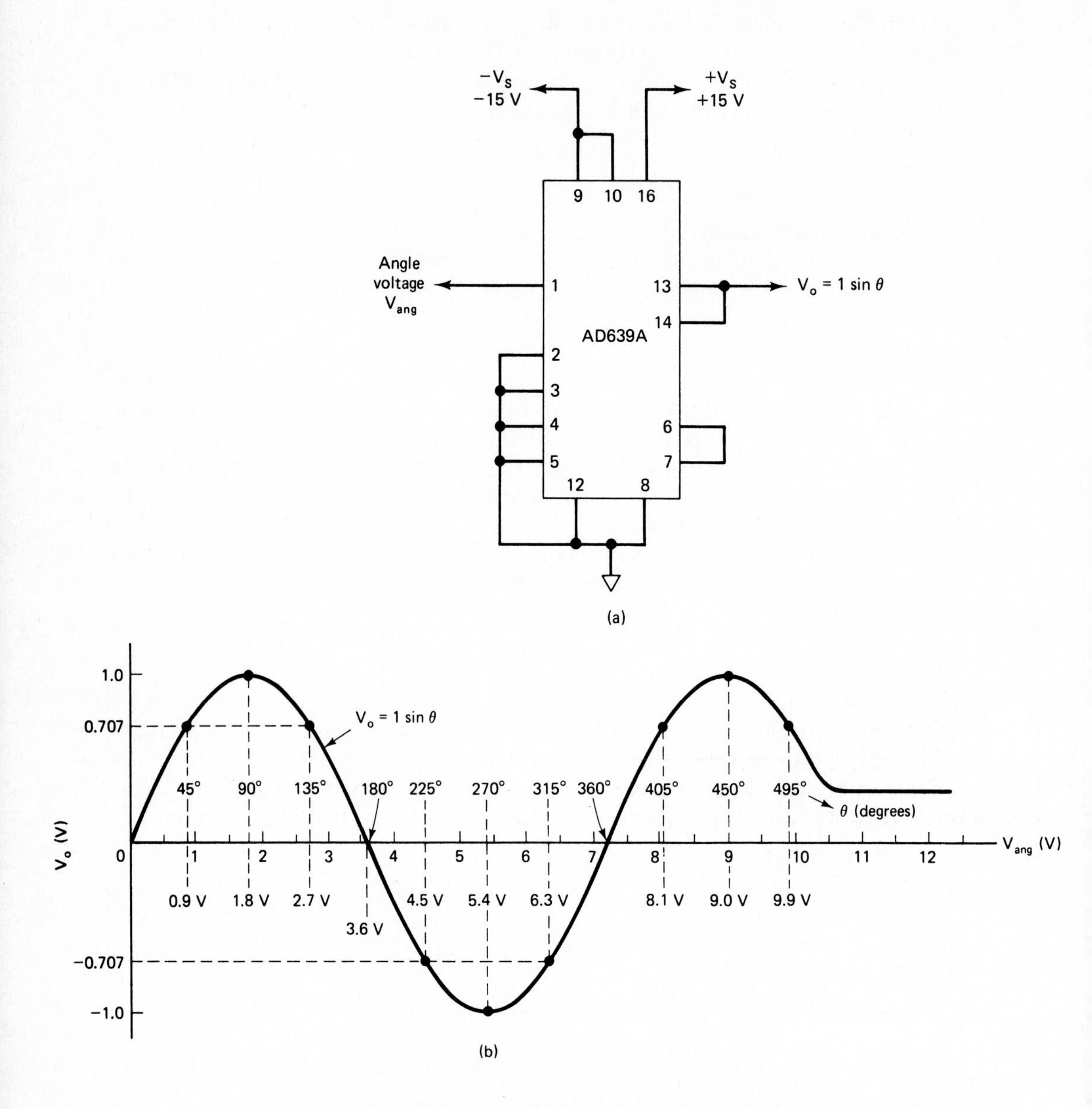

FIGURE 6-11 The AD639 is pin-programmed in (a) to act as a sine function generator. Each ±20 mV of input angle voltage corresponds to an input angle of $\theta = \pm 1°$. Output V_o equals $1 \times \sin \theta$. (a) The AD639A is pin-programmed to output the sine of the angle voltage; (b) output voltage V_o equals the sine of θ if θ is represented by an angle voltage of 20 mV per angular degree.

TABLE 6-1 AD639 SINE FUNCTIONS[a]

Input		Output (V)	
θ (angular degrees)	Angle voltage, V_{ang} (V)	$V_o = 1 \sin\theta$ (wire pin 10 to 9)	$V_o = 10 \sin\theta$ (wire pin 10 to 16)
0	0.00	0.000	0.000
±45	±0.90	±0.707	±7.07
±90	±1.80	±1.000	±10.07
±135	±2.70	±0.707	±7.07
±180	±3.60	0.000	0.000
±225	±4.50	∓0.707	∓7.07
±270	±5.40	∓1.000	∓10.00
±315	±6.30	∓0.707	∓7.07
±360	±7.20	0.000	0.00
±405	±8.10	±0.707	±7.07
±450	±9.00	±1.000	±10.00
±495	±9.90	±0.707	±7.07
±500	±10.00	±0.643	±6.43

[a] Connect terminal 10 to 9 to pin program $V_o = \sin\theta$; or connect pin 10 to 16 to pin program $V_o = 10 \sin\theta$. Input angle voltage $V_{ang} = (20\text{ mV}/1°\text{C})\ \theta$.

or ±10.000 V. We extend the results of Example 6-12 to summarize briefly the performance of the sine function generator in Table 6-1 and Fig. 6-11(b).

In Fig. 6-11(b), V_o is plotted against both V_{ang} and θ. A study of this figure shows that if V_{ang} *could be varied linearly* by a triangle wave, V_o *would vary sinusoidally*. Further, if the frequency of the triangle wave could be varied easily, the sine-wave frequency could *easily* be tuned, adjusted, or varied. We pursue this observation in the next section.

6-9 PRECISION SINE-WAVE GENERATOR

6-9.1 Circuit Operation

Connect the precision triangle-wave oscillator of Fig. 6-10 to the sine function generator of Fig. 6-11 to construct the *precision* sine-wave generator in Fig. 6-12. As a bonus, we also have precision triangle-wave and square-wave outputs. The 1.80-V reference voltage of the AD639 is connected to modulation inputs 16 and 17 of the AD630 modulator (Fig. 6-9). Circuit operation is now examined by reference to Fig. 6-12.

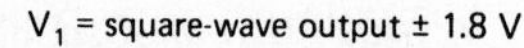

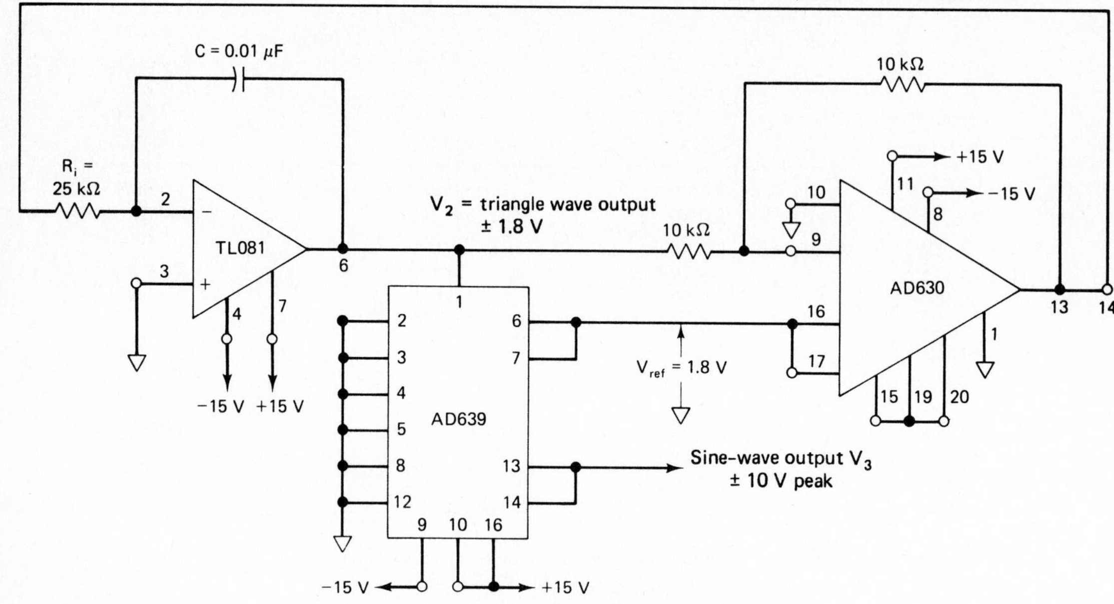

(a) Precision sine-wave generator circuit

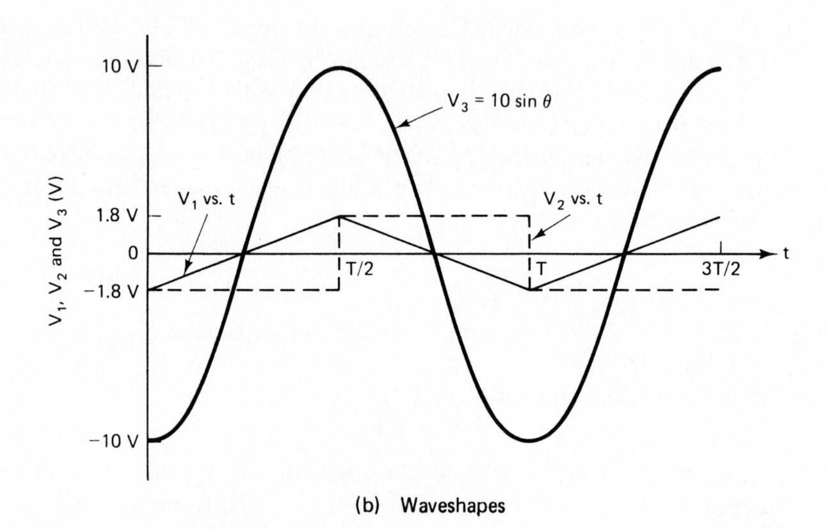

(b) Waveshapes

FIGURE 6-12 Frequency of the precision sine-square-triangle wave generator in (a) can be easily changed by adjusting R_i. Output waveshapes are shown in (b). Their amplitudes are independent of frequency.

Triangle-wave rise time, 0 to $T/2$ in Fig. 6-12(b)

1. AD630:
   a. Pin 13 is at $-V_{ref} = -1.8$ V, causing:
   b. Pin 9 to select gain = -1 to hold 13 at -1.8 V, and
   c. Op-amp output voltage to ramp up.
2. *Op amp:*
   a. Pin 6 ramps from $-V_{ref} = -1.8$ V *toward* $+V_{ref} = 1.8$ V to:
   b. Hold pin 9 of the AD630 negative, and
   c. Drive input 1 of the AD639 with an angle voltage linearly from -1.8 to 1.8 V.
3. *AD639:*
   a. Pin 1's input angle voltage corresponds to an input angle that varies linearly from -90 to $+90°$.
   b. Pin 13 outputs $V_o = 10 \sin \theta$ from -10 to $+10$ V.

When op amp pin 6 reaches +1.8V, pin 9 of the AD630 goes positive to select a gain of +1. Its output, in 13, snaps to +1.8 V. This begins the *fall time*.

Triangle-wave fall time, $T/2$ to T in Fig. 6-12(b)

1. *AD630:* Causes the triangle wave to ramp down from +1.8 V to -1.8 V. At -1.8 V, gain is switched to -1 and a new cycle begins.
2. *Op amp:* Applies an angle voltage to input pin 1 of the AD639 that varies linearly from +1.8 to -1.8 V.
3. *AD639:* Its input angle voltage corresponds to an input angle of $\theta = +90$ to $-90°$. Pin 13 outputs a sine wave that varies from +10 to -10 V.

6-9.2 Frequency of Oscillation

The frequency of oscillation, f, is determined by R_i, C, and the op amp in Fig. 6-12(a) from

$$f = \frac{1}{4R_iC} \tag{6-14}$$

Peak amplitudes of the triangle wave and square wave are precisely equal to ±1.8 V. The sine wave has peak amplitude of ±10 V and is synchronized to the triangle wave (for the ±1-V peak, change the AD639 pin 10 connection to $-V_S$).

Example 6-13

Let $C = 0.025\ \mu F$ in Fig. 6-12(a) (two 0.05-μF capacitors in series). How does frequency change as R_i is changed from 10 kΩ to 100 kΩ?

Solution From Eq. (6-14),

$$f = \frac{1}{4(10\ k\Omega)(0.025\ \mu F)} = \text{kHz}, \qquad f = \frac{1}{4(100\ k\Omega)(0.025\ \mu F)}$$

$$= 100\ \text{Hz} \qquad\qquad = 10\ \text{Hz}$$

Example 6-13 shows the overwhelming superiority of this multiwave generator. Frequency is easily tuned *and* with precision.

LABORATORY EXERCISES

Some of the circuits shown in this chapter are introductory in nature and were offered to simplify understanding of circuit operation. The remainder give excellent experience in the design and analysis of signal generators. The following suggestions are offered as starting points for some basic experiments.

6-1. **(a)** Build the multivibrator circuit of Fig. 6-1 with the component values given in Examples 6-1 to 6-3. Sketch the waveshapes for V_C and V_o.
(b) Plot frequency versus R_f as R_f is varied from 100 kΩ to 10 kΩ.

6-2. **(a)** Build the one-shot multivibrator of Fig. 6-4, except make $R_f = 10$ kΩ so that τ will be about 0.2 ms.
(b) Use a square-wave oscillator for E_i with a frequency of 500 Hz at 5 V peak. Use a CRO to monitor the output pulse. Vary the peak value of E_i to find the minimum value required to reliably trigger the one-shot.
(c) Monitor pin 2 to observe the recovery time before and after adding D_D and R_D of Fig. 6-5(a).

6-3. **(a)** Build the triangle-wave and square-wave oscillator of Fig. 6-6. Carefully measure the waveshape at V_A and V_B with a CRO. The positive and negative peak voltages will differ slightly, as will the time for each half-cycle.
(b) Connect a 10-kΩ resistor across the R resistor to double the resistor ratio p. Verify that the frequency of oscillation approximately doubles [see Eq. (6-6)]. Peak amplitudes for the triangle wave will approximately halve [see Eqs. (6-5)].
(c) Add a diode to your circuit in part (a) to observe the unipolar triangle wave of Fig. 6-7(b). Make sure that the CRO is dc coupled, to see that V_o goes to exactly 0 V at its bottom peak. Reverse the diode to see a negative unipolar triangle wave.

6-4. For a low-frequency voltage-controlled oscillator with a sawtooth output, build the circuit of Fig. 6-8 and Design Example 6-8. Vary E_i from -1 to -10 V and plot the ramp voltage frequency versus E_i. The frequency should increase linearly with E_i from about 100 to 1000 Hz.

6-5. The advanced integrated circuits of Figs. 6-9 to 6-12 are a joy to work with. The authors have presented these circuits in the order in which they should be used in a lab. To complete a precision sine/triangular/square-wave oscillator the steps involved are:

1. Build the circuit of Fig. 6-9(a). Set V_C for ± 1 V peak at 100 Hz. Measure the waveshapes of Fig. 6-9b to learn about operating the AD630 as a switched gain amplifier. Now remove V_{ref} and V_C and save the circuit.
2. Connect the AD630 to the TL081 (or 741) circuit in Fig. 6-10(a) to learn about an excellent precision triangle/square-wave generator. Note that oscillating frequency does *not* vary as you change V_{ref} to set the peak triangle- and square-wave voltages. Vary R_i to change frequency and note that the peaks do not vary (they depend only on V_{ref}). Now save this circuit.
3. Use the circuit of Fig. 6-11 and Example 6-12 to gain experience with the AD639. Save this circuit. Now you have all the ingredients for the final product.
4. Connect the AD639 (step 3) to the precision triangle-wave generator (step 2) to obtain the *precision* sine/square/triangle-wave generator of Fig. 6-12. Vary R_i from 100 Ω to 10 kΩ to see how easy it is to vary sinusoidal frequency by a decade from 10 to 100 Hz. Change C to 0.1 μF and the frequency will change from 100 to 1000 Hz for the same 100- to 10-kΩ variation in R_i. *Note:* The peak amplitudes of all waves stay the same and independent of frequency.

PROBLEMS

6-1. Make two drawings of a multivibrator circuit with $R_1 = 100$ kΩ, $R_2 = 86$ kΩ, $R_f = 10$ kΩ, and $C = 0.01$ μF. Show the direction of current through C and calculate both V_{UT} and V_{LT} for: **(a)** $V_o = +V_{sat} = 15$ V; **(b)** $V_o = -V_{sat} = -15$ V.

6-2. Calculate the frequency of oscillation for the multivibrator circuit in Problem 6-1.

6-3. In Problem 6-1, if C is changed to 0.1 μF, do you expect the output frequency to oscillate at 500 Hz? (See Example 6-3.) What could you do to R_f to increase frequency to 1000 Hz?

6-4. The monostable multivibrator of Figs. 6-4 and 6-5 generates a negative output pulse in response to a negative-going input signal. How would you change these circuits to get a positive output pulse for a positive-going input edge?

6-5. Explain what is meant by *monostable recovery time*.

6-6. Sketch a one-shot multivibrator circuit whose output will deliver a negative pulse lasting 1 ms with a recovery time of about 0.1 ms.

6-7. Assume for simplicity that saturation voltages in the triangle-wave oscillator of Fig. 6-6 are ± 15 V, $R_i = R = 10$ kΩ, $C = 0.1$ μF, and $pR = 50$ kΩ. Find the peak triangle-wave voltages and oscillating frequency.

6-8. Refer to the triangular-wave oscillator circuit of Fig. 6-6. What happens to peak output voltages and oscillating frequency if you **(a)** double pR only; **(b)** double R_i only; **(c)** double capacitor C only?

6-9. Change pR to 14 kΩ and C to 0.1 μF in the unipolar triangle-wave generator of Fig. 6-7. Find the resulting peak output voltage and frequency of oscillation. (See Example 6-7.)

6-10. In the sawtooth-wave generator of Fig. 6-8(a), let $V_{ref} = 1$ V, $R_i = 10$ kΩ, and $C = 0.1$ μF.

(a) Find an expression for frequency f in terms of E_i.

(b) Calculate f for $E_i = 1$ V and $E_i = 2$ V.

6-11. These questions concern the AD630 balanced modulator circuit in Fig. 6-9.

(a) Name the application for which the AD630 is wired.

(b) When pin 9 is at a positive voltage, which amplifier is selected, and what is the value of V_o?

(c) Suppose that V_{ref} is a ±1-V peak sine wave and pin 9 is at 1 V; what happens at V_o when pin 9 is changed to −1 V?

6-12. Figure 6-10 shows a precision triangle/square-wave oscillator. Three components control peak output voltages and oscillating frequency, R_i, C, and V_{ref}.

(a) Which does what?

(b) Can the oscillating frequency be adjusted independent of peak outputs, and vice versa?

(c) What must be done to change the frequency from 100 Hz to 500 Hz, and the peak voltages from ±5 V to ±1 V?

6-13. $V_o = 0.866$ V in the sine function generator circuit of Fig. 6-11.

(a) What angle does this represent?

(b) What is the value of the input angle voltage?

6-14. Calculate V_o in Fig. 6-11 when the input angle is 30° and pin 10 is wired to **(a)** pin 9 or **(b)** pin 10.

6-15. Design a sine-wave oscillator whose frequency can be varied from 0.5 Hz to 50 Hz with just a single variable resistor.

CHAPTER 7

Op Amps with Diodes

LEARNING OBJECTIVES

Upon completing this chapter on op amps with diodes, you will be able to:

- Draw the circuit for a precision (or linear) half-wave rectifier.
- Show current flow and circuit voltages in a precision half-wave rectifier for either positive or negative inputs.
- Do the same as above in objectives 1 and 2 for precision full-wave rectifiers.
- Sketch two types of precision full-wave rectifier circuits.
- Explain the operation of a peak detector circuit.
- Add one capacitor to a precision half-wave rectifier to make an ac-to-dc converter (mean-average-value) circuit.
- Explain the operation of dead-zone circuits.
- Draw the circuit for and explain the operation of precision clipper circuits.
- Name at least five application areas for precision rectifiers.

7-0 INTRODUCTION TO PRECISION RECTIFIERS

The major limitation of ordinary silicon diodes is that they cannot rectify voltages below 0.6 V. For example, Fig. 7-1(a) shows that V_o does not respond to positive inputs below 0.6 V in a half-wave rectifier built with an ordinary silicon diode. Figure 7-1(b) shows the waveforms for a half-wave rectifier built with an ideal diode. An output voltage occurs for all positive input voltages, even those below 0.6 V. A circuit that acts like an ideal diode can be designed using an op amp and two ordinary diodes. The result is a powerful circuit capable of rectifying input signals of only a few millivolts.

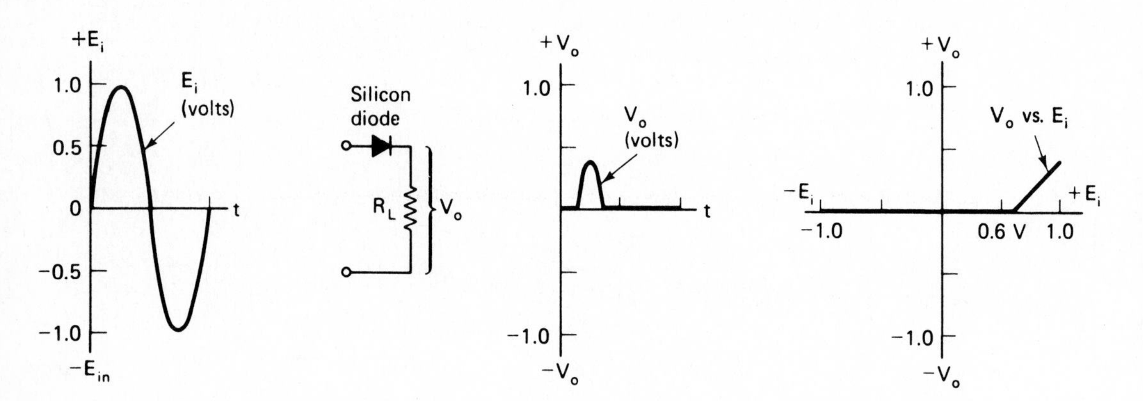

(a) Real diodes cannot rectify small ac voltages because of the diode's 0.6-V voltage drop

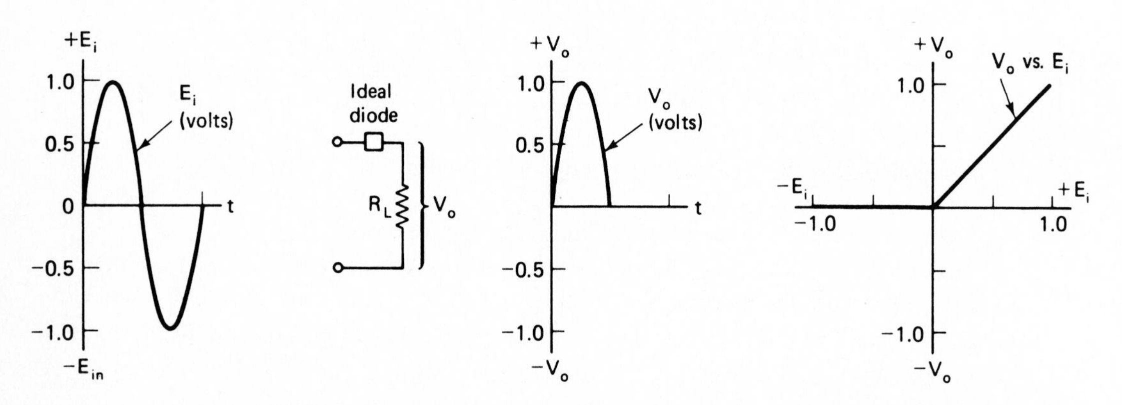

(b) A linear or precision half-wave rectifier circuit precisely rectifies any ac signal regardless of amplitude and acts as an ideal diode

FIGURE 7-1 The ordinary silicon diode requires about 0.6 V of forward bias in order to conduct. Therefore, it cannot rectify small ac voltages. A precision half-wave rectifier circuit overcomes this limitation.

The low cost of this equivalent ideal diode circuit allows it to be used routinely for many applications. They can be grouped loosely into the following classifications: linear half-wave rectifiers and precision full-wave rectifiers.

1. *Linear half-wave rectifiers.* The linear half-wave rectifier circuit delivers an output that depends on the magnitude *and polarity* of the input voltage. It will be shown that the linear half-wave rectifier circuit can be modified to perform a variety of signal-processing applications. The linear half-wave rectifier is also called a *precision half-wave rectifier* and acts as an ideal diode.
2. *Precision full-wave rectifiers.* The precision full-wave rectifier circuit delivers an output proportional to the magnitude but *not* the polarity of the input. For example, the output can be positive at 2 V for inputs of either +2 V or −2 V. Since the absolute value of +2 V and −2 V is equal to +2 V, the precision full-wave rectifier is also called an *absolute-value circuit*.

Applications for both linear half-wave and precision full-wave rectifiers include:

1. Detection of amplitude-modulated signals
2. Dead-zone circuits
3. Precision bound circuits or *clippers*
4. Current switches
5. Waveshapers
6. Peak-value indicators
7. Sample-and-hold circuits
8. Absolute-value circuits
9. Averaging circuits
10. Signal polarity detectors
11. Ac-to-dc converters

7-1 LINEAR HALF-WAVE RECTIFIERS

7-1.1 Introduction

Linear half-wave rectifier circuits transmit only one-half cycle of a signal and eliminate the other by *bounding* the output to zero volts. The input half-cycle that is transmitted can be either inverted or noninverted. It can also experience gain, attenuation, or remain unchanged in magnitude, depending on the choice of resistors and placement of diodes in the op amp circuit.

7-1.2 Inverting Linear Half-Wave Rectifier, Positive Output

The inverting amplifier is converted into an ideal (linear precision) half-wave rectifier by adding two diodes as shown in Fig. 7-2. When E_i is positive in Fig. 7-2(a), diode D_1 conducts causing the op amp's output voltage, V_{OA}, to go negative by one diode drop ($\simeq$0.6 V). This forces diode D_2 to be reverse biased. The circuit's output voltage V_o equals zero because input current I flows through D_1. For all practical purposes, no current flows through R_F and therefore $V_o = 0$.

Note the load is modeled by a resistor R_L and must always be resistive. If the load is a capacitor, inductor, voltage, or current source, then V_o will *not* equal zero.

In Fig. 7-2(b), negative input E_i forces the op amp output V_{OA} to go positive. This causes D_2 to conduct. The circuit then acts like an inverter since $R_F = R_i$ and $V_o = -(-E_i) = +E_i$. Since the (−) input is at ground potential, diode D_1 is reverse biased. Input current is set by E_i/R_i and gain by $-R_F/R_i$. Remember that this gain equation applies only for negative inputs, and V_o can only be positive or zero.

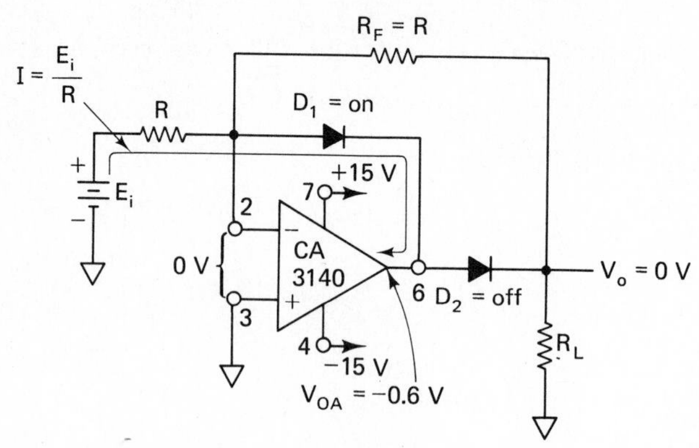

(a) Output V_o is bound at 0 V for all positive input voltages

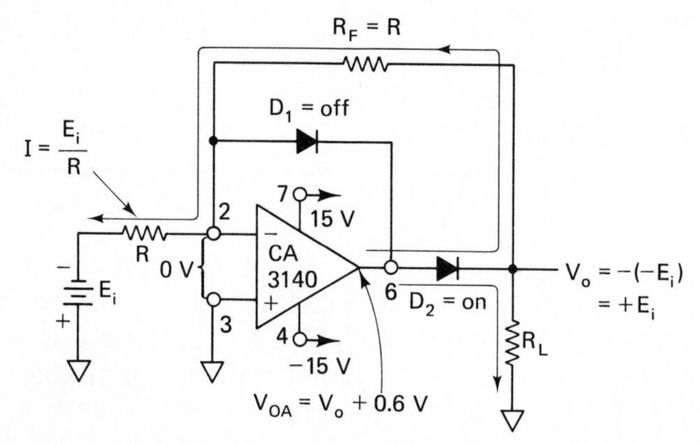

(b) Output V_o is positive and equal to the magnitude of E_i for all negative inputs

FIGURE 7-2 Two diodes convert an inverting amplifier into a positive-output, inverting, linear (ideal) half-wave rectifier. Output V_o is positive and equal to the magnitude of E_i for negative inputs, and V_o equals 0 V for all positive inputs. Diodes are 1N914 or 1N4154.

Circuit operation is summarized by the waveshapes in Fig. 7-3. V_o can only go positive in a linear response to negative inputs. The most important property of this linear half-wave rectifier will now be examined. An ordinary silicon diode or even a hot-carrier diode requires a few tenths of volts to become forward biased. Any signal voltage below this threshold voltage cannot be rectified. However, by connecting the diode in the feedback loop of an op amp, the threshold voltage of the diode is essentially eliminated. For example, in Fig. 7-2(b) let E_i be a low voltage of -0.1 V. E_i and R_i convert this low voltage to a current that is conducted through D_2. V_{OA} goes to whatever voltage is required to supply the necessary diode drop plus the voltage drop across R_f. Thus millivolts of input voltage can be rectified since the diode's forward bias is supplied automatically by the negative feedback action of the op amp.

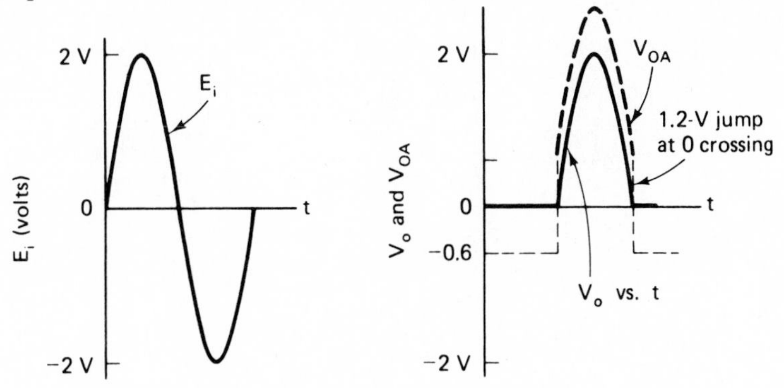

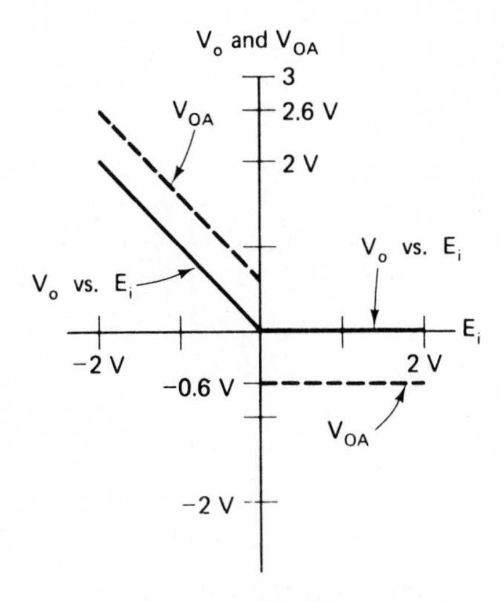

FIGURE 7-3 Input, output, and transfer characteristics of a positive-output, ideal, inverting half-wave rectifier.

Finally, observe the waveshape of op amp output V_{OA} in Fig. 7-3. When E_i crosses 0 V (going negative), V_{OA} jumps quickly from -0.6 V to $+0.6$ V as it switches from supplying the drop for D_2 to supplying the drop for D_1. This jump can be monitored by a differentiator to indicate the zero crossing. During the jump time the op amp operates open loop.

7-1.3 Inverting Linear Half-Wave Rectifier, Negative Output

The diodes in Fig. 7-2 can be reversed as shown in Fig. 7-4. Now only positive input signals are transmitted and inverted. The output voltage V_o equals 0 V for all negative inputs. Circuit operation is summarized by the plot of V_o and V_{OA} versus E_i in Fig. 7-4(b).

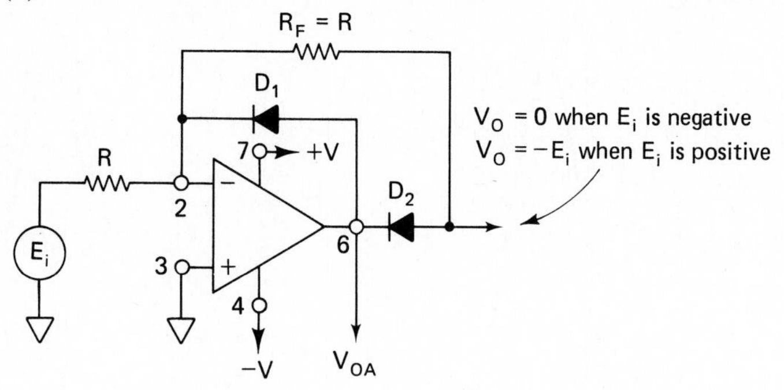

(a) Inverting linear half-wave rectifier: negative output

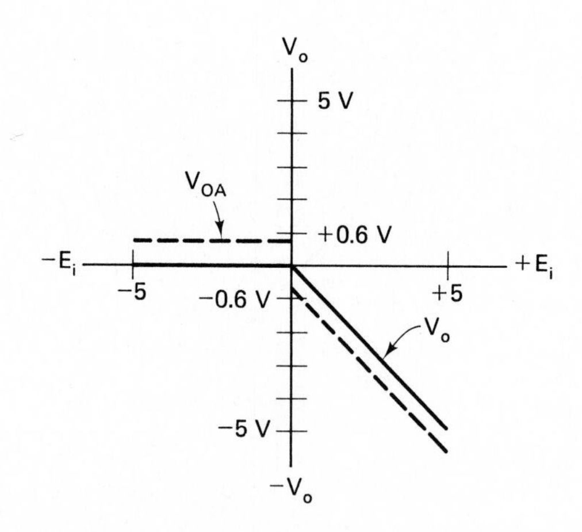

(b) Transfer characteristic V_o vs. E_i

FIGURE 7-4 Reversing the diodes in Fig. 7-2 gives an inverting linear half-wave rectifier. This circuit transmits and inverts only positive input signals.

7-1.4 Signal Polarity Separator

The circuit of Fig. 7-5 is an expansion of the circuits in Figs. 7-2 and 7-4. When E_i is positive in Fig. 7-5(a), diode D_1 conducts and an output is obtained only on output V_{o1}. V_{o2} is bound at 0 V. When E_i is negative, D_2 conducts, $V_{o2} = -(-E_i) = +E_i$, and V_{o1} is bound at 0 V. This circuit's operation is summarized by the waveshapes in Fig. 7-6.

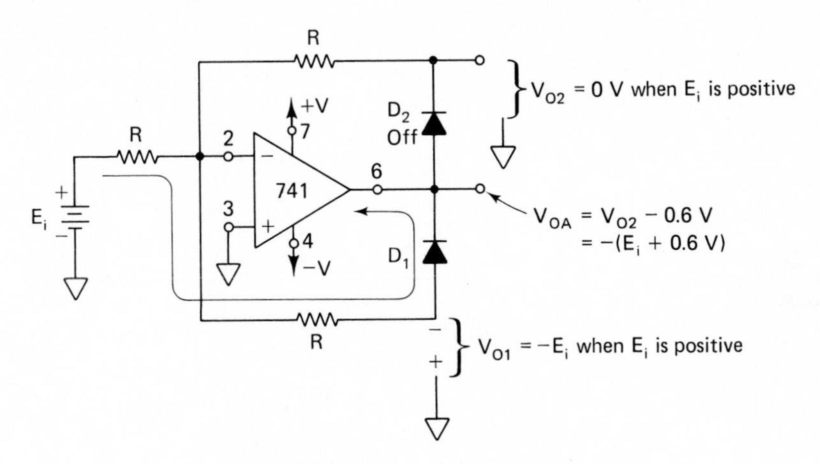

(a) When E_i is positive, V_{O1} is negative and V_{O2} is bound at 0 V

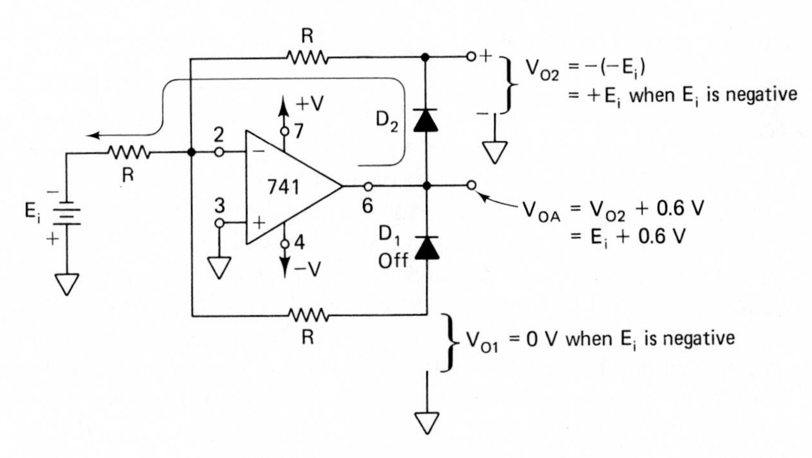

(b) When E_i is negative, V_{O1} = 0 V and V_{O2} goes positive

FIGURE 7-5 This circuit inverts and separates the polarities of input signal E_i. A positive output at V_{o2} indicates that E_i is negative, and a negative output at V_{o1} indicates that E_i is positive. These outputs should be buffered.

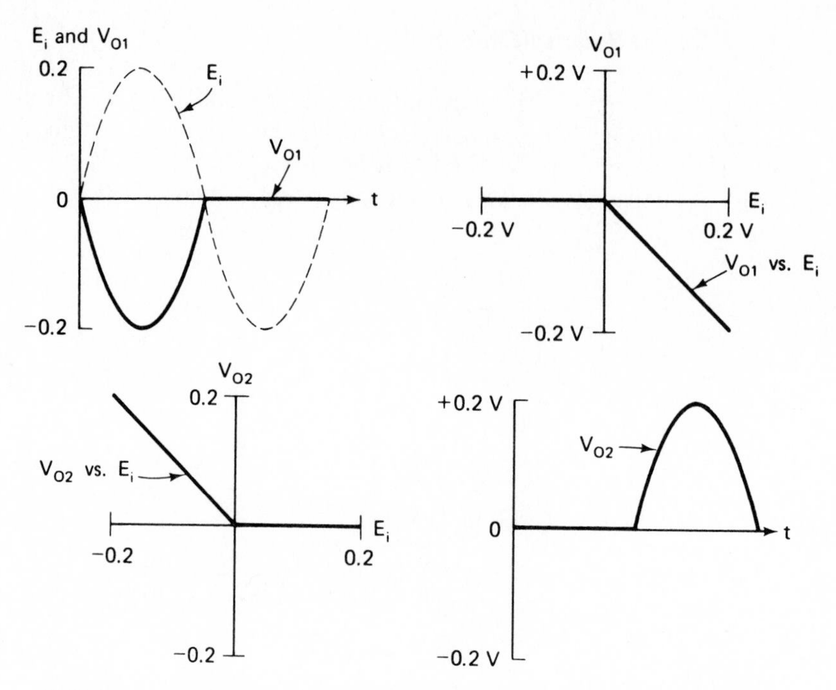

FIGURE 7-6 Input and output voltages for the polarity separator of Fig. 7-5.

7-2 PRECISION RECTIFIERS: THE ABSOLUTE-VALUE CIRCUIT

7-2.1 Introduction

The precision full-wave rectifier transmits one polarity of the input signal and inverts the other. Thus both half-cycles of an alternating voltage are transmitted but are converted to a single polarity of the circuit's output. The precision full-wave rectifier can rectify input voltages with millivolt amplitudes.

This type of circuit is useful to prepare signals for multiplication, averaging, or demodulation. The characteristics of an ideal precision rectifier are shown in Fig. 7-7.

The precision rectifier is also called an *absolute-value* circuit. The absolute value of a number (or voltage) is equal to its magnitude regardless of sign. For example, the absolute values of $|+2|$ and $|-2|$ are equal to $+2$. The symbol $|*|$ means "absolute value of." Figure 7-7 shows that the output equals the absolute value of the input. In a precision rectifier circuit the output is either negative or positive, depending on how the diodes are installed.

7-2.2 Types of Precision Full-Wave Rectifiers

Three types of precision rectifiers will be presented. The first is inexpensive because it uses two op amps, two diodes, and five *equal* resistors. Unfortunately, it does not

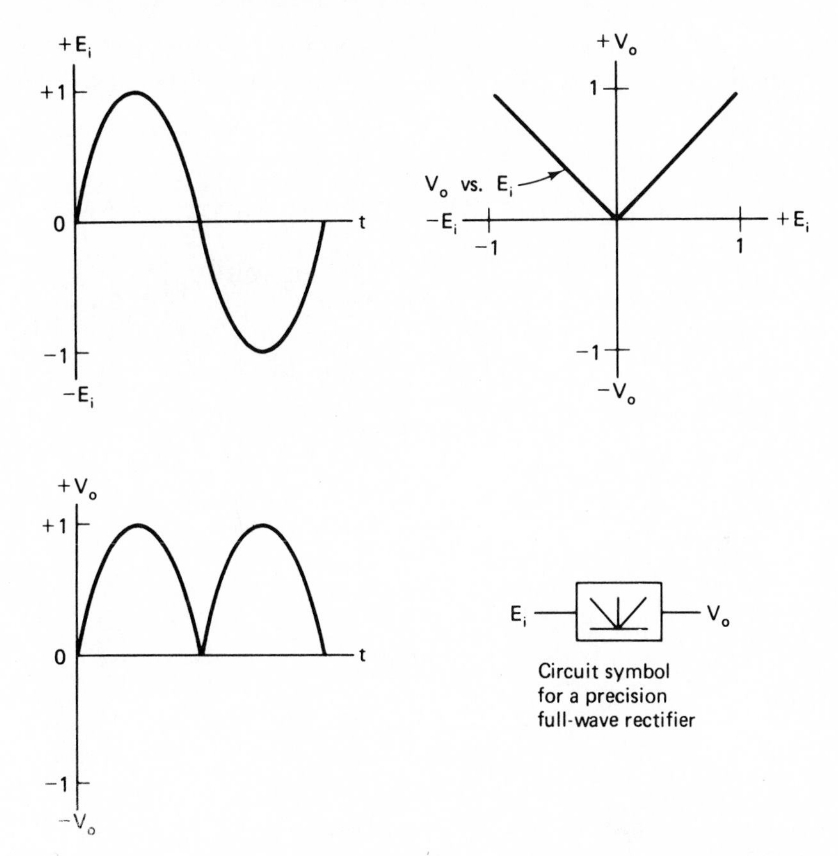

FIGURE 7-7 The precision full-wave rectifier fully rectifies input voltages, including those with values less than a diode threshold voltage.

have high input resistance, so a second type is given that does have high input resistance but requires resistors that are precisely proportioned but *not* all equal. Neither type has a summing node at virtual ground potential, so a third type will be presented in Section 7-4.2 to allow averaging.

Full-wave precision rectifier with equal resistors. The first type of precision full-wave rectifier or absolute-value circuit is shown in Fig. 7-8. This circuit uses equal resistors and has an input resistance equal to R. Figure 7-8(a) shows current directions and voltage polarities for positive input signals. Diode D_P conducts so that both op amps A and B act as inverters, and $V_o = +E_i$.

Figure 7-8(b) shows that for negative input voltages, diode D_N conducts. Input current I divides as shown, so that op amp B acts as an inverter. Thus output voltage V_o is positive for either polarity of input E_i and V_o is equal to the absolute value of E_i.

The waveshapes in Fig. 7-8(c) show that V_o is always of positive polarity and equal to the absolute value of the input voltage. To obtain negative outputs for either polarity of E_i, simply reverse the diodes.

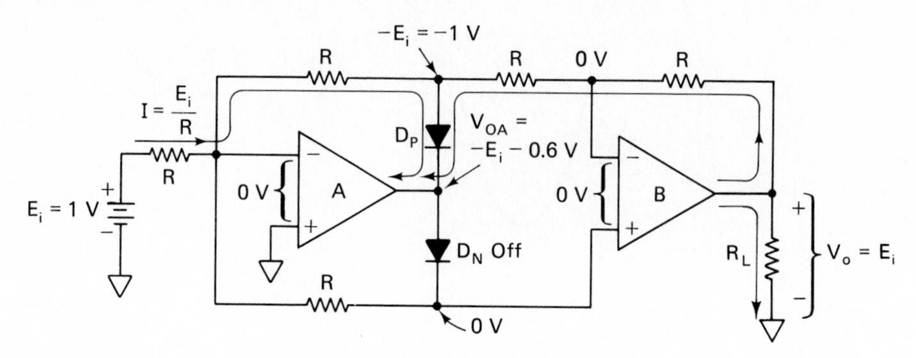

(a) For positive inputs D_P conducts; op amps A and B act as inverting amplifiers

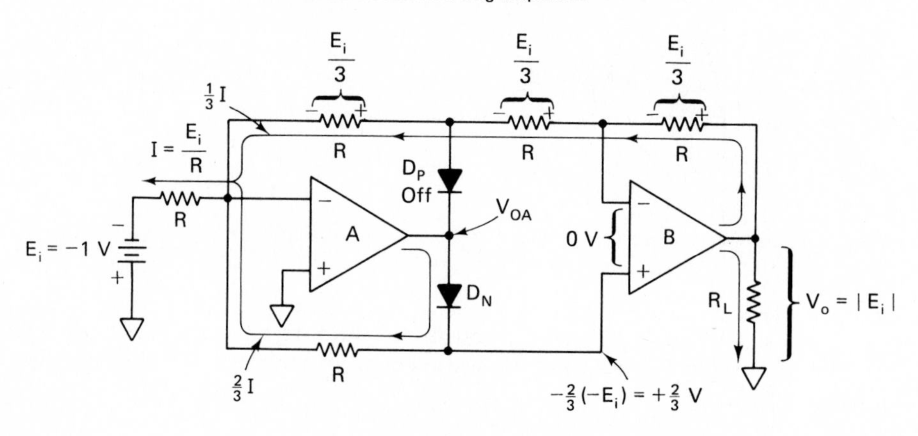

(b) For negative inputs, D_N conducts

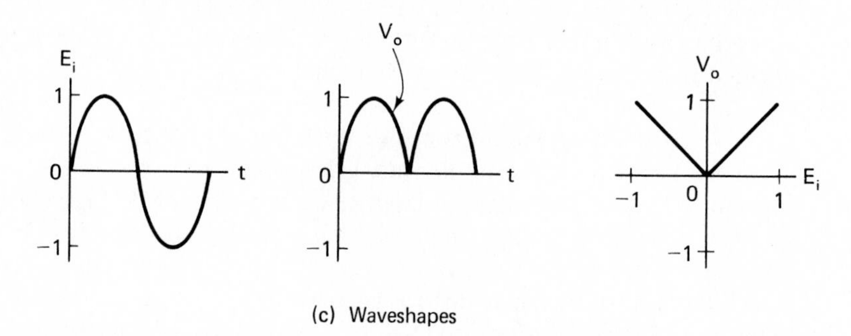

(c) Waveshapes

FIGURE 7-8 Absolute-value circuit or precision full-wave rectifier, $V_o = |E_i|$.

High-impedance precision full-wave rectifier. The second type of precision rectifier is shown in Fig. 7-9. The input signal is connected to the noninverting op amp inputs to obtain high input impedance. Figure 7-9(a) shows what happens for positive inputs. E_i and R_i set the current through diode D_P. The (−) inputs of

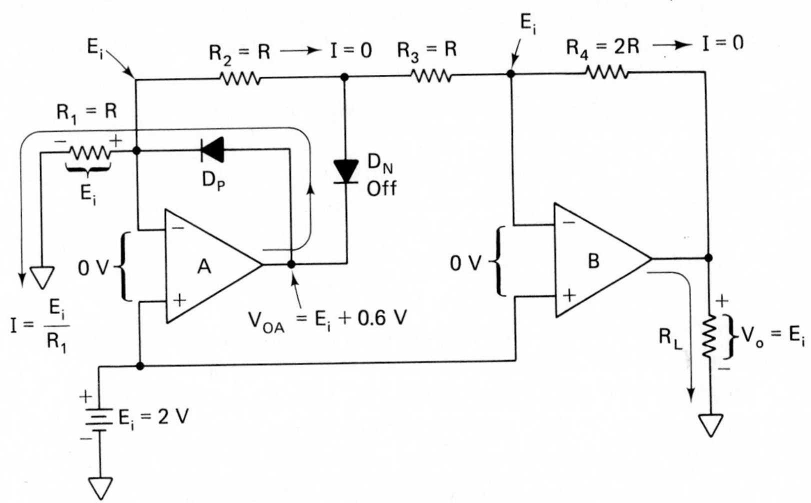

(a) Voltage levels for positive inputs: $V_o = +E_i$ for all positive E_i

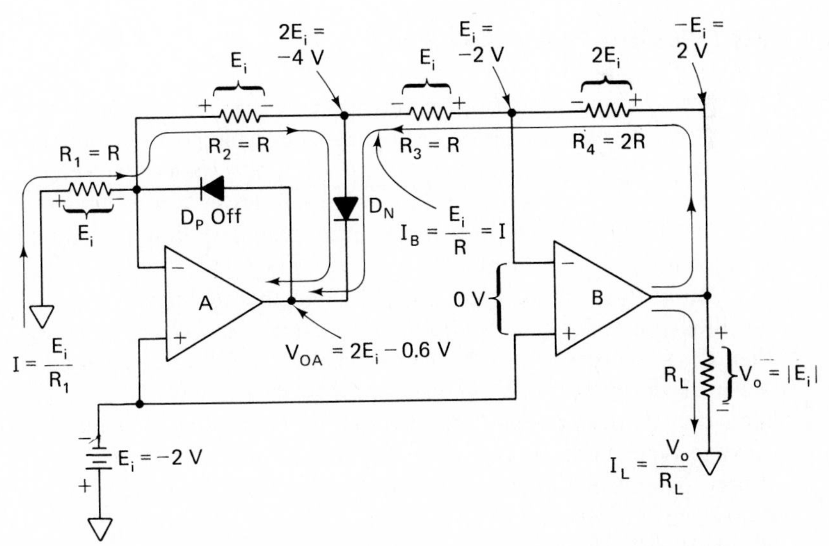

(b) Voltage levels for negative inputs: $V_o = -(-E_i) = |E_i|$

FIGURE 7-9 Precision full-wave rectifier with high input impedance. $R = 10\ \text{k}\Omega$, $2R = 20\ \text{k}\Omega$.

both op amps are at a potential equal to E_i so that no current flows through R_2, R_3, and R_4. Therefore, $V_o = E_i$ for all positive input voltages.

When E_i goes negative in Fig. 7-9(b), E_i and R_1 set the current through both R_1 and R_2 to turn on diode D_N. Since $R_1 = R_2 = R$, the anode of D_N goes to $2E_i$ or $2(-E_i) = -4$ V. The (−) input of op amp B is at $-E_i$. The voltage drop across R_3 is $2E_i - E_i$ or $(-4 \text{ V}) - (-2) = -2$ V. This voltage drop and R_3 establishes a current I_3 through both R_3 and R_4 equal to the input current I. Consequently, V_o is positive when E_i is negative. Thus V_o is always positive despite the polarity of E_i, so $V_o = |E_i|$.

The waveshapes for this circuit are the same as in Fig. 7-8(c). Note that the maximum value of E_i is limited by the negative saturation voltage of the op amps.

7-3 PEAK DETECTORS

In addition to rectifying a signal precisely, diodes and op amps can be interconnected to build a peak detector circuit. This circuit follows the voltage peaks of a signal and stores the highest value (almost indefinitely) on a capacitor. If a higher peak signal value comes along, this new value is stored. The highest peak voltage is stored until the capacitor is discharged by a mechanical or electronic switch. This peak detector circuit is also called a *follow-and-hold* or *peak follower*. We shall also see that reversing two diodes changes this circuit from a peak to a *valley follower*.

7-3.1 Positive Peak Follower and Hold

The peak *follower-and-hold* circuit is shown in Fig. 7-10. It consists of two op amps, two diodes, a resistor, a hold capacitor, and a reset switch. Op amp A is a precision half-wave rectifier that charges C only when input voltage E_i exceeds capacitor voltage V_C. Op amp B is a voltage follower whose output signal is equal to V_C. The follower's high input impedance does not allow the capacitor to discharge appreciably.

To analyze circuit operation, let us begin with Fig. 7-10(a). When E_i exceeds V_C, diode D_P is forward biased to charge hold capacitor C. As long as E_i is greater than V_C, C charges toward E_i. Consequently, V_C follows E_i as long as E_i exceeds V_C. When E_i drops below V_C, diode D_N turns on as shown in Fig. 7-10(b). Diode D_P turns off and disconnects C from the output of op amp A. Diode D_P must be a low-leakage-type diode or the capacitor voltage will discharge (droop). To minimize droop, op amp B should require small input bias currents (see Chapter 9). For that reason op amp B should be a metal-oxide-semiconductor (MOS) or bipolar-field-effect (BiFET) op amp.

Figure 7-11 shows an example of voltage waveshapes for a positive voltage follower-and-hold circuit. To reset the hold capacitor voltage to zero, connect a discharge path across it with a 2-kΩ resistor.

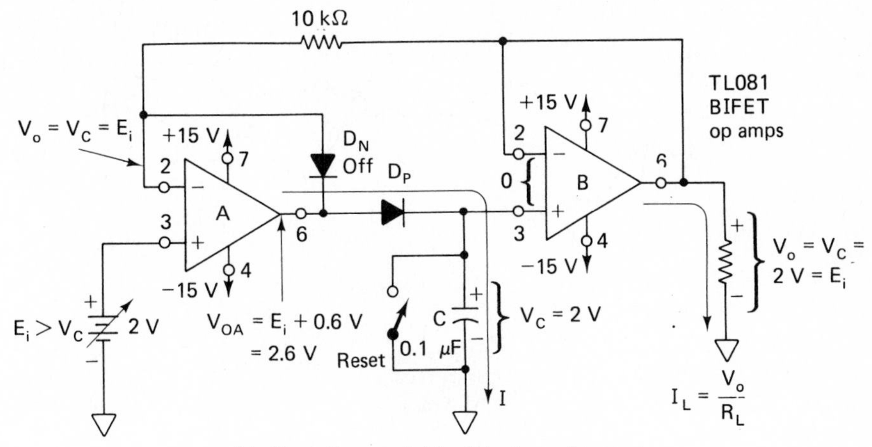

(a) When E_i exceeds V_C, C is charged toward E_i via D_P

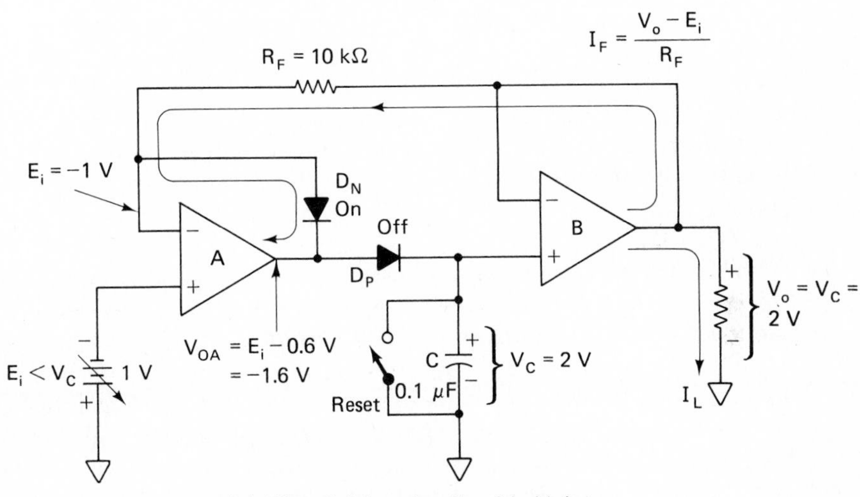

(b) When E_i is less than V_C, C holds its voltage at the highest previous value of E_i

FIGURE 7-10 Positive peak follower and hold or peak detector circuit. Op amps are TL081 BiFETs.

7-3.2 Negative Peak Follower and Hold

When it is desired to hold the lowest or most negative voltage of a signal, reverse both diodes in Fig. 7-10. For bipolar or negative input signals, V_o will store the most negative voltage. It may be desired to monitor a positive voltage and catch any negative dips of short duration. Simply connect a wire from V_C to the positive voltage to

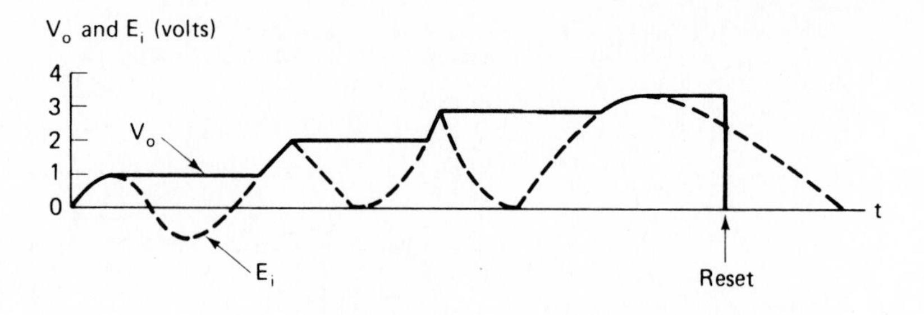

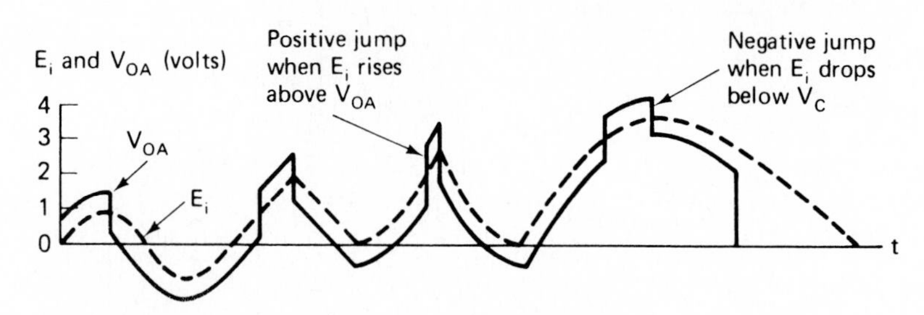

FIGURE 7-11 Waveshapes for the positive detector of Fig. 7-10(a).

be monitored to load C with an equal positive voltage. Then when the monitored voltage drops and recovers, V_C will follow the drop and store the lowest value.

7-4 AC-TO-DC CONVERTER

7-4.1 AC-to-DC Conversion or MAV Circuit

In this section we show how to design and build an op amp circuit that computes the average value of a rectified ac voltage. This type of circuit is called an ac-to-dc converter. Since a full-wave rectifier circuit is also known as an absolute-value circuit and since an average value is also called a mean value, the ac-to-dc converter is also referred to as a *mean-absolute-value* (MAV) circuit.

To see how the MAV circuit is useful, refer to Fig. 7-12. A sine, triangle, and square wave are shown with equal maximum (peak) values. Therefore, a peak detector could not distinguish between them. The positive half-cycle and negative half-cycles are equal for each particular wave. Therefore, the average value of each signal is zero, so you could not distinguish one from another with an averaging circuit or device such as a dc volt meter. However, the MAV of each voltage is different (see Fig. 7-12).

The MAV of a voltage wave is approximately equal to its rms value. Thus an inexpensive MAV circuit can be used as a substitute for a more expensive rms calculating circuit.

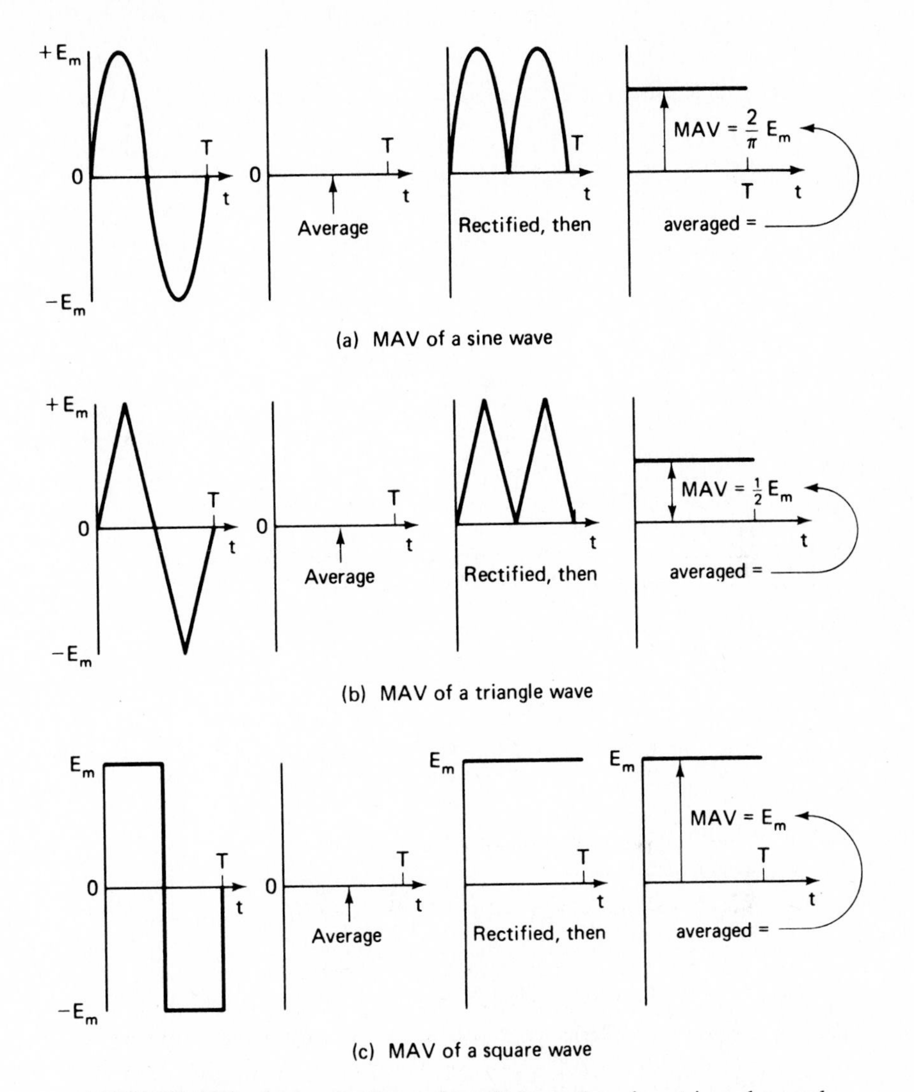

FIGURE 7-12 Mean absolute value of alternating sine, triangular, and square waves.

7-4.2 Precision Rectifier with Grounded Summing Inputs

To construct an ac-to-dc converter, we begin with the precision rectifier or absolute-value amplifier of Fig. 7-13. For positive inputs in Fig. 7-13(a), op amp A inverts E_i. Op amp B sums the output of A and E_i to give a circuit output of $V_o = E_i$. For negative inputs as shown in Fig. 7-13(b), op amp B inverts $-E_i$ and the circuit output V_o is $+E_i$. Thus the circuit output V_o is positive and equal to the rectified or absolute value of the input.

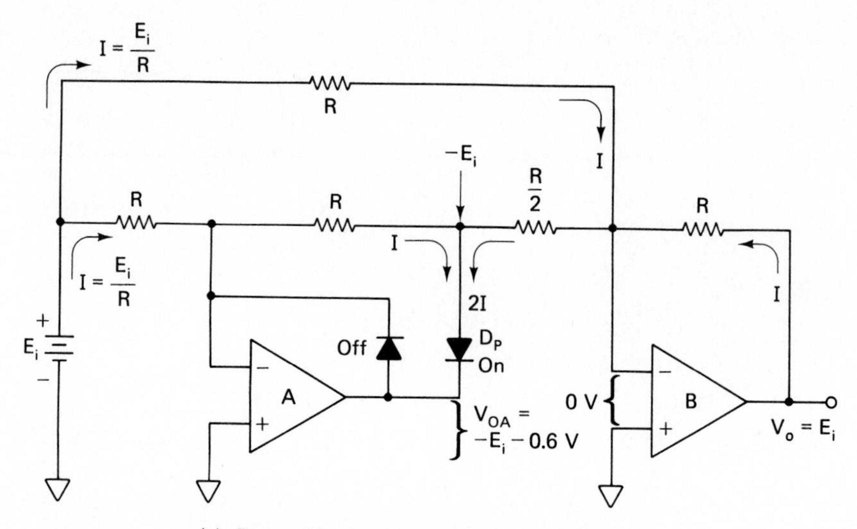

(a) For positive inputs, op amp A inverts E_i; op amp B is an inverting adder, so that $V_o = E_i$

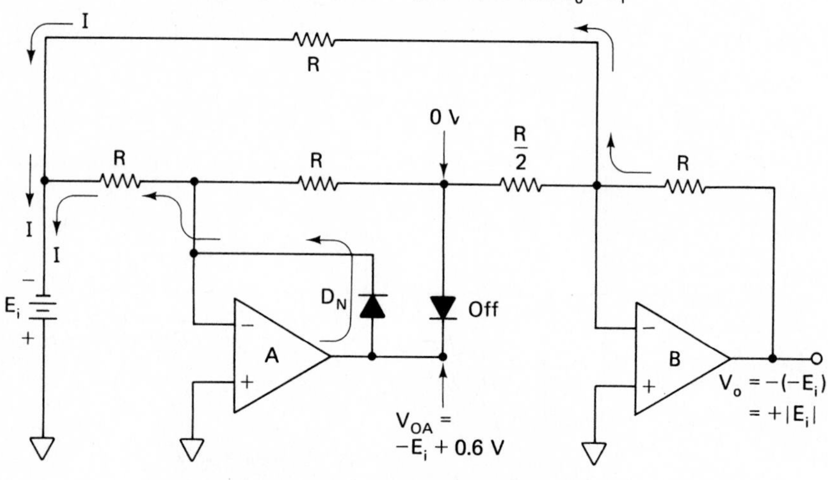

(b) For negative inputs, the output of A is rectified to 0; op amp B inverts E_i, so that $V_o = +E_i$

FIGURE 7-13 This absolute-value amplifier has both summing nodes at ground potential during either polarity of input voltage. $R = 20\ \text{k}\Omega$.

7-4.3 AC-to-DC Converter

A large-value low-leakage capacitor (10-μF tantalum) is added to the absolute-value circuit of Fig. 7-13. The resultant circuit is the MAV amplifier or ac-to-dc converter shown in Fig. 7-14. Capacitor C does the averaging of the rectified output of op amp

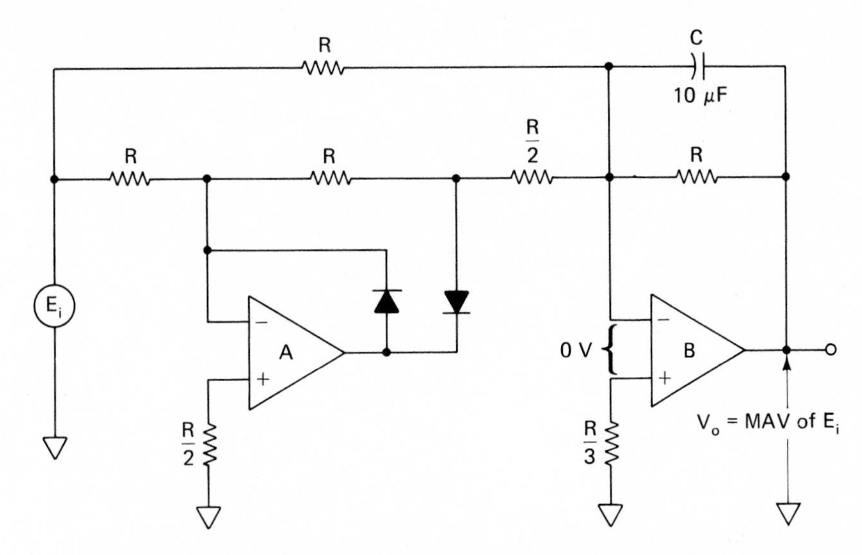

FIGURE 7-14 Add one capacitor to the absolute-value amplifier of Fig. 7-13 to get this ac-to-dc converter or mean-absolute-value amplifier.

B. It takes about 50 cycles of input voltage before the capacitor voltage settles down to its final reading. If the waveshapes of Fig. 7-12 are applied to the ac-to-dc converter, its output will be the MAV of each input signal.

7-5 DEAD-ZONE CIRCUITS

7-5.1 Introduction

Comparator circuits tell *if* a signal is below or above a particular reference voltage. In contrast with the comparator, a dead-zone circuit tells *by how much* a signal is below or above a reference voltage.

7-5.2 Dead-Zone Circuit with Negative Output

Analysis of a dead-zone circuit begins with the circuit of Fig. 7-15. A convenient regulated supply voltage $+V$ and resistor mR establish a reference voltage V_{ref}. V_{ref} is found from the equation $V_{ref} = +V/m$. As will be shown, the *negative* of V_{ref}, $-V_{ref}$, will establish the dead zone. In Fig. 7-15(a), current I is determined by $+V$ and resistor mR at $I = +V/mR$.

Diode D_N will conduct for all positive values of E_i, clamping V_{oA} and V_{oB} to 0 V. Therefore, all positive inputs are eliminated from affecting the output. In order to get any output at V_{oA}, E_i must go negative, as shown in Fig. 7-15(b). Diode D_P will conduct when the loop current E_i/R through E_i exceeds the loop current V/mR through resistor mR.

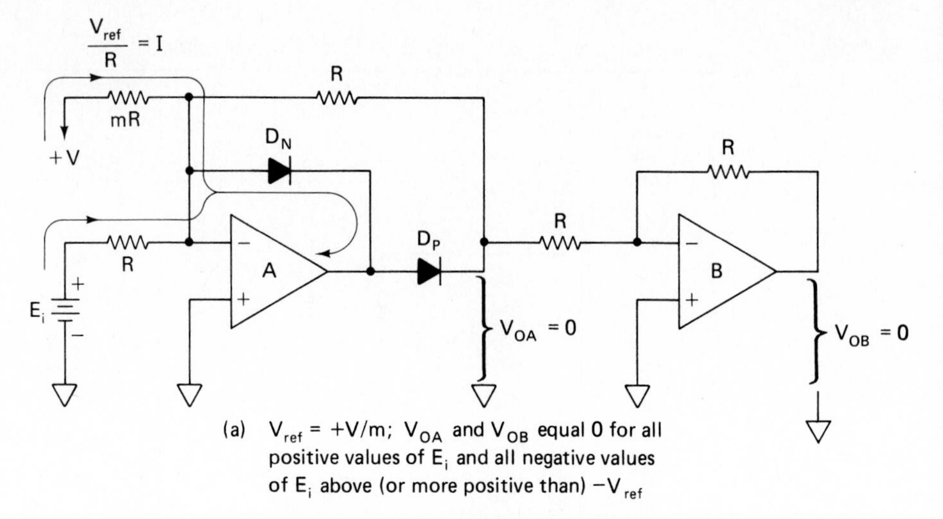

(a) $V_{ref} = +V/m$; V_{OA} and V_{OB} equal 0 for all positive values of E_i and all negative values of E_i above (or more positive than) $-V_{ref}$

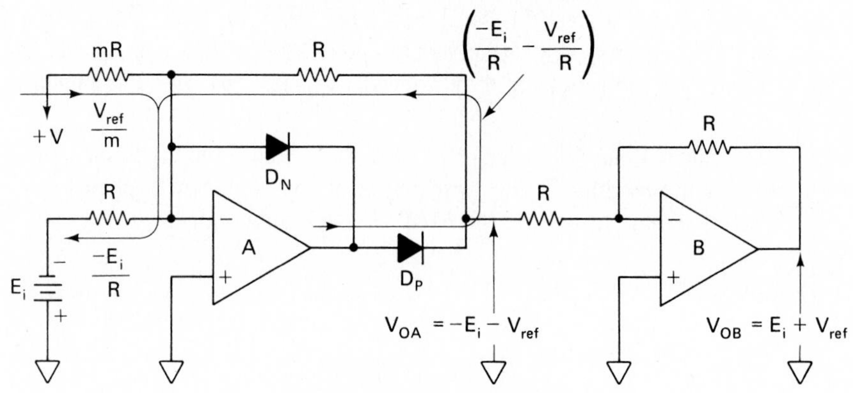

(b) When E_i is negative and below $-V_{ref}$, V_{OA} goes positive to a value of $-(E_i + V_{ref})$ and V_{OB} goes negative to $E_i + V_{ref}$

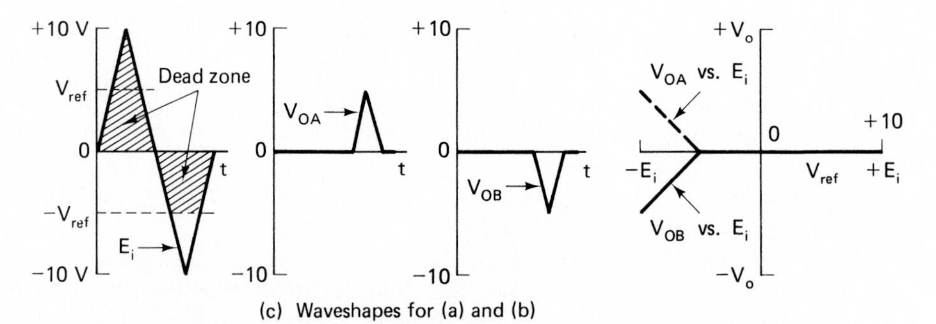

(c) Waveshapes for (a) and (b)

FIGURE 7-15 The dead-zone circuit output V_{oB} eliminates all portions of the signal above $-V_{ref}$ where $V_{ref} = +V/m$.

The value of E_i necessary to turn on D_P in Fig. 7-15(b) is equal to $-V_{ref}$. This conclusion is found by equating the currents

$$-\frac{E_i}{R} = \frac{+V}{mR}$$

and solving for E_i:

$$E_i = -\frac{+V}{m} = -V_{ref} \tag{7-1a}$$

where

$$V_{ref} = \frac{+V}{m} \tag{7-1b}$$

Thus all values of E_i above $-V_{ref}$ will lie in a dead zone where they will not be transmitted [see Fig. 7-15(c)]. Outputs V_{oA} and V_{oB} will be zero.

When E_i is below V_{ref}, E_i and V_{ref} will be added and their sum is inverted at output V_{oA}. V_{oA} is reinverted by op amp B. Thus V_{oB} only has an output when E_i goes below V_{ref}. V_{oB} tells you by how many volts E_i lies below V_{ref}.

Circuit operation is summarized by the waveshapes of Fig. 7-15(c). Circuit operation is illustrated by an example.

Example 7-1

In the circuit of Fig. 7-15, $+V = +15$ V, $mR = 30$ kΩ, and $R = 10$ kΩ, so that $m = 3$. Find (a) V_{ref}; (b) V_{oA} when $E_i = -10$ V; (c) V_{oB} when $E_i = -10$ V.

Solution (a) From Eq. (7-1b), $V_{ref} = +15\text{ V}/3 = 5$ V. (b) V_{oA} and V_{oB} will equal zero for all values of E_i above $-V_{ref} = -5$ V, from Eq. (7-1a). Therefore, $V_{oA} = -E_i - V_{ref} = -(-10\text{ V}) - 5\text{ V} = +5$ V. (c) Op amp B inverts the output of V_{oA} so that $V_{oB} = -5$ V. Thus, V_{oB} indicates how much E_i goes below $-V_{ref}$. All input signals *above* $-V_{ref}$ fall in a dead zone and are eliminated from the output.

7-5.3 Dead-Zone Circuit with Positive Output

If the diodes in Fig. 7-15 are reversed, the result is a positive-output dead-zone circuit as shown in Fig. 7-16. Reference voltage V_{ref} is found from Eq. (7-1b). $V_{ref} = -15\text{ V}/3 = -5$ V. Whenever E_i goes above $-V_{ref} = -(-5\text{ V}) = +5$ V, the

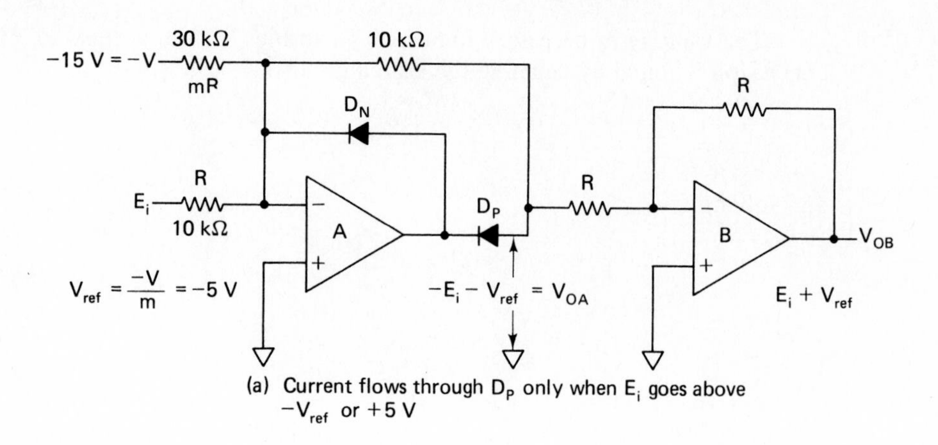

(a) Current flows through D_P only when E_i goes above $-V_{ref}$ or +5 V

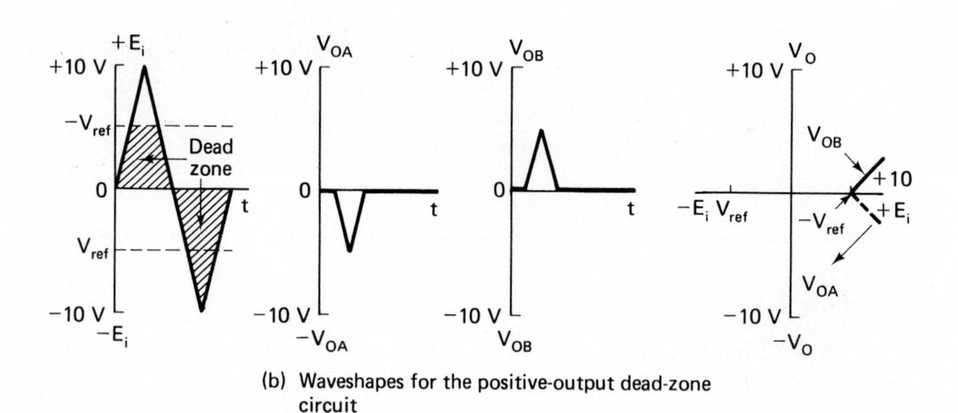

(b) Waveshapes for the positive-output dead-zone circuit

FIGURE 7-16 Positive-output dead-zone circuit.

output V_{oB} tells by how much E_i exceeds $-V_{ref}$. The dead zone exists for all values of *E_i below $-V_{ref}$*.

7-5.4 Bipolar-Output Dead-Zone Circuit

The positive and negative output dead-zone circuits can be combined as shown in Fig. 7-17 and discussed in Fig. 7-18. The V_{oA} outputs from Fig. 7-15 and 7-16 are connected to an inverting adder. The adder output V_{oB} tells how much E_i goes above one positive reference voltage and also how much E_i goes below a different negative reference voltage.

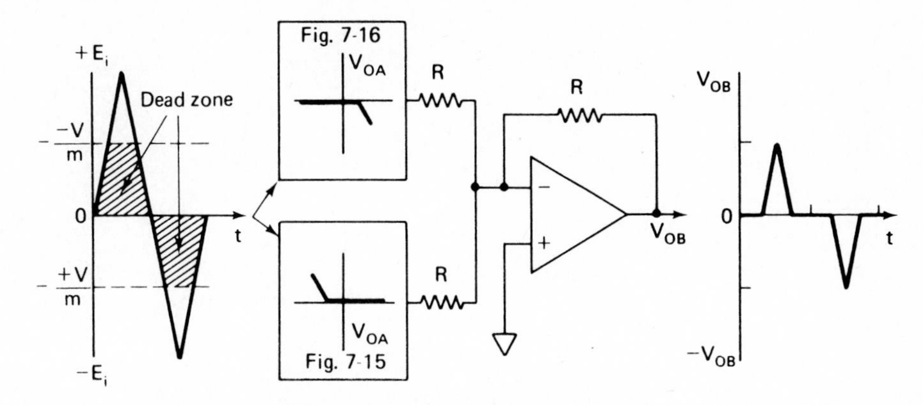

FIGURE 7-17 The V_{oA} outputs of Figs. 7-15 and 7-16 are combined by an inverting adder to give a bipolar output dead-zone circuit.

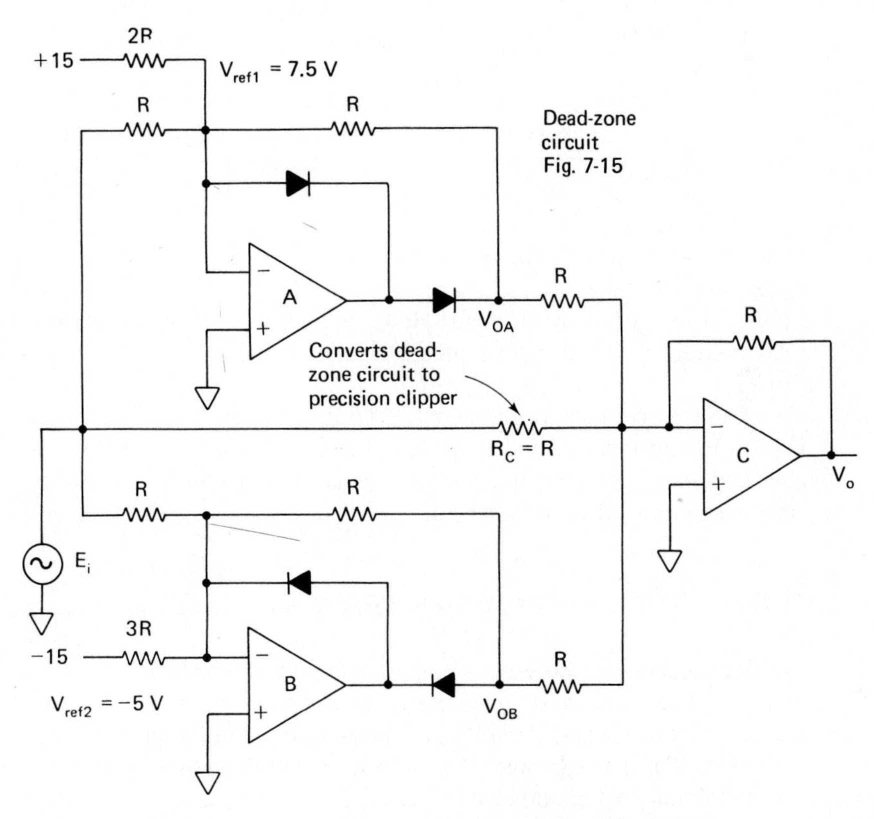

(a) Adding a resistor R_C to the dead-zone circuit of Fig. 7-17 gives a precision clipper

FIGURE 7-18 A precision clipper is made from a bipolar dead-zone circuit plus an added resistor R_C.

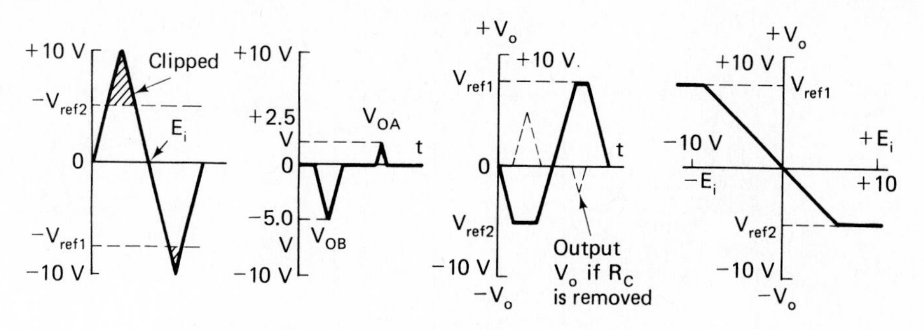

(b) Waveshapes for the precision clipper

FIGURE 7-18 *(cont.)*

7-6 PRECISION CLIPPER

A *clipper* or *amplitude limiter* circuit clips off all signals above a positive reference voltage and all signals below a negative reference voltage. The reference voltages can be made symmetrical or nonsymmetrical around zero. Construction of a precision clipper circuit is accomplished by adding a single resistor, R_C, to a bipolar output dead-zone circuit as shown in Fig. 7-18. The outputs of op amps A and B are each connected to the input of the inverting adder. Input signal E_i is connected to a third input of the inverting adder, via resistor R_C. If R_C is removed, the circuit would act as a dead-zone circuit. However, when R_C is present, input voltage E_i is subtracted from the dead-zone circuit's output and the result is an inverting precision clipper.

Circuit operation is summarized by the waveshapes in Fig. 7-18(b). Outputs V_{oA} and V_{oB} are inverted and added to $-E_i$. The plot of V_o versus time shows by solid lines how the clipped output appears. The dashed lines show how the circuit acts as a dead-zone circuit if R_C is removed.

7-7 TRIANGULAR-TO-SINE WAVE CONVERTER

Variable-frequency sine-wave oscillators are much harder to build than variable-frequency triangular-wave generators. The circuit of Fig. 7-19 converts the output of a triangular-wave generator into a sine wave that can be adjusted for less than 5% distortion. The triangle-sine converter is an amplifier whose gain varies inversely with amplitude of the output voltage.

R_1 and R_3 set the slope of V_o at low amplitudes near the zero crossings. As V_o increases, the voltage across R_3 increases to begin forward biasing D_1 and D_3 for positive outputs, or D_2 and D_4 for negative outputs. When these diodes conduct, they

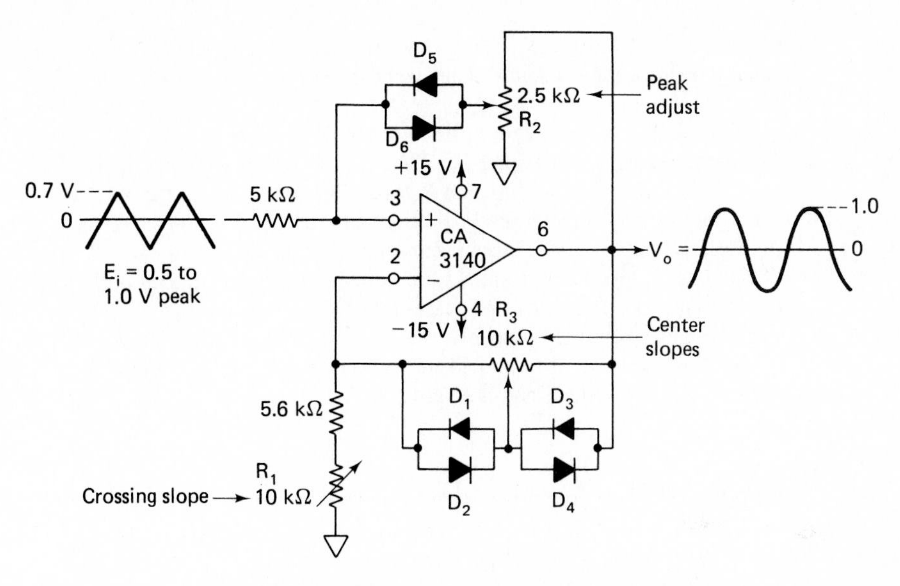

FIGURE 7-19 Triangular-to-sine wave shaper.

shunt feedback resistance R_3, lowering the gain. This tends to shape the triangular output above about 0.4 V into a sine wave. In order to get rounded tops for the sine-wave output, R_2 and diodes D_5 and D_6 are adjusted to make amplifier gain approach zero at the peaks of V_o.

The circuit is adjusted by comparing a 1-kHz sine wave and the output of the triangle/sine-wave converter on a dual-trace CRO. R_1, R_2, R_3, and the peak amplitude of E_i are adjusted in sequence for best sinusoidal shape. The adjustments interact, so they should be repeated as necessary.

LABORATORY EXERCISES

All of the circuits in this chapter can be used to gain operating experience in the laboratory. The authors offer a few practical suggestions.

1. Make the wiring as short as possible.
2. Use 1N914 or 1N4148 fast diodes. Even with fast diodes, the operating frequency will be limited. Use test frequencies of about 100 Hz.
3. Always dc couple the oscilloscope inputs for both time and *x-y* plots.

Some interesting experience can be gained from the peak follower circuit of Fig. 7-10. Momentarily short the capacitor with a jumper wire. Adjust E_i to 2.0 V. You should see 2 V at V_o. Reduce E_i to 0 V. V_o should remain at 2 V. Connect a dc-coupled CRO across capacitor

C and watch V_o drop to 0 V. This is because the 1-MΩ input resistance of the CRO gives a discharge path for the capacitor. Thus you can only measure capacitor voltage at the output of voltage follower B.

The TL081 or any other BiFET op amp has an extremely high input impedance and very low bias currents (Chapter 9). Thus capacitor C will retain its charge for a relatively long period of time. Replace op amp B with a 741 to see how the input bias currents of a general-purpose op amp rapidly discharge the voltage stored on the capacitor.

One last experience: charge a capacitor to about 25 V dc for several minutes. Connect a DVM across its terminals. Remove the battery and short the cap until the voltmeter reads zero. Remove the capacitor and leave the DVM connected to the cap (on the lowest dc voltage range). You will probably see the capacitor voltage slowly rise. This is due to "dielectric absorption." You should use capacitors selected for low dielectric absorption for peak followers, or sample-and-hold circuits.

PROBLEMS

7-1. What is the absolute value of +3 V and −3 V?

7-2. If the peak value of $E_i = 0.5$ V in Fig. 7-1, sketch the waveshapes of V_o vs. t and V_o vs. E_i for both a silicon and an ideal diode.

7-3. If E_i is a sine wave with a peak value of 1 V in Figs. 7-2 and 7-3, sketch the waveshapes of V_o vs. t and V_o vs. E_i.

7-4. If diodes D_1 and D_2 are reversed in Fig. 7-2, sketch V_o vs. E_i and V_o vs. t.

7-5. Sketch the circuit for a signal polarity separator.

7-6. Let both diodes be reversed in Fig. 7-8. What is the value of V_o if $E_i = +1$ V or $E_i = -1$ V?

7-7. What is the name of a circuit that follows the voltage peaks of a signal and stores the highest value?

7-8. How do you reset the hold capacitor's voltage to zero in a peak follower-and-hold circuit?

7-9. How do you convert the absolute-value amplifier of Fig. 7-13 to an ac-to-dc converter?

7-10. If resistor mR is changed to 50 kΩ in Example 7-1, find **(a)** V_{ref}; **(b)** V_{oA} when $E_i =$ 10 V; **(c)** V_{oB} when $E_i = 10$ V.

7-11. If resistor R_C is removed in Fig. 7-18, sketch V_o vs. E_i.

CHAPTER 8

Differential, Instrumentation, and Bridge Amplifiers

LEARNING OBJECTIVES

When you complete this chapter on differential instrumentation and bridge amplifiers, you will be able to:

- Draw the circuit for a basic differential amplifier, state its output–input equation, and explain why it is superior to a single-input amplifier.
- Define common-mode and differential input voltage.
- Draw the circuit for a differential input to differential output voltage amplifier and add a differential amplifier to make a three-op-amp instrumentation amplifier (IA).
- Calculate the output voltage of a three-op-amp instrumentation amplifier if you are given the input voltages and resistance values.
- Explain how the sense and reference terminals of an IA allow you to (1) eliminate the effects of connecting-wire resistance on load voltage, (2) obtain load current boost, or (3) make a differential voltage-to-current converter (ac current source.)

- Draw the circuit for a bridge amplifier and show how it converts a change in transducer resistance to an output voltage.
- Use the bridge amplifier to make a temperature-to-voltage converter.
- Explain how a strain gage converts tension or compression forces into a change in resistance.
- Connect strain gages into a passive bridge resistance network to convert gage resistance change into an output voltage.
- Amplify the strain gage bridge's differential output with an instrumentation amplifier.
- Measure pressure, force, or weight.

8-0 INTRODUCTION

The most useful amplifier for measurement, instrumentation, or control is the *instrumentation amplifier*. It is designed with several op amps and precision resistors, which make the circuit extremely stable and useful where accuracy is important. There are now many integrated circuits and modular versions available in single packages. Unfortunately, these packages are relatively expensive (from $5 to over $100). But when performance and precision are required, the instrumentation amplifier is well worth the price, because its performance cannot be matched by the average op amp.

An inexpensive first cousin to the instrumentation amplifier is the basic *differential amplifier*. This chapter begins with the differential amplifier to show in which applications it can be superior to the ordinary inverting or noninverting amplifier. The differential amplifier, with some additions, leads into the instrumentation amplifier, which is discussed in the second part of this chapter. The final sections consider *bridge amplifiers,* which involve both instrumentation and basic differential amplifiers.

8-1 BASIC DIFFERENTIAL AMPLIFIER

8-1.1 Introduction

The differential amplifier can measure as well as amplify small signals that are buried in much larger signals. How the differential amplifier accomplishes this task will be studied in Section 8-2. But first, let us build and analyze the circuit performance of the basic differential amplifier.

Four precision (1%) resistors and an op amp make up a differential amplifier, as shown in Fig. 8-1. There are two input terminals, labeled (−) input, and (+) input, corresponding to the closest op amp terminal. If E_1 is replaced by a short circuit, E_2 sees an inverting amplifier with a gain of $-m$. Therefore, the output voltage

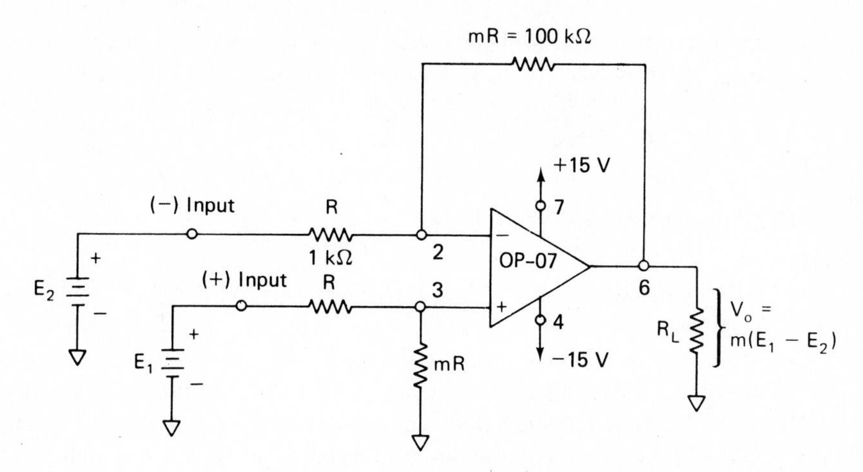

FIGURE 8-1 Basic differential amplifier.

due to E_2 is $-mE_2$. Now let E_2 be short-circuited; E_1 divides between R and mR to apply a voltage of $E_1m/(1 + m)$ at the op amp's (+) input. This divided voltage sees a noninverting amplifier with a gain of $(m + 1)$. The output voltage due to E_1 is the divided voltage, $E_1m/(1 + m)$, times the noninverting amplifier gain, $(1 + m)$, which yields mE_1. Therefore, E_1 is amplified at the output by the multiplier m to mE_1. When both E_1 and E_2 are present at the (+) and (−) inputs, respectively, V_o is $mE_1 - mE_2$, or

$$V_o = mE_1 - mE_2 = m(E_1 - E_2) \tag{8-1}$$

Equation (8-1) shows that the output voltage of the differential amplifier, V_o, is proportional to the *difference* in voltage applied to the (+) and (−) inputs. Multiplier m is called the *differential gain* and is set by the resistor ratios.

Example 8-1

In Fig. 8-1, the differential gain is found from

$$m = \frac{mR}{R} = \frac{100\text{ k}\Omega}{1\text{ k}\Omega} = 100$$

Find V_o for $E_1 = 10$ mV and (a) $E_2 = 10$ mV, (b) $E_2 = 0$ mV, and (c) $E_2 = -20$ mV.

Solution By Eq. (8-1), (a) $V_o = 100(10 - 10)$ mV $= 0$; (b) $V_o = 100(10 - 0)$ mV $=$ 1.0 V; (c) $V_o = 100[10 - (-20)]$ mV $= 100(30$ mV$) = 3$ V.

As expected from Eq. (8-1) and shown from part (a) of Example 8-1, when $E_1 = E_2$ the output voltage is 0. To put it another way, when a common (same) voltage is applied to the input terminals, $V_o = 0$. Section 8-1.2 examines this idea of a common voltage in more detail.

8-1.2 Common-Mode Voltage

The output of the differential amplifier should be 0 when $E_1 = E_2$. The simplest way to apply equal voltages is to wire inputs together and connect them to the voltage source (see Fig. 8-2). For such a connection, the input voltage is called the *common-mode input voltage, E_{CM}*. Now V_o will be 0 if the resistor ratios are equal (*mR* to *R* for the inverting amplifier gain equals *mR* to *R* of the voltage-divider network.) Practically, the resistor ratios are equalized by installing a potentiometer in series with one resistor, as shown in Fig. 8-2. The potentiometer is trimmed until V_o is reduced to a negligible value. This causes the *common-mode voltage gain, V_o/E_{CM}*, to approach 0. It is this characteristic of a differential amplifier that allows a small signal voltage to be picked out of a larger noise voltage. It may be possible to arrange the circuit so that the larger undesired signal is the common-mode input voltage and the small signal is the differential input voltage. Then the differential amplifier's output voltage will contain only an amplified version of the differential input voltage. This possibility is investigated in Section 8-2.

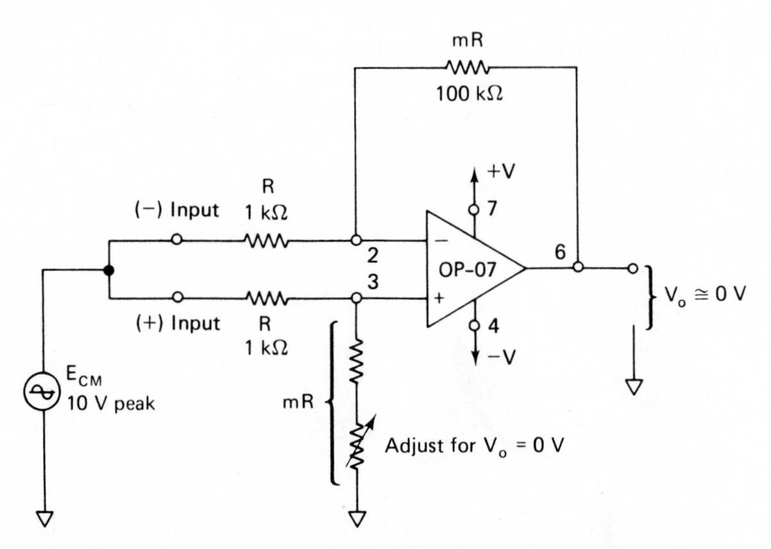

FIGURE 8-2 The common-mode voltage gain should be zero.

8-2 DIFFERENTIAL VERSUS SINGLE-INPUT AMPLIFIERS

8-2.1 Measurement with a Single-Input Amplifier

A simplified wiring diagram of an inverting amplifier is shown in Fig. 8-3. The power common terminal is shown connected to earth ground. Earth ground comes from a connection to a water pipe on the street side of the water meter. Ground is extended via conduit or a bare Romex wire to the third (green wire) of the instrument line cord and finally to the chassis of the amplifier. This equipment or chassis ground is made to ensure the safety of human operators. It also helps to drain off static charges or any capacitive coupled noise currents to earth.

The signal source is also shown to be connected to earth ground in Fig. 8-3. Even if it were not grounded, there would be a leakage resistance or capacitance coupling to earth, to complete a ground loop.

Inevitably, noise currents and noise voltages abound from a variety of sources that are often not easily identifiable. The net effect of all this noise is modeled by noise voltage source E_n in Fig. 8-3. It is evident that E_n is in series with signal

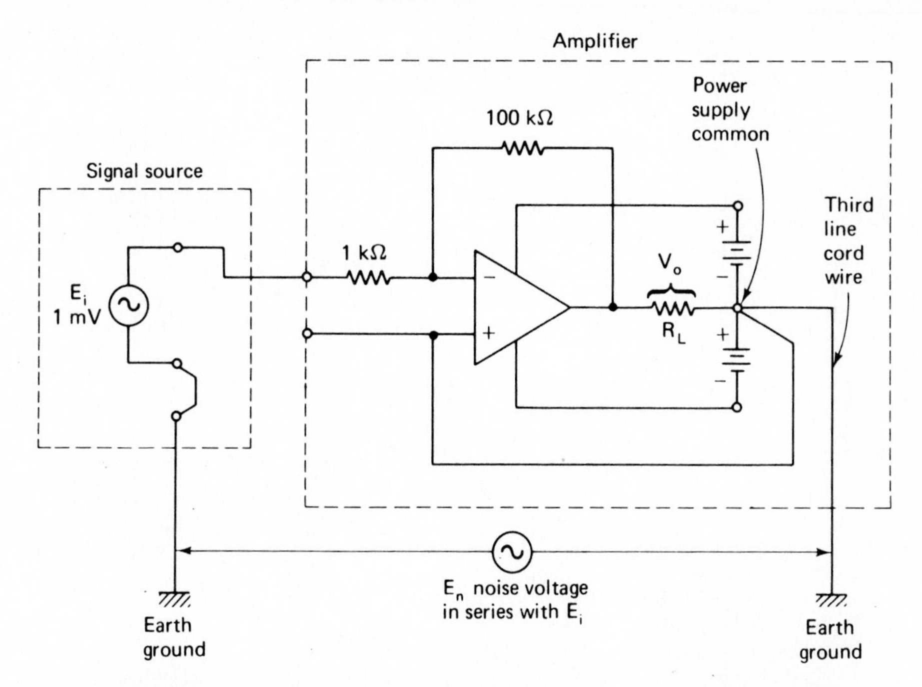

FIGURE 8-3 Noise voltages act as if they are in series with the input signal E_i. Consequently, both are amplified equally. This arrangement is unworkable if E_n is equal or greater than E_i.

voltage E_i, so that both are amplified by a factor of -100 due to the inverting amplifier. E_n may be much larger than E_i. For example, the skin signal voltage due to heart beats is less than 1 mV, whereas the body's noise voltage may be tenths of volts or more. So it would be impossible to make an EKG measurement with a single-input amplifier. What is needed is an amplifier that can distinguish between E_i and E_n and amplify only E_i. Such a circuit is the differential amplifier.

8-2.2 Measurement with a Differential Amplifier

A differential amplifier is employed to measure only the signal voltage (see Fig. 8-4). The signal voltage E_i is connected across the (+) and (−) inputs of the differential amplifier. Therefore, E_i is amplified by a gain of -100. Noise voltage E_n becomes the common-mode voltage input voltage to the differential amplifier as in Fig. 8-2. Therefore, the noise voltage is *not* amplified and has been effectively eliminated from having any significant effect on the output V_o.

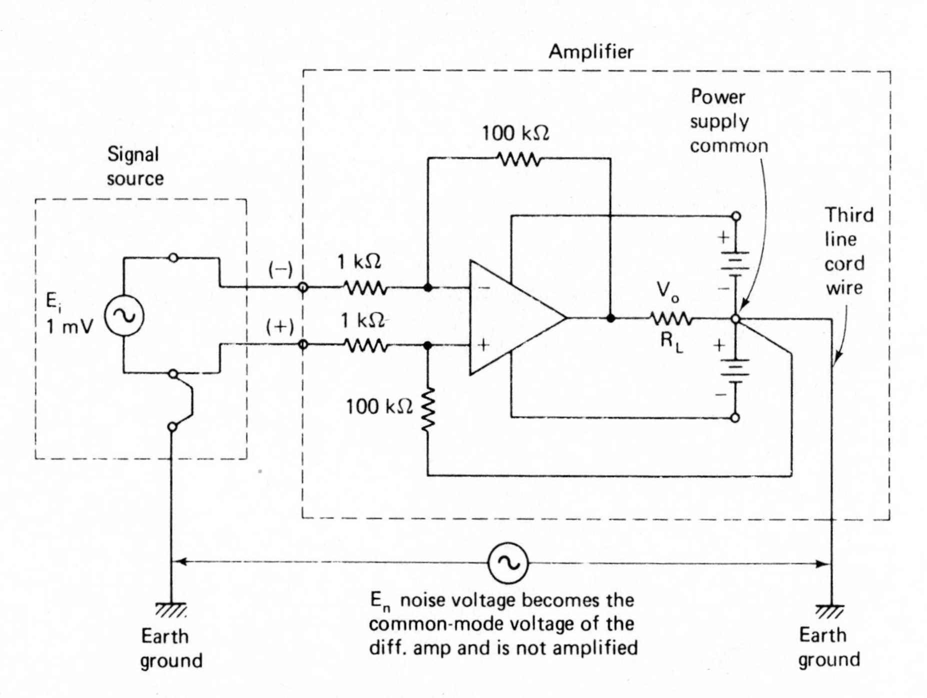

FIGURE 8-4 The differential amplifier is connected so that noise voltage becomes the common-mode voltage and is not amplified. Only the signal voltage E_i is amplified because it has been connected as the differential input voltage.

8-3 IMPROVING THE BASIC DIFFERENTIAL AMPLIFIER

8-3.1 Increasing Input Resistance

There are two disadvantages to the basic differential amplifier studied thus far: It has low input resistance, and changing gain is difficult, because the resistor ratios must be closely matched. The first disadvantage is eliminated by *buffering* or isolating the inputs with voltage followers. This is accomplished with two op amps connected as voltage followers in Fig. 8-5(a). The output of op amp A_1 with respect to ground is E_1, and the output of op amp A_2 with respect to ground is E_2. The differential output voltage V_o is developed across the load resistor R_L. V_o equals the difference between E_1 and E_2 ($V_o = E_1 - E_2$). Note that the output of the basic differential amplifier of Fig. 8-1 is a single-ended output; that is, one side of R_L is connected to ground, and V_o is measured from the output pin of the op amp to ground. The buffered differential amplifier of Fig. 8-5(a) is a differential output; that is, neither side of R_L is connected to ground, and V_o is measured only across R_L.

8-3.2 Adjustable Gain

The second disadvantage of the basic differential amplifier is the lack of adjustable gain. This problem is eliminated by adding three more resistors to the buffered amplifier. The resulting buffered, differential-input to differential-output amplifier, with adjustable gain is shown in Fig. 8-5(b). The high input resistance is preserved by the voltage followers.

Since the differential input voltage of each op amp is 0 V, the voltages at points 1 and 2 (with respect to ground) are respectively equal to E_1 and E_2. Therefore, the voltage across resistor aR is $E_1 - E_2$. Resistor aR may be a fixed resistor or a potentiometer that is used to adjust the gain. The current through aR is

$$I = \frac{E_1 - E_2}{aR} \tag{8-2}$$

When E_1 is above (more positive than) E_2, the direction of I is as shown in Fig. 8-5(b). I flows through both resistors labeled R, and the voltage across all three resistors establishes the value of V_o. In equation form,

$$V_o = (E_1 - E_2)\left(1 + \frac{2}{a}\right) \tag{8-3}$$

where

$$a = \frac{aR}{R}$$

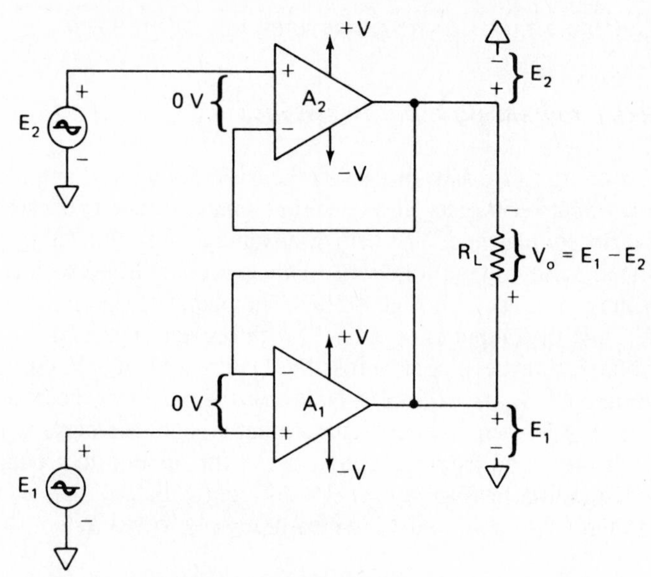

(a) Buffered differential-input to differential-output amplifier

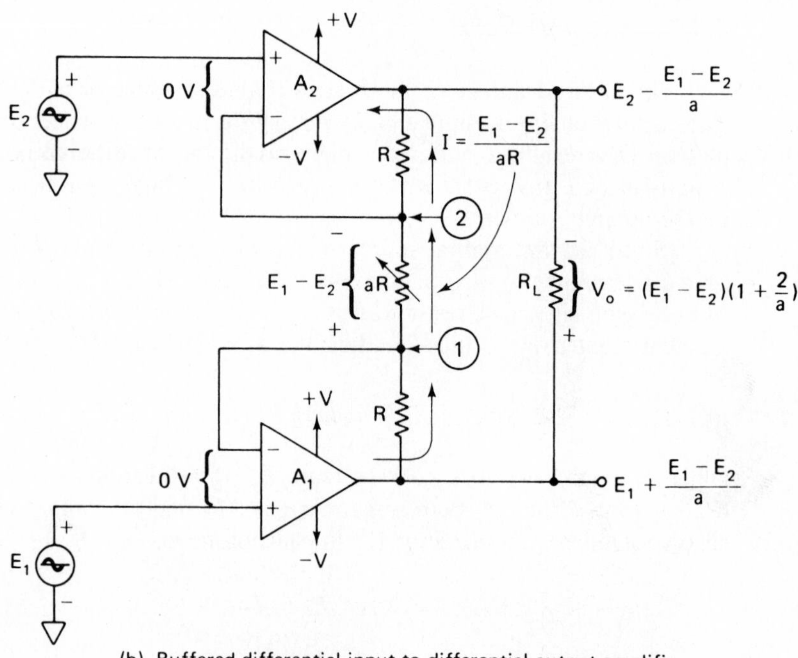

(b) Buffered differential-input to differential-output amplifier with adjustable gain

FIGURE 8-5 Improving the basic differential amplifier.

Example 8-2

In Fig. 8-5(b), $E_1 = 10$ mV and $E_2 = 5$ mV. If $aR = 2\ \text{k}\Omega$ and $R = 9\ \text{k}\Omega$, find V_o.

Solution Since $aR = 2\ \text{k}\Omega$ and $R = 9\ \text{k}\Omega$,

$$\frac{aR}{R} = \frac{2\ \text{k}\Omega}{9\ \text{k}\Omega} = \frac{2}{9} = a$$

From Eq. (8-3),

$$1 + \frac{2}{a} = 1 + \frac{2}{2/9} = 10$$

Finally,

$$V_o = (10\ \text{mV} - 5\ \text{mV})(10) = 50\ \text{mV}$$

Conclusion. To change the amplifier gain, *only a single resistor aR* now has to be adjusted. However, the buffered differential amplifier has one disadvantage: It can only drive floating loads. *Floating loads* are loads that have neither terminal connected to ground. To drive grounded loads, a circuit must be added that converts a differential input voltage to a single-ended output voltage. Such a circuit is the basic differential amplifier. The resulting circuit configuration, to be studied in Section 8-4, is called an *instrumentation amplifier*.

8-4 INSTRUMENTATION AMPLIFIER

8-4.1 Circuit Operation

The instrumentation amplifier is one of the most useful, precise, and versatile amplifiers available today. You will find at least one in every data acquisition unit. It is made from three op amps and seven resistors, as shown in Fig. 8-6. To simplify circuit analysis, note that the instrumentation amplifier is actually made by connecting a buffered amplifier [Fig. 8-5(b)] to a basic differential amplifier (Fig. 8-1). Op amp A_3 and its four equal resistors R form a differential amplifier with a gain of 1. Only the A_3 resistors have to be matched. The primed resistor R' can be made variable to

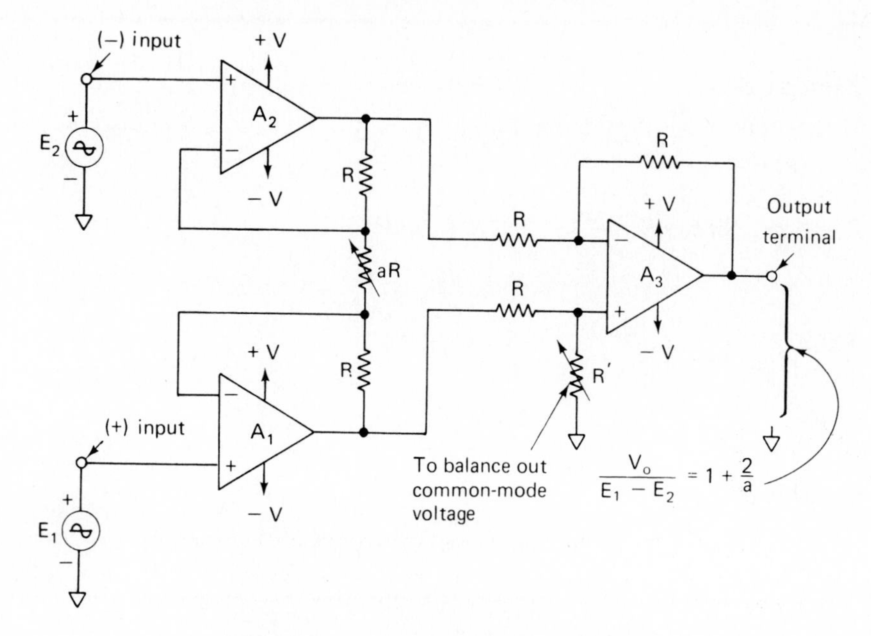

FIGURE 8-6 Instrumentation amplifier.

balance out any common-mode voltage, as shown in Fig. 8-2. Only *one resistor*, aR, is used to set the gain according to Eq. (8-3), repeated here for convenience:

$$\frac{V_o}{E_1 - E_2} = 1 + \frac{2}{a} \tag{8-3}$$

where $a = aR/R$.

E_1 is applied to the (+) input and E_2 to the (−) input. V_o is proportional to the difference between input voltages. *Characteristics of the instrumentation amplifier are summarized* as follows:

1. The voltage gain, from differential input ($E_1 - E_2$) to single-ended output, is set by *one* resistor.
2. The input resistance of both inputs is very high and does not change as the gain is varied.
3. V_o does *not* depend on the voltage common to both E_1 and E_2 (common-mode voltage), only on their difference.

Example 8-3

In Fig. 8-6, $R = 25\ \text{k}\Omega$ and $aR = 50\ \Omega$. Calculate the voltage gain.

Solution From Eq. (8-3),

$$\frac{aR}{R} = \frac{50}{25{,}000} = \frac{1}{500} = a$$

$$\frac{V_o}{E_1 - E_2} = 1 + \frac{2}{a} = 1 + \frac{2}{1/500} = 1 + (2 \times 500) = 1001$$

Example 8-4

If aR is removed in Fig. 8-6 so that $aR = \infty$, what is the voltage gain?

Solution $a = \infty$, so

$$\frac{V_o}{E_1 - E_2} = 1 + \frac{2}{\infty} = 1$$

Example 8-5

In Fig. 8-6, the following voltages are applied to the inputs. Each voltage polarity is given with respect to ground. Assuming the gain of 1001 from Example 8-3, find V_o for (a) $E_1 = 5.001$ V and $E_2 = 5.002$ V; (b) $E_1 = 5.001$ V and $E_2 = 5.000$ V; (c) $E_1 = -1.001$ V, $E_2 = -1.002$ V.

Solution (a)

$$V_o = 1001(E_1 - E_2) = 1001(5.001 - 5.002) \text{ V}$$
$$= 1001(-0.001) \text{ V} = -1.001 \text{ V}$$

(b) $V_o = 1001(5.001 - 5.000) \text{ V} = 1001(0.001) \text{ V} = 1.001 \text{ V}$
(c) $V_o = 1001[-1.001 - (-1.002)] \text{ V} = 1001(0.001) \text{ V} = 1.001 \text{ V}$

8-4.2 Referencing Output Voltage

In some applications, it is desirable to offset the output voltage to a reference level other than 0 V. For example, it would be convenient to position a pen on a chart recorder or oscilloscope trace from a control on the instrumentation amplifier. This can be done quite easily by adding a reference voltage in series with one resistor of the basic differential amplifier. Assume that E_1 and E_2 are set equal to 0 V in Fig. 8-7(a). The outputs of A_1 and A_2 will equal 0 V. Thus we can show the inputs of the differential amplifiers as 0 V in Fig. 8-7(a).

An offset voltage or reference voltage V_{ref} is inserted in series with *reference* terminal R. V_{ref} is divided by 2 and applied to the A_3 op amp's (+) input. Then the noninverting amplifier gives a gain of 2 so that V_o equals V_{ref}. Now V_o can be set to any desired offset value by adjusting V_{ref}. In practice V_{ref} is the output of a voltage-follower circuit as shown in Fig. 8-7(b).

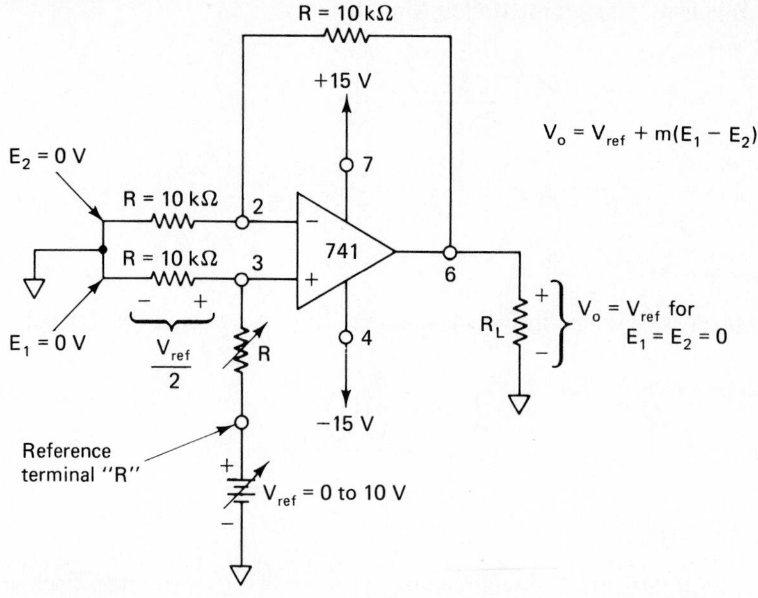

(a) Op amp 3 of the IA in Fig. 8-6 has its "normally grounded" terminal brought out; the new terminal is called "reference terminal." R

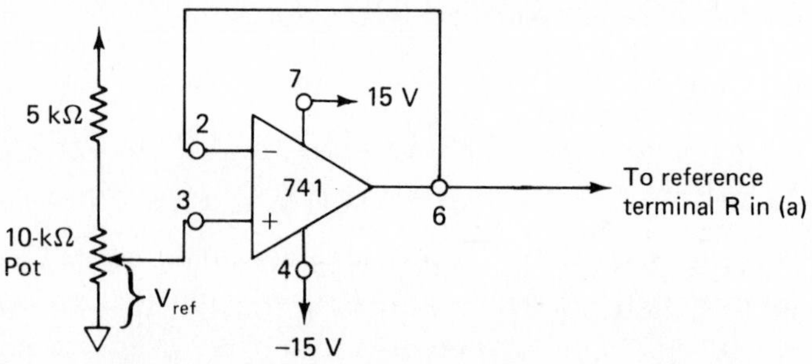

(b) Practically, the reference voltage in (a) must have a very low output impedance; a buffering op amp solves the problem

FIGURE 8-7 The output voltage of an instrumentation amplifier (IA) may be offset by connecting the desired offset voltage (+ or −) to the reference terminal.

8-5 SENSING AND MEASURING WITH THE INSTRUMENTATION AMPLIFIER

8-5.1 Sense Terminal

The versatility and performance of the instrumentation amplifier in Fig. 8-6 can be improved by breaking the negative feedback loop around op amp A_3 and bringing out three terminals. As shown in Fig. 8-8, these terminals are *output* terminal 0, *sense* terminal S, and *reference* terminal R. If long wires or a current-boost transistor

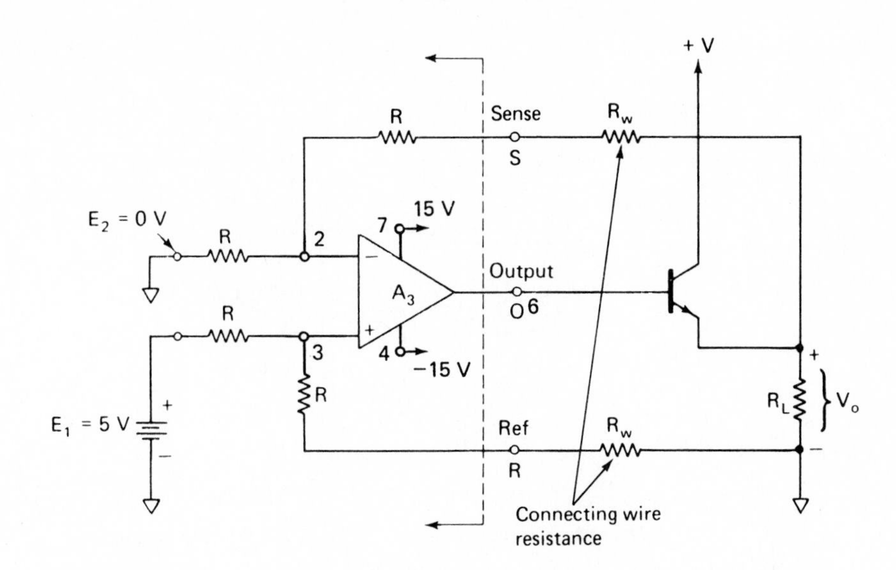

FIGURE 8-8 Extending the sense and reference terminals to the load terminals makes V_o depend on the amplifier gain and the input voltages, not on the load current or load resistance.

are required between the instrumentation amplifier and load, there will be voltage drops across the connecting wires. To eliminate these voltage drops, the sense terminal and reference terminal are wired *directly to the load*. Now, wire resistance is added equally to resistors in series with the sense and reference terminals to minimize any mismatch. Even more important, by sensing voltage at the load terminals and *not* at the amplifier's output terminal, feedback acts to hold load voltage constant. If only the basic differential amplifier is used, the output voltage is found from Eq. (8-1) with $m = 1$. If the instrumentation amplifier is used, the output voltage is determined from Eq. (8-3).

This technique is also called *remote voltage sensing;* that is, you sense and control the voltage at the remote load and *not* at the amplifier's output terminals.

8-5.2 Differential Voltage Measurements

The schematic drawing of a modern instrumentation amplifier is presented in Fig. 8-9. Five pins are brought out so that the user can choose gains of 1, 10, 100, or 1000 by making a single wire strap. (For any intermediate or higher gains, consult Analog Devices' *Data Acquisition Handbook,* Vol. 1.) No strap to either pin, 3 or 16, gives a gain of $V_o/(E_1 - E_2) = 1$ [see Fig. 8-9(a)].

The usual way to measure V_{CE} of a working common-emitter amplifier circuit is to (1) measure collector voltage (with respect to ground), (2) measure emitter voltage (with respect to ground), and (3) calculate the difference. The IA allows you

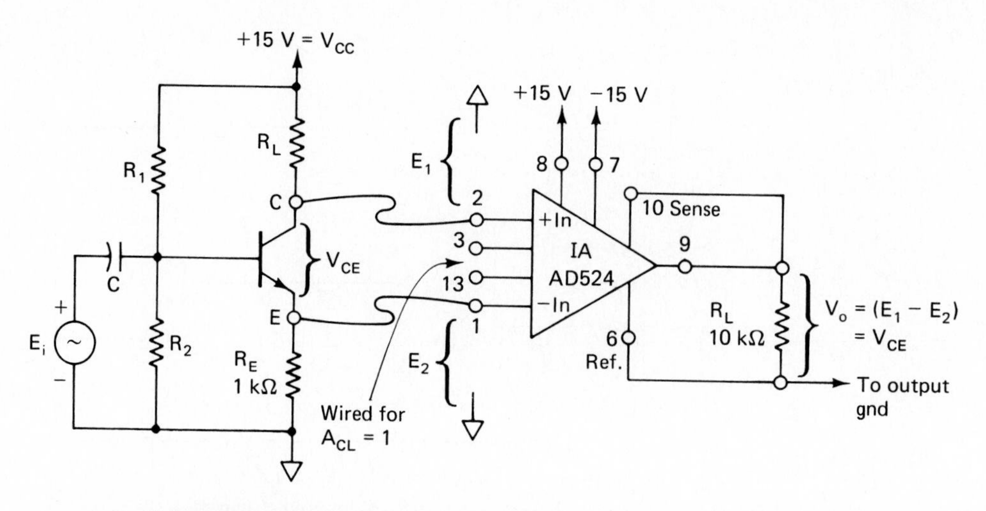

(a) Measuring a voltage that is not grounded at either terminal with an IA

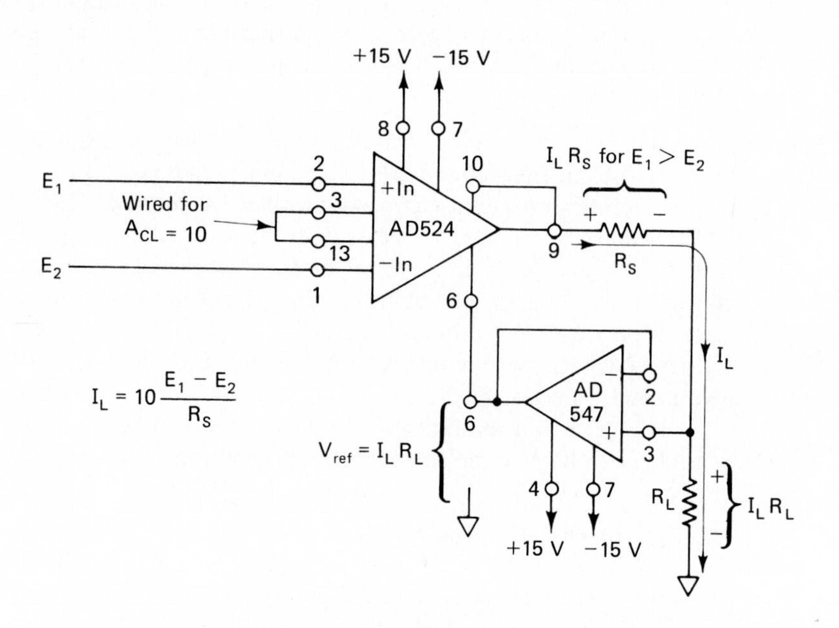

(b) Connect a 547 BiFET op amp to an IA together with a current-sense resistor to make a voltage-to-current converter

FIGURE 8-9 The IA is used to measure a floating differential voltage in (a). A differential voltage-to-current converter is made from an IA, op amp, and resistor in (b).

to make the measurement in one step, as shown in Fig. 8-9(a). Since $E_1 = V_{collector}$ and $E_2 = V_{emitter}$,

$$V_o = (1)(E_1 - E_2) = (1)(V_{collector} - V_{emitter}) = V_{CE} \tag{8-4}$$

Example 8-6

Given $V_o = 5$ V in Fig. 8-9(a), find V_{CE}.

Solution From Eq. (8-4),

$$5 \text{ V} = (E_1 - E_2) = V_{CE}$$

Example 8-7

Extend Example 8-6 as follows. Connect (+)in to the emitter and (−)in to ground and assume V_o measures 1.2 V and calculate (a) emitter current I_E; (b) the voltage across R_L or V_{RL}.

Solution (a) Since $V_o = 1.2$ V, $E_1 - E_2 = 1.2$ V, and therefore $V_{RE} = 1.2$ V. Use Ohm's Law to find I_E.

$$I_E = \frac{V_{RE}}{R_E} = \frac{1.2 \text{ V}}{1 \text{ k}\Omega} = 1.2 \text{ mA}$$

(b)

$$V_{collector} = V_{CE} + V_{RE} = 5 \text{ V} + 1.2 \text{ V} = 6.2 \text{ V}$$

$$V_{RL} = V_{CC} - V_{collector} = 15 \text{ V} - 6.2 \text{ V} = 8.8 \text{ V}$$

Part (a) of this example shows how to measure current in a working circuit by measuring the voltage drop across a known resistor.

8-5.3 Differential Voltage-to-Current Converter

Figure 8-9(b) shows how to make an excellent current source that can sink or source dc current into a grounded load. It can also be an ac current source.

To understand how this circuit operates, one must understand that the IA's output voltage at pin 9 depends on load current, I_L, load resistor, R_L, and *current set resistor*, R_s. In equation form

$$V_9 = I_L R_s + I_L R_L \tag{8-5a}$$

The output voltage of an IA can also be expressed generally by

$$V_9 = V_{ref} + \text{gain}(E_1 - E_2) \tag{8-5b}$$

The AD547 voltage follower forces the reference voltage to equal load voltage or $V_{ref} = I_L R_L$. Since the IA's gain is set for 10 in Fig. 8-9, we can rewrite Eq. (8-5b) as

$$V_9 = I_L R_L + 10(E_1 - E_2) \tag{8-5c}$$

Equate Eqs. (8-5a) and (8-5c) to solve for I_L, which yields

$$I_L = 10\left(\frac{E_1 - E_2}{R_s}\right) \tag{8-5d}$$

Equation (8-5d) indicates that load resistor, R_L, does *not* control load current; this is true as long as neither amplifier is forced to saturation. I_L is controlled by R_s and the difference between E_1 and E_2.

Example 8-8

In the circuit of Fig. 8-9(b), $R_s = 1\ \text{k}\Omega$, $E_1 = 100$ mV, $E_2 = 0$ V, and $R_L = 5\ \text{k}\Omega$. Find (a) I_L; (b) V_{Rs}; (c) V_{ref}; (d) V_9.

Solution (a) From Eq. (8-5d),

$$I_L = 10\left(\frac{0.1\text{ V} - 0\text{ V}}{1000\ \Omega}\right) = 1\text{ mA}$$

(b) $V_{Rs} = I_L R_s = (1\text{ mA})(1\ \text{k}\Omega) = 1\text{ V}$
(c) $V_{ref} = I_L R_L = (1\text{ mA})(5\ \text{k}\Omega) = 5\text{ V}$
(d) From Eq. (8-5a) or (8-5c),

$$V_9 = I_L R_s + I_L R_L = 1\text{ V} + 5\text{ V} = 6\text{ V}$$

or

$$V_9 = V_{ref} + \text{gain}(E_1 - E_2) = 5\text{ V} + 10(0.1\text{ V}) = 6\text{ V}$$

8-6 BASIC BRIDGE AMPLIFIER

8-6.1 Introduction

An op amp, four resistors, and a transducer form the basic bridge amplifier in Fig. 8-10(b). The transducer in this case is any device that converts an environmental change to a resistance change. For example, a *thermistor* is a transducer whose resis-

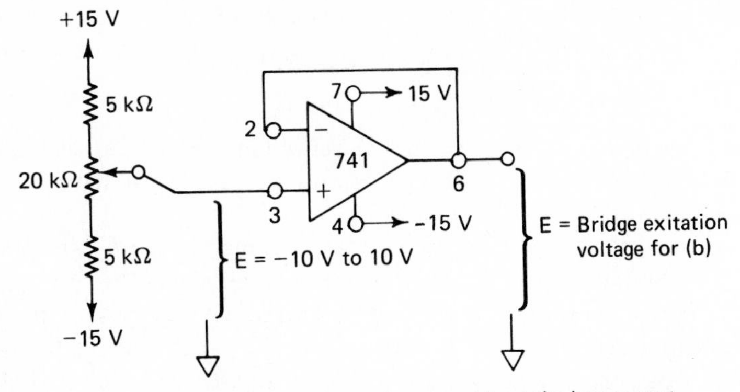

(a) A bridge exitation voltage must have the low output resistance of a voltage follower and be derived from a regulated, stable voltage source

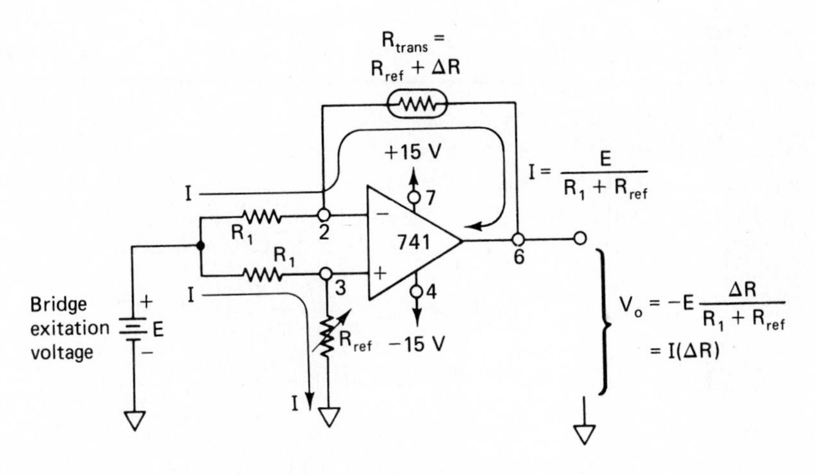

(b) Practical bridge amplifier

FIGURE 8-10 This bridge amplifier outputs a voltage that is directly proportional to the *change* in transducer resistance.

tance increases as its temperature decreases. A *photoconductive cell* is a transducer whose resistance decreases as light intensity increases. For circuit analysis, the transducer is represented by a resistor R plus a *change* in resistance ΔR. R is the resistance value at the desired reference, and ΔR is the amount of change in R. For example, a UUA 41J1 thermistor has a resistance of 10,000 Ω at a reference of 25°C. A temperature change of +1° to 26°C results in a thermistor resistance of 9573 Ω. ΔR is found to be negative from

$$R_{\text{transducer}} = R_{\text{reference}} + \Delta R$$

$$9573\ \Omega = 10{,}000\ \Omega + \Delta R$$

$$\Delta R = -427\ \Omega$$

Since we have defined 25°C to be the reference temperature, we define the reference resistance to be $R_{ref} = 10{,}000\ \Omega$. Our definitions force ΔR to have a negative sign if the transducer's resistance is less than R_{ref}.

To operate the bridge, we need a stable bridge voltage E, which may be either ac or dc. E should have an internal resistance that is small with respect to R. The simplest way to generate E is to use a voltage divider across the stable supply voltages as shown in Fig. 8-10(a). Then connect a simple voltage follower to the divider. For the resistor values shown, E can be adjusted between +10 and −10 V.

8-6.2 Basic Bridge Circuit Operation

A basic low-parts-count bridge circuit is presented in Fig. 8-10(b). Resistors R_1 are 1% low Tempco (metal film) resistors. Current I is constant and set by R_1, R_{ref}, and E. That is, $I = E/(R_1 + R_{ref})$. Note that transducer current is *constant* and equal to I because the voltage drops across both R_1 resistors are equal ($E_d = 0$ V).

The resistor from (+) input to ground is always chosen to equal the reference resistance of the transducer. We want V_o to be zero volts when $R_{trans} = R_{ref}$. This will allow us to calibrate or check operation of the bridge. For Fig. 8-10(b), ΔR is the input and V_o is the output. The output–input relation is given by

$$V_o = -E\left(\frac{\Delta R}{R_{ref} + R}\right) = -I\ \Delta R \qquad (8\text{-}6)$$

where $I = E/(R_{ref} + R)$, $\Delta R = R_{trans} - R_{ref}$.

Zeroing procedure

1. Place the transducer in the reference environment: for example, 25°C.
2. Adjust R_{ref} until $V_o = 0$ V.

Normally, it is too costly to control an environment for test or calibration of a single circuit. Therefore, (1) replace the transducer, R_{trans}, with a resistor equal to R_{ref}; (2) now, ΔR equals zero; (3) from Eq. (8-6), V_o should also be equal to zero.

Suppose that V_o is close to, but not equal to, zero. You want to adjust V_o to precisely zero volts.

1. Check that the R_1 resistors are equal to within 1%.
2. Check that the replacement transducer R_{ref} equals the value of R_{ref} [from (+) input to ground in Fig. 8-10(b)] within 1%.

3. Add an offset null circuit to the op amp and adjust V_o to zero volts (see Chapter 9 for null circuits).

8-6.3 Temperature Measurement with a Bridge Circuit

In this section we show how to design a low-parts-count temperature-measuring system to illustrate a design procedure.

Design Example 8-9

Design a temperature-to-voltage converter that will measure temperatures between 25 and 50°C.

Design Procedure

1. *Select any thermistor* on a trial basis. The thermistor converts a temperature change to a resistance change. Select the Fenwal UUA41J1 and list its corresponding temperature versus resistance as in Table 8-1. (Note the nonlinearity between temperature and resistance.)
2. *Select the reference temperature.* At the reference temperature, V_o must equal zero. Select either the low limit of 25°C or the high limit of 50°C. We shall select the low limit of 25°C for this example. *We have just defined* R_{ref}. R_{ref} is equal to the transducer's resistance at the reference temperature. Specifically, T_{ref} = 25°C; therefore, R_{ref} = 10,000 Ω. Now calculate ΔR for each temperature from

$$R_{trans} = R_{ref} + \Delta R$$

At 50°C,

$$3603\ \Omega = 10{,}000\ \Omega + \Delta R$$

$$\Delta R = -6397\ \Omega$$

Note the negative sign for ΔR.

TABLE 8-1 RESISTANCE VERSUS TEMPERATURE OF A UUA41J1 THERMISTOR

Temp. (°C)	R_{trans}	
25	10,000 } 1983	Ohmic change for 5°C change
30	8,057	
35	6,530	nonlinear
40	5,327	
45	4,370 } 767	
50	3,603	Ohmic change for 5°C change

3. *Predict the voltage–temperature characteristics.* We shall select the bridge circuit of Fig. 8-10 because it converts a resistance change ΔR into an output voltage [see Eq. (8-6)].
   a. Select resistors R_1 to equal 10 kΩ—1%.
   b. Make a trial choice for $E = 1.0$ V.

   If you are wondering why we should make these particular choices, the answers are: (1) 10-kΩ resistor sizes are readily available, and (2) a 1-V

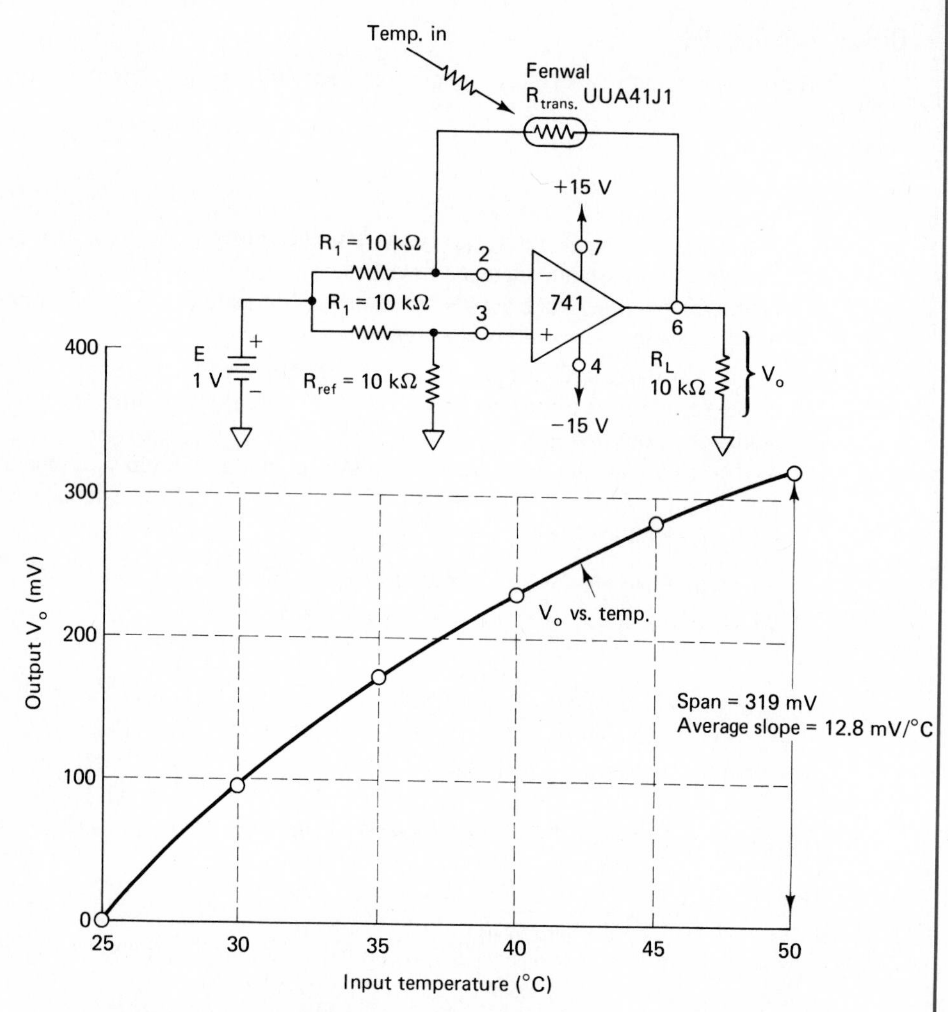

FIGURE 8-11 Solution to Design Example 8-9. An input temperature change of 25 to 50°C gives an output voltage change of 0 to 319 mV. The circuit is a temperature-to-voltage converter.

selection will give us an idea of the size of V_o. If later you want to double or triple V_o, simply double or triple E.

c. Calculate I from Eq. (8-6).

$$I = \frac{E}{R_{\text{ref}} + R_1} = \frac{1\text{ V}}{10\text{ k}\Omega + 10\text{ k}\Omega} = 0.050\text{ mA}$$

d. Calculate V_o for each value of R and tabulate the results (see Table 8-2). From Eq. (8-6),

$$V_o = -I\,\Delta R$$

For 50°C,

$$V_o = -(0.050\text{ mA})(-6397\ \Omega) = 319\text{ mV}$$

4. *Document performance.* V_o is plotted against temperature in Fig. 8-11, where the design circuit is also drawn.

TABLE 8-2 CALCULATIONS FOR TEMPERATURE-TO-VOLTAGE CONVERTER

Temp. (°C)	R_{trans} (Ω)	ΔR (Ω)	V_o (mV)
25	10000	0	0
30	8057	−1943	97
35	6530	−3470	173
40	5327	−4673	233
45	4370	−5630	281
50	3603	−6397	319

Summary review and comments. Example 8-9 shows how a bridge circuit converts the resistance change of a transducer into a voltage change. The circuit output voltage is linear with respect to ΔR [see Eq. (8-6)]. However, ΔR is *not* linear with respect to temperature (see Table 8-1 and Fig. 8-11). Therefore, V_o is *not* linear with respect to temperature. The bridge simply transmits the nonlinearity of the thermistor.

The sensitivity of the temperature-to-voltage converter can be increased easily by increasing E. The maximum value of E is set by the maximum thermistor current to avoid self-heating, typically 1 mA. Therefore, E has a maximum value of

$$E = I(R_{\text{ref}} + R_1) = (1\text{ mA})(10 + 10)\text{ k}\Omega = 20\text{ V}$$

If we wanted to increase the 319-mV output span ($E = 1$ V) to a 2.50-V span, simply increase E by 7.84 to 7.84 V [(2.50 V/0.319 V = 7.84)].

8-6.4 Bridge Amplifiers and Computers

Thus far, the bridge amplifier shown converts temperature changes to a voltage. In Chapter 2 you studied the principles of how a voltage change could be converted into a change of high time by a pulse-width modulator. Those principles can be used again to communicate between the analog world of continuous temperature variation and the digital world of the computer. The bridge amplifier provides measurement; the PWM provides an interface.

8-7 ADDING VERSATILITY TO THE BRIDGE AMPLIFIER

8-7.1 Grounded Transducers

In some applications it is necessary to have one terminal of the transducer connected to ground. The standard technique is shown in Fig. 8-12(a). Note that current I depends on transducer resistance (in Figs. 8-10 and 8-11, the current was constant). Note also that V_o is not linear with ΔR because ΔR appears in the denominator of the equation for V_o versus ΔR. Finally, in contrast with Design Example 8-9, if E_i is positive and T_{ref} is at the low end of the scale, V_o goes negative for negative values of ΔR. That is, if R_{trans} is a thermistor, V_o goes more negative as temperature increases.

8-7.2 High-Current Transducers

If the current required by the transducer is higher than the current capability of the op amp (5 mA), use the circuit of Fig. 8-12(b). Transducer current is furnished from

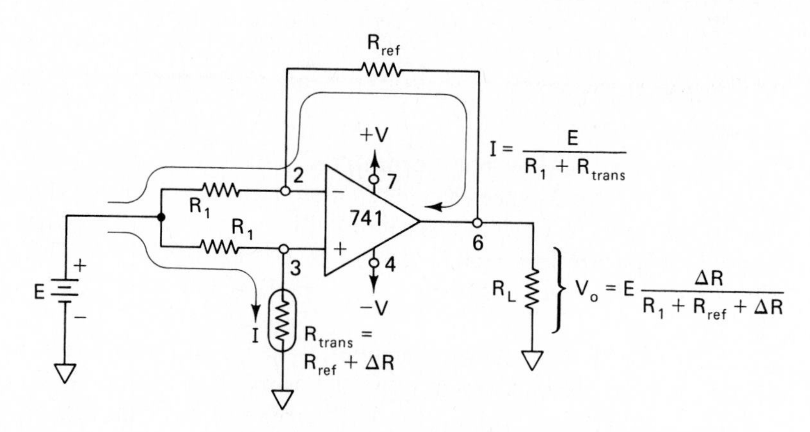

(a) Bridge amplifier with grounded transducer

FIGURE 8-12 The bridge amplifier is used with a grounded transducer in (a) and with a high current transducer in (b).

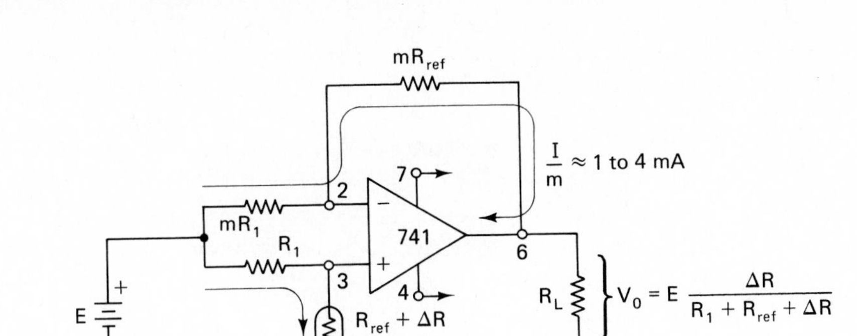

(b) Some transducers require currents larger than the op amp can supply; The transducer's current is scaled down by the multiplying factor m so that feedback current supplied by the op amp is I/m

FIGURE 8-12 *(cont.)*

E. Resistors *mR* are chosen to hold their currents to about 1 to 4 mA. Transducer current and output voltage may be found from the equation in Fig. 8-12(b). If the transducer current is very small (high-resistance transducers), the same circuit can be used except that the *mR* resistors will be smaller than *R* to hold output current of the op amp at about 1 mA. Also BiFET op amps such as the TL081, CA3140, AD547, or LF355 should be used. They have small bias currents (see Chapter 9).

8-8 THE STRAIN GAGE AND MEASUREMENT OF SMALL RESISTANCE CHANGES

8-8.1 Introduction to the Strain Gage

A strain gage is a conducting wire whose resistance changes by a small amount when it is lengthened or shortened. The change in length is small, a few millionths of an inch. The strain gage is bonded to a structure so that the percent change in length of the strain gage and structure are identical.

A foil-type gage is shown in Fig. 8-13(a). The active length of the gage lies along the transverse axis. The strain gage must be mounted so that its transverse axis lies in the same direction as the structure motion that is to be measured [see Fig. 8-13(b) and (c)]. Lengthening the bar by tension lengthens the strain gage conductor and increases its resistance. Compression reduces the gage's resistance because the normal length of the strain gage is reduced.

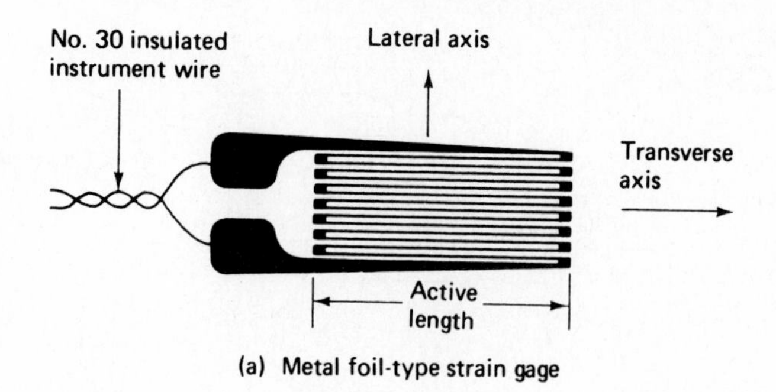

(a) Metal foil-type strain gage

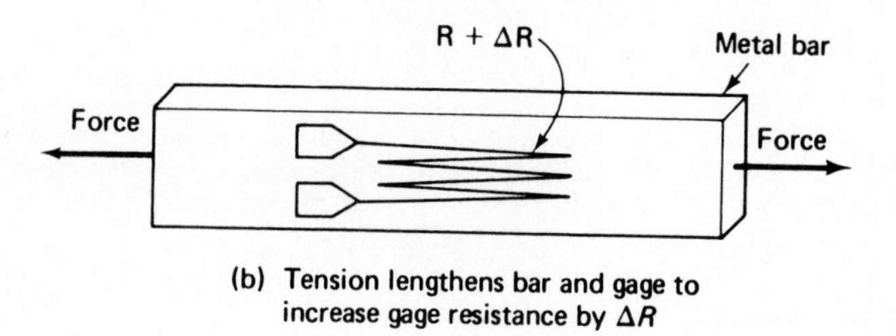

(b) Tension lengthens bar and gage to increase gage resistance by ΔR

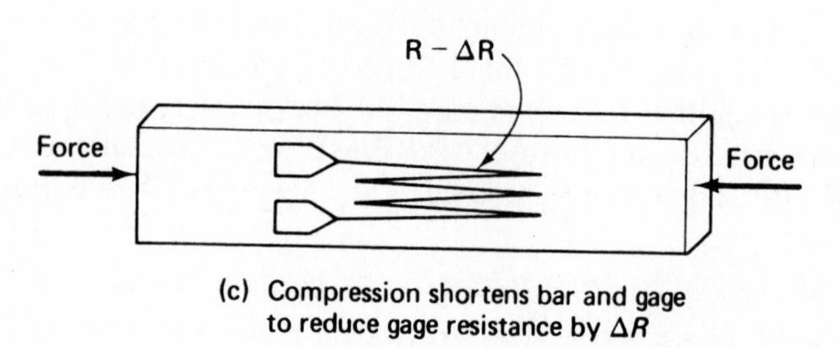

(c) Compression shortens bar and gage to reduce gage resistance by ΔR

FIGURE 8-13 Using a strain gage to measure the change in length of a structure.

8-8.2 Strain-Gage Material

Strain gages are made from metal alloy such as constantan, Nichrome V, Dynaloy, Stabiloy, or platinum alloy. For high-temperature work they are made of wire. For moderate temperature, strain gages are made by forming the metal alloy into very thin sheets by a photoetching process. The resultant product is called a foil-type strain gage and a typical example is shown in Fig. 8-13(a).

8-8.3 Using Strain-Gage Data

In the next section we show that our instrumentation measures only the gage's *change* in resistance ΔR. The manufacturer specifies the unstrained gage's resistance R. Once ΔR has been measured, the ratio $\Delta R/R$ can be calculated. The manufacturer also furnishes a specified *gage factor* (GF) for each gage. The gage factor is the ratio of the percent change in resistance of a gage to its percent change in length. These percent changes may also be expressed as decimals. If the ratio $\Delta R/R$ is divided by gage factor G, the result is the ratio of the *change* in length of the gage ΔL to its original length L. Of course the structure where the gage is mounted has the same $\Delta L/L$. An example will show how gage factor is used.

Example 8-10

A 120-Ω strain gage with a gage factor of 2 is affixed to a metal bar. The bar is stretched and causes a ΔR of 0.001 Ω. Find $\Delta L/L$.

Solution

$$\frac{\Delta L}{L} = \frac{\Delta R/R}{GF} = \frac{0.001\ \Omega/120\ \Omega}{2}$$

$$\simeq 4.1 \text{ microinches per inch}$$

The ratio $\Delta L/L$ has a name. It is called *unit strain*. It is the unit strain data (we have developed from a measurement of ΔR) that mechanical engineers need. They can use this unit strain data together with known characteristics of the structural material (modulus of elasticity) to find the *stress* on the beam. *Stress is the amount of force acting on a unit area*. The unit for stress is pounds per square inch (psi). If the bar in Example 8-10 were made of mild steel, its stress would be about 125 psi. *Strain is the deformation of a material resulting from stress*, or $\Delta L/L$.

8-8.4 Strain-Gage Mounting

Before mounting a strain gage the surface of the mounting beam must be cleaned, sanded, and rinsed with alcohol, Freon, or methyl ethyl ketone (MEK). The gage is then fastened permanently to the cleaned surface by Eastman 910, epoxy, polymide adhesive, or ceramic cement. The manufacturer's procedures should be followed carefully.

8-8.5 Strain-Gage Resistance Changes

It is the *change* of resistance in a strain gage ΔR that must be measured and this change is *small.* ΔR has values of a few milliohms. The technique employed to measure small resistance change is discussed next.

8-9 MEASUREMENT OF SMALL RESISTANCE CHANGES

8-9.1 Need for a Resistance Bridge

To measure resistance we must first find a technique to convert the *resistance change* to a current or voltage for display on an ammeter or voltmeter. If we must measure a small *change* of resistance, we will obtain a very small voltage *change*. For example, if we passed 5 mA of current through a 120-Ω strain gage, the voltage across the gage would be 0.600 V. If the resistance *changed* by 1 mΩ, the voltage *change* would be 5 μV. To display the 5-μV change, we would need to amplify it by a factor of, for example, 1000 to 5 mV. However, we would also amplify the 0.6 V by 1000 to obtain 600 V plus 5 mV. It is difficult to detect a 5-mV difference in a 600-V signal. Therefore, we need a circuit that allows us to amplify only the *difference* in voltage across the strain gage caused by a *change* in resistance. The solution is found in the bridge circuit.

8-9.2 Basic Resistance Bridge

The strain gage is placed in one arm of a resistance bridge, as shown in Fig. 8-14. Assume that the gage is unstrained, so that its resistance = R. Also assume that R_1, R_2, and R_3 are all precisely equal to R. (This unlikely assumption is dealt with in Section 8-10.) Under these conditions $E_1 = E_2 = E/2$ and $E_1 - E_2 = 0$. The bridge is said to be *balanced*. If the strain gage is *compressed,* R would decrease by ΔR and the differential voltage $E_1 - E_2$ would be given by

$$E_1 - E_2 = E\frac{\Delta R}{4R} \tag{8-7}$$

This approximation is valid because $2\,\Delta R \ll 4R$ for strain gages.

Equation (8-7) shows that E should be made large to maximize the bridge differential output voltage, $E_1 - E_2$.

Example 8-11

If $\Delta R = 0.001\ \Omega$, $R = 120\ \Omega$, and $E = 1.0$ V in Fig. 8-14, find the output of the bridge, $E_1 - E_2$.

Solution From Eq. (8-7),

$$E_1 - E_2 = 1.0 \text{ V} \times \frac{0.001\ \Omega}{(4)(120)\ \Omega} = 2.2\ \mu\text{V}$$

If E is increased to 10 V, then $E_1 - E_2$ will be increased to 22 μV.

An instrumentation amplifier can then be used to amplify the differential voltage $E_1 - E_2$ by 1000 to give an output of about 22 mV per milliohm of ΔR.

We conclude that a voltage E and bridge circuit plus an instrumentation amplifier can convert a change in resistance of 1 mΩ to an output voltage change of 22 mV.

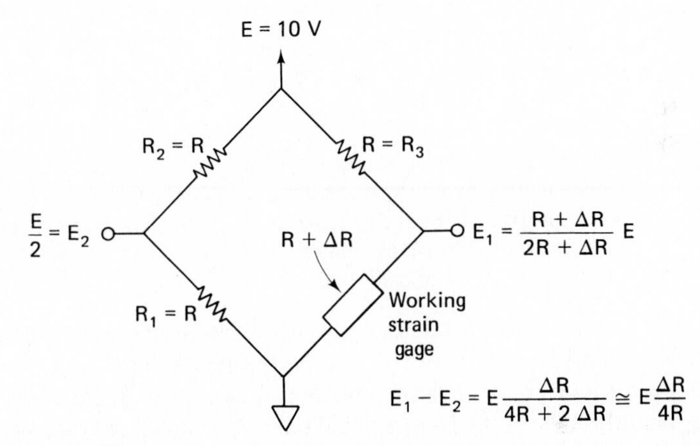

FIGURE 8-14 The resistor bridge arrangement and supply voltage E convert a resistance change in the strain gage ΔR to a differential output voltage $E_1 - E_2$. If $R = 120\ \Omega$, $E = 10$ V, and $R = 1$ mΩ, $E_1 - E_2 =$ 22 μV.

8-9.3 Thermal Effects on Bridge Balance

Even if you succeed in balancing the bridge circuit of Fig. 8-14, it will not stay in balance because slight temperature changes in the strain gage cause resistance change equal to or greater than those caused by strain. This problem is solved by mounting another identical strain gage immediately adjacent to the working strain gage so that both share the same thermal environment. Therefore, as temperature changes, the added gage's resistance changes exactly as the resistance of the working gage. The added gage provides automatic temperature compensation, and is appropriately called the *temperature-compensation* or *dummy* gage.

The *temperature-compensation gage* is mounted with its transverse axis perpendicular to the transverse axis of the working gage, as shown in Fig. 8-15. This

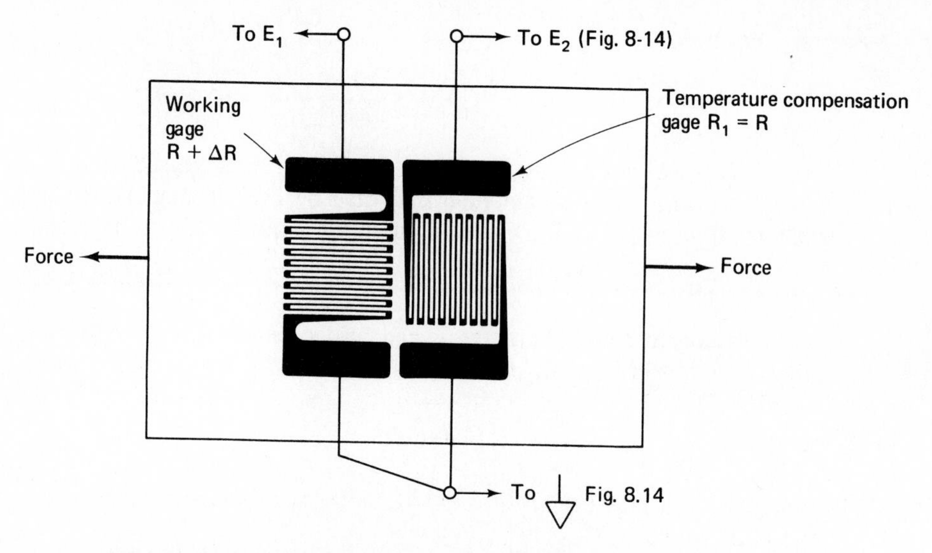

FIGURE 8-15 The temperature-compensation gage has the same resistance changes as the working gage with changes in temperature. Only the working gage changes resistance with strain. By connecting in the bridge circuit of Fig. 8-14 as shown, resistance changes due to temperature changes are automatically balanced out.

type of standard gage arrangement is available from manufacturers. The new gage is connected in place of resistor R_1 in the bridge circuit of Fig. 8-14. Once the bridge has been balanced, R of the temperature-compensation gage and working gage track one another to hold the bridge in balance. Any unbalance is caused strictly by ΔR of the working gage due to strain.

8-10 BALANCING A STRAIN-GAGE BRIDGE

8-10.1 The Obvious Technique

Suppose that you had a working gage and temperature-compensation gage in Fig. 8-16 that are equal to within 1 mΩ. To complete the bridge, you install two 1%, 120-Ω resistors. One is high by 1% at 121.200 Ω and one is low by 1% at 118.800 Ω. They must be equalized to balance the bridge. To do so, a 5-Ω 20-turn balancing pot is installed, as shown in Fig. 8-16. Theoretically, the pot should be set as shown to equalize resistances in the top branches of the bridge at 122.500 Ω.

Further assume that an instrumentation amplifier with a gain of 1000 is connected to the bridge of Fig. 8-16. From Example 8-11, the *output* of the instrumentation amplifier (IA) will be about 22 mV per milliohm of unbalance. This means that the 5-Ω pot must be adjusted to within 1 mΩ of the values shown, so that $E_1 - E_2$ and consequently, V_o of the IA will equal 0 V $\pm$22 mV.

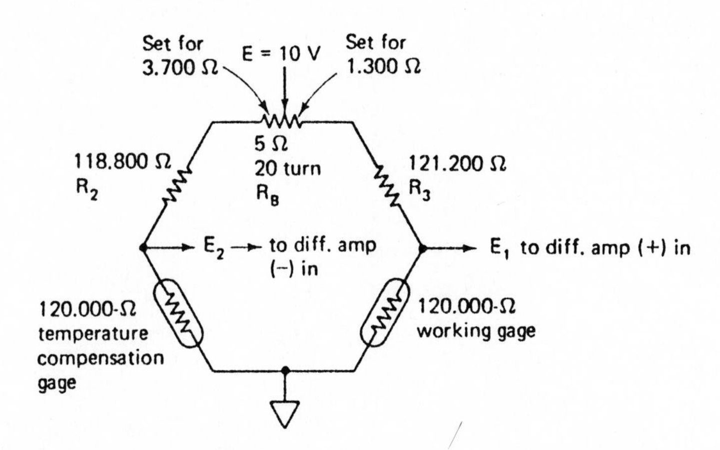

FIGURE 8-16 Balance pot R_B is adjusted in an attempt to make $E_1 - E_2 = 0$ V.

Unfortunately, it is very difficult in practice to adjust for balance. This is because each turn of the pot is worth 5 Ω/20 turns = 250 mΩ. When you adjust the pot it is normal to expect a backlash of $\pm\frac{1}{50}$ of a turn. Therefore, your best efforts result in an unbalance at the pot of about ±5 mΩ. You observe this unbalance at the IA's output, where V_o changes by ±0.1 V on either side of zero as you fine-tune the 20-turn pot. It turns out there is a better technique that uses an ordinary linear potentiometer ($\frac{3}{4}$ turn) and a single resistor.

8-10.2 The Better Technique

To analyze operation of the balance network in Fig. 8-17, assume that the R_2 and R_3 bridge resistors are reasonably equal, to within ±1%. The strain gage's resistance should have equal resistances within several milliohms if the working gage is not under strain.

Resistor R_{B1} is an ordinary $\frac{3}{4}$-turn linear pot. Its resistance should be about $\frac{1}{10}$ or less than resistor R_{B2} so that the voltage fE depends only on E and the decimal fraction f. Values of f vary from 0 to 1.0 as the pot is adjusted from one limit to the other. R_{B1} should be 10 or more times the gage resistance.

Resistor R_{B2} is chosen to be greater than 10 or more times R_{B1}. Under these conditions R_{B2} does not load down the voltage-divider action of R_{B1}. Also, the size of R_{B2} determines the maximum balancing current that can be injected into, or extracted from, the E_2 node. The pot setting f determines how much of that maximum current is injected or extracted.

Balancing action is summarized by observing that if $f > 0.5$, a small current is injected into the E_2 node and flows through the temperature gage to ground, this makes E_2 more positive. If $f < 0.5$ current is extracted from the E_2 node, this increases current through R_2 to make E_2 less positive.

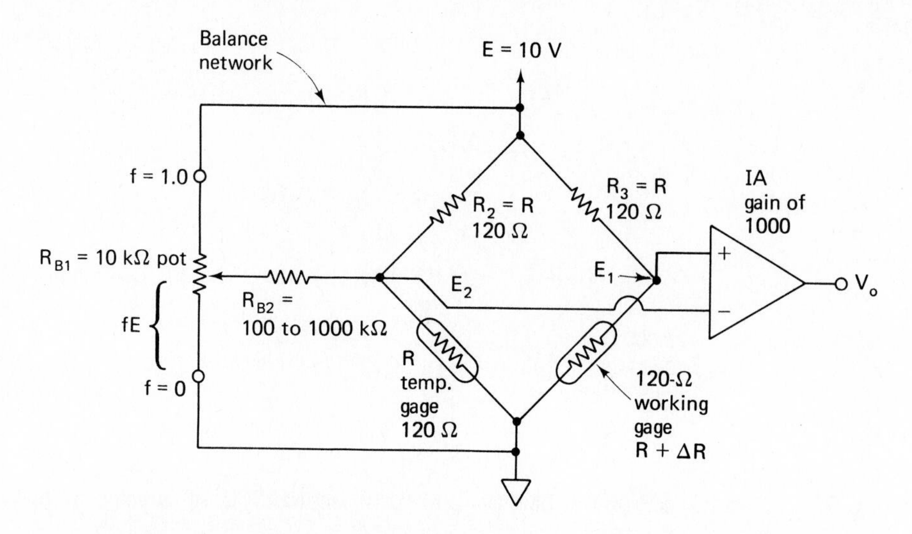

FIGURE 8-17 Improved balance network R_{B1} and R_{B2} allow easy adjustment of V_o to 0 V.

In a real bridge setup, begin with $R_{B2} = 100\text{ k}\Omega$ and $R_{B1} = 10\text{ k}\Omega$. Monitor V_o of the IA and check the balancing action. If the variation in V_o is larger than you want, increase R_{B2} to $1000\text{ k}\Omega$ and recheck the balance action. The final value of R_{B2} is selected by experiment and depends on the magnitude of unbalance between R_2 and R_3.

8-11 INCREASING STRAIN-GAGE BRIDGE OUTPUT

A single working gage and temperature-compensation gage were shown to give a differential bridge output in Fig. 8-14 of

$$E_1 - E_2 = E\frac{\Delta R}{4R} \tag{8-7}$$

This bridge circuit and placement of the gages is shown again in Fig. 8-18(a).

The bridge output voltage $E_1 - E_2$ can be doubled by doubling the number of working gages, as in Fig. 8-18(b). Gages 1–2 and 5–6 are the working gages and will increase resistance (tension) if force is applied as shown. By arranging the working gages in opposite arms of the bridge and the temperature gages in the other arms, the bridge output is

$$E_1 - E_2 = E\frac{\Delta R}{2R + \Delta R} \simeq E\frac{\Delta R}{2R}$$

If the structural member experiences bending as shown in Fig. 8-18(c), even greater bridge sensitivity can be obtained. The upper side of the bar will lengthen

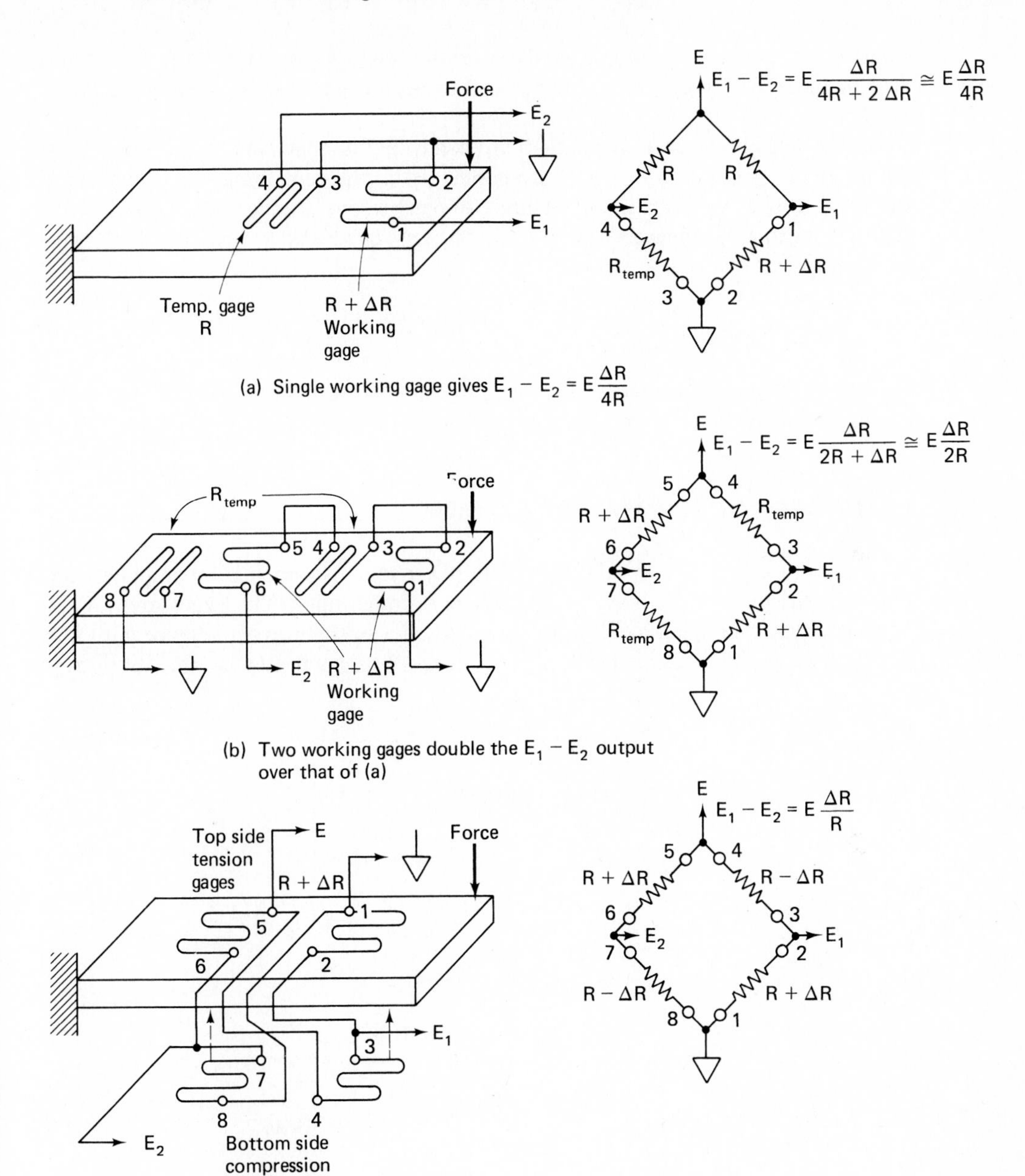

(c) Four working gages quadruple the $E_1 - E_2$ over that of (a)

FIGURE 8-18 Comparison of sensitivity for three stain-gage bridge arrangements. (ΔR is small with respect to R for foil strain gages.)

(tension) to increase resistance of the working strain gages by $(+)\Delta R$. The lower side of the bar will shorten (compression) to decrease the working strain gages by $(-)\Delta R$.

The tension gages 1–2 and 5–6 are connected in opposite arms of the bridge. Compression gages 3–4 and 7–8 are connected in the remaining opposite arms of the bridge. The gages also temperature-compensate one another. The output of the four-strain-gage arrangement in Fig. 8-18(c) is quadrupled over the single-gage bridge to

$$E_1 - E_2 = E\frac{\Delta R}{R}$$

Of course, each bridge arrangement in Fig. 8-18 should be connected to a balance network [which, for clarity, was not shown (see Fig. 8-17 and Section 8-10)].

8-12 A PRACTICAL STRAIN-GAGE APPLICATION

As shown in Fig. 8-19, an AD521 (Analog Devices) instrumentation amplifier (IA) is connected to a bridge arrangement of four strain gages. The gages are 120-Ω, SR4, foil-type strain gages. They are mounted on a steel bar in accordance with Fig. 8-18(c). Also the balance network of Fig. 8-17 is connected to the strain-gage

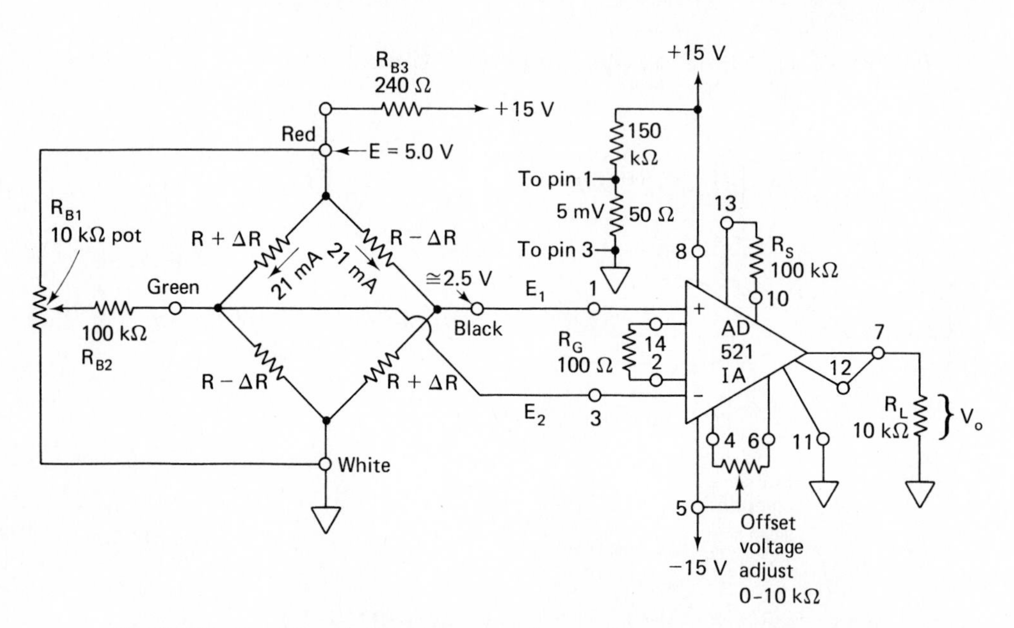

FIGURE 8-19 The AD521 instrumentation amplifier is used to amplify output of the four working strain gages [see Fig. 8-18(c)].

bridge. R_{B2} was selected, after experiment, as 100 kΩ. Strain gages were mounted in strict accordance with the manufacturer's instructions (BLH Electronics, Inc.)

The calibration procedure for the instrumentation is as follows:

1. *Gain resistors*. Select gain setting resistors $R_{scale}(R_S)$ and $R_{gain}(R_G)$ to give a gain of 1000. Gain is set by the ratio of R_S/R_G, or 100 kΩ/100 Ω.
2. *Offset voltage adjust*. With R_S and R_G installed, ground inputs 1 and 3. Adjust the IA's offset voltage pot for $V_o = 0$ V (see Chapter 9).
3. *Gain measurement*. Connect the 5-mV signal from the 150-kΩ and 50-Ω voltage divider as $E_1 - E_2$ to pins 1 and 3. Measure V_o; calculate gain = $V_o/(E_1 - E_2)$.
4. *Zeroing the bridge network*. Connect E_1 and E_2 to pins 1 and 3 of the IA. Adjust R_{B1} for $V_o = 0$ V when the gages are *not* under strain.

Example 8-12

The test setup of Fig. 8-19 is used to measure the strain resulting from deflection of a steel bar. V_o is measured to be 100 mV. Calculate (a) ΔR; (b) $\Delta R/R$; (c) $\Delta L/L$. Assume that the calibration procedure has been followed and that the gain from step 3 is 1000. The gage factor is 2.0.

Solution (a) Find $E_1 - E_2$ from

$$E_1 - E_2 = \frac{V_o}{\text{gain}} = \frac{100 \text{ mV}}{1000} = 0.1 \text{ mV}$$

From ΔR from Fig. 8-18(c):

$$\Delta R = \frac{R(E_1 - E_2)}{E} = \frac{120\ \Omega(0.1 \times 10^{-3}\text{ V})}{5.0\text{ V}}$$

$$= 0.0024\ \Omega = 2.4\text{ m}\Omega$$

(b)

$$\frac{\Delta R}{R} = \frac{0.0024\ \Omega}{120\ \Omega} = 0.000020 = 20 \times 10^{-6}\ \mu\Omega/\Omega$$

(c) From gage factor = $(\Delta R/R)/(\Delta L/L)$, we obtain

$$\frac{\Delta L}{L} = \frac{20 \times 10^{-6}}{2} = 10 \times 10^{-6} = 10\ \mu\text{in./in.}$$

Note: Resistor R_{B3} is selected to restrict gage current *below* 25 mA to limit self-heating.

Since we now know the value of strain from $\Delta L/L$, we can look up the *modulus of elasticity* for steel, $E = 30 \times 10^6$. The *stress* can be calculated from

$$\text{stress} = E \times \text{strain} = (30 \times 10^6)(10 \times 10^{-6})$$
$$= 300 \text{ psi}$$

8-13 MEASUREMENT OF PRESSURE, FORCE, AND WEIGHT

Example 8-12 illustrated how pressure could be measured by a strain-gage system. The mechanical engineers can be given $\Delta L/L$ by electrical personnel, who can measure $\Delta R/R$ and look up the gage factor. From the value of $\Delta L/L$, the mechanical engineers and technicians can calculate pressure on a structure. Since pressure is force per unit area, they can calculate force by measuring the structure's area.

Furthermore, the weight of an object exerts a force on any supporting structure. By installing a strain gage on the supporting structure, you can weigh very heavy objects such as a gravel-filled truck or a 747 aircraft.

LABORATORY EXERCISES

8-1. *Basic differential amplifier*. Figures 8-1 and 8-2 can be used to gain experience with a basic differential amplifier. Select $R = 10\text{ k}\Omega$, 1% and $mR = 10\text{ k}\Omega$, 1% resistors. E_1 and E_2 must have low-impedance outputs. Signal generators with 50 Ω impedance or lab-type power supplies are satisfactory. Any voltage derived from a resistor voltage divider should be buffered with a voltage follower. Hold signal frequencies at or below 1 kHz (see Chapter 10 for the reason).

We suggest that you do *not* breadboard an instrumentation amplifier with three op amps. Rather, use a commercial IA and put the circuit on a pc board. Be sure to use shielded leads and bypass the supply voltage with 0.01- or 0.1-μF capacitors mounted at the power supply pins of the IA.

8-2. *Thermistor*. Section 8-3 can be used to make an excellent laboratory exercise. Get the resistance temperature data on any thermistor. Pick R_{ref} approximately equal to the thermistor's resistance at the low end of the temperature scale you wish to measure. (Keep the temperature spread below 50°C.) For Fig. 8-10 or 8-11, pick $R_1 = R_{ref} \pm 20\%$. Use a decade box to dial in the transducer resistances for each temperature (because few can afford a $10,000 precision-controlled temperature oven for one experiment).

8-3. *Strain gage*. A strain-gage experiment first requires obtaining a mounting kit from the strain-gage manufacturer and mounting the gages *exactly* as instructed. Do *not* skip a cleaning step. Buy gages *with* the solder dots, it makes soldering the leads a lot easier.

Make a strain gage set up with the IAs recommended. Now bang the structure with a hammer and watch the output voltage on a CRO; use a slow sweep speed. You will observe the natural resonant frequency of the structure.

For your guidance, the authors recommend the following instrumentation amplifiers. We have included only three manufacturers because we have had experience using their devices. They also furnish superb application notes, a data sheet, and guidance on how to use them.

Manufacturer	Instrumentation amplifier
Analog Devices	AD521, AD524, AD624
National Semiconductor	LH0036, LH0037
Burr-Brown	INA104, 3626, 3629

PROBLEMS

8-1. In Fig. 8-1, $m = 20$, $E_1 = 0.2$ V, and $E_2 = 0.25$ V, Find V_o.

8-2. If $V_o = 10$ V in Fig. 8-1, $E_1 = 7.5$ V, and $E_2 = 7.4$ V, find m.

8-3. If $E_{cm} = 5.0$ V in Fig. 8-2, find V_o.

8-4. In Fig. 8-3, $E_i = 2$ mV and $E_n = 50$ mV. What is the output voltage due to **(a)** E_i; **(b)** E_n?

8-5. In Fig. 8-4, $E_i = 2$ mV and $E_n = 50$ mV. What is the output voltage due to **(a)** E_i; **(b)** E_n?

8-6. What is the main advantage of a differential amplifier over an inverting amplifier with respect to an input noise signal voltage?

8-7. Find V_o in Fig. 8-5(b) if $E_1 = -5$ V and $E_2 = -3$ V.

8-8. In Fig. 8-5(b), $R = 10$ kΩ and $aR = 2$ kΩ. If $E_1 = 1.5$ V and $E_2 = 0.5$ V, find V_o.

8-9. In Fig. 8-6 the overall gain is 21 and $V_o = 3$ V. Determine **(a)** $E_1 - E_2$; **(b)** a.

8-10. In Fig. 8-6, $R = 25$ kΩ, $aR = 100$ Ω, $E_1 = 1.01$ V, and $E_2 = 1.02$ V. Find V_o.

8-11. If $V_{ref} = 5.0$ V in Fig. 8-7, find **(a)** V_o; **(b)** the voltage at the (+) input with respect to ground.

8-12. Refer to the circuit of Fig. P8-12. Complete the table below for each input condition.

	E_1(V)	E_2(V)	E_3(V)	V_o	V at (+) input
(a)	−2	−2	0		
(b)	−2	−2	2		
(c)	2	−2	−2		
(d)	2	0	2		

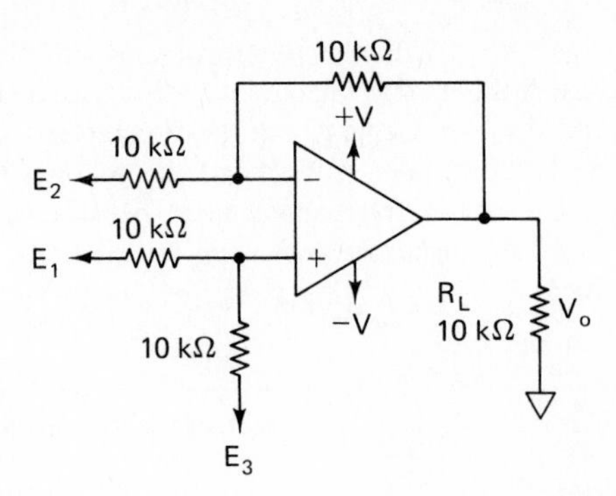

FIGURE P8-12

8-13 Refer to the voltage-to-current converter of Fig. 8-9(b). Assume that the AD524 is wired for a gain of 1 [no wires on pins 16 and 3 per Fig. 8-9(a)]. The load current is now $I_L = (E_1 - E_2)/R_s$. Let $R_s = 1\ \text{k}\Omega$, $E_2 = 0$ V or ground, and $E_1 = 1$ V. **(a)** Will the direction of I_L be up or down in Fig. 8-9(b)? **(b)** Find I_L. **(c)** Find the voltage across R_L if $R_L = 100\ \Omega$. **(d)** Find the output voltage of the IA ($V_{\text{pin }9}$) if $R_L = 3\ \text{k}\Omega$.

8-14. Repeat Problem 8-13 except change E_2 = ground and $E_1 = 1$ V. (Note that Problems 8-13 and 8-14 tell you how to make an ac voltage-controlled current source for a grounded load.)

8-15. Change E_1 to -1 V in Problem 8-13. **(a)** Would V_o be positive or negative with respect to ground? **(b)** Would V_o decrease or increase in *magnitude* as temperature increased?

8-16. Refer to Section 8-6.3. To gain experience with this type of instrumentation circuit, repeat Design Example 8-9 except change only your reference temperature to 50°C. R_1 remains at 10 kΩ and $E = 1$ V. Present your solution in the same format as shown in Table 8-2 and Fig. 8-11. Redraw the new design schematic like that of Fig. 8-11. [Remember that R_{ref} will now be 3603 Ω so that $I = 1\ \text{V}/(10{,}000 + 3603)\ \Omega = 73.51\ \mu\text{A}$.]

8-17. You want a circuit that has an increasing magnitude of output voltage as temperature of a thermistor increases. You put the thermistor in the feedback loop [see Fig. 8-10(b)]. Would you choose R_{ref} at the low or the high end of the temperature scale? (*Hint*: Compare V_o vs. V_{temp} of Design Example 8-9 with the solution of Problem 8-16.) This problem forces you to face briefly the issue of "human engineering." People want to see a voltage get higher as temperature gets higher.

8-18. In Fig. 8-18, the value for $R = 120.00\ \Omega$, $\Delta R = 1.2\ \text{m}\Omega$, and $E = 10.0$ V. Find $(E_1 - E_2)$ for the strain-gage arrangement of **(a)** Fig. 8-18(a); **(b)** Fig. 8-18(b); **(c)** Fig. 8-18(c).

8-19. Assume that an IA with a gain of 1000 is wired to the bridges of Problem 8-18. Find V_o for each of the three bridge arrangements.

8-20. Consider a gage factor of 2 in Problems 8-18 and 8-19 and calculate $\Delta L/L$ for each bridge arrangement.

CHAPTER 9

DC Performance: Bias, Offsets, and Drift

LEARNING OBJECTIVES

Upon completion of this chapter on dc performance, you will be able to:

- Name the op amp characteristics that add dc error components to the output voltage.
- Show how an op amp requires a small bias current at both (−) and (+) inputs to activate its internal transistors.
- Give the definition for input offset voltage and show how it is modeled in an op amp circuit.
- Write the equation for input offset current in terms of the bias currents.
- Calculate the effect of input offset voltage on the output voltage of either an inverting or noninverting amplifier.

- Calculate the effects of bias currents on the output voltage of an inverting or non-inverting amplifier.
- Calculate the value of and install a compensating resistor to minimize the errors in output voltage caused by bias currents.
- Connect a nulling circuit to null out any errors due to bias currents and input offset voltage.
- Measure offset voltage and bias currents.

9-0 INTRODUCTION

The op amp is widely used in amplifier circuits to amplify dc or ac signals or combinations of them. In dc amplifier applications, certain electrical characteristics of the op amp can cause large errors in the output voltage. The ideal output voltage should be equal to the product of the dc input signal and the amplifier's closed-loop voltage gain. However, the output voltage may have an added error component. This error is due to differences between an ideal op amp and a real op amp. If the ideal value of output voltage is large with respect to the error component, then we can usually ignore the op amp characteristic that causes it. But if the error component is comparable to or even larger than the ideal value, we must try to minimize the error. Op amp characteristics that add error components to the dc output voltage are:

1. Input bias currents
2. Input offset current
3. Input offset voltage
4. Drift

When the op amp is used in an ac amplifier, coupling capacitors eliminate dc output-voltage error. Therefore, characterisitics 1 to 4 above are often unimportant in ac applications. However, there are new problems for ac amplifiers. They are:

5. Frequency response
6. Slew rate

Frequency response refers to how voltage gain varies as frequency changes. The most convenient way to display such data is by a plot of voltage gain versus frequency. Op amp manufacturers give such a plot for open-loop gain versus frequency. A glance at the plot quickly shows how much gain is obtainable at a particular frequency.

If the op amp has sufficient gain at a particular frequency, there is still a possibility of an error being introduced in V_o. This is because there is a fundamental limit imposed by the op amp (and certain circuit capacitors) on *how fast the output voltage can change*. If the input signal tells the op amp output to change faster than it can, distortion is introduced in the output voltage. The op amp characteristic responsible for this type of error is its internal capacitance. This type of error is called *slew-rate limiting*.

Op amp characteristics and the circuit applications that each type of error *may* affect are summarized in Table 9-1. The first four characteristics can limit dc performance; the last two can limit ac performance.

Op amp characteristics that cause errors primarily in dc performance are studied in this chapter. Those that cause errors in ac performance are studied in Chapter 10. We begin with input bias currents and ways in which they cause errors in the dc output voltage of an op amp circuit.

TABLE 9-1 OP AMP APPLICATIONS AND CHARACTERISTICS THAT AFFECT OPERATION

Op-amp characteristic that may affect performance	Op-amp application: Dc amplifier, Small output	Op-amp application: Dc amplifier, Large output	Op-amp application: Ac amplilfier, Small output	Op-amp application: Ac amplilfier, Large output
1. Input bias current	Yes	Maybe	No	No
2. Offset current	Yes	Maybe	No	No
3. Input offset voltage	Yes	Maybe	No	No
4. Drift	Yes	No	No	No
5. Frequency response	No	No	Yes	Yes
6. Slew rate	No	Yes	No	Yes

9-1 INPUT BIAS CURRENTS

Transistors within the op amp must be *biased* correctly before any signal voltage is applied. Biasing correctly means that the transistor has the right value of base and collector current as well as collector-to-emitter voltage. Until now, we have considered that the input terminals of the op amp conduct no signal or bias current. This is the ideal condition. Practically, however, the input terminals do conduct a small value of dc current to bias the op amps' transistors (see Appendices 1 and 2). A simplified diagram of the op amp is shown in Fig. 9-1(a). To discuss the effect of input bias currents, it is convenient to model them as current sources in series with each input terminal, as shown in Fig. 9-1(b).

The (−) input's bias current, I_{B-}, will usually not be exactly equal to the (+) input's bias current, I_{B+}. Manufacturers specify an *average* input bias current I_B, which is found by adding the *magnitudes* of I_{B+} and I_{B-} and dividing this sum by 2. In equation form,

$$I_B = \frac{|I_{B+}| + |I_{B-}|}{2} \tag{9-1}$$

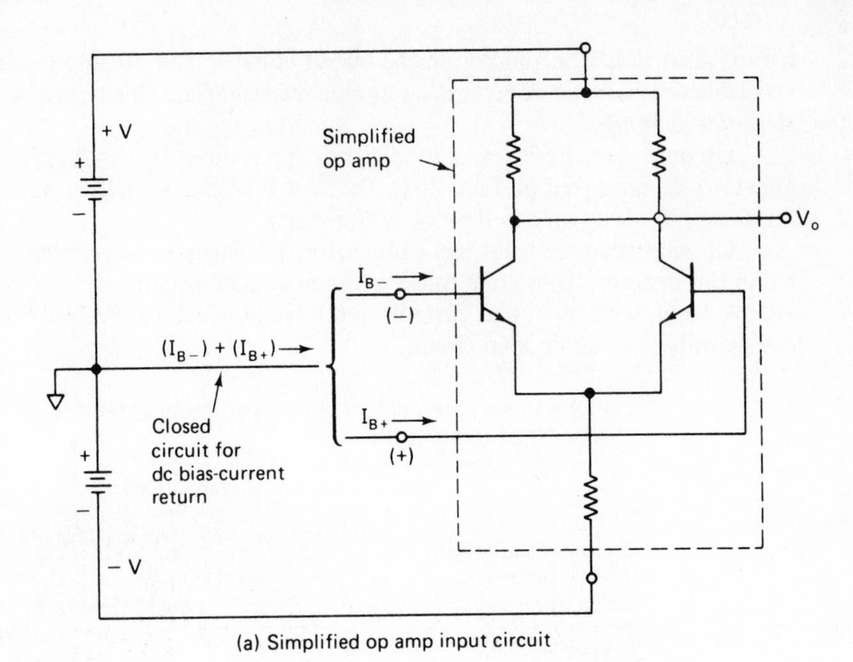

(a) Simplified op amp input circuit

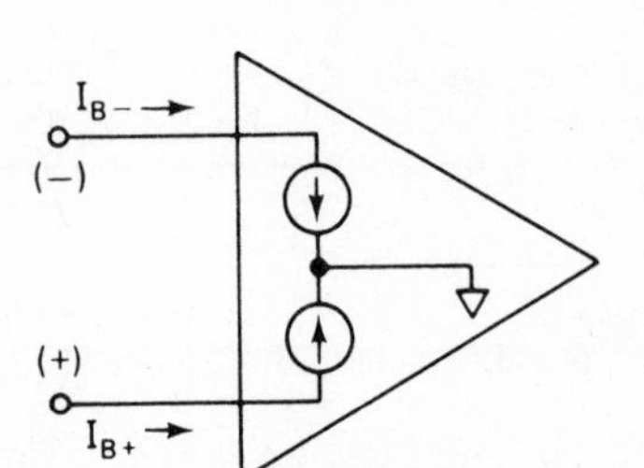

(b) Model for bias currents

FIGURE 9-1 Origin and model of dc input bias currents.

where $|I_{B+}|$ is the magnitude of I_{B+} and $|I_{B-}|$ is the magnitude of I_{B-}. The range of I_B is from 1 μA or more for general-purpose op amps to 1 pA or less for op amps that have field-effect transistors at the input.

9-2 INPUT OFFSET CURRENT

The difference in magnitudes between I_{B+} and I_{B-} is called the *input offset current* I_{os}:

$$I_{os} = |I_{B+}| - |I_{B-}| \tag{9-2}$$

Manufacturers specify I_{os} for a circuit condition where the output is at 0 V and the temperature is 25°C. The typical I_{os} is less than 25% of I_B, for the average input bias current (see Appendices 1 and 2).

Example 9-1

If $I_{B+} = 0.4\ \mu A$ and $I_{B-} = 0.3\ \mu A$, find (a) the average bias current I_B; (b) the offset current I_{os}.

Solution (a) By Eq. (9-1),

$$I_B = \frac{(0.4 + 0.3)\ \mu A}{2} = 0.35\ \mu A$$

(b) By Eq. (9-2),

$$I_{os} = (0.4 - 0.3)\ \mu A = 0.1\ \mu A$$

9-3 EFFECT OF BIAS CURRENTS ON OUTPUT VOLTAGE

9-3.1 Simplification

In this section it is assumed that bias currents are the only op amp characteristic that will cause an undesired component in the output voltage. The effects of other op amp characteristics on V_o will be dealt with individually.

9-3.2 Effect of (−) Input Bias Current

Output voltage should ideally equal 0 V in each circuit of Fig. 9-2, because input voltage E_i is 0 V. The fact that a voltage component will be measured is due strictly to I_{B-}. (Assume for simplicity that V_{io}, input offset voltage, is zero. V_{io} is discussed in Section 9-5.) In Fig. 9-2(a), the bias current is furnished from the output terminal. Since negative feedback forces the differential input voltage to 0 V, V_o must rise to supply the voltage drop across R_f. Thus, the output voltage error due to I_{B-} is found from $V_o = R_f I_{B-}$. I_{B+} flows through 0 Ω, so it causes no voltage error. Signal source E_i must contain a dc path to ground.

The circuit of Fig. 9-2(b) has the same output-voltage error expression, $V_o = R_f I_{B-}$. No current flows through R_i, because there is 0 V on each side of R_i. Thus all of I_{B-} flows through R_f. [Recall that an ideal amplifier with negative feedback has 0 voltage between the (+) and (−) inputs.]

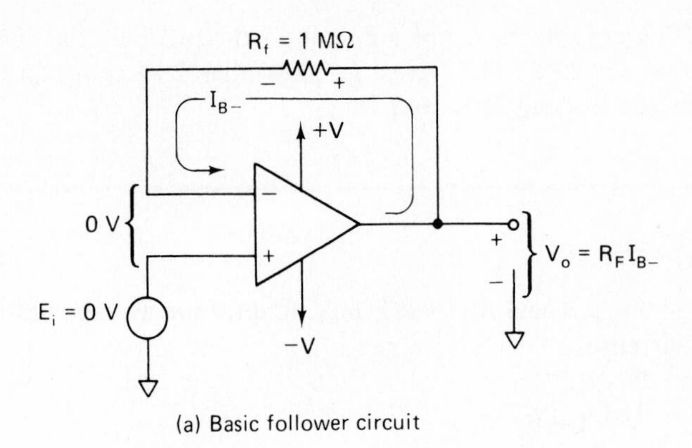

(a) Basic follower circuit

R_f = R_i = 1 MΩ

I = 0

$V_o = R_f I_{B-}$

(b) Basic inverting circuit

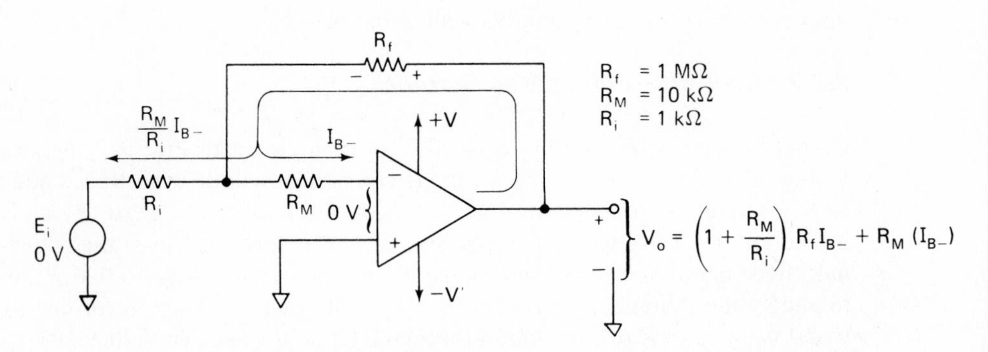

(c) Multiplier resistor R_M increases effect of I_{B-} on V_o

FIGURE 9-2 Effects of (−) input bias current on output voltages.

Example 9-2

In Fig. 9-2(a), $V_o = 0.4$ V. Find I_{B-}.

Solution

$$I_{B-} = \frac{V_o}{R_f} = \frac{0.4 \text{ V}}{1 \text{ M}\Omega} = 0.4 \ \mu\text{A}$$

Placing a multiplying resistor R_M in series with the (−) input in Fig. 9-2(c) multiplies the effect of I_{B-} on V_o. I_{B-} sets up a voltage drop across R_M that establishes an equal drop across R_i.

Now both the R_i current and I_{B-} must be furnished through R_f. Thus the error in V_o will be much larger. R_M would be undesirable in a normal circuit; however, if we want to measure low values of the bias current, Fig. 9-2(c) shows a way of doing it. For the resistor values shown, $V_o \simeq 11R_fI_{B-}$. I_{B-} acts to drive the output positive.

9-3.3 Effect of (+) Input Bias Current

Since $E_i = 0$ V in Fig. 9-3, V_o should ideally equal 0 V. However, the positive input bias current I_{B+} flows through the internal resistance of the signal generator. Internal generator resistance is modeled by resistor R_G in Fig. 9-3. I_{B+} sets up a voltage

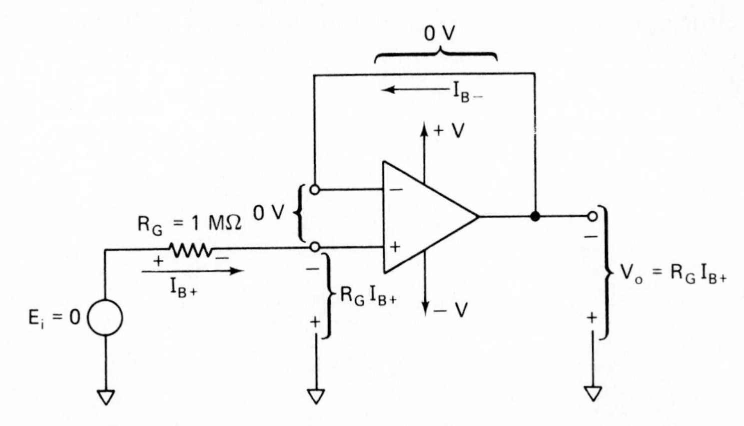

FIGURE 9-3 Effect of (+) input bias current on output voltages.

drop of R_GI_{B+} across R_G and applies it to the (+) input. The differential input voltage of 0 V, so the (−) input is also at R_GI_{B+} in Fig. 9-3. Since there is 0 resistance in the feedback loop, V_o equals R_GI_{B+}. (The return path for I_{B+} is through $-V$ supply and back to ground.) I_{B+} acts to drive the output negative.

Example 9-3

In Fig. 9-3, $V_o = -0.3$ V. Find I_{B+}.

Solution

$$I_{B+} = -\frac{V_o}{R_G} = -\frac{-0.3\text{ V}}{1\text{ M}\Omega} = 0.3\ \mu\text{A}$$

9-4 EFFECT OF OFFSET CURRENT ON OUTPUT VOLTAGE

9-4.1 Current Compensating the Voltage Follower

If I_{B+} and I_{B-} were always equal, it would be possible to compensate for their effects on V_o. For example, in the voltage follower of Fig. 9-4(a), I_{B+} flows through the signal generator resistance R_G. If we insert $R_f = R_G$ in the feedback loop, I_{B-} will develop a voltage drop across R_f of $R_f I_{B-}$. If $R_f = R_G$ and $I_{B+} = I_{B-}$, their voltage drops will cancel each other and V_o will equal 0 V when $E_i = 0$ V. Unfortunately, I_{B+} is seldom equal to I_{B-}. V_o will then be equal to R_G times the difference between I_{B+} and $I_{B-}(I_{B+} - I_{B-} = I_{os})$. Therefore, by making $R_f = R_G$, we have reduced the error in V_o from $R_G I_{B+}$ in Fig. 9-3 to $-R_G I_{os}$ in Fig. 9-4(a). Recall that I_{os} is typically 25% of I_B. If the value of I_{os} is too large, an op amp with a smaller value of I_{os} is needed.

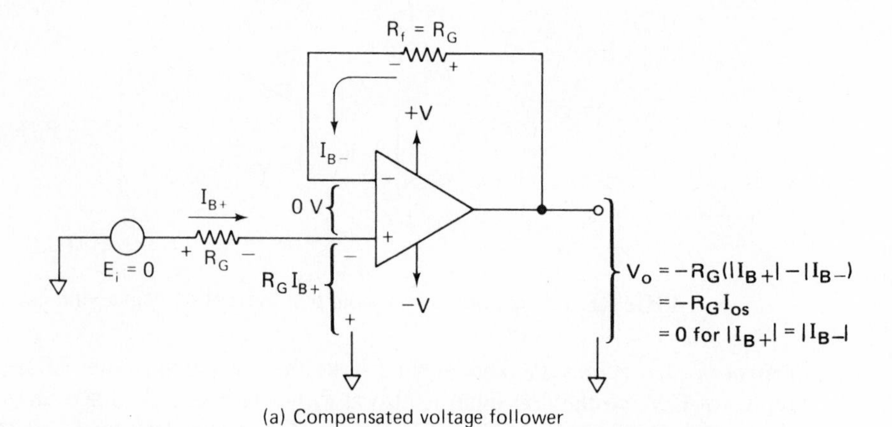

(a) Compensated voltage follower

FIGURE 9-4 Balancing-out effects of bias current in V_o.

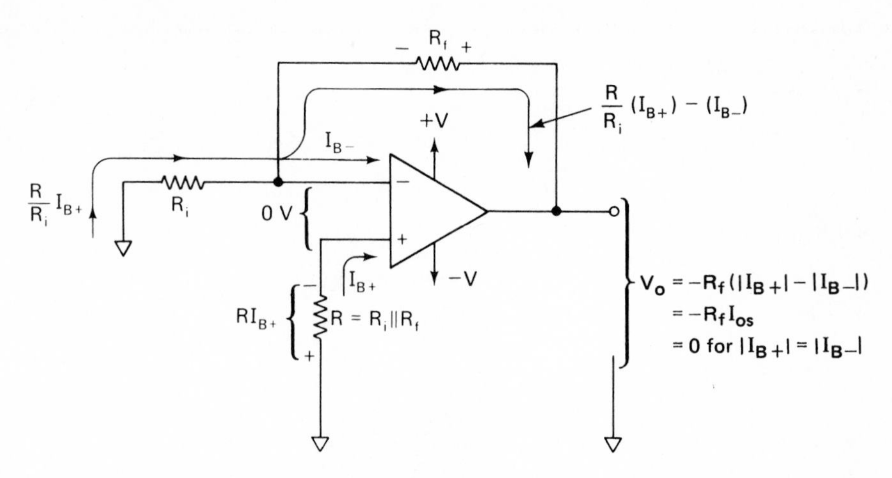

(b) Compensation for inverting or noninverting amplifiers

FIGURE 9-4 *(cont.)*

9-4.2 Current Compensating Other Amplifiers

To minimize errors in V_o due to bias currents for either inverting or non-inverting amplifiers, resistor R as shown in Fig. 9-4(b), must be added to the circuit. With no input signal applied, V_o depends on R_f times I_{os} [where I_{os} is given by Eq. (9-2)]. Resistor R is called the *current-compensating resistor* and is equal to the parallel combination of R_i and R_f, or

$$R = R_i \parallel R_f = \frac{R_i R_f}{R_i + R_f} \tag{9-3}$$

R_i and R should include any signal generator resistance. By inserting resistor R, the error voltage in V_o will be reduced more than 25% from $R_f I_{B-}$ in Fig. 9-2(b) to $-R_f I_{os}$ in Fig. 9-4(b). In the event that $I_{B-} = I_{B+}$, then $I_{os} = 0$ and $V_o = 0$.

9-4.3 Summary on Bias-Current Compensation

Always add a bias-current compensating resistor R in series with the (+) input terminal (except for FET input op amps). The value of R should equal the parallel combination of all resistance branches connected to the (−) terminal. Any internal resistance in the signal source should also be included in the calculations.

In circuits where more than a single resistor is connected to the (+) input, bias-current compensation is accomplished by observing the following principle. *The dc resistance seen from the (+) input to ground should equal the dc resistance seen from the (−) input to ground.* In applying this principle, signal sources are replaced by their internal dc resistance and the op amp output terminal is considered to be at ground potential.

Example 9-4

(a) In Fig. 9-4(b), $R_f = 100\ \text{k}\Omega$ and $R_i = 10\ \text{k}\Omega$. Find R. (b) If $R_f = 100\ \text{k}\Omega$ and $R_i = 100\ \text{k}\Omega$, find R.

Solution (a) By Eq. (9-3),

$$R = \frac{(100\ \text{k}\Omega)(10\ \text{k}\Omega)}{100\ \text{k}\Omega + 10\ \text{k}\Omega} = 9.1\ \text{k}\Omega$$

(b) By Eq. (9-3),

$$R = \frac{(100\ \text{k}\Omega)(100\ \text{k}\Omega)}{100\ \text{k}\Omega + 100\ \text{k}\Omega} = 50\ \text{k}\Omega$$

9-5 INPUT OFFSET VOLTAGE

9-5.1 Definition and Model

In Fig. 9-5(a), the output voltage V_o should equal 0 V. However, there will be a small error-voltage component present in V_o. Its value can range from microvolts to millivolts and is caused by very small but unavoidable imbalances inside the op amp. The easiest way to study the *net effect* of all these internal imbalances is to visualize a small dc voltage in *series* with one of the input terminals. This dc voltage

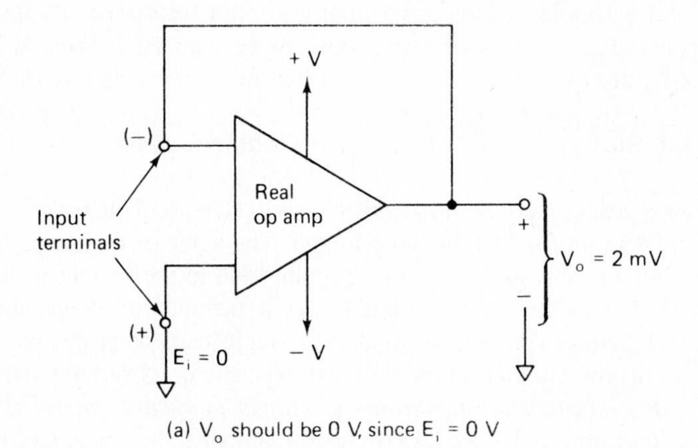

FIGURE 9-5 Effect of input offset voltage in the real op amp of (a) is modeled by an ideal op amp plus battery V_{io} in (b).

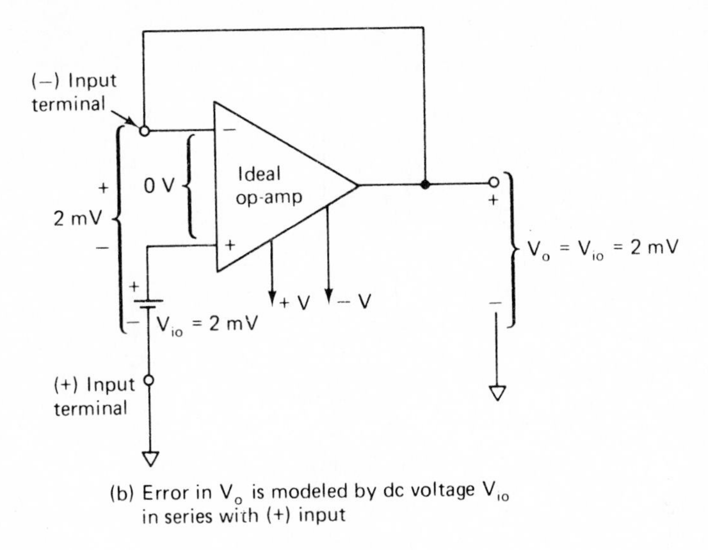

(b) Error in V_o is modeled by dc voltage V_{io} in series with (+) input

FIGURE 9-5 *(cont.)*

is modeled by a battery in Fig. 9-5(b) and is called *input offset voltage,* V_{io} (see Appendices 1 and 2 for typical values). Note that V_{io} is shown in series with the (+) input terminal of the op amp. It makes no difference whether V_{io} is modeled in series with the (−) input or the (+) input. But it is easier to determine the polarity of V_{io} if it is placed in series with the (+) input. For example, if the output terminal is positive (with respect to ground) in Fig. 9-5(b), V_{io} should be drawn with its (+) battery terminal connected to the ideal op amp's (+) input.

9-5.2 Effect of Input Offset Voltage on Output Voltage

Fig. 9-6(a) shows that V_{io} and the large value of the open-loop gain of the op amp act to drive V_o to negative saturation. Contrast the polarity of V_{io} in Figs. 9-5(b) and 9-6(a). If you buy several op amps and plug them into the test circuit of Fig. 9-6(a), some will drive V_o to $+V_{sat}$ and the remainder will drive V_o to $-V_{sat}$. Therefore, the magnitude and polarity of V_{io} varies from op amp to op amp. To learn how V_{io} affects amplifiers with negative feedback, we study how to measure V_{io}.

9-5.3 Measurement of Input Offset Voltage

For simplicity, the effects of bias currents are neglected in the following discussion. Figure 9-6(b) shows how to measure V_{io}. It also shows how to predict the magnitude of error that V_{io} will cause in the output voltage. Since $E_i = 0$ V, V_o should equal 0 V. But V_{io} acts exactly as would a signal in series with the noninverting input. Therefore, V_{io} is amplified exactly as any signal applied to the (+) input of a noninverting amplifier (see Section 3-6). The error in V_o due to V_{io} is given by

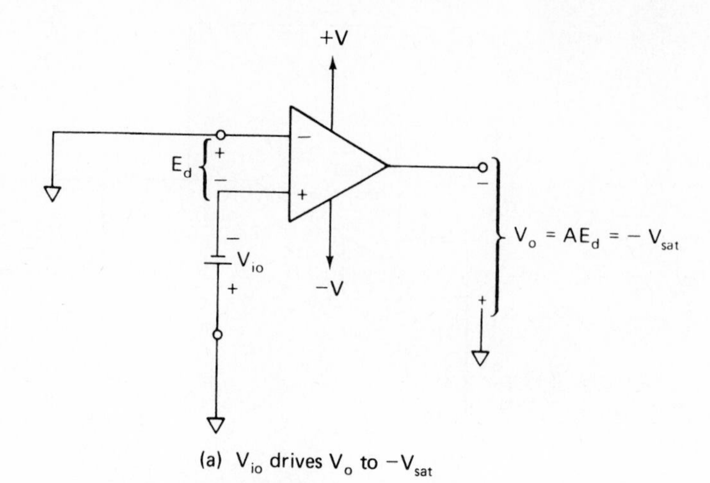

(a) V_{io} drives V_o to $-V_{sat}$

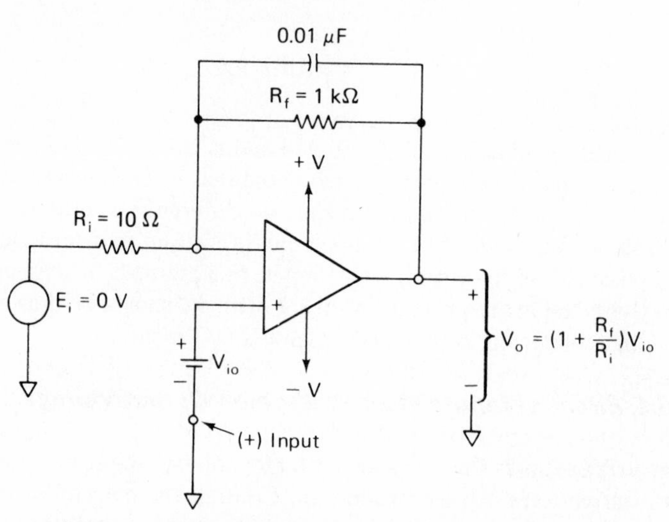

(b) V_{io} is amplified, causing a large error in V_o

FIGURE 9-6 V_o should be 0 V in (a) and (b) but contains a dc error voltage due to V_{io}. (Error component due to bias current is neglected.)

$$V_o = \text{error voltage due to } V_{io} = V_{io}\left(1 + \frac{R_f}{R_i}\right) \tag{9-4}$$

The output error voltage in Fig. 9-6(b) is given by Eq. (9-4) whether the circuit is used as an inverting or as a noninverting amplifier. That is, E_i could be inserted in series with R_i (inverting amplifier) for a gain of $-(R_f/R_i)$ or in series with the (+) input (noninverting amplifier) for a gain of $1 + (R_f/R_i)$. A bias-current compensating resistor (a resistor in series with the (+) input) has no effect on this type of error in the output voltage due to V_{io}.

Conclusion. To measure V_{io}, set up the circuit of Fig. 9-6(b). The capacitor is installed across R_f to minimize noise in V_o. Measure V_o, R_f and R_i. Calculate V_{io} from

$$V_{io} = \frac{V_o}{1 + R_f/R_i} \tag{9-5}$$

Note that R_f is made small to minimize the effect of input bias current.

Example 9-5

V_{io} is specified to be 1 mV for a 741-type op amp. Predict the value of V_o that would be measured in Fig. 9-6(b).

Solution From Eq. (9-5),

$$V_o = \left(1 + \frac{1000}{10}\right)(1 \text{ mV}) = 101 \text{ mV}$$

9-6 INPUT OFFSET VOLTAGE FOR THE ADDER CIRCUIT

9-6.1 Comparison of Signal Gain and Offset Voltage Gain

In both inverting and noninverting amplifier applications, the input offset voltage V_{io} is multiplied by $(1 + R_f/R_i)$. The input signal in either circuit is multiplied by a different gain. R_f/R_i is the gain for the inverter and $(1 + R_f/R_i)$ for the noninverter. In the inverting adder circuit of Fig. 9-7(a) (neglecting bias currents), V_{io} is multiplied by a larger number than the signal at each input.

For example, in Fig. 9-7(a) signals E_1 and E_2 are each larger than V_{io} but E_1 is multiplied by $-R_f/R_1 = -1$ and develops a component of -5 mV in V_o. E_2 is likewise multiplied by -1 and adds a -5-mV component to V_o. Thus the correct value of V_o should be -10 mV. Since E_3 is 0 its contribution to V_o is 0 (see Section 3-2).

If we temporarily let E_1 and $E_2 = 0$ V in Fig. 9-7(a), the $(-)$ input sees three equal resistors forming parallel paths to ground. The single equivalent series resistance, R_i, is shown in Fig. 9-7(b). For three equal 10-kΩ resistors in parallel, the equivalent resistance R_i is found by 10 kΩ/3 = 3.33 kΩ. V_{io} is amplified just as in Fig. 9-6(b) to give an output error of $+10$ mV. Therefore, the total output voltage in Fig. 9-7(a) is 0 instead of -10 mV.

Conclusions. In an adder circuit, the input offset voltage has a gain of *1 plus the number of inputs*. The more inputs, the greater the error component in the

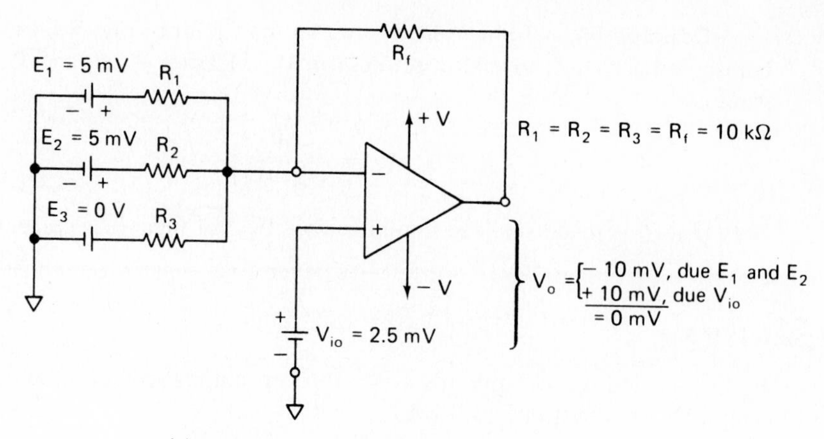

(a) V_o has a -10 mV component due to E_1 and E_2 plus a 10 mV error component due to V_{io}

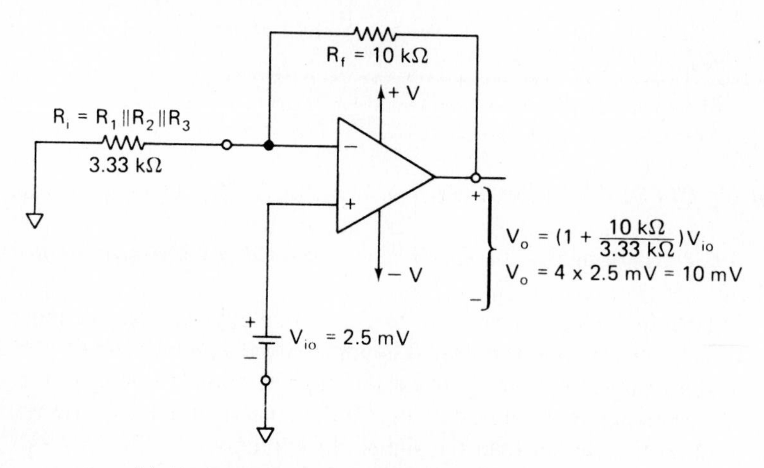

(b) V_{io} is multiplied by a gain of 4 to generate a 10 mV error component in V_o

FIGURE 9-7 Each input voltage of the inverting adder in (a) is multiplied by a gain of -1. V_{io} is multiplied by a gain of $+4$.

output voltage. Since the gain for the inputs is -1, *the offset voltage gain always exceeds the signal voltage gain.*

9-6.2 How Not to Eliminate the Effects of Offset Voltage

One might be tempted to add an adder input such as E_3 in Fig. 9-7(a) to balance out the effect of V_{io}. For example, if E_3 is made equal to 10 mV, then E_3, R_3, and R_f will add a -10 mV component to V_o and balance out the $+10$ mV due to V_{io}. There are two disadvantages to this approach. First, such a small value of E_3 would have to be obtained from a resistor-divider network between the power supply terminals of $+V$ and $-V$. The second disadvantage is that any resistance added between the $(-)$ input and ground raises the *noise gain*. This situation is treated in Sections 10-5.3 and 10-5.4. In Section 9-7 we show how to minimize the output voltage errors caused by both bias currents and input offset voltage.

9-7 NULLING-OUT EFFECT OF OFFSET VOLTAGE AND BIAS CURRENTS

9-7.1 Design or Analysis Sequence

To minimize dc error voltages in the output voltage, follow this sequence:

1. Select a bias-current compensating resistor in accordance with the principles set forth in Section 9-4.3.
2. Get a circuit for minimizing effects of the input offset voltage from the op amp manufacturer's data sheet. This principle is treated in more detail in Section 9-7.2 and in Appendices 1 and 2.
3. Go through the output-voltage nulling procedure given in Section 9-7.3.

9-7.2 Null Circuits for Offset Voltage

It is possible to imagine a fairly complex resistor-divider network that would inject a small variable voltage into the $(+)$ or $(-)$ input terminal. This would compensate for the effects of *both* input offset voltage and offset current. However, the extra components are more costly and bulky than necessary. It is far better to go to the op amp manufacturer for guidance. The data sheet for your op amp will have a *voltage offset null circuit* recommended by the manufacturer. The op amp will have *null* terminals brought out for connection to the null circuit. Experts have designed the null circuit to minimize offset errors at the lowest cost to the user (see Appendices 1 and 2).

Some typical output-voltage null circuits are shown in Fig. 9-8. In Fig. 9-8(a), one variable resistor is connected between the $+V$ supply and a *trim* terminal. For an expensive op amp, the manufacturer may furnish a metal film resistor selected especially for that op amp. In Fig. 9-8(b), a 10-kΩ pot is connected between terminals called *offset null*. More complicated null circuits are shown in Fig. 9-8(c) and (d). Note that only the offset-voltage compensating resistors are shown by the manufacturer. They assume that a current-compensating resistor will be installed in series with the $(+)$ input.

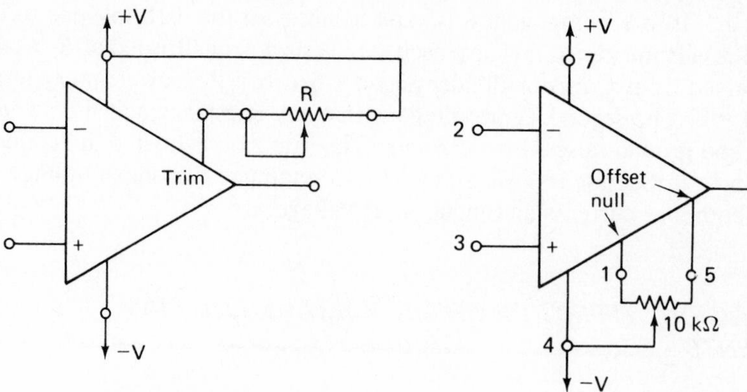

(a) Trim resistor 0 to 50 kΩ or 25 kΩ fixed for discrete op amp

(b) 741 Offset voltage adjustment (minidip)

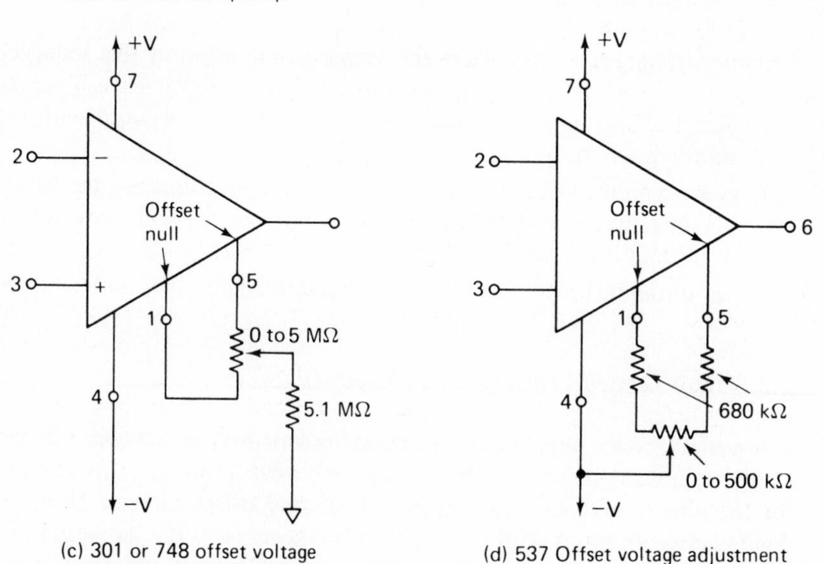

(c) 301 or 748 offset voltage adjustment (TO 99 case)

(d) 537 Offset voltage adjustment

FIGURE 9-8 Typical circuits to minimize errors in output voltage due to input offset voltage and offset current (see also Appendices 1 and 2).

9-7.3 Nulling Procedure for Output Voltage

1. Build the circuit. Include (a) the current-compensating resistor (see Section 9-4.3) and (b) the voltage offset null circuit (see Section 9-7.2).
2. Reduce all generator signals to 0. If their output cannot be set to 0, replace them with resistors equal to their internal resistance. This step is unnecessary if their internal resistance is negligible with respect to (more than about 1% of) any series resistor R_i connected to the generator.

3. Connect the load to the output terminal.
4. Turn on the power and wait a few minutes for things to settle down.
5. Connect a dc voltmeter or a CRO (dc coupled) across the load to measure V_o. (The voltage sensitivity should be capable of reading down to a few millivolts.)
6. Vary the offset voltage adjustment resistor until V_o reads 0 V. Note that output voltage errors due to *both* input offset voltage and input offset current are now minimized.
7. Install the signal sources and do *not* touch the offset-voltage adjustment resistor again.

9-8 DRIFT

It has been shown in this chapter that dc error components in V_o can be minimized by installing a current-compensating resistor in series with the (+) input and by trimming the offset-voltage adjustment resistor. It must also be emphasized that the zeroing procedure holds only at one temperature and at one time.

The offset current and offset voltage change with time because of aging of components. The offsets will also be changed by temperature changes in the op amp. In addition, if the supply voltage changes, bias currents, and consequently the offset current, change. By use of a well-regulated power supply, the output changes that depend on supply voltage can be eliminated. However, the offset changes with temperature can only be minimized by (1) holding the temperature surrounding the circuit constant, or (2) selecting op amps with offset current and offset voltage ratings that change very little with temperature changes.

The changes in offset current and offset voltage due to temperature are described by the term *drift*. Drift is specified for offset current in nA/°C (nanoamperes per degree Celsius). For offset voltage, drift is specified in μV/°C (microvolts per degree Celsius). Drift rates may differ at different temperatures and may even reverse; that is, at low temperatures V_{io} may drift by +20 μV/°C (increase), and at high temperatures V_{io} may change by −10 μV/°C (decrease). For this reason, manufacturers may specify either an average or maximum drift between two temperature limits. Even better is to have a plot of drift vs. temperature. An example is shown to calculate the effects of drift.

Example 9-6

A 301 op amp in the circuit of Fig. 9-9 has the following drift specifications. As temperature changes from 25°C to 75°C, I_{os} changes by a *maximum* of 0.3 nA/°C and V_{io} changes by a *maximum* of 30 μV/°C. Assume that V_o has been zeroed at 25°C; then the surrounding temperature is raised to 75°C. Find the maximum error in output voltage due to drift in (a) V_{io}; (b) I_{os}.

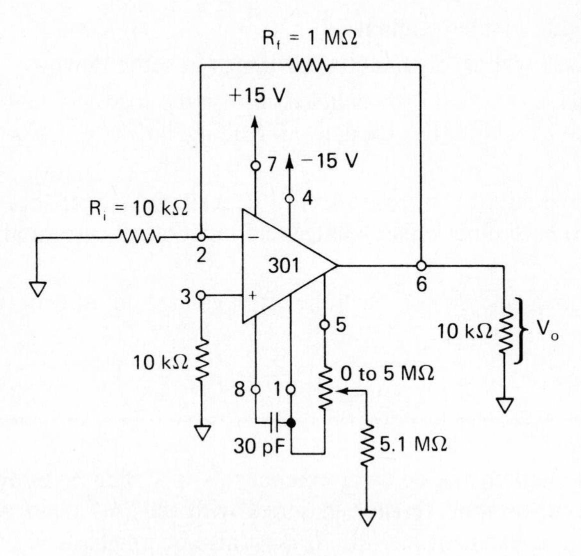

FIGURE 9-9 Circuit for Example 9-6.

Solution (a) V_{io} will change by

$$\pm\frac{30\ \mu\text{V}}{°C} \times (75 - 25)°\text{C} = \pm 1.5 \text{ mV}$$

From Eq. (9-4), the change in V_o due to the change in V_{io} is

$$1.5 \text{ mV}\left(1 + \frac{R_f}{R_i}\right) = 1.5 \text{ mV}(101) \simeq \pm 150 \text{ mV}$$

(b) I_{os} will change by

$$\pm\frac{0.3 \text{ nA}}{°\text{C}} \times 50°\text{C} = \pm 15 \text{ nA}$$

From Section 9-4, the change in V_o due to the change in I_{os} is ± 15 nA $\times$ $R_f = \pm 15$ nA(1 MΩ) $= \pm 15$ mV.

The changes in V_o due to both V_{io} and I_{os} can either add or subtract from one another. Therefore, the worst possible change in V_o is either $+165$ mV or -165 mV, from the 0 value at 25°C.

9-9 MEASUREMENT OF OFFSET VOLTAGE AND BIAS CURRENTS

The effects of offset voltage and bias currents have been discussed separately to simplify the problem of understanding how error voltage components appear in the dc output voltage of an op amp. However, their effects are *always* present simultaneously.

In order to measure V_{io}, I_{B+}, and I_{B-} of general-purpose op amps as inexpensively as possible, the following procedure is recommended.

1. As shown in Fig. 9-10(a), measure V_o with a digital voltmeter and calculate input offset voltage V_{io}:

$$V_{io} = \frac{V_o}{(R_f + R_i)/R_i} = \frac{V_o}{101} \tag{9-6}$$

   Note that R_i and R_f are small. Therefore, by adding the 50-Ω current-compensation resistor, we force the output voltage error component due to I_{os} to be negligible.

2. To measure I_{B-}, set up the circuit of Fig. 9-10(b). Measure V_o. Using the value of V_{io} found in step 1, calculate I_{B-} from

$$I_{B-} = \frac{V_o - V_{io}}{R_f} \tag{9-7}$$

3. To measure I_{B+}, measure V_o in Fig. 9-10(c) and calculate I_{B+} from

$$I_{B+} = -\left(\frac{V_o - V_{io}}{R_f}\right) \tag{9-8}$$

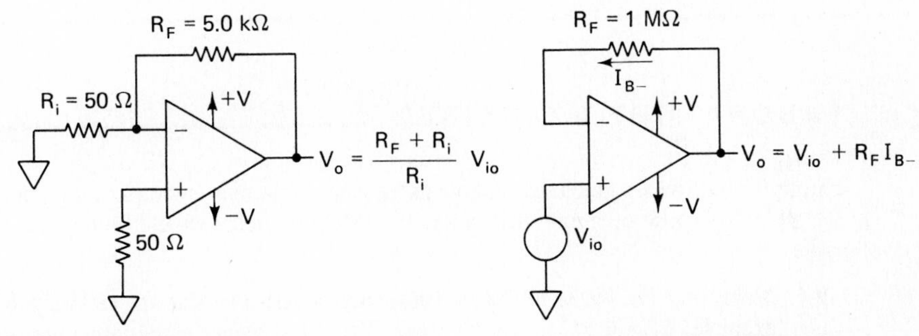

(a) Circuit to measure V_{io}; effect of I_{os} is minimized

(b) Circuit to measure I_{B-}

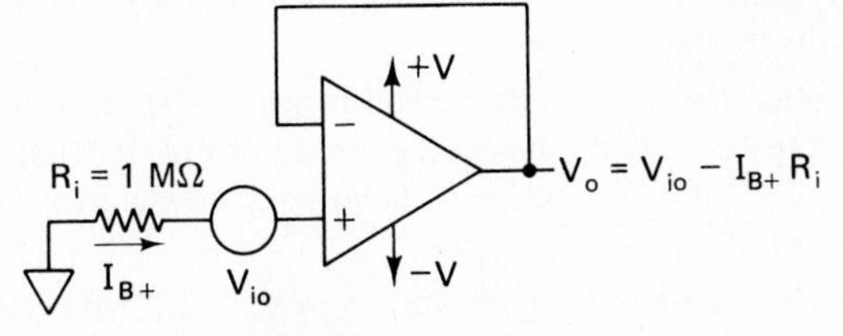

(c) Circuit to measure I_{B+}

FIGURE 9-10 Procedure to measure offset voltage, then bias currents for a general-purpose op amp.

Example 9-7

The circuit of Fig. 9-10 is used with the resistance values shown for a 741 op amp. Results are $V_o = +0.421$ V for Fig. 9-10(a), $V_o = 0.097$ V for Fig. 9-10(b), and $V_o = -0.082$ V for Fig. 9-10(c). Find (a) V_{io}; (b) I_{B-}; (c) I_{B+}.

Solution (a) From Eq. (9-6),

$$V_{io} = \frac{0.421 \text{ V}}{101} = 4.1 \text{ mV}$$

(b) From Eq. (9-7),

$$I_{B-} = \frac{(97 - 4.1) \text{ mV}}{1 \text{ M}\Omega} = 93 \text{ nA}$$

(c) From Eq. (9-8),

$$I_{B+} = -\frac{(-82 - 4.1) \text{ mV}}{1 \text{ M}\Omega} = 86 \text{ nA}$$

Note that I_{os} is found to be $I_{B+} - I_{B-} = -7$ nA.

LABORATORY EXERCISES

Figure 9-10 provides guidance on how to measure input bias currents and input offset voltage for general-purpose op amps such as the 741. Measurements should be taken in the following order:

9-1. *Measuring V_{io}.* Refer to Fig. 9-10(a). Select resistor values to yield a gain of 100. For example, $R_f = 1$ kΩ and $R_i = 10$ Ω. These low values essentially eliminate the output voltage errors due to bias currents. Measure V_o and calculate V_{io} from Eq. (9-6). If the dc value of V_o appears to change inexplicably, (1) connect a 0.1-μF bypass capacitor from each power supply terminal to ground; and (2) if you are breadboarding, keep all lead lengths as short as possible.

9-2. *Measuring I_{B-} and I_{B+}.* In the circuit of Fig. 9-10(b), with $R_f = 2.2$ MΩ, measure V_o and calculate I_{B-}. Then connect the circuit of Fig. 9-10(c) and measure V_o to calculate I_{B+} ($R_f = 2.2$ MΩ). Calculate I_{os} from your measured values of I_{B+} and I_{B-} [see Eq. (9-2)].

9-3. *Measuring I_{os}.* Refer to Fig. 9-4(a). Let $R_f = 2.2$ MΩ and install a 2.2 MΩ resistor from (+) input to ground. Measure V_o and calculate I_{os} from $I_{os} = -V_o/2.2$ MΩ. Your results may not agree with I_{os} calculated in Laboratory Exercise 9-2. This is because I_{B+} and I_{B-} are almost equal in modern op amps. Their difference is very small and therefore I_{os} found from Laboratory Exercise 9-3 will be inaccurate.

PROBLEMS

9-1. Which op amp characteristics normally have the most effect on **(a)** dc amplifier performance; **(b)** ac amplifier performance?

9-2. If $I_{B+} = 0.2\ \mu A$ and $I_{B-} = 0.1\ \mu A$, find **(a)** the average bias current I_B; **(b)** the offset current I_{os}.

9-3. In Example 9-2, $V_o = 0.2$ V. Find I_{B-}.

9-4. In Example 9-3, $V_o = -0.2$ V. Find I_{B+}.

9-5. I_{B-} is 0.2 μA in Fig. 9-2(c). Find V_o.

9-6. In Fig. 9-4(a), $R_f = R_G = 100\ k\Omega$. $I_{B+} = 0.3\ \mu A$ and $I_{B-} = 0.2\ \mu A$. Find V_o.

9-7. In Fig. 9-4(b), $R_f = R_i = 25\ k\Omega$ and $R = 12.5\ k\Omega$. If $I_{os} = 0.1\ \mu A$, find V_o.

9-8. In Fig. 9-4(b), $R_i = R_f = 25\ k\Omega$ and $R = 12.5\ k\Omega$. If $I_{os} = -0.1\ \mu A$, find V_o.

9-9. $V_o = 200$ mV in Fig. 9-6(b). Find V_{io}.

9-10. Resistors R_1, R_2, R_3, and R_f all equal 20 kΩ in Fig. 9-7(a). $E_1 = E_2 = E_3 = V_{io} =$ 2 mV. Find **(a)** the actual value of V_o; **(b)** V_o assuming that $V_{io} = 0$.

9-11. What value of current-compensating resistor should be added in Problem 9-10?

9-12. What is the general procedure to null the dc output voltage of an op amp to 0 V?

9-13. In Fig. 9-9, V_{io} changes by ±1 mV when the temperature changes by 50°C. What is the change in V_o due to the change in V_{io}?

9-14. I_{os} changes by ±20 nA in Fig. 9-9 for a temperature change of 50°C. What is the resulting change in V_o?

9-15. $V_o = 101$ mV in the circuit of Fig. 9-10(a), $V_o = 201$ mV in Fig. 9-10(b), and $V_o = -99$ mV in Fig. 9-10(c). Find **(a)** V_{io}; **(b)** I_{B-}; **(c)** I_{B+}.

9-16. Refer to Fig. P9-16. $V_{io} = 3$ mV, $I_{B-} = 0.4\ \mu A$, and $I_{B+} = 0.1\ \mu A$. **(a)** What is the best value for resistor R? Calculate the individual error in the output voltage, V_o, due to **(b)** V_{io} only; **(c)** I_{B+} only; **(d)** I_{B-} only; **(e)** I_{os} only. The ideal value of V_o should be 1.00 V because of E_i. **(f)** What is the actual value of V_o when both input offset voltage and current are present along with E_i?

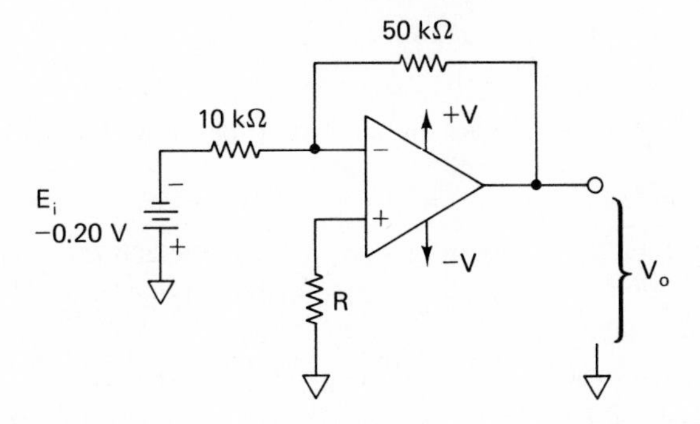

FIGURE P9-16

CHAPTER 10

AC Performance: Bandwidth, Slew Rate, Noise, and Frequency Compensation

LEARNING OBJECTIVES

Upon completion of this chapter on ac performance of an op amp, you will be able to:

- Recognize an op amp's frequency response graph in the manufacturer's data sheet and from it determine (1) dc open-loop gain A_{OL}, (2) small-signal unity-gain bandwidth B, and (3) read the magnitude of A_{OL} at any frequency.
- Calculate the unity-gain bandwidth if rise time is given, and vice versa.
- Predict the open-loop gain of an op amp at any frequency if you know the unity-gain bandwidth.
- Measure the rise time.
- Show how closed loop gain, A_{CL}, of either an inverting or noninverting amplifier depends on open-loop gain, A_{OL}.

- Measure the frequency response of an inverting or noninverting amplifier.
- Predict the bandwidth or upper cutoff frequency for an inverting amplifier if you know the external resistor values and the op amp's small-signal unity.
- Calculate the maximum sinusoidal frequency that can be obtained from an op amp at a given peak output voltage if you know its slew rate.
- Calculate the maximum peak output voltage at any given sine frequency if the op amp's slew rate is known.
- Install a frequency compensating capacitor on an op amp if it is not internally compensated like the 741.

10.0 INTRODUCTION

When the op amp is used in a circuit that amplifies only ac signals, we must consider whether ac output voltages will be small signals (below about 1 V peak) or large signals (above 1 V peak). If only small ac *output* signals are present, the important op amp characteristics that limit performance are *noise* and *frequency response*. If large ac output signals are expected, then an op amp characteristic called *slew-rate limiting* determines whether distortion will be introduced by the op amp, and may further limit frequency response.

Bias currents and offset voltages affect dc performance and usually do not have to be considered with respect to ac performance. This is true especially if a coupling capacitor is in the circuit to pass ac signals and block dc currents and voltages. We begin with an introduction to the frequency response of an op amp.

10-1 FREQUENCY RESPONSE OF THE OP AMP

10-1.1 Internal Frequency Compensation

Many types of general-purpose op amps and specialized op amps are *internally compensated;* that is, the manufacturer has installed within such op amps a small capacitor, usually 30 pF. This *internal frequency compensation capacitor* prevents the op amp from oscillating at high frequecies. Oscillations are prevented by decreasing the op amp's gain as frequency increases. Otherwise, there would be sufficient gain and phase shift at some high frequency where enough output signal could be fed back to the input and cause oscillations (see Appendix 1).

From basic circuit theory it is known that the reactance of a capacitor goes down as frequency goes up: $X_C = 1/(2\pi fC)$. For example, if the frequency is increased by 10, the capacitor reactance decreases by 10. Thus, it is no accident that the voltage gain of an op amp goes down by 10 as the frequency of the input signal is increased by 10. A change in frequency of 10 is called a *decade*. Manufacturers show how the open-loop gain of the op amp is related to the frequency of the differential input signal by a curve called *open-loop voltage gain versus frequency*. The curve may also be called *small-signal response*.

10-1.2 Frequency-Response Curve

A typical curve is shown in Fig. 10-1 for internally compensated op amps such as the 741 and 747. At low frequencies (below 0.1 Hz), the open-loop voltage gain is very high. A typical value is 200,000 (106 dB), and it is this value that is specified on data sheets where a curve is not given. See "Large-Signal Voltage Gain" equals 200,000 in Appendix 1 and 160 V/mV in Appendix 2.

Point *A* in Fig. 10-1 locates the *break frequency* where the open-loop voltage gain of the op amp is 0.707 times its value at very low frequencies. Therefore, the voltage gain at point *A* (where the frequency of E_d is 5 Hz) is about 140,000, or 0.707 × 200,000.

Points *C* and *D* show how gain drops by a factor of 10 as frequency rises by a factor of 10. Changing frequency or gain by a factor of 10 is expressed more

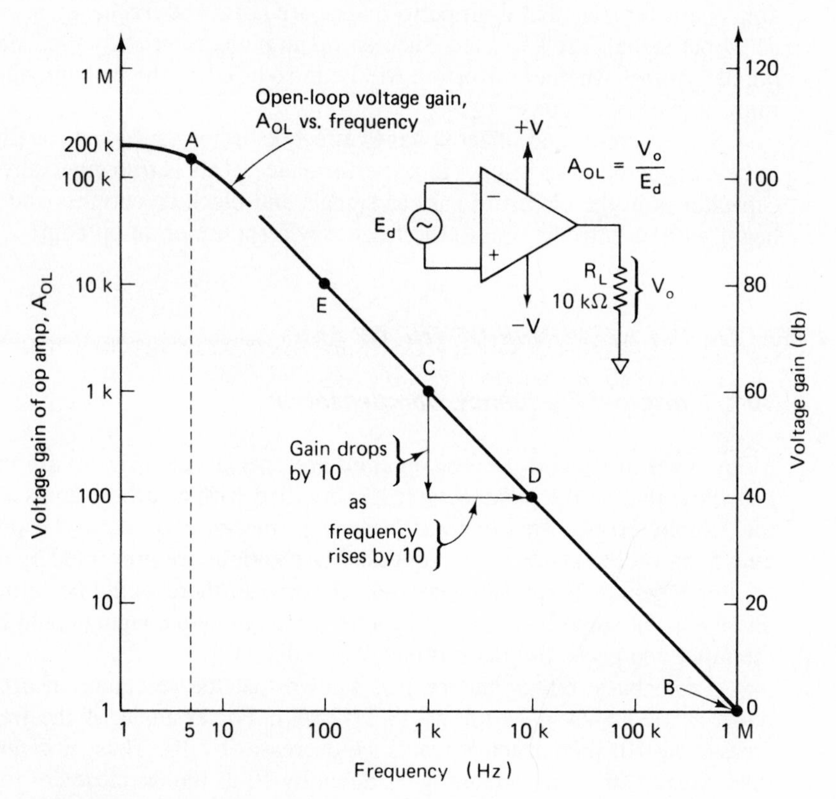

FIGURE 10-1 Open-loop voltage gain of a 741 op amp versus frequency.

efficiently by the term *per decade* ("decade" signifies 10). The right-hand vertical axis of Fig. 10-1 is a plot of voltage gain in decibels (dB). The voltage gain decreases by 20 dB for an increase in frequency of 1 decade. This explains why the frequency-response curve from A to B is described as *rolling off at* 20 *dB/decade*. An alternative description is 6 *dB/octave rolloff* ("octave" signifies a frequency change of 2). Therefore, each time the frequency doubles, the voltage gain decreases by 6 dB.

10-1.3 Unity-Gain Bandwidth

When an amplifier is made from an op amp and a few resistors, the frequency response of the amplifier depends on the frequency response of the op amp. The key op amp characteristic is defined as that frequency where the op amp's gain equals unity. We will use the symbol B for this op amp characteristic. It is called *small-signal unity-gain bandwidth*. Later in this chapter we will need a value for B of the op amp to predict the high-frequency response of an amplifier constructed with this op amp.

Three ways to obtain B from a manufacturer's data sheet are presented in this section. First, if you have the manufacturer's plot of A_{OL} versus frequency, look for that frequency where $A_{OL} = 1$ (see point B in Fig. 10-1, $B = 1$ MHz). Second, some data sheets may not give a specification called unity-gain bandwidth or a curve like Fig. 10-1. Instead, they give a specification called *transient response rise time (unity gain)*. For a 741 op amp it is typically 0.25 μs and 0.8 μs at maximum. The bandwidth B is calculated from the rise-time specification by

$$B = \frac{0.35}{\text{rise time}} \tag{10-1}$$

where B is in hertz and rise time is in seconds. Rise time is defined in Section 10-1.4. (See Appendix 1, "Transient Response—Unity Gain" = 0.3 μs typical.)

Example 10-1

A 741 op amp has a rise time of 0.35 μs. Find the small-signal or unity-gain bandwidth.

Solution From Eq. (10-1),

$$B = \frac{0.35}{0.35\ \mu\text{s}} = 1\ \text{MHz}$$

Example 10-2

What is the open-loop voltage gain for the op amp of Example 10-1 at 1 MHz?

Solution From the definition of B, the voltage gain is 1.

Example 10-3

What is the open-loop voltage gain at 100 kHz for the op amp in Examples 10-1 and 10-2?

Solution By inspection of Fig. 10-1, if the frequency goes down by 10, the gain goes up by 10. Therefore, since the frequency goes down a decade (from 1 MHz to 100 kHz), the gain must go up by a decade from 1 at 1 MHz to 10 at 100 kHz.

Example 10-3 leads to the conclusion that if you divide the frequency of the signal, f, into the unity-gain bandwidth, B, the result is the op amp's gain at the signal frequency. Expressed mathematically,

$$\text{open-loop gain at } f = \frac{\text{bandwidth at unity gain}}{\text{input signal frequency, } f} \tag{10-2}$$

Example 10-4

What is the open-loop gain of an op amp that has a unity-gain bandwidth of 1.5 MHz for a signal of 1 kHz?

Solution From Eq. (10-2), the open-loop gain at 1 kHz is

$$\frac{1.5 \text{ MHZ}}{1 \text{ kHz}} = 1500$$

Equation (10-2) gives a third way to find B. If you know the op amp's open-loop gain at one frequency (in the roll-off region), simply multiply the two values to obtain B. Let's consider Example 10-4 again. If $A_{OL} = 1000$ at a frequency of 1500 Hz, then $B = 1500 \times 1000 = 1.5$ MHz.

The data shown in Fig. 10-1 are useful for learning but may probably not apply to your op amp. For example, while 200,000 is a specified typical open-loop gain, the manufacturer guarantees only a minimum gain of 20,000 for general-purpose op amps. Still, 20,000 may be enough to do the job. Section 10-2 deals with this question.

10-1.4 Rise Time

Assume that the input voltage E_i of a unity-gain amplifier is changed very rapidly by a square wave or pulse signal. Ideally, E_i should be changed from 0 V + 20 mV in 0 time; practically, a few nano-seconds are required to make this change (see Appendix 1, "Transient Response Curve"). At unity gain, the output should change from 0 to +20 mV in the same few nanoseconds. However, it takes time for the signal to propagate through all the transistors in the op amp. It also takes time for the output voltage to rise to its final value. *Rise time* is defined as the time required for the output voltage to rise from 10% of its final value to 90% of its final value. From Section 10-1.3, the rise time of a 741 is 0.35 μs. Therefore, it would take 0.35 μs for the output voltage to change from 2 mV to 18 mV.

10-2 AMPLIFIER GAIN AND FREQUENCY RESPONSE

10-2.1 Effect of Open-Loop Gain on Closed-Loop Gain of an Amplifier, DC Operation

It is necessary to learn how open-loop gain A_{OL} affects the actual closed-loop gain of an amplifier with dc signal voltages (zero frequency). First, we must define *ideal* closed-loop gain of an amplifier as that gain which should be determined only by external resistors. However, the *actual* dc closed loop of an amplifier is determined by *both* the external resistors and open-loop gain of an op amp.

The actual dc closed-loop gain of a *noninverting* amplifier is

$$\text{actual } A_{CL} = \frac{(R_f + R_i)/R_i}{1 + \dfrac{1}{A_{OL}}\left(\dfrac{R_f + R_i}{R_i}\right)} \tag{10-3a}$$

where

$$\frac{R_f + R_i}{R_i} = \text{ideal } A_{CL} \text{ for } \textit{noninverting} \text{ amplifiers} \tag{10-3b}$$

If A_{OL} is very large, the denominator of Eq. (10-3a) approaches unity. Then the amplifier gain will *not* depend on the open-loop gain of the op amp but rather only on the external resistors and can be calculated from Eq. (10-3b).

The actual dc gain of an *inverting amplifier* depends on A_{OL} according to

$$\text{actual } A_{CL} = \frac{-R_f/R_i}{1 + \dfrac{1}{A_{OL}}\left(\dfrac{R_f + R_i}{R_i}\right)} \tag{10-3c}$$

where

$$\frac{-R_f}{R_i} = \text{ideal } A_{CL} \text{ for } \textit{inverting} \text{ amplifiers} \tag{10-3d}$$

Equation (10-3d) is valid if A_{OL} is large with respect to $(R_f + R_i)/R_i$.

Example 10-5

Find the actual gain for a dc noninverting amplifier if ideal $A_{CL} = 100$ and A_{OL} is (a) 10,000; (b) 1000; (c) 100; (d) 10; (e) 1. Repeat for a dc inverting amplifier with an ideal gain of -100.

Solution (a) For the noninverting amplifier: $(R_f + R_i)/R_i = 100 =$ ideal gain. From Eq. (10-3a),

$$\text{actual } A_{CL} = \frac{100}{1 + \left(\dfrac{1}{10{,}000}\right)100} = 99.0099$$

For the inverting amplifier, $R_f/R_i = |\text{ideal gain}| = 100$. Therefore, $(R_f + R_i)/R_i = 101$. From Eq. (10-3c),

$$\text{actual } A_{CL} = \frac{-100}{1 + \left(\dfrac{1}{10{,}000}\right)101} = -99.0000$$

If these steps are repeated for parts (b) through (e), the results may be tabulated as follows:

	A_{OL}					
	1	10	100	10^3	10^4	10^5
Actual A_{CL}, noninverting	0.99	9.09	50	90.9	99.0	99.9
Actual A_{CL}, inverting	−0.98	−9.01	−49.7	−90.8	−99.0	−99.9

The results of Example 10-5 are shown by the plot of A_{CL} versus A_{OL} in Fig. 10-2. There are two important lessons to be learned from Example 10-5 and Fig. 10-2. First the actual gains of both noninverting and inverting amplifiers are of approximately the same magnitudes for the same value of open-loop gain. Second, we would like the actual closed-loop gain to be equal to the ideal closed-loop gain. An examination of Eqs. (10-3a) and (10-3c) shows that this will be true *if* the open-loop gain of the op amp A_{OL} is large with respect to the ideal closed-loop gain of the op amp. Practically, we would like A_{OL} to be 100 or more times the ideal A_{CL}, so that the external precision resistors and *not* the op amp's A_{OL} determine the actual gain, to within 1%.

We already learned in Section 10-1.2 that A_{OL} depends on frequency. Since A_{OL} of the op amp also determines A_{CL} of an amplifier, then A_{CL} of the amplifier will

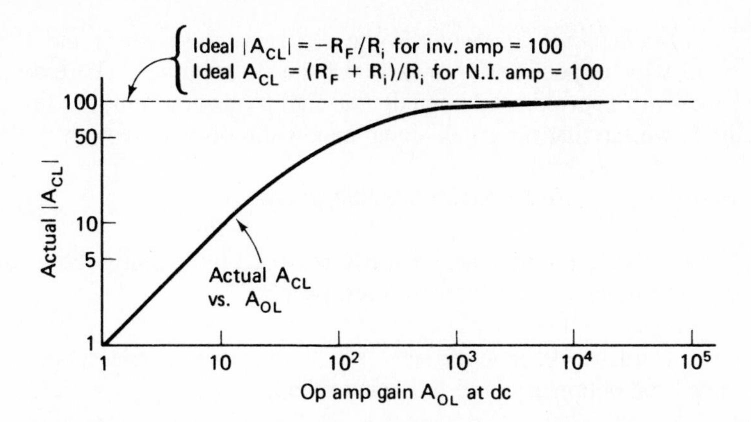

FIGURE 10-2 The actual closed-loop gain of an inverting or noninverting amplifier depends on both the ideal gain that is set by resistor ratios, and the open-loop gain of the op amp at dc (see Example 10-5).

also depend on frequency. But before we look at the amplifier's frequency response we must define it and also define bandwidth.

10-2.2 Small-Signal Bandwidth, Low- and High-Frequency Limits

The useful frequency range of any amplifier (closed- or open-loop) is defined by a high-frequency limit f_H and a low-frequency limit f_L. At f_L and f_H, the voltage gain is down to 0.707 times its maximum value in the middle of the useful frequency range. In terms of decibels, the voltage gain is down 3 dB at both f_L and f_H. These statements are summarized on the general frequency response curve in Fig. 10-3 and in Appendices 1 and 2.

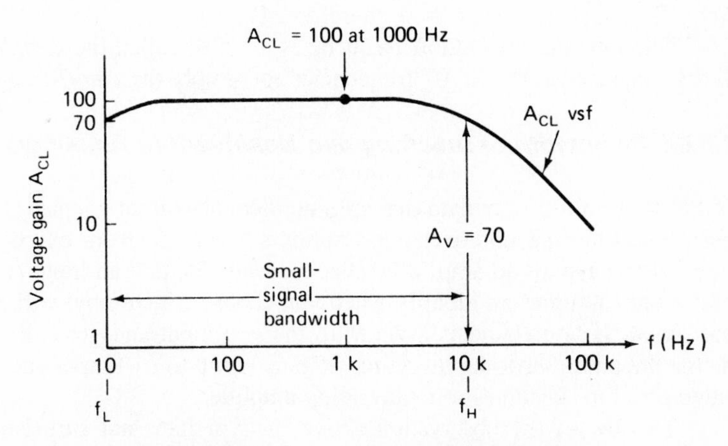

FIGURE 10-3 Small-signal bandwidth.

Small-signal bandwidth is the difference between f_H and f_L. Often f_L is very small with respect to f_H, or f_L is 0 for a dc amplifier. Therefore, the small-signal bandwidth approximately equals the high-frequency limit f_H. From point *A* of Fig. 10-1, we see that the small-signal bandwidth of an op amp is 5 Hz.

10-2.3 Measuring Frequency Response

You can learn a lot about frequency response by learning the techniques of how to measure frequency response at a test bench.

Laboratory procedure. The frequency-response curve of Fig. 10-3 would have been obtained in the following manner:

1. Adjust the input voltage E_i of an op amp to some convenient value, say 30 mV rms.
2. Set the sinusoidal frequency of E_i to some convenient midband value, say 1000 Hz.
3. Measure the midband output voltage; assume that it equals 3.0 V.
4. Calculate the midband voltage gain $A_{CL} = 3\text{ V}/0.030\text{ V} = 100$.
5. Calculate the expected value of V_o at f_L and f_H, $V_o = (0.707)$ (V_o midband). Thus $V_o = (0.707)\ 3\text{ V} = 2.1\text{ V}$ rms where $A_{CL} = 70.7$.
6. Hold E_i *constant* in magnitude at 30 mV. Reduce the oscillator frequency until $V_o = 2.1$ V. Read the oscillator dial frequency to obtain the *lower cutoff* frequency f_L.
7. Hold E_i constant in amplitude at 30 mV. Increase the oscillator frequency (beyond 1 kHz) until V_o again drops to 2.1 V. Read f_H from the oscillator dial.
8. Calculate bandwidth B from $B = f_H - f_L$.

Note: For dc amplifiers, $f = 0$; therefore, $B = f_H$.

The low and high cutoff frequencies are also called the *corner* frequencies, the *3*-dB frequencies, the 0.707 frequencies, or simply the *cutoff* frequencies.

10-2.4 Bandwidth of Inverting and Noninverting Amplifiers

In this section let's stipulate that all amplifiers are direct coupled. Next, observe that both inverting and noninverting amplifiers are made from exactly the same structure. They have an op amp, a feedback resistor R_f, and an input resistor R_i. An amplifier only assumes an identity when you choose which input will experience the input signal. If you connect E_i via R_i to the (−) input and ground the (+) input, *you* define the amplifier to be inverting. If E_i is wired to (+) input and ground to R_i, the same structure becomes a noninverting amplifier.

In view of the observation above, it is perhaps not surprising that the *upper cutoff frequency* f_H *for both inverting and noninverting amplifiers is given by*

$$f_H = \frac{B}{(R_f + R_i)/R_i} \tag{10-4}$$

where B = op amp small-signal bandwidth
R_f = feedback resistance
R_i = input resistor

Example 10-6

Given that $R_f = R_i = 10\ \text{k}\Omega$ for an inverting amplifier and also for a noninverting amplifier. Find the gain and bandwidth of (a) the inverting amplifier; (b) the noninverting amplifier. (c) What is the gain and bandwidth of a voltage follower? The op amp is a 741 with a small-signal gain–bandwidth product of $B = 1$ MHz.

Solution (a) From Eq. (3-2b) or (10-3d), $A_{CL} = -R_f/R_i = -1$. From Eq. (10-4),

$$f_H = \frac{1 \times 10^6\ \text{Hz}}{(10\ \text{k}\Omega + 10\ \text{k}\Omega)/10\ \text{k}\Omega} = 500\ \text{kHz}$$

(b) From Eq. (3-11b) or (10-3b), $A_{CL} = (R_f + R_i)/R_i = 2$. f_H is the same as in part (a). The noninverting amplifier has a higher gain times bandwidth product than the inverting amplifier.
(c) The voltage follower has a gain of 1 [see Eq. (3-9b)]. In Eq. (3-11b), $R_f = 0$ and R_i is an open circuit approaching an infinite resistance for a voltage follower. Therefore, $(R_f + R_i)/R_i = 1$. Hence the upper cutoff frequency f_H is calculated from Eq. (10-4) as

$$f_H = \frac{1 \times 10^6\ \text{Hz}}{(R_f + R_i)/R_i} = \frac{10^6\ \text{Hz}}{1} = 1\ \text{MHz}$$

10-2.5 Finding Bandwidth by a Graphical Method

There is a graphical technique for obtaining the frequency response of a noninverting amplifier. An example is shown in Fig. 10-4. Let the amplifier gain equal 1000 at low and mid frequencies. From Eq. (10-4), $f_H = 999\ \text{Hz} \simeq 1\ \text{kHz}$. At f_H, the amplifier gain is approximately 700 ($0.707 \times 1000 \simeq 700$). For all frequencies above f_H, the frequency response of the amplifier and op amp coincide. For another example, use Fig. 10-4 and draw a horizontal line starting at $A_{CL} = 100$. The ending point where it intercepts the curve of A_{OL} versus f shows the amplifier's bandwidth. For this case, $f_H \simeq 10$ kHz. The conclusion is that the gain–bandwidth product of a noninverting amplifier is equal to B of the op amp. There is a direct trade-off. If you want more closed-loop gain, you must sacrifice bandwidth.

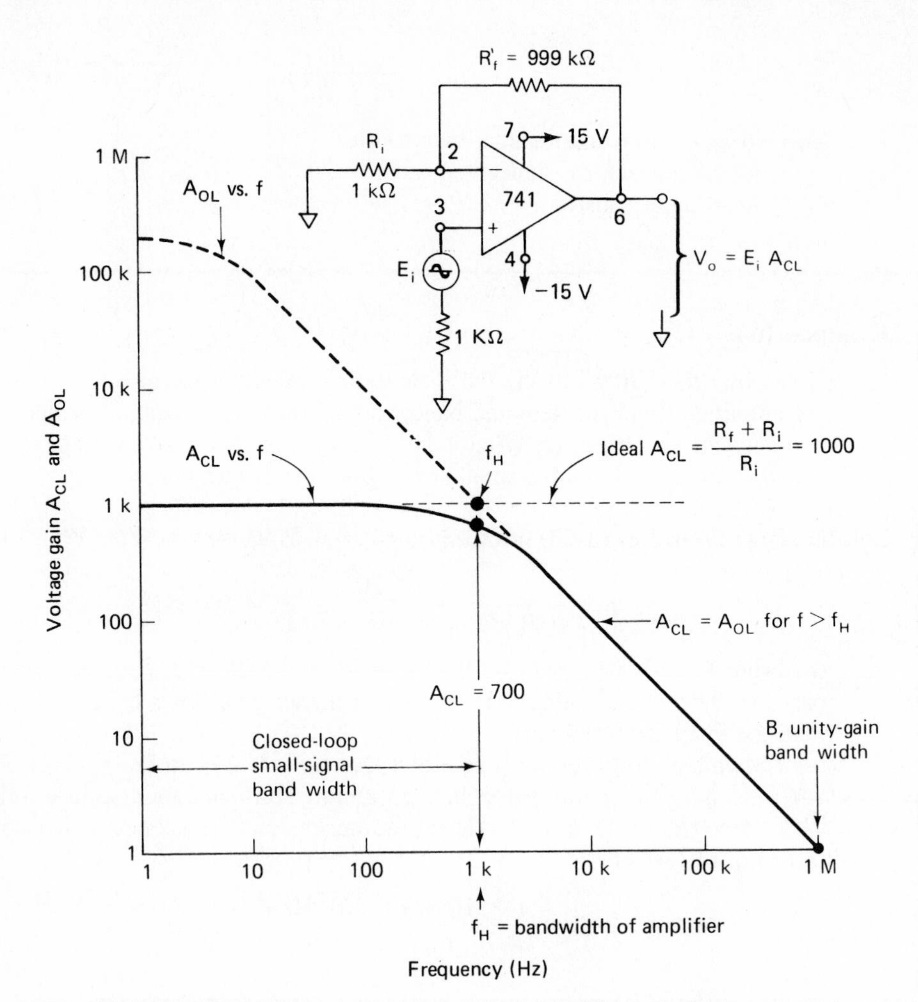

FIGURE 10-4 Op amp small-signal bandwidth and amplified closed-loop bandwidth.

10-3 SLEW RATE AND OUTPUT VOLTAGE

10-3.1 Definition of Slew Rate

The slew rate of an op amp tells how fast its output voltage can change. For a general-purpose op amp such as the 741, the maximum slew rate is 0.5 V/μs. This means that the output voltage can change a maximum of $\frac{1}{2}$ V in 1 μs. Slew rate depends on many factors: the amplifier gain, compensating capacitors, and even whether the output voltage is going positive or negative. The worst case, or slowest slew rate, occurs at unity gain. Therefore, slew rate is usually specified at unity gain (see Appendices 1 and 2).

10-3.2 Cause of Slew-Rate Limiting

Either within or outside the op amp there is at least one capacitor required to prevent oscillation (see Section 10-1.1). Connected to this capacitor is a portion of the op amp's internal circuitry that can furnish a maximum current that is limited by op amp design. The ratio of this maximum current I to the compensating capacitor C is the *slew rate*. For example, a 741 can furnish a maximum of 15 μA to its internal 30-pF compensating capacitor (see Appendix 1). Therefore,

$$\text{slew rate} = \frac{\text{output voltage change}}{\text{time}} = \frac{I}{C} = \frac{15\ \mu\text{A}}{30\ \text{pF}} = 0.5\frac{\text{V}}{\mu\text{s}} \qquad (10\text{-}5)$$

From Eq. (10-5), a faster slew rate requires the op amp to have either a higher maximum current or a smaller compensating capacitor. For example, the AD518 has a slew rate of 80 V/μs with $I = 400\ \mu\text{A}$ and $C = 50$ pF.

Example 10-7

An instantaneous input change of 10 V is applied to a unity-gain inverting amplifier. If the op amp is a 741, how long will it take for the output voltage to change by 10 V?

Solution By Eq. (10-5),

$$\text{slew rate} = \frac{\text{output voltage change}}{\text{time}}$$

$$\frac{0.5\ \text{V}}{\mu\text{s}} = \frac{10\ \text{V}}{\text{time}}, \qquad \text{time} = \frac{10\ \text{V} \times \mu\text{s}}{0.5\ \text{V}} = 20\ \mu\text{s}$$

10-3.3 Slew-Rate Limiting of Sine Waves

In the voltage follower of Fig. 10-5, E_i is a sine wave with peak amplitude E_p. The maximum rate of change of E_i depends on both its frequency f and the peak amplitude. It is given by $2\pi f E_p$. If this rate of change is larger than the op amp's slew rate, the output V_o will be distorted. That is, output V_o tries to follow E_i but cannot do so because of slew-rate limiting. The result is distortion, as shown by the triangular shape of V_o in Fig. 10-5. The maximum frequency f_{max} at which we can obtain an undistorted output voltage with a peak value of V_{op} is determined by the slew rate in accordance with

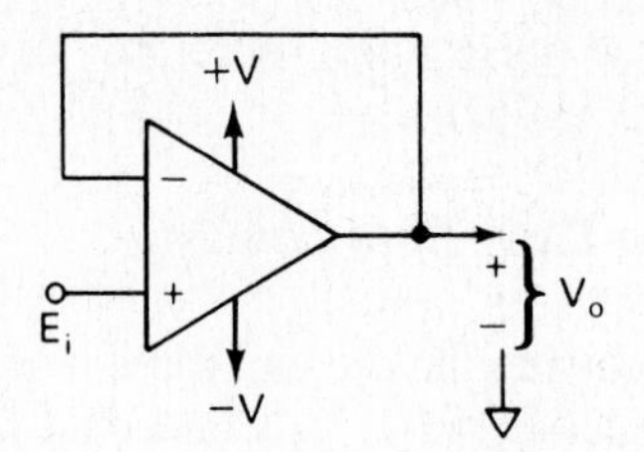

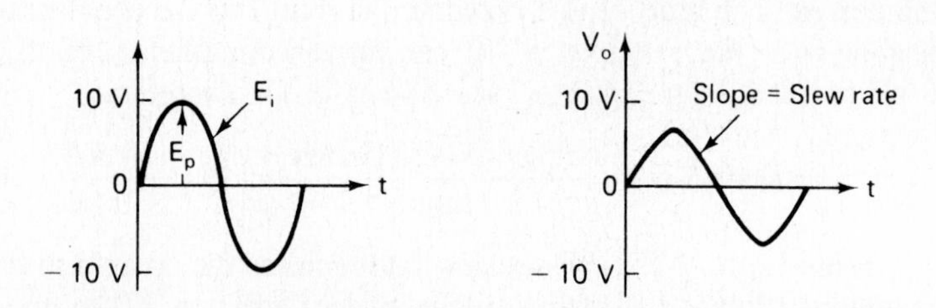

FIGURE 10-5 Example of slew-rate limiting of output voltage V_o.

$$f_{max} = \frac{\text{slew rate}}{6.28 \times V_{op}} \tag{10-6a}$$

where f_{max} is the maximum frequency in Hz, V_{op} is the maximum undistorted output voltage in volts, and the slew rate is in volts per microsecond.

The maximum peak sinusoidal output voltage $V_{op\,max}$ that can be obtained at a given frequency f is found from

$$V_{op\,max} = \frac{\text{slew rate}}{6.28 \times f} \tag{10-6b}$$

Example 10-8

The slew rate for a 741 is 0.5 V/μs. At what maximum frequency can you get an undistorted sine-wave output voltage of (a) 10 V peak; (b) 1 V peak?

Solution (a) From Eq. (10-6a),

$$f_{max} = \frac{1}{6.28 \times 10\text{ V}} \times \frac{0.5\text{ V}}{\mu\text{s}} = 8\text{ kHz}$$

(b) From Eq. (10-6a),

$$f_{max} = 80\text{ kHz}$$

In the next example, we learn that the slew rate *and* bandwidth must *both* be considered before we can predict the highest frequency at which we can obtain an undistorted output voltage.

Example 10-9

In Example 10-6, the small-signal bandwidth was 500 kHz for *both* an inverting amplifier with a gain of −1 and a noninverting amplifier with a gain of 2. Find (a) the maximum peak and undistorted sine wave output voltage at $f_H = 500$ kHz; (b) the maximum frequency where you can obtain a peak output voltage of 10 V.

Solution Since the op amp is a 741, its maximum slew rate is 0.5 V/μs. (a) From Eq. (10-6b),

$$V_{op\,max} = \frac{0.5\ \text{V}/\mu\text{s}}{6.28(500 \times 10^3)\ \text{Hz}} = 160\ \text{mV}$$

(b) From Eq. (10-6a),

$$f_{max} = \frac{0.5\ \text{V}/\mu\text{s}}{(6.28)(10\ \text{V})} \simeq 8\ \text{kHz}$$

f_{max} is defined as *full-power output frequency* at *full-power output*. The meaning of these new terms will become clear after a brief introduction.

A prudent amplifier design would restrict V_o to limits of ±10 V. Then you have a safety margin of ±20% if the amplifier is overdriven at ±12 V (almost into $\pm V_{sat}$). Manufacturers of op amps specify the ±10 V output voltage level as *full-power output*. Note a *full-power output frequency* specification is often supplied by the manufacturer (see Appendix 1, "Output Voltage Swing as a Function of Frequency"). Examples 10-8 and 10-9 showed that the op amp's slew rate limits the upper frequency of large-amplitude output voltages. As the peak output voltage required from the op amp is reduced, the upper-frequency limitation imposed by the slew rate increases.

Recall that the upper-frequency limitation imposed by small-signal response increases as the closed-loop gain decreases. For each amplifier application, the upper-frequency limit imposed by slew-rate limiting (Section 10-3.3) and small-signal bandwidth (Section 10-2.3) must be calculated. The *smaller* value determines the actual upper-frequency limit. In general, the slew rate is a large-signal frequency limitation and small-signal frequency response is a small-signal frequency limitation.

10-3.4 Slew Rate Made Easy

Figure 10-6 simplifies the problem of finding f_{max} at any peak output voltage for slew rates between 0.5 and 5 V/μs. For example, to do part (b) of Example 10-8, locate

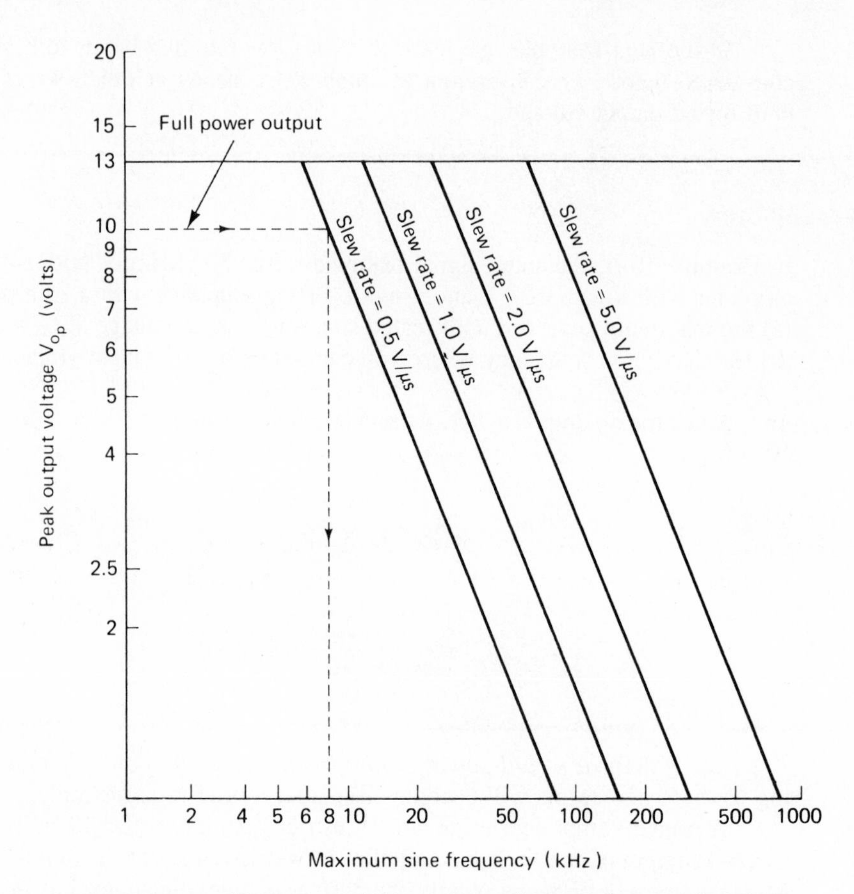

FIGURE 10-6 Slew rate made easy. Any point on a slew-rate line shows the maximum sinusoidal frequency allowed for the corresponding peak output voltage.

where the horizontal line of V_{op} = 10 V intersects the slew-rate line 0.5 V/μs. Below the intersection, read f_{max} = 8 kHz.

10-4 NOISE IN THE OUTPUT VOLTAGE

10-4.1 Introduction

Undesired electrical signals present in the output voltage are classified as *noise*. Drift (see Chapter 9) and offsets can be considered as very low frequency noise. If you view the output voltage of an op amp amplifier with a sensitive CRO (1 mV/cm), you will see a random display of noise voltages called *hash*. The frequencies of these noise voltages range from 0.01 Hz to megahertz.

Noise is generated in any material that is above absolute zero (−273°C). Noise is also generated by all electrical devices and their controls. For example, in an automobile, the spark plugs, voltage regulator, fan motor, air conditioner, and generator all generate noise. Even when headlights are switched on (or off), there is a sudden change in current that generates noise. This type of noise is external to the op amp. Effects of external noise can be minimized by proper construction techniques and circuit selection (see Sections 10-4.3 to 10-4.5).

10-4.2 Noise in Op Amp Circuits

Even if there were no external noise, there would still be noise in the output voltage caused by the op amp. This internal op amp noise is modeled most simply by a noise voltage source E_n. As shown in Fig. 10-7, E_n is placed in series with the (+) input. On data sheets, noise voltage is specified in microvolts (rms) for different values of source resistance over a particular frequency range. For example, the 741 op amp has 2 μV of *total* noise over a frequency of 10 Hz to 10 kHz. This noise voltage is valid for source resistors (R_i) between 100 Ω and 20 kΩ. The noise voltage goes up directly with R_i, once R_i exceeds 20 kΩ. Thus R_i should be kept below 20 kΩ to minimize noise in the output (see Appendix 1).

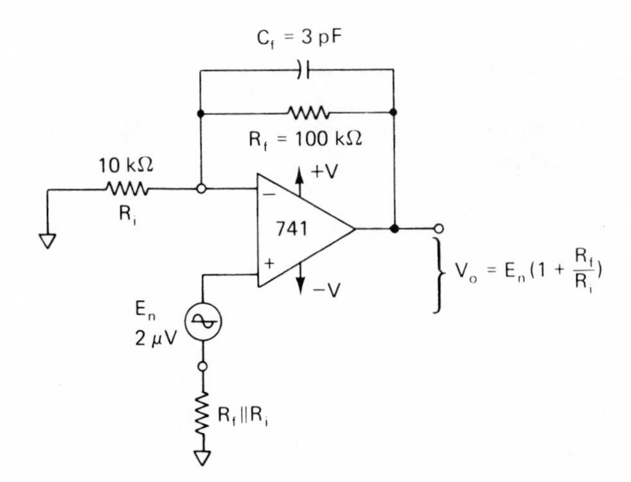

FIGURE 10-7 Op amp noise is modeled by a noise voltage in series with the (+) input.

10-4.3 Noise Gain

Noise voltage is amplified just as offset voltage is. That is, *noise voltage gain* is the same as the gain of a noninverting amplifier:

$$\text{noise gain} = 1 + \frac{R_f}{R_i} \tag{10-7}$$

What can you do about minimizing output voltage errors due to noise? First, avoid, if possible, large values of R_i and R_f. Install a small capacitor (3-pF) across R_f to shunt it at high noise frequencies. Then the higher noise frequencies will not be amplified so much. Next, do not shunt R_i with a capacitor; otherwise, the $R_i C$ combination will have a smaller impedance at higher noise frequencies than R_i alone, and gain will increase with frequency and aggravate the situation. Finally, try to keep R_i at about 10 kΩ or below.

Noise currents, like bias currents, are also present at each op amp input terminal. If a bias-current compensation resistor is installed (see Chapter 9), the effect of noise currents on output voltage will be reduced. As with offset current, the effects of noise currents also depend on the feedback resistor. So if possible, reduce the size of R_f to minimize the effects of noise currents.

10-4.4 Noise in the Inverting Adder

In the inverting adder (see Section 3-2), each signal input voltage has a gain of 1. However, the noise gain will be 1 plus the number of inputs; for example, a four-input adder would have a noise gain of 5. Thus noise voltage has five times as much gain as each input signal. Therefore, low-amplitude signals should be preamplified before connecting them to an adder.

10-4.5 Summary

To reduce the effects of op amp noise:

1. *Never* connect a capacitor across the input resistor or from (−) input to ground. There will always be a few picofarads of stray capacitance from (−) input to ground due to wiring, so
2. *Always* connect a small capacitor (3 pF) across the feedback resistor. This reduces the noise gain at high frequencies.
3. If possible, avoid large resistor values.

10-5 EXTERNAL FREQUENCY COMPENSATION

10-5.1 Need for External Frequency Compensation

Op amps with internal frequency compensation (see Section 10-1.1 and Appendix 1) are very stable with respect to signal frequencies. They do not burst into spontaneous oscillation or wait to oscillate occasionally when a signal is applied. However, the trade-offs for frequency stability are limited small-signal bandwidth, slow slew rate, and reduced-power bandwidth. Internally compensated op amps are useful at audio frequencies but not at higher frequencies.

A 741 that has a 1-MHz gain–bandwidth product will give a useful gain of 1000 only up to frequencies of about 1 kHz. To obtain more gain from the op amp at higher frequencies, the internal frequency-compensating capacitor of the op amp must be removed. If this is done, the resulting op amp structure has a higher slew rate and greater power bandwidth. But these improvements would be cancelled out because the op amp probably would oscillate continually. As usual, there is a trade-off: frequency stability for a larger bandwidth and slew rate.

In order to be able to make these trade-offs, manufacturers of op amps bring out from one to three *frequency-compensating terminals*. Such terminals allow the user to choose the best allowable combination of stability and bandwidth. This choice is made by connecting external capacitors and resistors to the compensating terminals. Accordingly, this versatile type of op amp is classified as *externally frequency-compensated* (see Appendix 2).

10-5.2 Single-Capacitor Compensation

The frequency response of the 101 general-purpose op amp can be tailored by connecting a single capacitor, C_1, to pins 1 and 8. As shown in Fig. 10-8 and Appendix

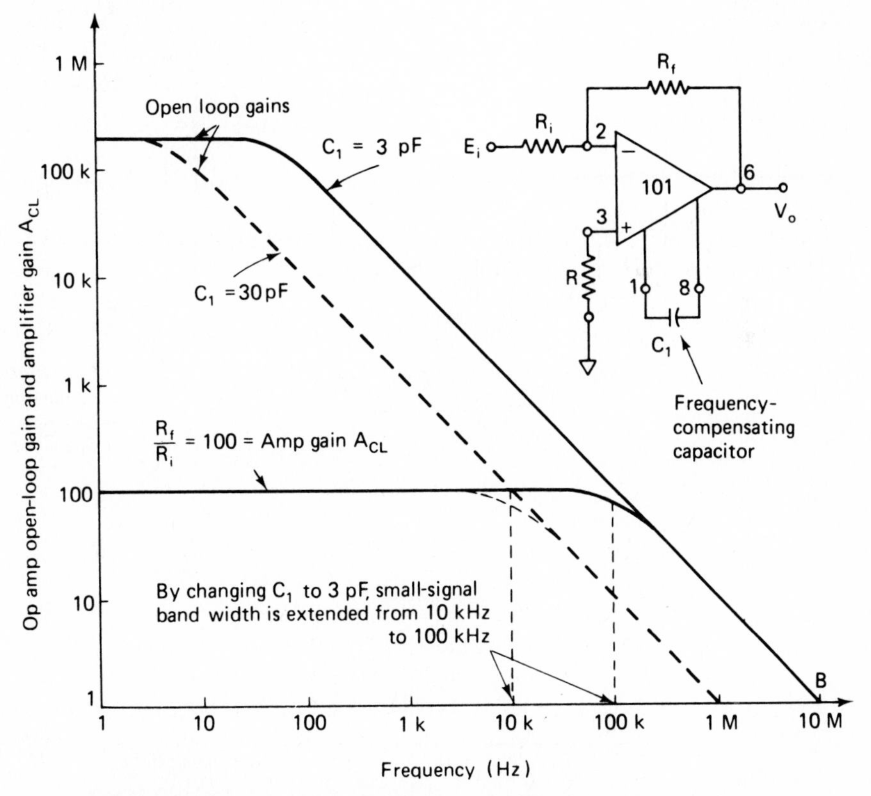

FIGURE 10-8 Extending frequency response with an external compensating capacitor.

2, by making $C_1 = 3$ pF, the 101 has an open-loop frequency-response curve with a small-signal bandwidth of 10 MHz. Increasing C_1 by a factor of 10 (to 30 pF) reduces the small-signal bandwidth by a factor of 10 (to 1 MHz). Therefore, the 101 can be externally compensated to have the same small-signal bandwidth as the 741.

When the 101 is used in an amplifier circuit, the amplifier's useful frequency range now depends on the compensating capacitor. For example, with R_f/R_i set for an amplifier gain of 100, its small-signal bandwidth would be 10 kHz for $C_1 =$ 30 pF. By reducing C_1 to 3 pF, the small-signal bandwidth is increased to 100 kHz. The full-power bandwidth is also increased, from about 6 kHz to 60 kHz.

10-5.3 Feed-Forward Frequency Compensation

There are many other types of frequency compensation. Among the more popular are *two-capacitor,* or *two-pole compensation* and *feed-forward compensation.* Manufacturers' data sheets give precise instructions on the type best suited for your application.

Feed-forward compensation for the 101 is illustrated in Fig. 10-9. Feed-

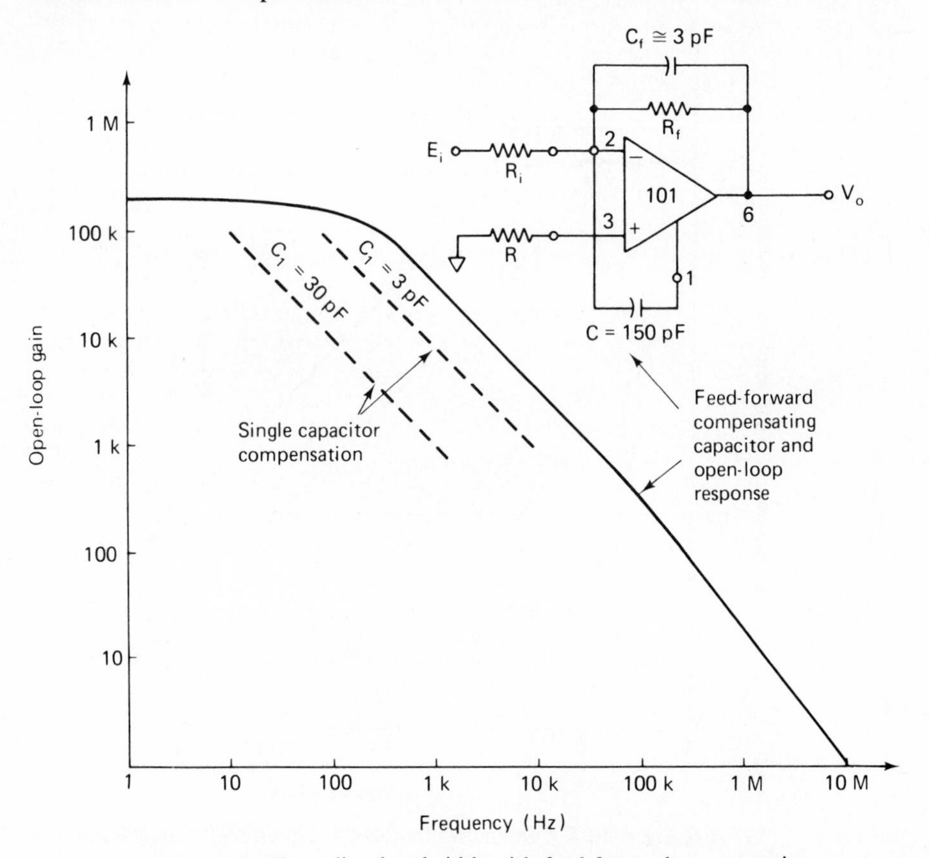

FIGURE 10-9 Extending bandwidth with feed-forward compensation.

forward capacitor C is wired from the (−) input to compensating terminal 1. A small capacitor C_f is needed across R_f to ensure frequency stability. The slew rate is increased to 10 V/μs and the full-power bandwidth to over 200 kHz. Of course, the added high-frequency gain will also amplify high-frequency noise.

We should conclude that frequency-compensation techniques must be applied only to the extent required for the circuit. Do not use any more high-frequency gain than is absolutely necessary; otherwise, there will be a needless amount of high-frequency noise in the output.

LABORATORY EXERCISES

Theoretical calculations of bandwidth and slew rate have been shown to be straightforward. However, when you build an amplifier on an ordinary breadboard, the actual measured frequency response will probably not agree with predicted performance. The following laboratory experience illustrates this point.

10-1. Consider that you use a 741 op amp and build a noninverting amplifier with a gain of 2. You restrict E_i and V_o to 100 mV peak to avoid distortion caused by slew-rate limiting. You expect V_o to stay at 100 mV peak until the oscillator frequency is increased to about 500 kHz. There V_{op} equals 70 mV when E_{ip} = 100 mV.

The frequency response will be as predicted if $R_i = R_f = 10\ k\Omega$. However, if $R_i = R_f = 1\ M\Omega$, the actual frequency response is very different from the predicted values. As you increase frequency of E_i (holding the amplitude constant), V_{op} remains at 100 mV up to about 20 kHz. Then V_{op} increases as frequency increases until the range of about 100 to 200 kHz. Above 200 kHz, V_{op} begins to decrease. At 300 to 500 kHz, V_{op} has decreased to 70 mV.

The reason for this apparent failure of the theory is that the equations did not include the effect of unpredictable input capacitance. Any time two conductors are separated by an insulator, a capacitor is formed. Therefore, capacitors exist between the (−) input and (1) ground, (2) all op amp terminals, (3) all component leads, and (4) breadboard terminals. The net capacitance from (−) input to ground acts as *if* you connected a 10- to 15-pF parasitic capacitor across R_i; see Fig. LE10-1. (Note in the figure that parasitic capacitance C_{in} is present from (−) in to ground. X_{Cin} shunts R_i and increases gain for frequencies above 10 to 30 kHz. The cure is to add C_f.)

Amplifier gain is now seen to be $(R_f + Z_i) / Z_i$, where $Z_i = R_i \| X_{Cin}$. As frequency increases, X_{Cin} decreases, reducing Z_i and increasing gain.

The cure is to bypass R_f with a capacitor C_f. C_f should bypass R_f at the same rate as C_{in} bypasses R_i. The value of C_f is chosen by trial and errror. Theoretically, if C_{in} is known, calculate C_f from

$$R_i C_{in} = R_f C_f \qquad (10\text{-}8)$$

Practical Laboratory Tips

If an amplifier oscillates or exhibits an unpredicted frequency response, try any or all of the following:

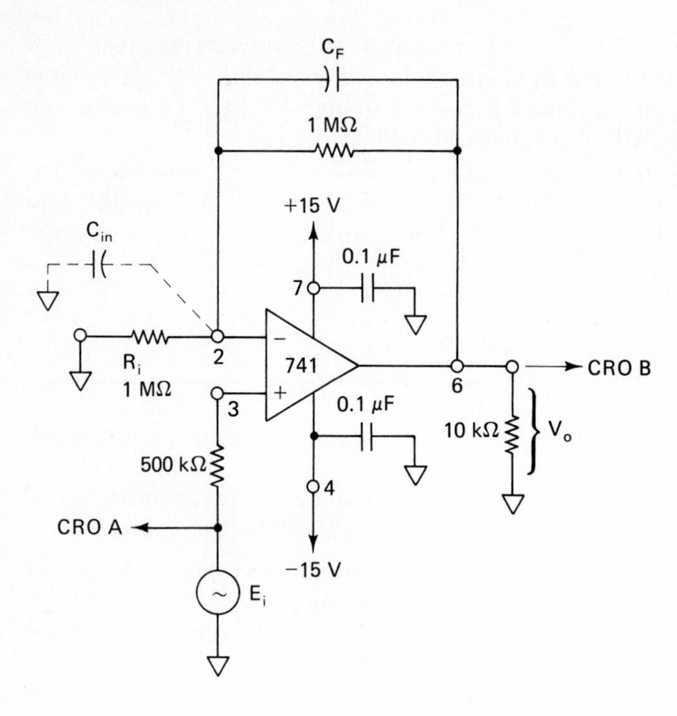

FIGURE LE10-1

1. Shorten leads. Mount components close to the op amp. Keep output components away from input components.
2. Bypass the op amp supply pins with 0.01- to 0.1-μF disk capacitors as shown in Fig. LE10-1. (Then try 1- to 10-μF tantalums.)
3. Arrange the circuit layout so that the input grounds are separate from output grounds and are connected only at the power supply common.
4. If a universal breadboard does not give satisfactory results, use a pc board with a ground plane.

PROBLEMS

10-1. What is the typical open-loop gain of a 741 op amp at very low frequencies?

10-2. The dc open-loop gain of an op amp is 100,000. Find the open-loop gain at its break frequency.

10-3. The transient response rise time (unity gain) of an op amp is 0.07 μs. Find the small-signal bandwidth.

10-4. An op amp has a small signal unity-gain bandwidth of 2 MHz. Find its open-loop gain at 200 kHz.

10-5. What is the difference between the open-loop and closed loop gain of an op amp?

10-6. What is the open-loop gain for the op amp of Problem 10-4 at 2 MHz?

10-7. What is *rise time?*

10-8. An op amp has a dc open-loop gain of 100,000. It is used in an inverting amplifier circuit with $R_f = 100\ k\Omega$, $R_i = 10\ k\Omega$. Find the actual dc closed-loop gain.

10-9. The op amp of Problem 10-8 is used in a noninverting amplifier with the same R_f and R_i. Find the amplifier's actual dc closed-loop gain.

10-10. What is the small-signal bandwidth of the op amp whose frequency response is given in Fig. 10-1?

10-11. The unity-gain bandwidth of an op amp is 10 MHz. It is used to make a noninverting amplifier with an ideal closed-loop gain of 100. Find the amplifier's **(a)** small-signal bandwidth; **(b)** A_{CL} at f_H.

10-12. How fast can the output of an op amp change by 10 V if its slew rate is 1 V/μs?

10-13. Find the maximum frequency for a sine-wave output voltage of 10 V peak with an op amp whose slew rate is 1 V/μs.

10-14. Find the noise gain for an inverting amplifier with a gain of $R_f/R_i = -10$.

10-15. What is the noise gain for a five-input inverting adder?

10-16. The op amp in Example 10-9 is changed to one with a slew rate of 1 V/μs. Find its maximum full-power output frequency. Assume that $V_{o\,max} = 10$ V peak.

10-17. Does increasing the compensating capacitor increase or decrease unity-gain bandwidth?

CHAPTER 11

Active Filters

LEARNING OBJECTIVES

Upon completion of this chapter on active filters, you will be able to:

- Name the four general classifications of filters and sketch a frequency-response curve that shows the band of frequencies that they pass and stop.
- Design or analyze circuits for three types of low-pass filters: −20 dB/decade, −40 dB/decade, or −60 dB/decade rolloff.
- Design or analyze circuits for three types of high-pass filters: 20 dB, 40 dB, and 60 dB per decade of roll-off.
- Cascade a low-pass filter with a high-pass filter to make a wide bandpass filter.
- Calculate the lower and upper cutoff frequencies of either a bandpass or a notch

filter if you are given (1) bandwidth and resonant frequency, (2) bandwidth and quality factor, or (3) resonant frequency and quality factor.

- Calculate quality factor, bandwidth, and resonant frequency of a bandpass or notch filter for a given lower and upper cutoff frequency.
- Design a bandpass filter that uses only one op amp.
- Make a notch filter by (1) designing a bandpass filter circuit with the same bandwidth and a resonant frequency equal to the notch frequency, and (2) properly connecting the bandpass circuit to an inverting adder.
- Explain the operation of a stereo equalizer circuit.

11-0 INTRODUCTION

A *filter* is a circuit that is designed to pass a specified band of frequencies while attenuating all signals outside this band. Filter networks may be either active or passive. *Passive filter networks* contain only resistors, inductors, and capacitors. *Active filters,* which are the only type covered in this text, employ transistors or op amps plus resistors, inductors, and capacitors. Inductors are not often used in active filters, because they are bulky and costly and may have large internal resistive components.

There are four types of filters: *low-pass, high-pass, bandpass,* and *band-elimination* (also referred to as *band-reject* or *notch*) filters. Figure 11-1 illustrates frequency-response plots for the four types of filters. A low-pass filter is a circuit that has a constant output voltage from dc up to a *cutoff frequency* f_c. As the frequency increases above f_c, the output voltage is attenuated (decreases). Figure 11-1(a) is a plot of the magnitude of the output voltage of a low-pass filter versus frequency. The solid line is a plot for the ideal low-pass filter, while the dashed lines indicate the curves for practical low-pass filters. The range of frequencies that are *transmitted* is known as the *passband*. The range of frequencies that are *attenuated* is known as the *stop band*. The cutoff frequency, f_c, is also called the 0.707 frequency, the −3-dB frequency, the corner frequency, or the break frequency.

High-pass filters attenuate the output voltage for all frequencies below the cutoff frequency f_c. Above f_c, the magnitude of the output voltage is constant. Figure 11-1(b) is the plot for ideal and practical high-pass filters. The solid line is the ideal curve, while the dashed curves show how practical high-pass filters deviate from the ideal.

Bandpass filters pass only a band of frequencies while attenuating all frequencies outside the band. Band-elimination filters perform in an exactly opposite way; that is, band-elimintion filters reject a specified band of frequencies while passing all frequencies outside the band. Typical frequency-response plots for bandpass and band-elimination filters are shown in Fig. 11-1(c) and (d). Once again, the solid line represents the ideal plot, while dashed lines show the practical curves.

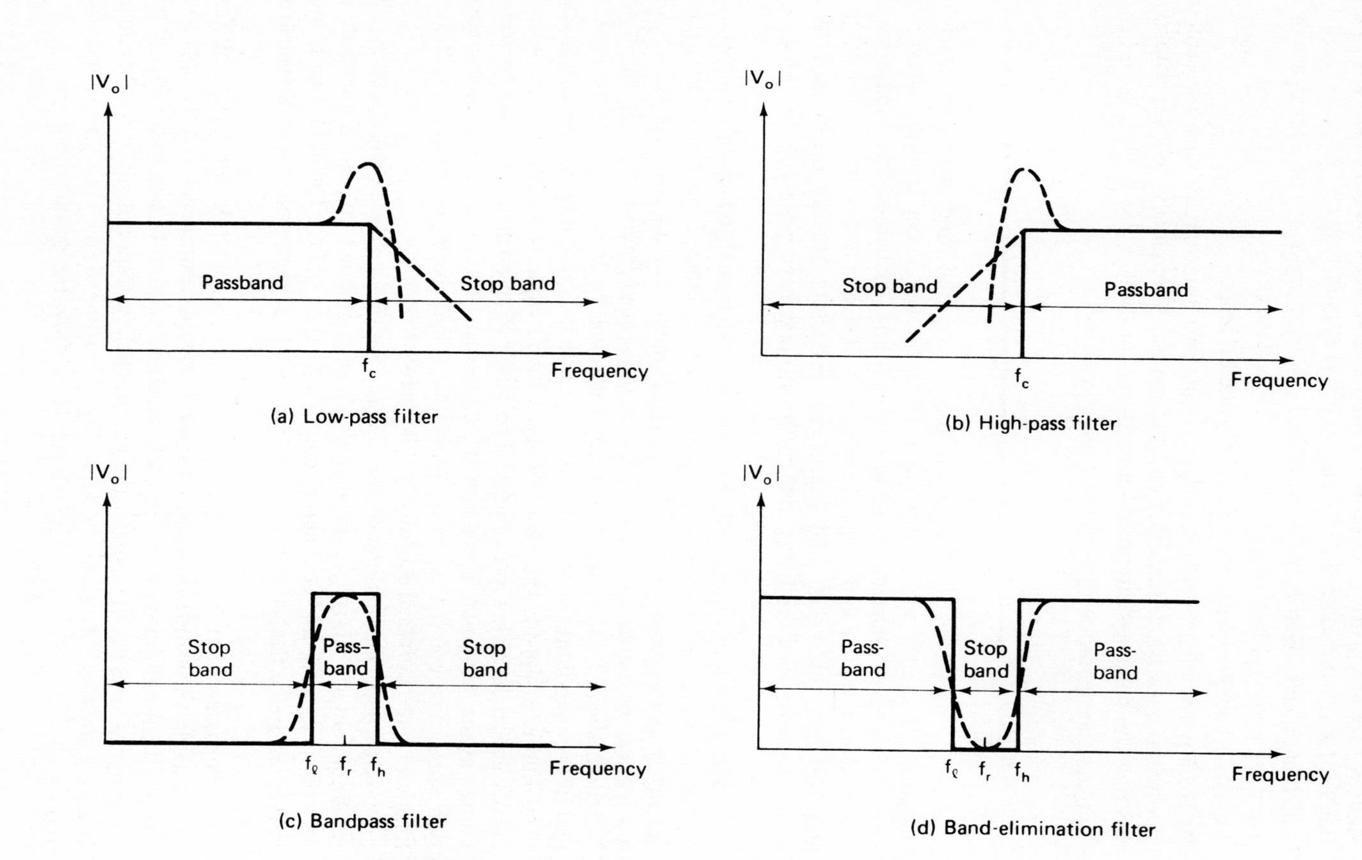

FIGURE 11-1 Frequency response for four categories of filters.

11-1 BASIC LOW-PASS FILTER

11-1.1 Introduction

The circuit of Fig. 11-2(a) is a commonly used low-pass active filter. The filtering is done by the *RC* network, and the op amp is used as a unity-gain amplifier. The resistor R_f is equal to R and is included for dc offset. [At dc, the capacitive reactance is infinite and the dc resistance path to ground for both input terminals should be equal (see Section 9-4).]

The differential voltage between pins 2 and 3 is essentially 0 V. Therefore, the

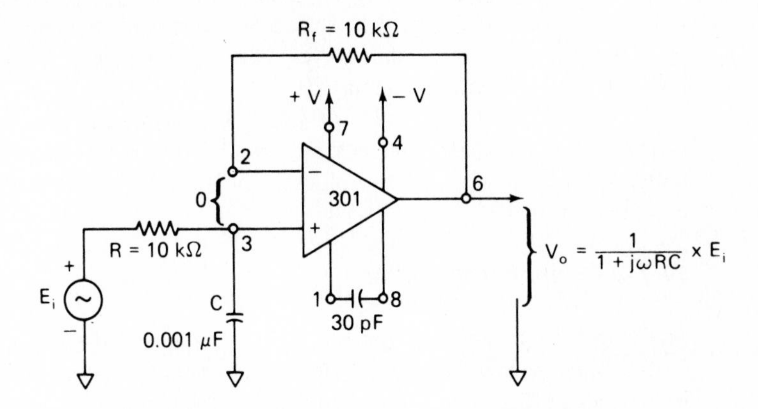

(a) Low-pass filter for a roll-off of − 20 db/decade

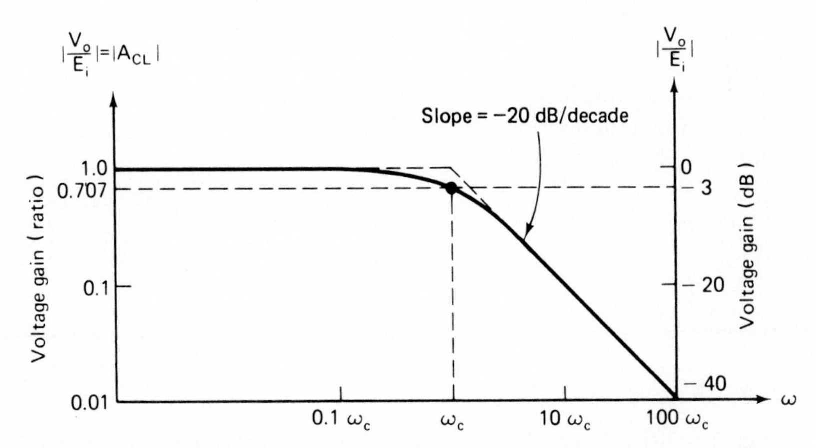

(b) Frequency-response plot for the circuit of part (a)

FIGURE 11-2 Low-pass filter and frequency-response plot for a filter with a −20-dB/decade roll-off.

voltage across capacitor C equals output voltage V_o, because this circuit is a voltage follower. E_i divides between R and C. The capacitor voltage equals V_o and is

$$V_o = \frac{1/j\omega C}{R + 1/j\omega C} \times E_i \qquad (11\text{-}1a)$$

where ω is the frequency of E_i in radians per second ($\omega = 2\pi f$) and j is equal to $\sqrt{-1}$. Rewriting Eq. (11-1a) to obtain the closed-loop voltage gain A_{CL}, we have

$$A_{CL} = \frac{V_o}{E_i} = \frac{1}{1 + j\omega RC} \qquad (11\text{-}1b)$$

To show that the circuit of Fig. 11-2(a) is a low-pass filter, consider how A_{CL} in Eq. (11-1b) varies as frequency is varied. At very low frequencies, that is, as ω approaches 0, $|A_{CL}| = 1$, and at very high frequencies, as ω approaches infinity, $|A_{CL}| = 0$. ($|\cdot|$ indicates magnitude.)

Figure 11-2(b) is a plot of $|A_{CL}|$ versus ω and shows that for frequencies *greater* than the cutoff frequency ω_c, $|A_{CL}|$ decreases at a rate of 20 dB/decade. This is the same as saying that the voltage gain is divided by 10 when the frequency of ω is increased by 10.

11-1.2 Designing the Filter

The cutoff frequency ω_c is defined as that frequency of E_i where $|A_{CL}|$ is reduced to 0.707 times its low-frequency value. This important point will be discussed further in Section 11-1.3. The cutoff frequency is evaluated from

$$\omega_c = \frac{1}{RC} = 2\pi f_c \qquad (11\text{-}2a)$$

where ω_c is the cutoff frequency in radians per second, f_c is the cutoff frequency in hertz, R is in ohms, and C is in farads. Equation (11-2a) may be rearranged to solve for R:

$$R = \frac{1}{\omega_c C} = \frac{1}{2\pi f_c C} \qquad (11\text{-}2b)$$

Example 11.1

Let $R = 10\text{ k}\Omega$ and $C = 0.001\ \mu\text{F}$ in Fig. 11-2(a); what is the cutoff frequency?

Solution By Eq. (11-2a),

$$\omega_c = \frac{1}{(10 \times 10^3)(0.001 \times 10^{-6})} = 100\text{ krad/s}$$

or

$$f_c = \frac{\omega_c}{6.28} = \frac{100 \times 10^3}{6.28} = 15.9 \text{ kHz}$$

Example 11-2

For the low-pass filter of Fig. 11-2(a), calculate R for a cutoff frequency of 2 kHz and $C = 0.005\ \mu\text{F}$.

Solution From Eq. (11-2b),

$$R = \frac{1}{\omega_c C} = \frac{1}{(6.28)(2 \times 10^3)(5 \times 10^{-9})} = 15.9 \text{ k}\Omega$$

Example 11-3

Calculate R for Fig. 11-2(a) for a cutoff frequency of 30 krad/s and $C = 0.01\ \mu\text{F}$.

Solution From Eq. (11-2b),

$$R = \frac{1}{\omega_c C} = \frac{1}{(30 \times 10^3)(1 \times 10^{-8})} \simeq 3.3 \text{ k}\Omega$$

Design Procedure The design of a low-pass filter similar to Fig. 11-2(a) is accomplished in three steps:

1. Choose the cutoff frequency—either ω_c or f_c.
2. Choose the capacitance C, usually between 0.001 and 0.1 μF.
3. Calculate R from Eq. (11-2b).

11-1.3 Filter Response

The value of A_{CL} at ω_c is found by letting $\omega RC = 1$ in Eq. (11-1b):

$$A_{CL} = \frac{1}{1 + j1} = \frac{1}{\sqrt{2}\,\underline{/45^\circ}} = 0.707\,\underline{/-45^\circ}$$

Therefore, the magnitude of A_{CL} at ω_c is

$$|A_{CL}| = \frac{1}{\sqrt{2}} = 0.707 = -3 \text{ dB}$$

and the phase angle is $-45°$.

The solid curve in Fig. 11-2(b) shows how the magnitude of the actual frequency response deviates from the straight dashed-line approximation in the vicinity of ω_c. At $0.1\ \omega_c$, $|A_{CL}| \simeq 1(0\ \text{dB})$, and at $10\omega_c$, $|A_{CL}| \simeq 0.1(-20\ \text{dB})$. Table 11-1 gives both the magnitude and the phase angle for different values of ω between $0.1\ \omega_c$ and $10\omega_c$.

Many applications require steeper roll-offs after the cutoff frequency. One common filter configuration that gives steeper roll-offs is the *Butterworth filter*.

TABLE 11-1 MAGNITUDE AND PHASE ANGLE FOR THE LOW-PASS FILTER OF FIG. 11-2(a)

ω	$\lvert A_{CL}\rvert$	Phase angle (deg)
$0.1\omega_c$	1.0	−6
$0.25\omega_c$	0.97	−14
$0.5\omega_c$	0.89	−27
ω_c	0.707	−45
$2\omega_c$	0.445	−63
$4\omega_c$	0.25	−76
$10\omega_c$	0.1	−84

11-2 INTRODUCTION TO THE BUTTERWORTH FILTER

In many low-pass filter applications, it is necessary for the closed-loop gain to be as close to 1 as possible within the passband. The *Butterworth filter* is best suited for this type of application. The Butterworth filter is also called a *maximally flat* or *flat-flat* filter, and all filters in this chapter will be of the Butterworth type. Figure 11-3 shows the ideal (solid line) and the practical (dashed lines) frequency response for three types of Butterworth filters. As the roll-offs become steeper, they approach the ideal filter more closely.

Two active filters similar to Fig. 11-2(a) could be coupled together to give a roll-off of −40 dB/decade. This would not be the most economical design, because it would require two op amps. In Section 11-3.1, it is shown how one op amp can be used to build a Butterworth filter with a single op amp to give a −40-dB/decade roll-off. Then in Section 11-4, a −40-dB/decade filter will be cascaded with a −20-dB/decade filter to produce a −60-dB/decade filter.

Butterworth filters are not designed to keep a constant phase angle at the cutoff frequency. A basic low-pass filter of −20 dB/decade has a phase angle of −45° at ω_c. A −40-dB/decade Butterworth filter has a phase angle of −90° at ω_c and a −60-dB/decade filter has a phase angle of −135° at ω_c. Therefore, for each increase of −20 dB/decade, the phase angle will increase by −45° at ω_c. We now proceed to a Butterworth filter that has a roll-off steeper than −20 dB/decade.

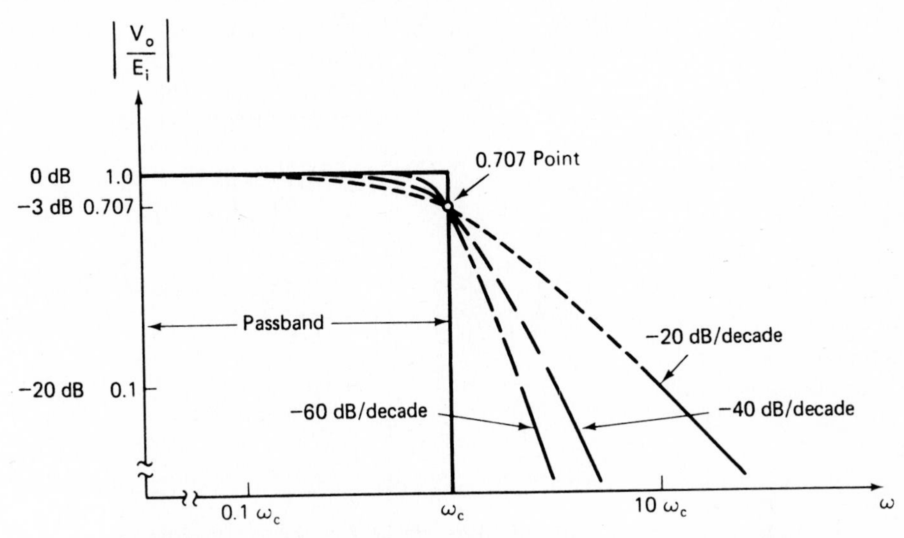

FIGURE 11-3 Frequency-response plots for three types of low-pass Butterworth filters.

11-3 −40-DB/DECADE LOW-PASS BUTTERWORTH FILTER

11-3.1 Simplified Design Procedure

The circuit of Fig. 11-4(a) is one of the most commonly used low-pass filters. It produces a roll-off of −40 dB/decade; that is, after the cutoff frequency, the magnitude of A_{CL} decreases by 40 dB as ω increases to $10\omega_c$. The solid line in Fig. 11-4(b) shows the actual frequency-response plot, which is explained in more detail in Section 11-3.2. The op amp is connected for dc unity gain. Resistor R_f is included for dc offset, as explained in Section 9-4. Since the op amp circuit is basically a voltage follower (unity-gain amplifier), the voltage across C_1 equals output voltage, V_o.

The design of the low-pass filter of Fig. 11-4(a) is greatly simplified by making resistors $R_1 = R_2 = R$. Then there are only five steps in the design procedure.

Design procedure

1. Choose the cutoff frequency, ω_c or f_c.
2. Pick C_1; choose a convenient value between 100 pF and 0.1 μF.
3. Make $C_2 = 2C_1$.
4. Calculate

$$R = \frac{0.707}{\omega_c C} \tag{11-3}$$

5. Choose $R_f = 2R$.

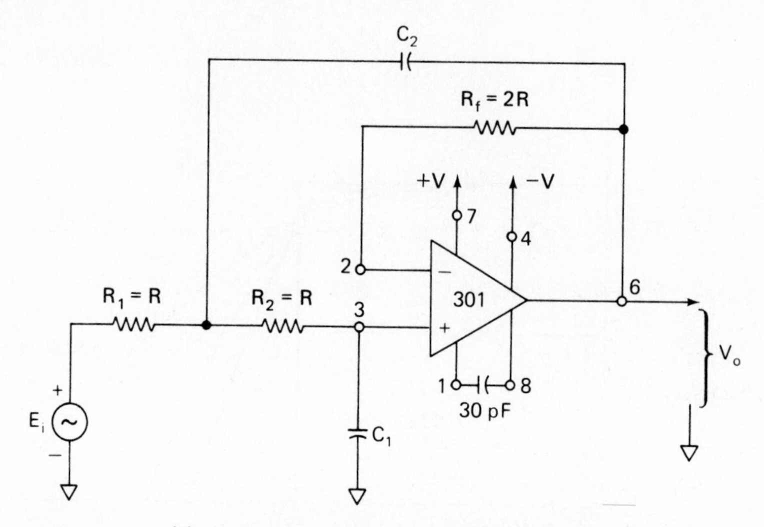

(a) Low-pass filter for a roll-off of −40 dB/decade

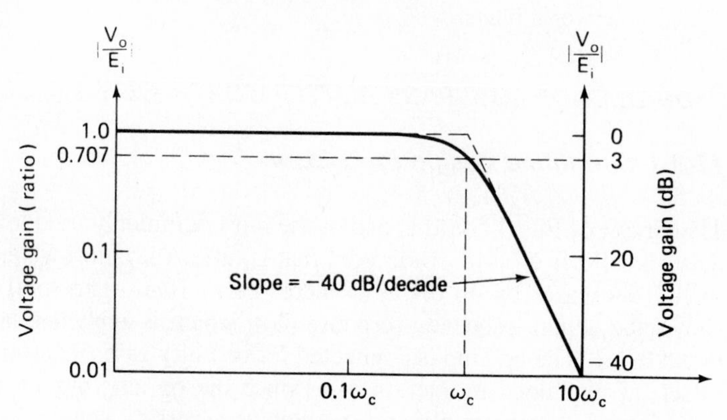

(b) Frequency-response plot for the low-pass filter of part (a)

FIGURE 11-4 Circuit and frequency plot for a low-pass filter of −40 dB/decade.

Example 11-4

Determine R_1 and R_2 in Fig 11-4(a) for a cutoff frequency of 1 kHz. Let $C_1 = 0.01\ \mu F$.

Solution Pick $C_2 = 2C_1 = 2(0.01\ \mu F) = 0.02\ \mu F$. Select $R_1 = R_2 = R$ from Eq. (11.3):

$$R = \frac{0.707}{(6.28)(1 \times 10^3)(0.01 \times 10^{-6})} = 11{,}258\ \Omega$$

and

$$R_f = 2(11{,}258\ \Omega) = 22{,}516\ \Omega$$

11-3.2 Filter Response

The dashed curve in Fig. 11-4(b) shows that the filter of Fig. 11-4(a) not only has a steeper roll-off after ω_c than does Fig. 11-2(a) but also remains at 0 dB almost up to about 0.25 ω_c. The phase angles for the circuit of Fig. 11-4(a) range from 0° at $\omega = 0$ rad/s (dc condition) to $-180°$ as ω approaches ∞ (infinity). Table 11-2 compares magnitude and phase angle for the low-pass filters of Figs. 11-2(a) and 11-4(a) from $0.1\omega_c$ to $10\omega_c$.

The next low-pass filter cascades the filter of Fig. 11-2(a) with the filter of Fig. 11-4(a) to form a roll-off of -60 dB/decade. As will be shown, the resistors are the only values that have to be calculated.

TABLE 11-2 MAGNITUDE AND PHASE ANGLE FOR FIGS. 11-2(a) AND 11-4(a)

	$\lvert A_{CL} \rvert$		Phase angle (deg)	
ω	-20 dB/decade; Fig. 11-2(a)	-40 dB/decade; Fig. 11-4(a)	Fig. 11-2(a)	Fig. 11-4(a)
$0.1\omega_c$	1.0	1.0	-6	-8
$0.25\omega_c$	0.97	0.998	-14	-21
$0.5\omega_c$	0.89	0.97	-27	-43
ω_c	0.707	0.707	-45	-90
$2\omega_c$	0.445	0.24	-63	-137
$4\omega_c$	0.25	0.053	-76	-143
$10\omega_c$	0.1	0.01	-84	-172

11-4 -60-DB/DECADE LOW-PASS BUTTERWORTH FILTER

11-4.1 Simplified Design Procedure

The low-pass filter of Fig. 11-5(a) is built using one low-pass filter of -40 dB/decade cascaded with another of -20 dB/decade to give an overall roll-off of -60 dB/decade. The overall closed-loop gain A_{CL} is the gain of the first filter times the gain of the second filter, or

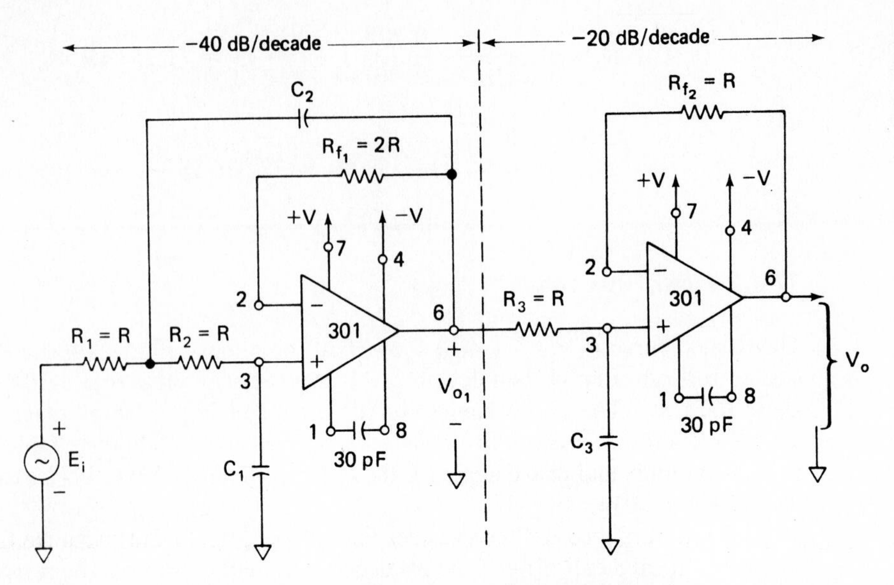

(a) Low-pass filter for a roll off of −60 dB/decade

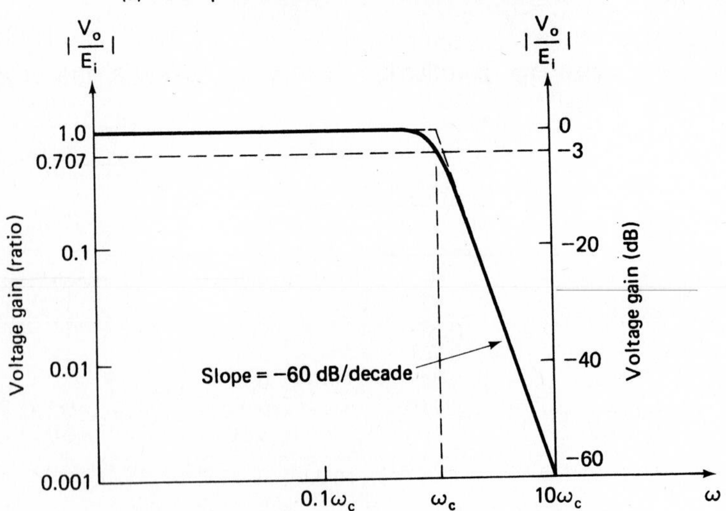

(b) Plot of frequency response for the circuit of part (a)

FIGURE 11-5 Low-pass filter designed for a roll-off of −60 dB/decade and corresponding frequency-response plot.

$$A_{CL} = \frac{V_o}{E_i} = \frac{V_{o1}}{E_i} \times \frac{V_o}{V_{o1}} \tag{11-4}$$

For a Butterworth filter, the magnitude of A_{CL} must be 0.707 at ω_c. To guarantee that the frequency response is flat in the passband, use the following design steps.

Design procedure

1. Choose the cutoff frequency, ω_c or f_c.
2. Pick C_3; choose a convenient value between 0.001 and 0.1 μF.
3. Make

$$C_1 = \tfrac{1}{2}C_3 \qquad \text{and} \qquad C_2 = 2C_3 \tag{11-5}$$

4. Calculate

$$R = \frac{1}{\omega_c C_3} \tag{11-6}$$

5. Make $R_1 = R_2 = R_3 = R$.
6. $R_{f1} = 2R$ and $R_{f2} = R$. For best results the value of R should be between 10 and 100 kΩ. If the value of R is outside this range, you should go back and pick a new value of C_3.

Example 11-5

For the -60-dB/decade low-pass filter of Fig. 11-5(a), determine the values of C_1, C_2, and R for a cutoff frequency of 1 kHz. Let $C_3 = 0.01$ μF.

Solution From Eq. (11-5),

$$C_1 = \tfrac{1}{2}C_3 = \tfrac{1}{2}(0.01\ \mu\text{F}) = 0.005\ \mu\text{F}$$

and

$$C_2 = 2C_3 = 2(0.01\ \mu\text{F}) = 0.02\ \mu\text{F}$$

From Eq. (11-6),

$$R = \frac{1}{(6.28)(1 \times 10^3)(0.01 \times 10^{-6})} = 15{,}915\ \Omega$$

Example 11-5 shows that the value of R in Fig. 11-5(a) is different from those of Fig. 11-4(a), although the cutoff frequency is the same. This is necessary so that $|A_{CL}|$ remains at 0 dB in the passband until the cutoff frequency is nearly reached; then $|A_{CL}| = 0.707$ at ω_c.

11-4.2 Filter Response

The solid line in Fig. 11-5(b) is the actual plot of the frequency response for Fig. 11-5(a). The dashed curve in the vicinity shows the straight-line approximation. Table 11-3 compares the magnitudes of A_{CL} for the three low-pass filters presented

TABLE 11-3 $|A_{CL}|$ FOR THE LOW-PASS FILTERS OF FIGS. 11-2(a), 11-4(a), AND 11-5(a)

ω	−20 dB/decade; Fig. 11-2(a)	−40 dB/decade; Fig. 11-4(a)	−60 dB/decade; Fig. 11-5(a)
$0.1\omega_c$	1.0	1.0	1.0
$0.25\omega_c$	0.97	0.998	0.999
$0.5\omega_c$	0.89	0.97	0.992
ω_c	0.707	0.707	0.707
$2\omega_c$	0.445	0.24	0.124
$4\omega_c$	0.25	0.053	0.022
$10\omega_c$	0.1	0.01	0.001

in this chapter. Note that the $|A_{CL}|$ for Fig. 11-5(a) remains quite close to 1 (0 dB) until the cutoff frequency, ω_c; then the steep roll-off occurs.

The phase angles for the low-pass filter of Fig. 11-5(a) range from 0° at $\omega = 0$ (dB condition) to −270° as ω approaches ∞. Table 11-4 compares the phase angles for the three low-pass filters.

TABLE 11-4 PHASE ANGLES FOR THE LOW-PASS FILTERS OF FIGS. 11-2(a), 11-4(a), AND 11-5(a)

ω	−20 dB/decade; Fig. 11-2(a)	−40 dB/decade; Fig. 11-4(a)	−60 dB/decade; Fig. 11-5(a)
$0.1\omega_c$	−6°	−8°	−12°
$0.25\omega_c$	−14°	−21°	−29°
$0.5\omega_c$	−27°	−43°	−60°
ω_c	−45°	−90°	−135°
$2\omega_c$	−63°	−137°	−210°
$4\omega_c$	−76°	−143°	−226°
$10\omega_c$	−84°	−172°	−256°

11-5 HIGH-PASS BUTTERWORTH FILTERS

11-5.1 Introduction

A high-pass filter is a circuit that attenuates all signals below a specified cutoff frequency ω_c and passes all signals whose frequency is above the cutoff frequency. Thus a high-pass filter performs the opposite function of the low-pass filter.

Figure 11-6 is a plot of the magnitude of the closed-loop gain versus ω for three types of Butterworth filters. The phase angle for a circuit of 20 dB/decade is +45° at ω_c. Phase angles at ω_c *increase* by +45° for each *increase* of 20 dB/decade. The phase angles for these three types of high-pass filters are compared in Section 11-5.5.

In this book the design of high-pass filters will be similar to that of the low-

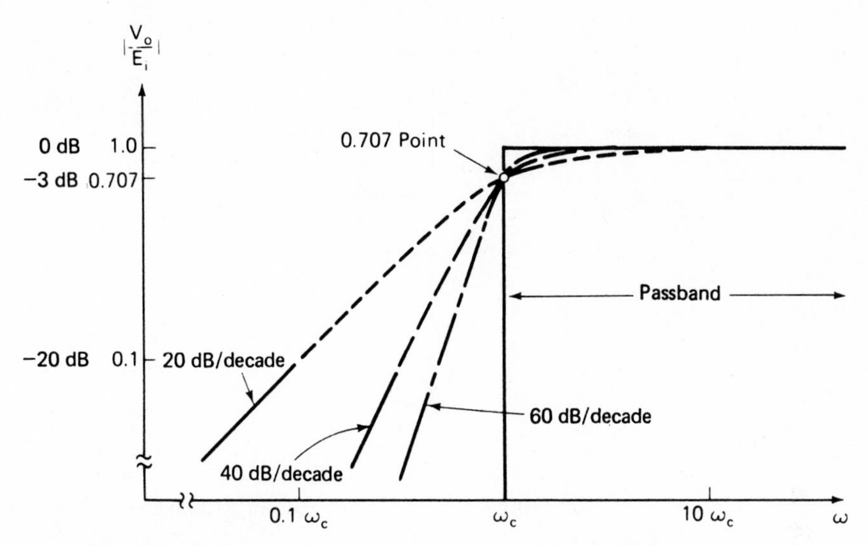

FIGURE 11-6 Comparison of frequency response for three high-pass Butterworth filters.

pass filters. In fact, the only difference will be the position of the filtering capacitors and resistors.

11-5.2 20-dB/Decade Filter

Compare the high-pass filter of Fig. 11-7(a) with the low-pass filter of Fig. 11-2(a) and note that C and R are interchanged. The feedback resistor R_f is included to minimize dc offset. Since the op amp is connected as a unity-gain follower in Fig. 11-7(a), the output voltage V_o equals the voltage across R and is expressed by

$$V_o = \frac{1}{1 - j(1/\omega RC)} \times E_i \tag{11-7}$$

When ω approaches 0 rad/s in Eq. (11-7), V_o approaches 0 V. At high frequencies, as ω approaches infinity, V_o equals E_i. Since the circuit is not an ideal filter, the frequency response is not ideal, as shown by Fig. 11-7(b). The solid line is the actual response, while the dashed lines show the straight-line approximation. The magnitude of the closed-loop gain equals 0.707 when $\omega RC = 1$. Therefore, the cutoff frequency ω_c is given by

$$\omega_c = \frac{1}{RC} = 2\pi f_c \tag{11-8a}$$

or

$$R = \frac{1}{\omega_c C} = \frac{1}{2\pi f_c C} \tag{11-8b}$$

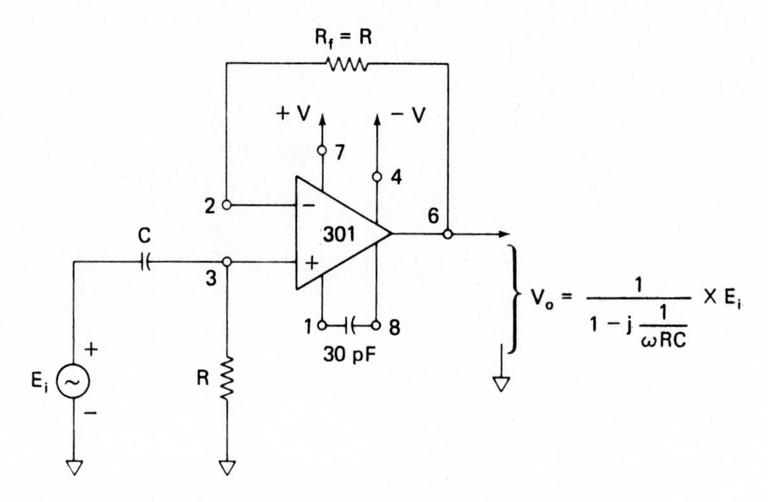

(a) High-pass filter with a roll-off of 20 dB/decade

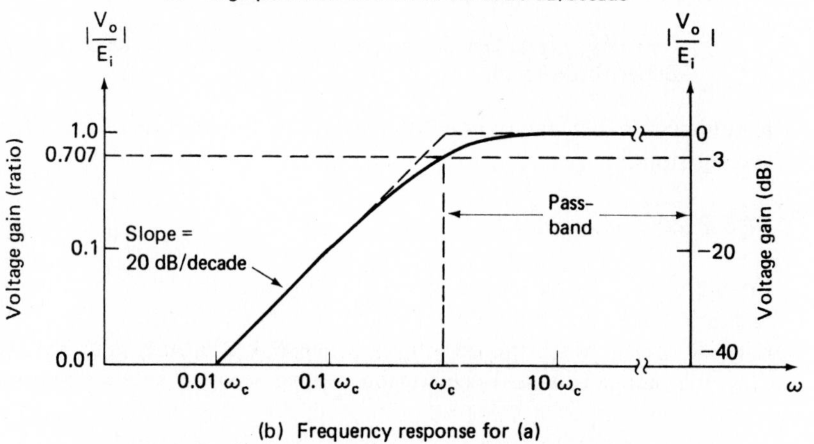

(b) Frequency response for (a)

FIGURE 11-7 Basic high-pass filter, 20 dB/decade.

The reason for solving for R and not C in Eq. (11-8b) is that it is easier to adjust R than it is C. The steps needed in designing Fig. 11-7(a) are as follows:

Design procedure for 20-dB/decade high-pass

1. Choose the cutoff frequency, ω_c or f_c.
2. Choose a convenient value of C, usually between 0.001 and 0.1 μF.
3. Calculate R from Eq. (11-8b).
4. Choose $R_f = R$.

Example 11-6

Calculate R in Fig. 11-7(a) if $C = 0.002\ \mu\text{F}$ and $f_c = 10$ kHz.

Solution From Eq. (11-8b),

$$R = \frac{1}{(6.28)(10 \times 10^3)(0.002 \times 10^{-6})} = 8\ \text{k}\Omega$$

Example 11-7

In Fig 11-7(a) if $R = 22\ \text{k}\Omega$ and $C = 0.01\ \mu\text{F}$, calculate (a) ω_c; (b) f_c.

Solution (a) From Eq. (11-8a),

$$\omega_c = \frac{1}{(22 \times 10^3)(0.01 \times 10^{-6})} = 4.54\ \text{krad/s}$$

(b)

$$f_c = \frac{\omega_c}{2\pi} = \frac{4.54 \times 10^3}{6.28} = 724\ \text{Hz}$$

11-5.3 40-dB/Decade Filter

The circuit of Fig. 11-8(a) is to be designed as a high-pass Butterworth filter with a roll-off of 40 dB/decade below the cutoff frequency, ω_c. To satisfy the Butterworth criteria, the frequency response must be 0.707 at ω_c and be 0 dB in the pass band. These conditions will be met if the following design procedure is followed:

Design procedure for 40-dB/decade high-pass

1. Choose a cutoff frequency, ω_c or f_c.
2. Let $C_1 = C_2 = C$ and choose a convenient value.
3. Calculate R_1 from

$$R_1 = \frac{1.414}{\omega_c C} \tag{11-9}$$

4. Select

$$R_2 = \tfrac{1}{2} R_1 \tag{11-10}$$

5. To minimize dc offset, let $R_f = R_1$.

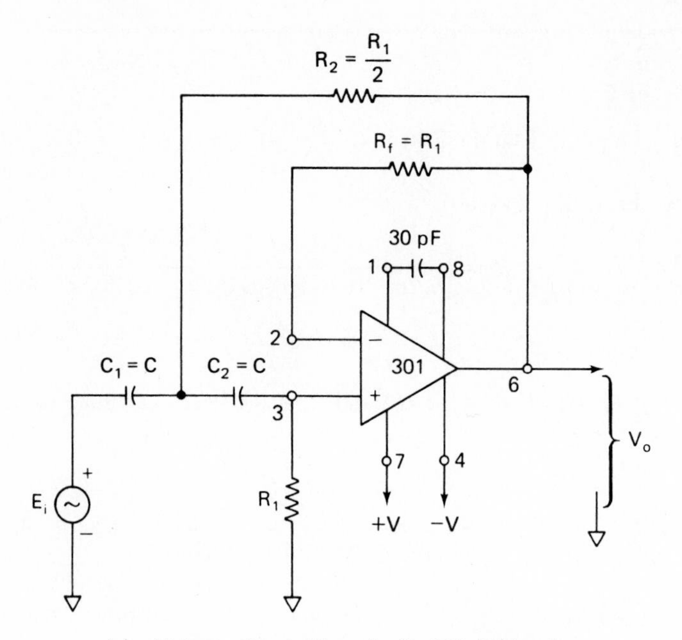

(a) High-pass filter will a roll-off of 40 dB/decade

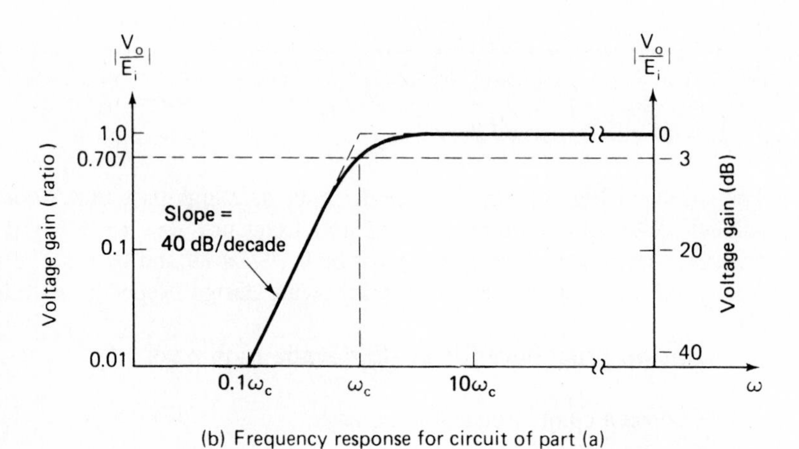

(b) Frequency response for circuit of part (a)

FIGURE 11-8 Circuit and frequency response for a 40-dB/decade high-pass Butterworth filter.

Example 11-8

In Fig. 11-8(a), let $C_1 = C_2 = 0.01\ \mu$F. Calculate (a) R_1 and (b) R_2 for a cutoff frequency of 1 kHz.

Solution (a) From Eq. (11-9),

$$R_1 = \frac{1.414}{(6.28)(1 \times 10^3)(0.01 \times 10^{-6})} = 22.5 \text{ k}\Omega$$

(b) $R_2 = \frac{1}{2}(22.5 \text{ k}\Omega) = 11.3 \text{ k}\Omega$.

Example 11-9

Calculate (a) R_1 and (b) R_2 in Fig. 11-8(a) for a cutoff frequency of 80 krad/s. $C_1 = C_2 = 125$ pF.

Solution (a) From (11-9),

$$R_1 = \frac{1.414}{(80 \times 10^3)(125 \times 10^{-12})} = 140 \text{ k}\Omega$$

(b) $R_2 = \frac{1}{2}(140 \text{ k}\Omega) = 70 \text{ k}\Omega$.

11-5.4 60-dB/Decade Filter

As with the low-pass filter of Fig. 11-5, a high-pass filter of 60 dB/decade can be constructed by cascading a 40-dB/decade filter with a 20-dB/decade filter. This circuit (like the other high- and low-pass filters) is designed as a Butterworth filter to have the frequency response in Fig. 11-9(b). The design steps for Fig. 11-9(a) are as follows:

Design procedure for 60-dB/decade high-pass

1. Choose the cutoff frequency, ω_c or f_c.
2. Let $C_1 = C_2 = C_3 = C$ and choose a convenient value between 100 pF and 0.1 μF.
3. Calculate R_3 from

$$R_3 = \frac{1}{\omega_c C} \tag{11-11}$$

4. Select

$$R_1 = 2R_3 \tag{11-12}$$

5. Select

$$R_2 = \frac{1}{2}R_3 \tag{11-13}$$

6. To minimize dc offset current, let $R_{f1} = R_1$ and $R_{f2} = R_3$.

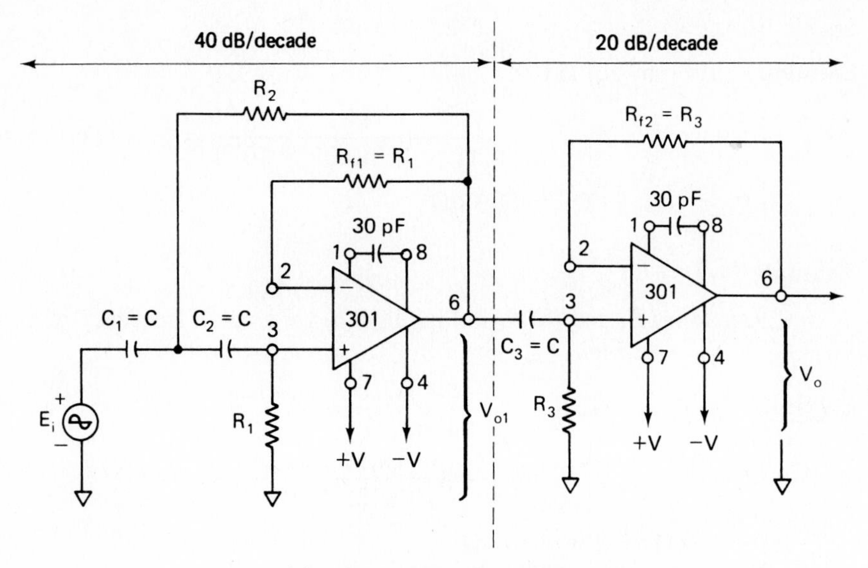

(a) High-pass filter for a 60 dB/decade slope

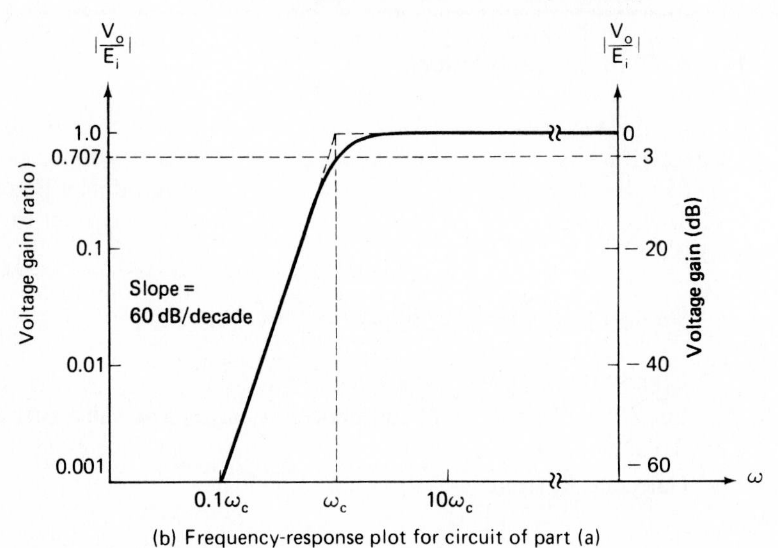

(b) Frequency-response plot for circuit of part (a)

FIGURE 11-9 Circuit and frequency response for a 60-dB/decade Butterworth high-pass filter.

Example 11-10

For Fig. 11-9(a), let $C_1 = C_2 = C_3 = C = 0.1\ \mu\text{F}$. Determine (a) R_3, (b) R_1, and (c) R_2 for $\omega_c = 1$ krad/s. ($f_c = 159$ Hz.)

Solution (a) By Eq. (11-11),

$$R_3 = \frac{1}{(1 \times 10^3)(0.1 \times 10^{-6})} = 10 \text{ k}\Omega$$

(b) $R_1 = 2R_3 = 2(10 \text{ k}\Omega) = 20 \text{ k}\Omega$.
(c) $R_2 = \frac{1}{2}R_3 = \frac{1}{2}(10 \text{ k}\Omega) = 5 \text{ k}\Omega$.

Example 11-11

Determine (a) R_3, (b) R_1, and (c) R_2 in Fig. 11-9(a) for a cutoff frequency of 60 kHz. Let $C_1 = C_2 = C_3 = C = 220$ pF.

Solution (a) From Eq. (11-11),

$$R_3 = \frac{1}{(6.28)(60 \times 10^3)(220 \times 10^{-12})} = 12 \text{ k}\Omega$$

(b) $R_1 = 2R_3 = 2(12 \text{ k}\Omega) = 24 \text{ k}\Omega$.
(c) $R_2 = \frac{1}{2}R_3 = \frac{1}{2}(12 \text{ k}\Omega) = 6 \text{ k}\Omega$.

If desired, the 20-dB/decade section can come before the 40-dB/decade section, because the op amps provide isolation and do not load one another.

11-5.5 Comparison of Magnitudes and Phase Angles

Table 11-5 compares the magnitudes of the closed-loop gain for the three high-pass filters. For each increase of 20 dB/decade, the circuit not only has a steeper roll-off below ω_c but also remains closer to 0 dB or a gain of 1 above ω_c.

The phase angle for a 20-dB/decade Butterworth high-pass filter is 45° at ω_c. For a 40-dB/decade filter it is 90°, and for a 60-dB/decade filter it is 135°. Other phase angles in the vicinity of ω_c for the three filters are given in Table 11-6.

TABLE 11-5 COMPARISON OF $|A_{CL}|$ FOR FIGS. 11-7(a), 11-8(a), AND 11-9(a)

ω	20 dB/decade; Fig. 11-7(a)	40 dB/decade; Fig. 11-8(a)	60 dB/decade; Fig. 11-9(a)
$0.1\omega_c$	0.1	0.01	0.001
$0.25\omega_c$	0.25	0.053	0.022
$0.5\omega_c$	0.445	0.24	0.124
ω_c	0.707	0.707	0.707
$2\omega_c$	0.89	0.97	0.992
$4\omega_c$	0.97	0.998	0.999
$10\omega_c$	1.0	1.0	1.0

TABLE 11-6 COMPARISON OF PHASE ANGLES FOR FIGS. 11-7(a), 11-8(a), AND 11-9(a)

ω	20 dB/decade; Fig. 11-7(a)	40 dB/decade; Fig. 11-8(a)	60 dB/decade; Fig. 11-9(a)
$0.1\omega_c$	84°	172°	256°
$0.25\omega_c$	76°	143°	226°
$0.5\omega_c$	63°	137°	210°
ω_c	45°	90°	135°
$2\omega_c$	27°	43°	60°
$4\omega_c$	14°	21°	29°
$10\omega_c$	6°	8°	12°

11-6 INTRODUCTION TO BANDPASS FILTERS

11-6.1 Frequency Response

A bandpass filter is a frequency selector. It allows one to select or pass only one particular band of frequencies from all other frequencies that may be present in a circuit. Its normalized frequency response is shown in Fig. 11-10. This type of filter has a maximum gain at a resonant frequency f_r. In this chapter all bandpass filters will have a gain of 1 or 0 dB at f_r. There is one frequency below f_r where the gain falls to 0.707. It is the *lower cutoff frequency*, f_L. At *higher cutoff frequency*, f_H, the gain also equals 0.707, as in Fig. 11-10.

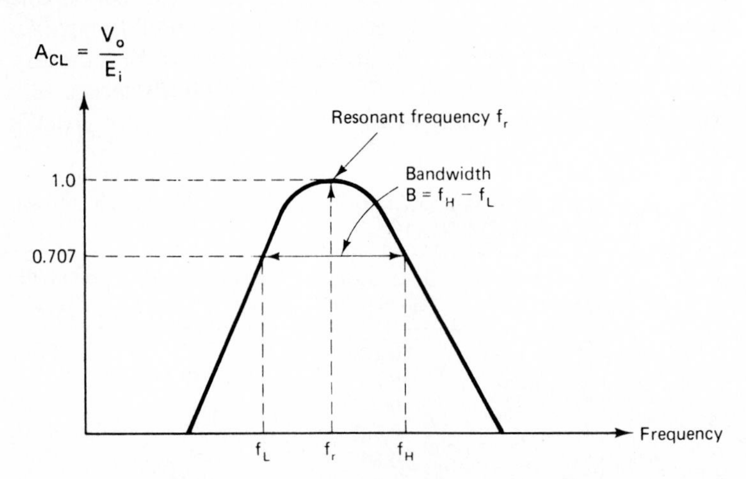

FIGURE 11-10 A bandpass filter has a maximum gain at resonant frequency f_r. The band of frequencies transmitted lies between f_L and f_H.

11-6.2 Bandwidth

The range of frequencies between f_L and f_H is called *bandwidth B,* or

$$B = f_H - f_L \tag{11-14}$$

The bandwidth is not exactly centered on the resonant frequency. (It is for this reason that we use the historical name "resonant frequency" rather than "center frequency" to describe f_r.)

If you know the values for f_L and f_H, the resonant frequency can be found from

$$f_r = \sqrt{f_L f_H} \tag{11-15}$$

If you know the resonant frequency, f_r, and bandwidth, B, cutoff frequencies can be found from

$$f_L = \sqrt{\frac{B^2}{4} + f_r^2} - \frac{B}{2} \tag{11-16a}$$

$$f_H = f_L + B \tag{11-16b}$$

Example 11-12

A bandpass voice filter has lower and upper cutoff frequencies of 300 and 3000 Hz. Find (a) the bandwidth; (b) the resonant frequency.

Solution (a) From Eq. (11-14),

$$B = f_H - f_L = (3000 - 300) = 2700 \text{ Hz}$$

(b) From Eq. (11-15),

$$f_r = \sqrt{f_L f_H} = \sqrt{(300)(3000)} = 948.7 \text{ Hz}$$

[Note that f_r is always *below* the center frequency of $(3000 + 300)/2 = 1650$ Hz.]

Example 11-13

A bandpass filter has a resonant frequency of 950 Hz and a bandwidth of 2700 Hz. Find its lower and upper cutoff frequencies.

Solution From Eq. (11-16a),

$$f_L = \sqrt{\frac{B^2}{4} + f_r^2} - \frac{B}{2} = \sqrt{\frac{(2700)^2}{4} + (950)^2} - \frac{2700}{2}$$

$$= 1650 - 1350 = 300 \text{ Hz}$$

From Eq. (11-16b), $f_H = 300 + 2700 = 3000$ Hz.

11-6.3 Quality Factor

The *quality factor* Q is defined as the ratio of resonant frequency to bandwidth, or

$$Q = \frac{f_r}{B} \tag{11-17}$$

Q is a measure of the bandpass filter's *selectivity*. A high Q indicates that a filter selects a smaller band of frequencies (more selective).

11-6.4 Narrowband and Wideband Filters

A *wideband* filter has a bandwidth that is two or more times the resonant frequency. That is, $Q \leq 0.5$ for wideband filters. In general, wideband filters are made by cascading a low-pass filter circuit with a high-pass filter circuit. This topic is covered in the next section. A narrowband filter ($Q > 0.5$) can usually be made with a single stage. This type of filter is presented in Section 11-8.

Example 11-14

Find the quality factor of a voice filter that has a bandwidth of 2700 Hz and a resonant frequency of 950 Hz (see Examples 11-12 and 11-13).

Solution From Eq. (11-7),

$$Q = \frac{f_r}{B} = \frac{950}{2700} = 0.35$$

This filter is classified as wideband because $Q < 0.5$.

11-7 BASIC WIDEBAND FILTER

11-7.1 Cascading

When the output of one circuit is connected in series with the input of a second circuit, the process is called *cascading* gain stages. In Fig. 11-11, the first stage is a 3000-Hz low-pass filter (Section 11-3). Its output is connected to the input of a 300-Hz high-pass filter (Section 11-5.3). The cascaded pair of active filters now form a bandpass filter from input E_i to output V_o. Note that it makes no difference if the high-pass is connected to the low-pass, or vice versa.

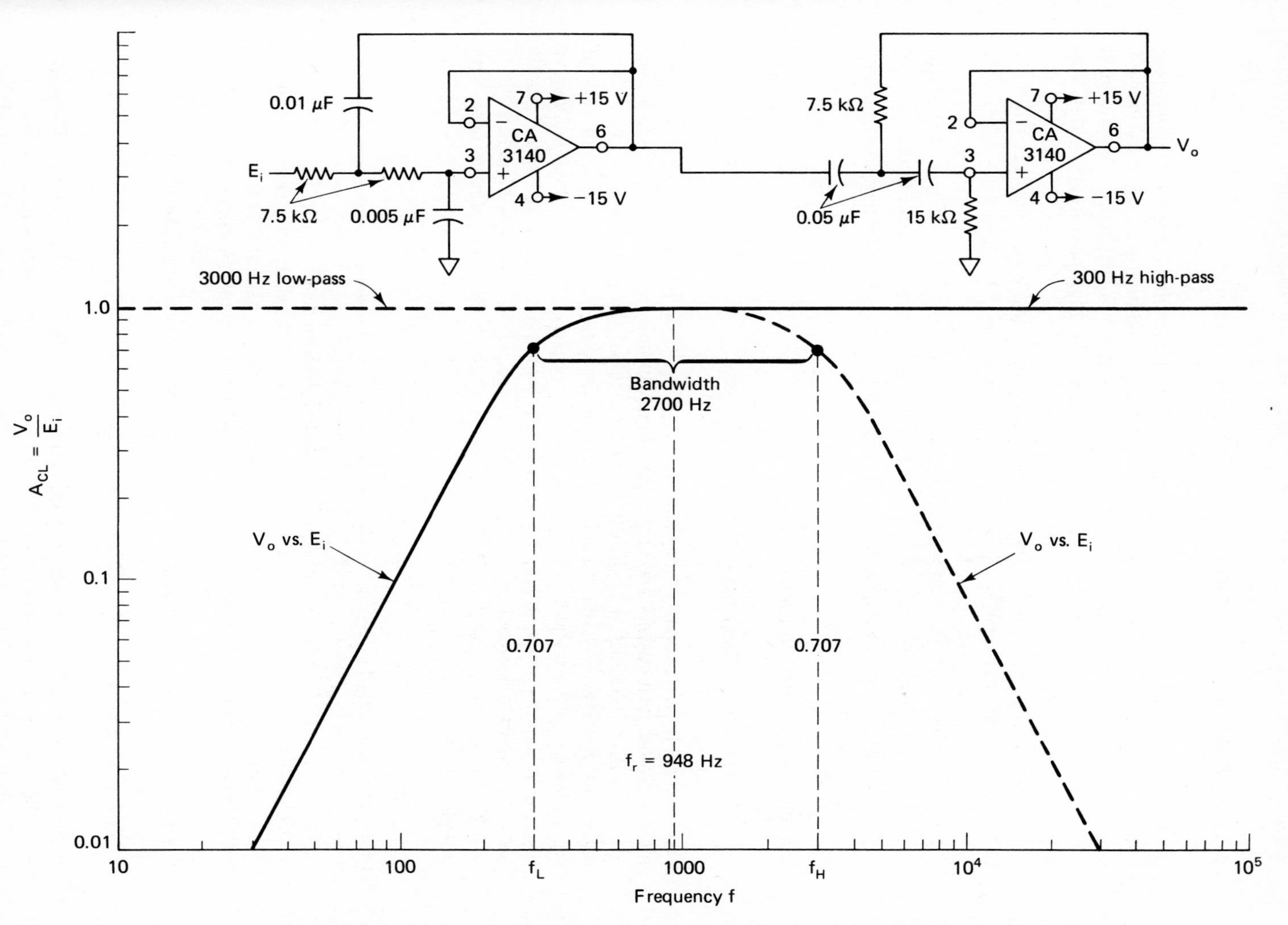

FIGURE 11-11 A 3000-Hz second-order low-pass filter is cascaded with a 300-Hz high-pass filter to form a 300- to 3000-Hz bandpass voice filter.

11-7.2 Wideband Filter Circuit

In general, a wideband filter ($Q \leq 0.5$) is made by cascading a low- and a high-pass filter (see Fig. 11-11). Cutoff frequencies of the low- and high-pass sections *must not overlap,* and each must have the same passband gain. Furthermore, the low-pass filter's cutoff frequency must be 10 or more times the high-pass filter's cutoff frequency.

For cascaded low- and high-pass filters, the resulting wideband filter will have the following characteristics:

1. The lower cutoff frequency, f_L, will be determined only by the high-pass filter.
2. The high cutoff frequency, f_H, will be set only by the low-pass filter.
3. Gain will be maximum at resonant frequency, f_r and equal to the passband gain of either filter.

These principles are illustrated next.

11-7.3 Frequency Response

In Fig. 11-11 the frequency response of a basic −40-dB/decade 3000-Hz low-pass filter is plotted as a dashed line. The frequency response of a 300-Hz high-pass filter is plotted as a solid line. The 40-dB/decade roll-off of the high-pass filter is seen to determine f_L. The −40-dB/decade roll-off of the low-pass sets f_H. Both roll-off curves make up the frequency response of the bandpass filter, V_o versus f. Observe that the resonant, low, and high cutoff frequencies plus bandwidth agree exactly with the values calculated in Examples 11-12 and 11-13. Narrow bandpass filters will be introduced in Section 11-8. Discussion of notch filters is deferred until Sections 11-9 and 11-10.

11-8 NARROWBAND BANDPASS FILTERS

Narrowband filters exhibit the typical frequency response shown in Fig. 11-12(a). The analysis and construction of narrowband filters is considerably simplified if we stipulate that the narrowband filter will have a maximum gain of 1 or 0 dB at the resonant frequency f_r. Equations (11-14) to (11-17) and bandpass terms were presented in Section 11-6. They gave an introduction to (cascaded pair) wideband filters. These equations and terms also apply to the narrowband filters that follow.

11-8.1 Narrowband Filter Circuit

A narrowband filter circuit uses only one op amp, as shown in Fig. 11-12. (Compare with the two-op-amp wideband filters in Fig. 11-11.) The filter's input resistance is established approximately by resistor R. If the feedback resistor ($2R$) is made two

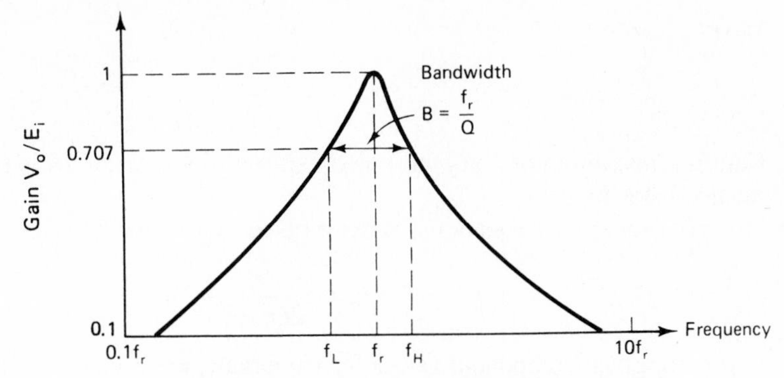

(a) Typical frequency response curve of a narrowband filter

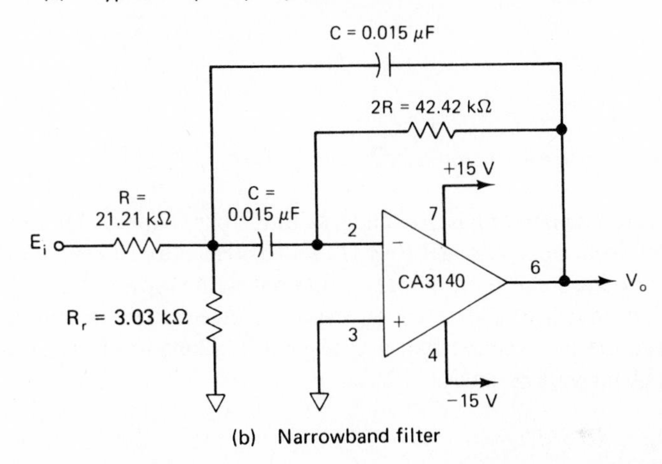

(b) Narrowband filter

FIGURE 11-12 Narrow bandpass filter circuit and its frequency response for the component values shown; $f_r = 100$ Hz, $B = 500$ Hz, $Q = 2$, $f_L = 780$ Hz, and $f_H = 1280$ Hz. (a) Typical frequency response of a bandpass filter; (b) narrow bandpass filter circuit.

times the input resistor R, the filter's maximum gain will be 1 or 0 dB at resonant frequency f_r. By adjusting R_r one can change (or exactly trim) the resonant frequency *without changing the bandwidth or gain.*

11-8.2 Performance

The performance of the *unity*-gain narrowband filter in Fig. 11-12 is determined by only a few simple equations. The bandwidth B in hertz is determined by resistor R and the two (matched) capacitors C by

$$B = \frac{0.1591}{RC} \tag{11-18a}$$

where

$$B = \frac{f_r}{Q} \tag{11-18b}$$

Gain is a maximum of 1 at f_r provided that feedback resistor $2R$ is twice the value of input resistor R.

The resonant frequency f_r is determined by resistor R_r according to

$$R_r = \frac{R}{2Q^2 - 1} \tag{11-19}$$

If you are given component values for the circuit, its resonant frequency can be calculated from

$$f_r = \frac{0.1125}{RC}\sqrt{1 + \frac{R}{R_r}} \tag{11-20}$$

11-8.3 Stereo-Equalizer Octave Filter

A stereo equalizer has 10 bandpass filters per channel. They separate the audio spectrum from approximately 30 Hz to 16 kHz into 10 separate octaves of frequency. Each octave can then be cut or boosted with respect to the other to achieve special sound effects, equalize room response, or equalize an automotive compartment to make the radio sound like it is playing in a large hall. The construction of one such equalizer will be analyzed by an example.

Example 11-15

Octave equalizers have resonant frequencies at approximately 32, 64, 128, 250, 500, 1000, 2000, 4000, 8000, and 16,000 Hz. Q of each filter is chosen to have values between 1.4 and 2. Let's make a unity-gain narrowband filter to select the sixth octave. Specifically, make a filter with $f_r = 1000$ Hz and $Q = 2$.

Solution From Eq. (11-18b),

$$B = \frac{f_r}{Q} = \frac{1000}{2} = 500 \text{ Hz}$$

[*Note:* From Eq. (11-16), $f_L = 80$ and $f_H = 1280$ Hz.] Choose $C = 0.015\ \mu$F. Find R from Eq. (11-18a).

$$R = \frac{0.1591}{BC} = \frac{0.1591}{(500)(0.015 \times 10^{-6}\text{ F})} = 21.21\text{ k}\Omega$$

The feedback resistor will be $2R = 42.42\ \text{k}\Omega$. Find R_r from Eq. (11-19).

$$R_r = \frac{R}{2Q^2 - 1} = \frac{21.21\ \text{k}\Omega}{2(2)^2 - 1} = \frac{21.21\ \text{k}\Omega}{7} = 3.03\ \text{k}\Omega$$

Example 11-16

Given a bandpass filter circuit with the component values in Fig. 11-12, find (a) the resonant frequency; (b) the bandwidth.

Solution (a) From Eq. (11-12),

$$f_r = \frac{0.1125}{RC}\sqrt{1 + \frac{R}{R_r}} = \frac{0.1125}{(21.21 \times 10^3)(0.015 \times 10^{-6})}\sqrt{1 + \frac{21.21\ \text{k}\Omega}{3.03\ \text{k}\Omega}}$$

$$= (353.6\ \text{Hz})\sqrt{1 + 7} = 353.6\ \text{Hz} \times 2.83 \simeq 1000\ \text{Hz}$$

(b) From Eq. (11-18a),

$$B = \frac{0.1591}{RC} = \frac{0.1591}{(21.21 \times 10^3)(0.015 \times 10^{-6})} = 500\ \text{Hz}$$

11-9 NOTCH FILTERS

11-9.1 Introduction

The notch or band-reject filter is named for the characteristic shape of its frequency-response curve in Fig. 11-13. Unwanted frequencies are attenuated in the stopband B. The desired frequencies are transmitted in the passband that lies on either side of the notch.

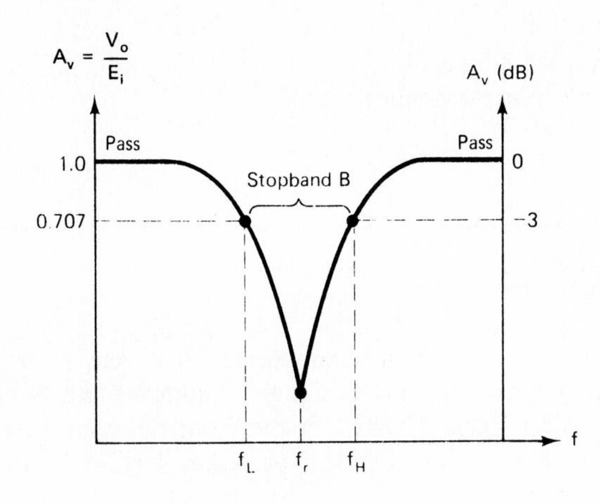

FIGURE 11-13 A notch filter transmits frequencies in the passband and rejects undesired frequencies in the stopband.

Notch filters usually have a passband gain of unity or 0 dB. The equations for Q, B, f_L, f_H, and f_r are identical to those of its associated bandpass filter. The reasons for this last statement are presented next.

11-9.2 Notch Filter Theory

As shown in Fig. 11-14, a notch filter is made by subtracting the output of a bandpass filter from the original signal. For frequencies in the notch filter's passband, the output of the bandpass filter section approaches zero. Therefore, input E_i is transmitted via adder input resistor R_1 to drive V_o to a value equal to $-E_i$. Thus $V_o = -E_i$ in both lower and upper passbands of the notch filter.

Suppose that the frequency of E_i is adjusted to resonant frequency f_r of the narrow bandpass filter component. (*Note:* f_r of the bandpass sets the notch frequency.) E_i will exit from the bandpass as $-E_i$ and then is inverted by R_1 and R to drive V_o to $+E_i$. However, E_i is transmitted via R_2 to drive V_o to $-E_i$. Thus V_o responds to both inputs of the adder and becomes $V_o = E_i - E_i = 0$ V at f_r.

In practice, V_o approaches zero only at f_r. The depth of the notch depends on how closely the resistors and capacitors are matched in the bandpass filter and judicious fine adjustment of resistor R_1 at the inverting adder's onput. This procedure is explained in Section 11-10.3.

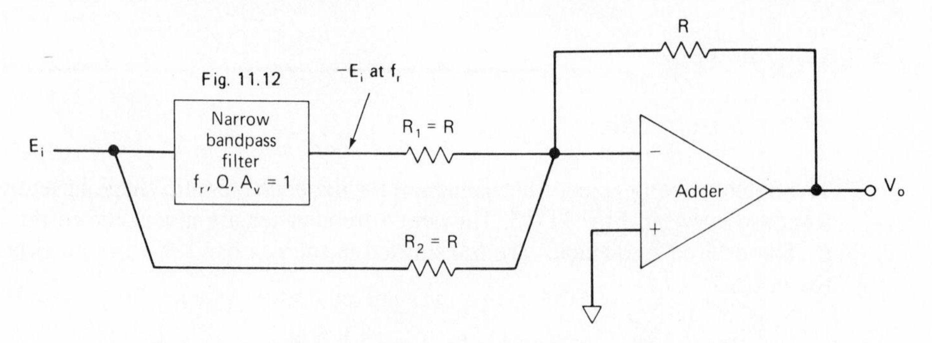

FIGURE 11-14 A notch filter is made by a circuit that subtracts the output of a bandpass filter from the original signal.

11-10 120-HZ NOTCH FILTER

11-10.1 Need for a Notch Filter

In applications where low-level signals must be amplified, there may be present one or more of an assortment of unwanted noise signals. Examples are 50-, 60-, or 400 Hz frequencies from power lines, 120-Hz, ripple from full-wave rectifiers, or even higher frequencies from regulated switching-type power supplies or clock oscillators. If both signals and a signal-frequency noise component are passed through a notch filter, only the desired signals will exit from the filter. The noise frequency is "notched out." As an example, let us make a notch filter to eliminate 120-Hz hum.

11-10.2 Statement of the Problem

The problem is to make a notch filter with a notch (resonant) frequency of $f_r =$ 120 Hz. Let us select a stopband of $B = 12$ Hz. Gain of the notch filter in the passband will be unity (0 dB), so that the desired signals will be transmitted without attenuations. We use Eq. (11-17) to determine a value for Q that is required by the notch filter:

$$Q = \frac{f_r}{B} = \frac{120}{12} = 10$$

This high value of Q means that (1) the notch and component bandpass filter will have narrow bands with very sharp frequency-response curves, and (2) the bandwidth is essentially centered on the resonant frequency. Accordingly, this filter will transmit all frequencies from 0 to $(120 - 6) = 114$ Hz, and also all frequencies above $(120 + 6) = 126$ Hz. The notch filter will stop all frequencies between 114 and 126 Hz.

11-10.3 Procedure to Make a Notch Filter

The procedure to make a notch filter is performed in two steps:

1. Make a bandpass filter that has the same resonant frequency, bandwidth, and consequently Q as the notch filter.
2. Connect the inverting adder of Fig. 11-15 by selecting equal resistors for R. Usually, $R = 10$ kΩ. (A practical fine-tuning procedure is presented in the next section).

11-10.4 Bandpass Filter Components

The first step in making a 120-Hz notch filter is best illustrated by an example (see Fig. 11-15).

Example 11-17

Design a bandpass filter with a resonant frequency of $f_r = 120$ Hz and a bandwidth of 12 Hz so that $Q = 10$. Thus gain of the bandpass section will be 1 at f_r and approach zero at the output of the notch labeled V_o.

Solution Choose $C = 0.33\ \mu F$. From Eq. (11-18a),

$$R = \frac{0.1591}{BC} = \frac{0.1591}{(12)(0.33 \times 10^{-6})} = 40.2\ k\Omega$$

Then the bandpass feedback resistor will be $2R$ equals 80.4 kΩ. From Eq. (11-19),

$$R_r = \frac{R}{2Q^2 - 1} = \frac{40.2\ k\Omega}{2(10)^2 - 1} = \frac{40.2\ k\Omega}{199} = 201\ \Omega$$

This bandpass filter component is built first and f_r is fine-tuned by adjusting R_r (see Section 11-8.2).

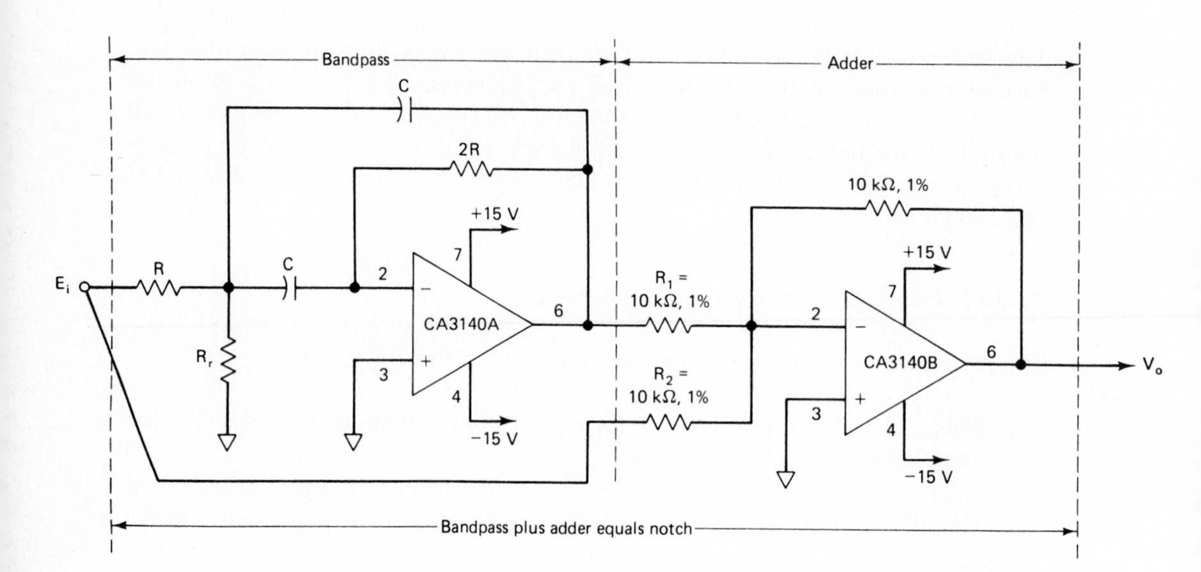

FIGURE 11-15 This two-op-amp notch filter is made from a bandpass filter plus an inverting adder. If $C = 0.33\ \mu F$, $R = 40.2\ k\Omega$, and $R_f = 201\ \Omega$, the notch frequency will be 120 Hz and reject a bandwidth of 12 Hz.

11-10.5 Final Assembly

Refer to Fig. 11-15. Simply connect an inverting adder (CA3140B or TL081) with equal 1% input and feedback 10-kΩ resistors as shown. The resultant notch filter (from E_i to V_o) exhibits a respectable performance that is an acceptable solution to the problem. The notch depth can be increased by fine trimming R_1 or R_2.

LABORATORY EXERCISES

A critical factor in designing any filter is the tolerance of the components. The filters that you studied in this chapter require measurement of all component values before building the circuits. Careful consideration must be given to match the values to less than 1%. Failure to carry out this tedious step only results in circuits that never meet design criteria. The capacitors should be type npo (zero-temperature coefficient).

11-1. Refer to the low-pass filter shown in Fig. LE11-1 and measure all component values before building the network. Using the measured component values, calculate the cutoff frequencies. If you have access to a function generator with sweep capabilities, use it for E_i.

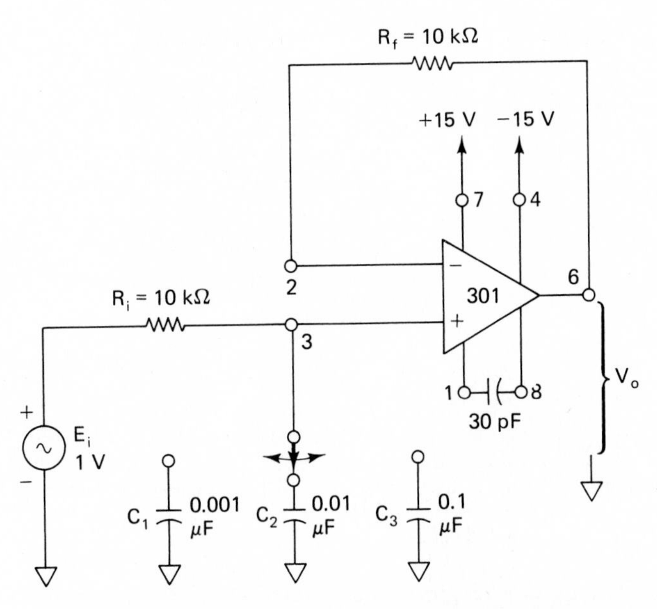

FIGURE LE11-1

(a) Set the switch to C_1 and monitor V_o on the CRO. Plot the envelope of V_o vs. time on log-log graph paper.
(b) Switch to C_2 and plot V_o vs. time on the same graph as in part (a).
(c) Repeat part (a) for the C_3 capacitor. As the value of C increases by a decade (times 10), the lower cutoff frequency should decrease by a decade.

11-2. This lab exercise can be done by two lab groups:

Group 1. Design a 40-dB/decade low-pass filter for a cutoff frequency of 10kHz. Choose a convenient value of C_1. Build and test your design. Plot V_o vs. time on log-log graph paper.

Group 2. Design a 40-dB/decade high-pass filter for a cutoff frequency of 1 kHz. Choose a convenient value of C. Build and test your design. Plot V_o vs. time on log-log graph paper.

Groups 1 and 2. Connect the output of the high-pass filter to the input of the low-pass filter. Set $E_i = 1$ V. If possible, use a sweep function generator. Sweep the frequency and plot V_o vs. time on log-log graph paper. (See Fig. 11-11.)

11-3. A stereo equalizer allows an operator to add, boost, or decrease the signal at certain frequencies throughout the audio spectrum. A 10-band stereo equalizer would have 10 controls for each channel. The 10 controls divide the audio spectrum into 10 octaves. The center frequency of the octave ranges are 32, 64, 128, 256, 512, 1024, 2048, 4096, 8192, and 16,384 Hz. Therefore, the key circuit in a stereo equalizer is a bandpass filter with the above-mentioned center frequencies. If the circuit is designed with a $Q = 1.5$, the 0.707 points on the frequency spectrum for each filter will intersect with its neighbor. See Fig. LE11-3a for the 128-, 256-, 512-, and 1024-Hz frequency plots.

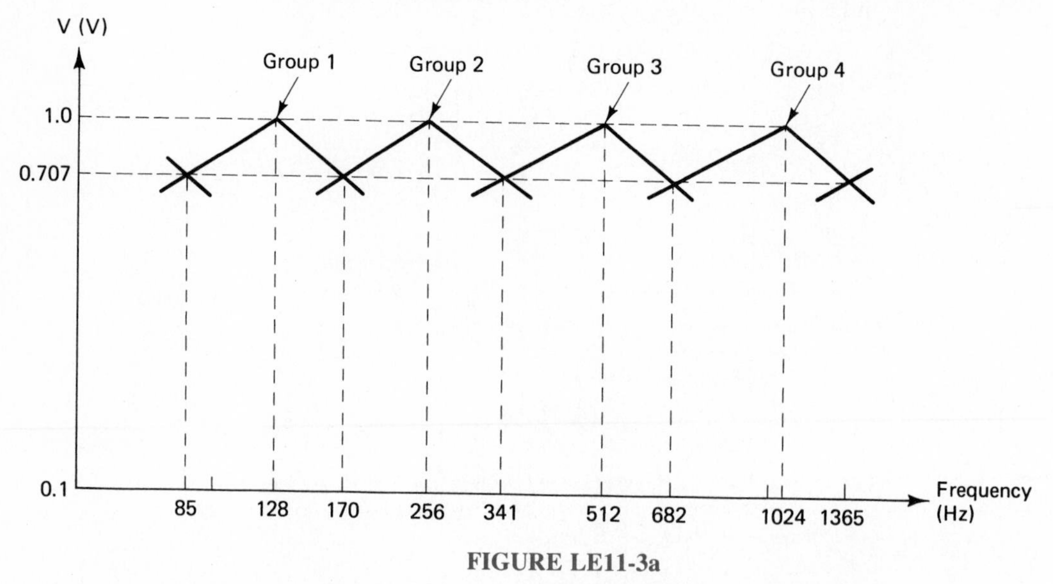

FIGURE LE11-3a

This lab exercise can be done by four lab groups. Each group is to build and test individually a bandpass filter. The resonant frequencies are: group 1, 128 Hz; group 2, 256 Hz; group 3, 512 Hz; and group 4, 1024 Hz. Each circuit is to have a $Q = 1.5$. To obtain the same resistor values, choose the following capacitor values:

f_r (Hz)	$C_1 = C_2 = C$ (μF)
128	0.04
256	0.02
512	0.01
1024	0.005

Each lab group is to plot V_o vs. time on log-log graph paper for their filter. Then each group can build a volume control and buffer (see Fig. LE11-3b).

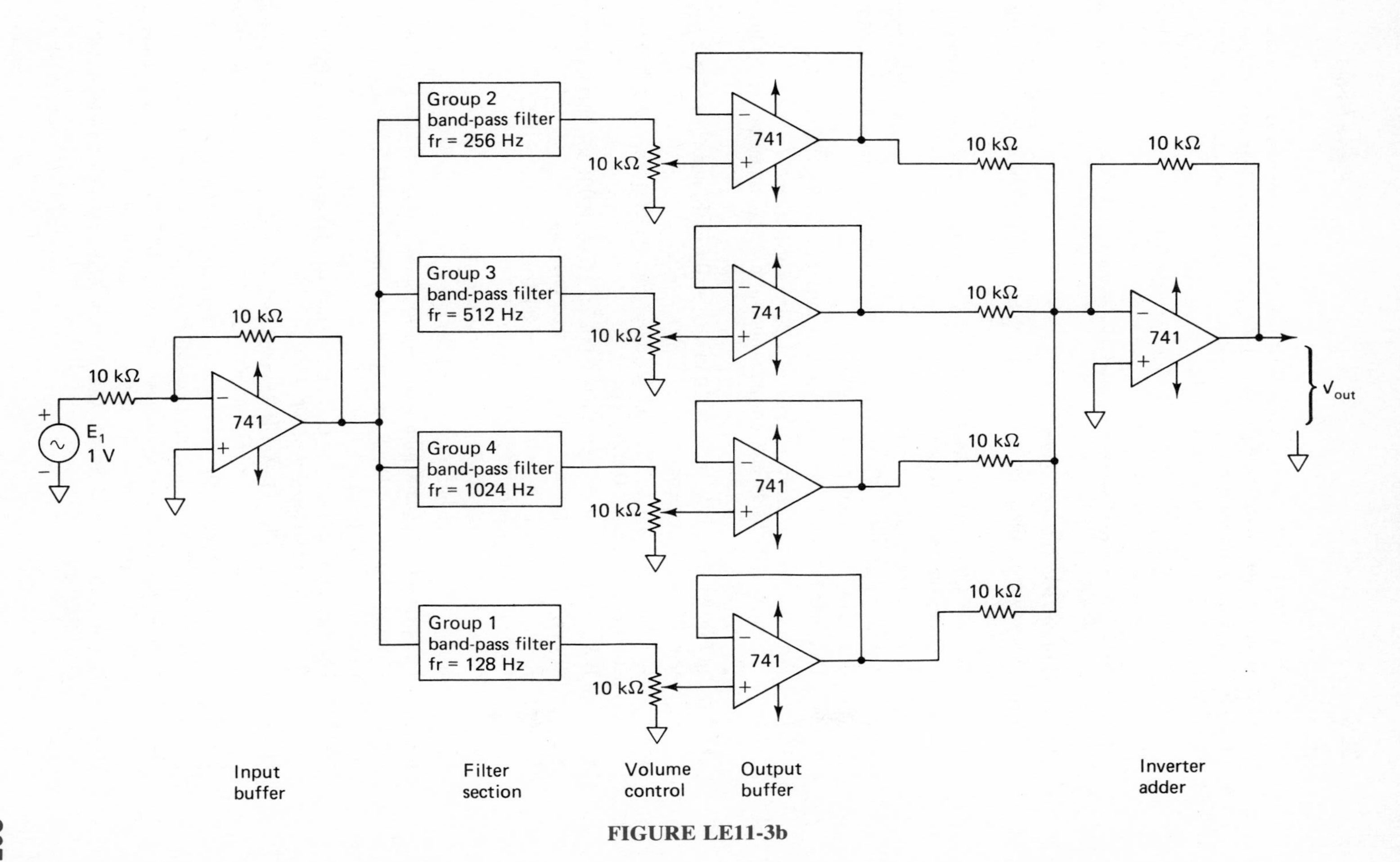

FIGURE LE11-3b

The first lab group to finish testing their bandpass filter and buffer network can build the inverter adder and input buffer. Connect all the networks together as shown in Fig. LE11-3b. Use a CRO to measure V_{out}. Sweep the input signal E_i and show the output voltage for different settings of the volume control.

PROBLEMS

11-1. List the four types of filters.

11-2. What type of filter has a constant output voltage from dc up to the cutoff frequency?

11-3. What is a filter called that passes a band of frequencies while attenuating all frequencies outside the band?

11-4. In Fig. 11-2(a), if R = 100 kΩ and C = 0.02 μF, what is the cutoff frequency?

11-5. The low-pass filter of Fig. 11-2(a) is to be designed for a cutoff frequency of 4.5 kHz. If C = 0.005 μF, calculate R.

11-6. Calculate the cutoff frequency for each value of C in Fig. LE11-1.

11-7. What are the two characteristics of a Butterworth filter?

11-8. Design a −40-dB/decade low-pass filter at a cutoff frequency of 10 krad/s. Let C_1 = 0.02 μF.

11-9. In Fig. 11-4(a), if $R_1 = R_2$ = 10 kΩ, C_1 = 0.01 μF, and C_2 = 0.002 μF, calculate the cutoff frequency f_c.

11-10. Calculate **(a)** R_3, **(b)** R_1, and **(c)** R_2 in Fig. 11-5(a) for a cutoff frequency of 10 krad/s. Let C_3 = 0.005 μF.

11-11. If $R_1 = R_2 = R_3$ = 20 kΩ, C_1 = 0.002 μF, C_2 = 0.008 μF, and C_3 = 0.004 μF in Fig. 11-5(a), determine the cutoff frequency ω_c.

11-12. In Fig. 11-5(a), C_1 = 0.01 μF, C_2 = 0.04 μF, and C_3 = 0.02 μF. Calculate R for a cutoff frequency of 1 kHz.

11-13. Calculate R in Fig. 11-7(a) if C = 0.04 μF and f_c = 500 Hz.

11-14. In Fig. 11-7(a) calculate **(a)** ω_c and **(b)** f_c if R = 10 kΩ and C = 0.01 μF.

11-15. Design a 40-dB/decade high-pass filter for ω_c = 5 krad/s. $C_1 = C_2$ = 0.02 μF.

11-16. Calculate **(a)** R_1 and **(b)** R_2 in Fig. 11-8(a) for a cutoff frequency of 40 krad/s. $C_1 = C_2$ = 250 pF.

11-17. For Fig. 11-9(a), let $C_1 = C_2 = C_3$ = 0.05 μF. Determine **(a)** R_3, **(b)** R_1, and **(c)** R_2 for a cutoff frequency of 500 Hz.

11-18. The circuit of Fig. 11-9(a) is designed with the values $C_1 = C_2 = C_3$ = 400 pF, R_1 = 100 kΩ, R_2 = 25 kΩ, and R_3 = 50 kΩ. Calculate the cutoff frequency f_c.

11-19. Find the **(a)** bandwidth, **(b)** resonant frequency, and **(c)** quality factor of a bandpass filter with lower and upper cutoff frequencies of 55 and 65 Hz.

11-20. A bandpass filter has a resonant frequency of 1000 Hz and a bandwidth of 2500 Hz. Find the lower and upper cutoff frequencies.

11-21. Use the capacitor and resistor values of the high-pass filter in Fig. 11-11 to prove f_C = 3000 Hz.

11-22. Use the capacitor and resistor values of the high-pass filter in Fig. 11-11 to prove that $f_C = 300$ Hz.

11-23. Find Q for the bandpass filter of Fig. 11-11.

11-24. Design a narrow bandpass filter using one op-amp. The resonant frequency is 128 Hz and $Q = 1.5$. Select $C = 0.1\ \mu F$ in Fig. 10-12.

11-25. **(a)** How would you convert the bandpass filter of Problem 11-24 above into a notch filter with the same resonant frequency and Q?
(b) Calculate f_L and f_H for the notch filter.

CHAPTER 12

Modulating, Demodulating, and Frequency Changing with the Multiplier

LEARNING OBJECTIVES

Upon completion of this chapter on multiplier ICs, you will be able to:

- Write the output–input equation of a multiplier IC and state the value of its scale factor.
- Multiply two dc voltages or divide one dc voltage by another.
- Square the value of a dc voltage or take its square root.
- Double the frequency of any sine wave.
- Measure the phase angle between two sine waves of exactly equal frequency.
- Show that amplitude modulation is actually a multiplication process.
- Multiply a carrier sine wave by a modulating sine wave and express the output voltage by either a product term or a term containing sum and difference frequencies.
- Calculate the amplitude and frequency of each output frequency term.
- Make either a balanced amplitude modulator or a standard amplitude modulator.
- Show how a multiplier can be used to shift frequencies.

12-0 INTRODUCTION

Analog multipliers are complex arrangements of op amps and other circuit elements now available in either integrated circuit or functional module form. Multipliers are easy to use; some of their applications are (1) measurement of power, (2) frequency doubling and shifting, (3) detecting phase-angle difference between two signals of equal frequency, (4) multiplying two signals, (5) dividing one signal by another, (6) taking the square root of a signal, and (7) squaring a signal. Another use for multipliers is to demonstrate the principles of amplitude modulation and demodulation. The schematic of a typical multiplier is shown in Fig. 12-1(a). There are two input terminals, x and y, which are used for connecting the two voltages to be multiplied. Typical input resistance of each input terminal is 10 kΩ or greater. One output terminal furnishes about the same current as an op amp to a grounded load (5 to 10 mA). The output voltage equals the product of the input voltages reduced by a scale factor. The *scale factor* is explained in Section 12-1.

12-1 MULTIPLYING DC VOLTAGES

12-1.1 Multiplier Scale Factor

The schematic of a multiplier shown in Fig. 12-1(a) may have a × to symbolize multiplication. Another type of schematic shows the inputs and the output voltage equation, as in Fig. 12-1(b). In general terms, the output voltage V_o is the product of input voltages x and y and is expressed by

$$V_o = kxy \tag{12-1a}$$

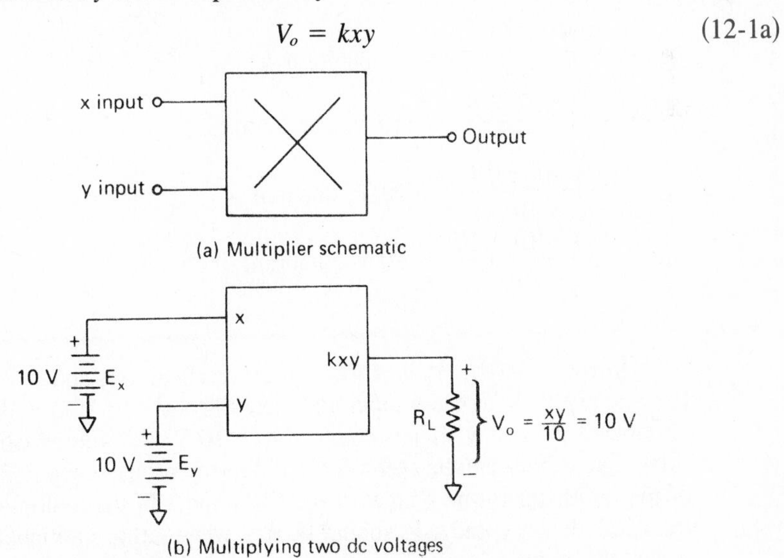

FIGURE 12-1 Introduction to the multiplier.

The constant k is called a *scale factor* and is usually equal to 1/10 V. This is because multipliers are designed for the same type of power supplies used for op amps, namely ±15 V. For best results, the voltages applied to either x or y inputs should not exceed +10 V or −10 V with respect to ground. This ±10-V limit also holds for the output, so the scale factor is usually the reciprocal of the voltage limit, or 1/10 V. If both input voltages are at their positive limits of +10 V, the output will be at its positive limit of 10 V. Thus Eq. (12-1a) is expressed for most multipliers by

$$V_o = \frac{xy}{10\text{ V}} = \frac{E_x E_y}{10\text{ V}} \tag{12-1b}$$

12-1.2 Multiplier Quadrants

Multipliers are classified by quadrants; for example, there are one-quadrant, two-quadrant, and four-quadrant multipliers. The classification is explained in two ways in Fig. 12-2. In Fig. 12-2(a), the input voltages can have four possible polarity combinations. If both x and y are positive, operation is in quadrant 1, since x is the horizontal and y the vertical axis. If x is positive and y is negative, quadrant 4 operation results, and so forth.

Example 12-1

Find V_o for the following combination of inputs: (a) $x = 10$ V, $y = 10$ V; (b) $x = -10$ V, $y = 10$ V; (c) $x = 10$ V, $y = -10$ V; (d) $x = -10$ V, $y = -10$ V.

Solution From Eq. (12-1b),

(a) $V_o = \dfrac{(10)(10)}{10} = 10$ V, quadrant 1

(b) $V_o = \dfrac{(-10)(10)}{10} = -10$ V, quadrant 2

(c) $V_o = \dfrac{(10)(-10)}{10} = -10$ V, quadrant 4

(d) $V_o = \dfrac{(-10)(-10)}{10} = 10$ V, quadrant 3

In Fig. 12-2(b), V_o is plotted on the vertical axis and x on the horizontal axis. If we apply 10 V to the y input and vary x from −10 V to +10 V, we plot the line *ab* labeled $y = 10$ V. If y is changed to −10 V, the line *cd* labeled $y = -10$ V results. These lines can be seen on a cathode-ray oscilloscope (CRO) by connecting V_o of the multiplier to the y input of the CRO and x of the multiplier to the $+x$ input of the CRO. For accuracy, V_o should be 0 V when either multiplier input is 0 V. If this is not the case, a zero trim adjustment should be made, as described in Section 12-1.3.

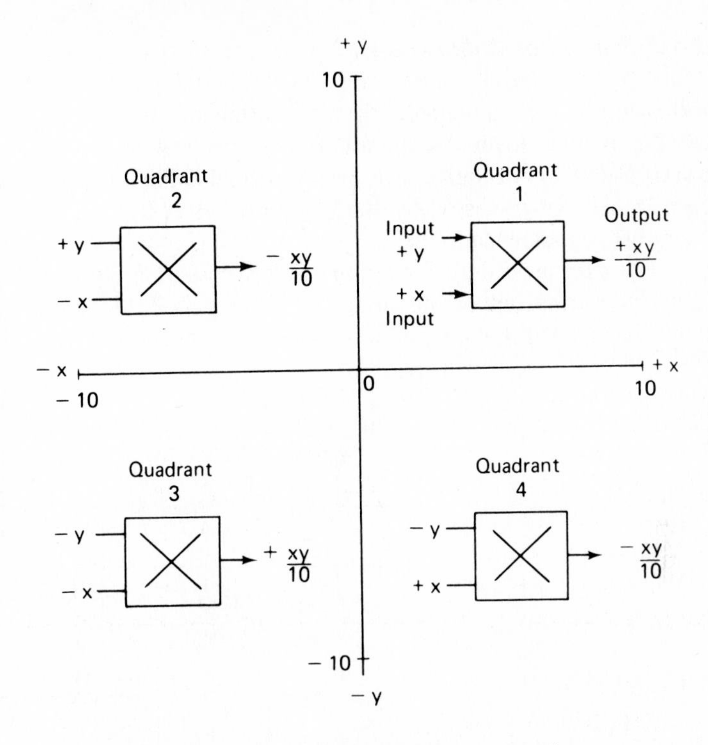

(a) y vs x plot shows location of input operating point in one of four quadrants

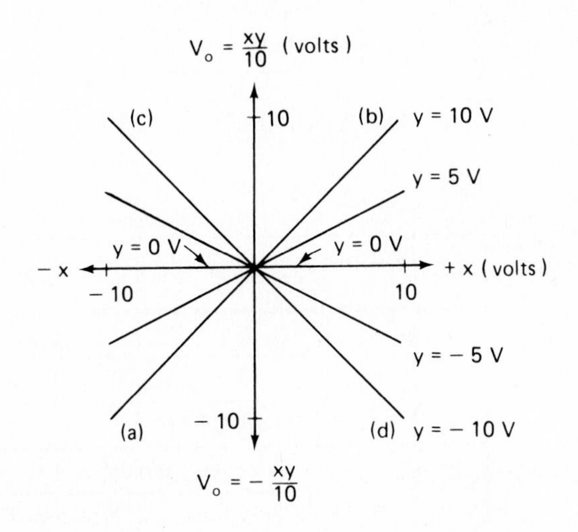

(b) Multiplier output xy/10 versus input x

FIGURE 12-2 Multiplying two dc voltages, x and y.

12-1.3 Multiplier Calibration

Ordinary low-cost integrated-circuit multipliers such as the AD533, 4200, and XR2208 require manual calibration with external circuitry. Precision multipliers such as the AD534 require little or no manual calibration. Any internal imbalances are trimmed out precisely by the manufacturer, using computer-controlled lasers. Their cost is somewhat higher.

The trim procedure for a popular low-cost IC multiplier is given in Fig. 12-3 together with the supporting circuitry. Three pots Z_o, X_o, and Y_o are adjusted in sequence to give (1) 0 V out when both x and y inputs are 0 V, (2) 0 V out when the x

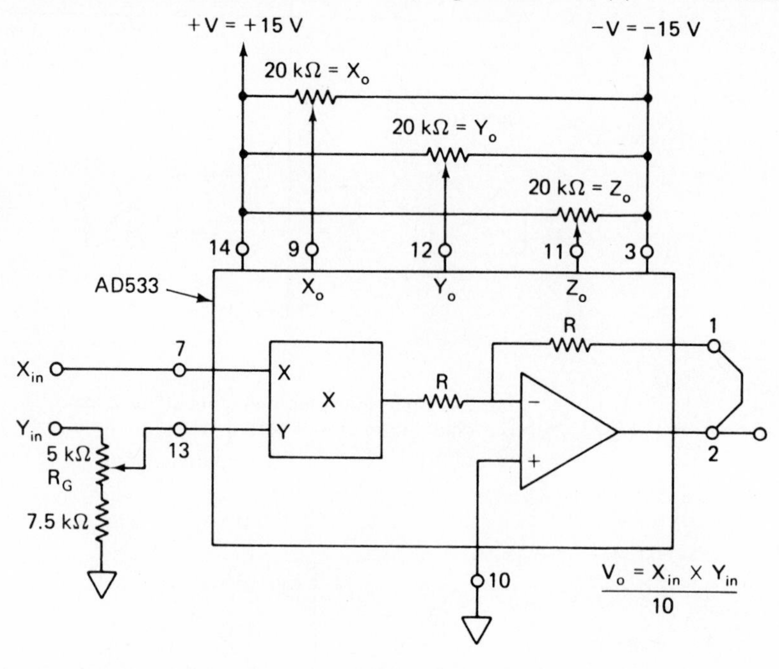

Trim Procedure

Step	Adjust pot	X_{in} at	Y_{in} at	For V_o =
1	Z_o	0 V	0 V	0 V dc
2	X_o	0 V	20 V, p-p, 50 Hz	Min. ac
3	Y_o	20 V, p-p, 50 Hz	0 V	Min. ac
4	Repeat steps 1 through 3, as required			
5	R_G	+10 V dc	20 V, p-p, 50 Hz	Y_{in}

FIGURE 12-3 Calibration procedure and circuit to trim an integrated-circuit multiplier.

input equals 0 V, and (3) 0 V out when the y input equals 0 V. Finally, the scale factor is adjusted to 0.1 by pot R_G (in step 5) to ensure that the output is at 10 V when both inputs are at 10 V. Multiturn pots are particularly useful for precise adjustment.

12-2 SQUARING A NUMBER OR DC VOLTAGE

Any positive or negative number can be squared by a multiplier, providing that the number can be represented by a voltage between 0 and 10 V. Simply connect the voltage E_i to *both* inputs as shown in Fig. 12-4. This type of connection is known as a *squaring circuit*.

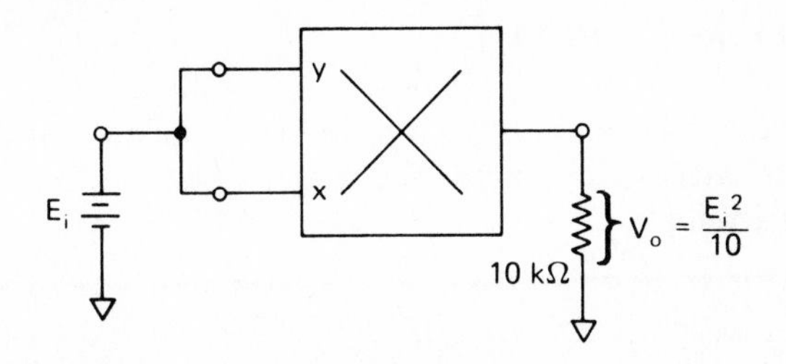

FIGURE 12-4 Squaring a dc voltage.

Example 12-2

Find V_o in Fig. 12-4 if (a) $E_i = +10$ V; (b) $E_i = -10$ V.

Solution From Fig. 12-4,

(a) $V_o = \dfrac{10^2}{10} = 10$ V.

(b) $V_o = \dfrac{(-10)(-10)}{10} = 10$ V.

Example 12-2 shows that the output of the multiplier follows the rules of algebra; that is, when either a positive or negative number is squared, the result is a positive number.

12-3 FREQUENCY DOUBLING

12-3.1 Principle of the Frequency Doubler

An ideal sinusoidal-wave frequency doubler would give an output voltage whose frequency is twice the frequency of the input voltage. The doubler circuit should not incorporate a tuned circuit, since the tuned circuit can be tuned only to one frequency.

A true doubler should double any frequency. The multiplier is very nearly an ideal doubler if only one frequency is applied to both inputs. The output voltage for a doubler circuit is given by the trigonometric identity*

$$(\sin 2\pi ft)^2 = \frac{1}{2} - \frac{\cos 2\pi(2f)t}{2} \tag{12-2}$$

Equation (12-2) predicts that squaring a sine wave with a frequency of (for example) $f = 10$ kHz gives a negative cosine wave with a frequency of $2f$ or 20 kHz plus a dc term of $\frac{1}{2}$. Note that *any* input frequency f will be doubled when passed through a squaring circuit.

12-3.2 Squaring a Sinusoidal Voltage

In Fig. 12-5(a), sine-wave voltage E_i is applied to both multiplier inputs. E_i has a peak value of 5 V and a frequency of 10 kHz. The output voltage V_o is predicted by the calculations shown in Example 12-3.

Example 12-3

Calculate V_o in the squaring circuit or frequency doubler of Fig. 12-5.

Solution Input $E_x = E_y = E_i$ and is expressed in volts by

$$E_i = E_x = E_y = 5 \sin 2\pi\, 10{,}000t$$

Substituting into Eq. (12-1b) yields

$$V_o = \frac{E_i^2}{10} = \frac{5^2}{10}(\sin 2\pi\, 10{,}000t)^2 \tag{12-3}$$

Applying Eq. (12-2), we obtain

$$V_o = 2.5\left[\frac{1}{2} - \frac{\cos 2\pi\, 20{,}000t}{2}\right]\text{V}$$

$$= \underbrace{1.25}_{} \qquad - \underbrace{1.25 \cos 2\pi\, 20{,}000t}_{}$$

$$= \text{dc term of } 1.25\text{ V} - \text{frequency doubled to } 20{,}000\text{ Hz, } 1.25\text{ V peak}$$

Both E_i and V_o are shown in Fig. 12-5. If you want to remove the dc voltage, simply

* Equation (12-2) is a special case of the general equation shown in math tables:

$$(\sin A)(\sin B) = \tfrac{1}{2}[(\cos(A - B) - \cos(A + B)]$$

Here $A = B = 2\pi ft$.

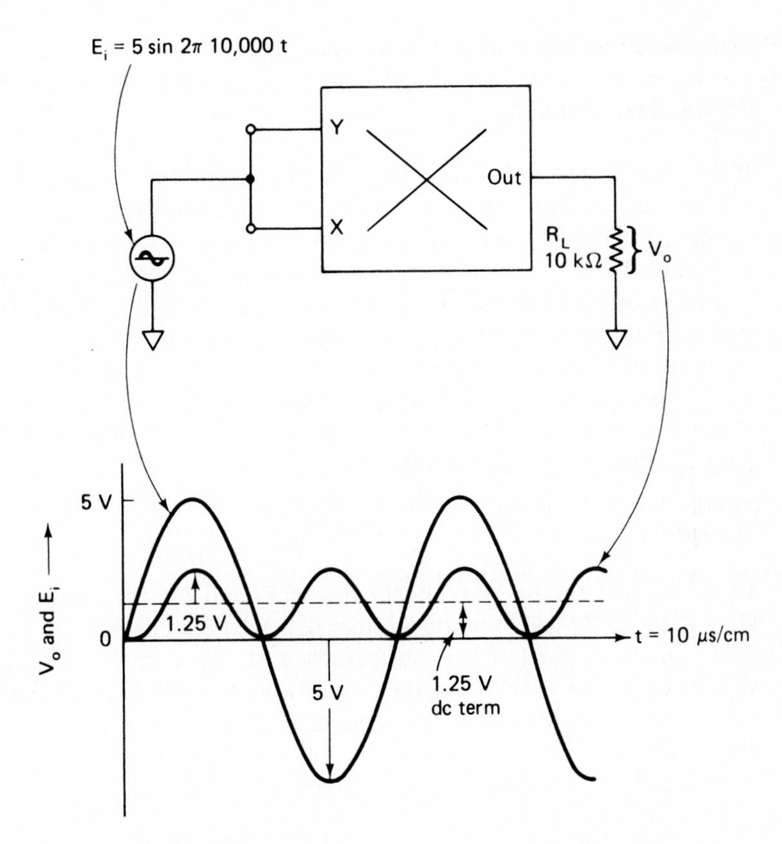

FIGURE 12-5 Squaring circuit as a frequency doubler.

install a 1-μF coupling capacitor between R_L and the output terminal. If you want to measure the dc term simply connect a dc voltmeter across V_o, without the capacitor.

Conclusion. V_o has two voltage components: (1) a dc voltage equal to $\frac{1}{20}(E_{ip})^2$, and (2) an ac sinusoidal wave whose peak value is $\frac{1}{20}(E_{ip})^2$ and whose frequency is double that of E_i.

Example 12-4

What are the dc and ac output voltage components of Fig. 12-5 if (a) $E_i = 10$ V peak at 1 kHz; (b) $E_i = 2$ V peak at 2.5 kHz?

Solution (a) dc value $= (10)^2/20 = 5$ V; peak ac value $= (10)^2/20 = 5$ V at 2 kHz. (b) dc value $= (2)^2/20 = 0.2$ V; peak ac value $= (2)^2/20 = 0.2$ V at 5 kHz.

12-4 PHASE-ANGLE DETECTION

12-4.1 Basic Theory

If two sine waves of the *same* frequency are applied to the multiplier inputs in Fig. 12-6(a), the output voltage V_o has a dc voltage component and an ac component whose frequency is twice that of the input frequency. This conclusion was developed in Section 12-3.2. The dc voltage is actually proportional to the difference in phase angle θ between E_x and E_y. For example, in Fig. 12-5, $\theta = 0°$, because there was no phase difference between E_x and E_y. Figure 12-6(b) shows two sine waves of identical frequency but a phase difference of 90°; therefore, $\theta = 90°$.

If one input sine wave differs in phase angle from the other, it is possible to calculate or measure the phase-angle difference from *the dc voltage component in* V_o. This dc component $V_{o\,dc}$ is given by*

$$V_{o\,dc} = \frac{E_{xp}E_{yp}}{20}(\cos\theta) \tag{12-4a}$$

where E_{xp} and E_{yp} are peak amplitudes of E_x and E_y. For example, if $E_{xp} = 10$ V, $E_{yp} = 5$ V, *and they are in phase,* then $V_{o\,dc}$ would indicate 2.5 V on a dc voltmeter. This voltmeter point would be marked as a phase angle of 0° ($\cos 0° = 1$). If $\theta = 45°$ ($\cos 45° = 0.707$), the dc meter would read $0.707 \times 2.5\text{ V} \simeq 1.75$ V. Our dc voltmeter can be calibrated as a *phase angle meter* 0° at 2.5 V, 45° at 1.75 V, and 90° at 0 V.

Equation (12-4a) may also be expressed by*

$$\cos\theta = \frac{20V_{o\,dc}}{E_{xp}E_{yp}} \tag{12-4b}$$

If we could arrange for the product $E_{xp}E_{yp}$ to equal 20, we could use a 0- to 1-V dc voltmeter to read $\cos\theta$ directly from the meter face and calibrate the meter face in degrees from a cosine table. That is, Eq. (12-4b) reduces to

$$V_{o\,dc} = \cos\theta \qquad \text{for } E_{xp} = E_{yp} = 4.47\text{ V} \tag{12-4c}$$

This point is explored further in Section 12-4.2.

* Trigonometric identity:

$$\sin A \sin B = \tfrac{1}{2}[\cos(A - B) - \cos(A + B)]$$

For equal frequencies, different phase angle:

$$A = 2\pi ft + \theta \text{ for } E_x, \qquad B = 2\pi ft \text{ for } E_y$$

Therefore,

$$[\sin(2\pi ft + \theta)][\sin 2\pi ft] = \tfrac{1}{2}[\cos\theta - \cos(4\pi ft + \theta)]$$
$$= \tfrac{1}{2}(\text{dc} + \text{double frequency term})$$

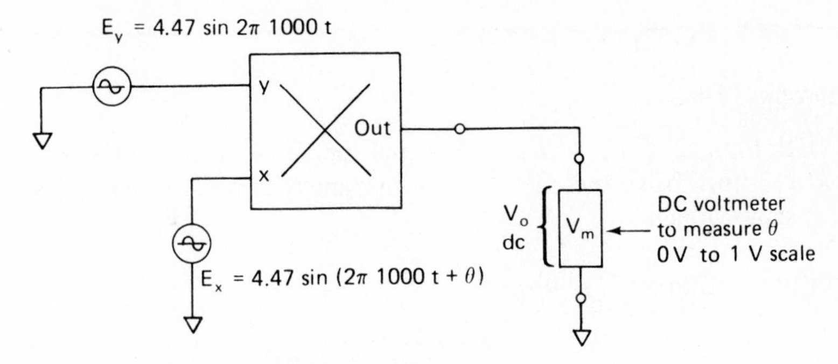

(a) Phase angle measurement

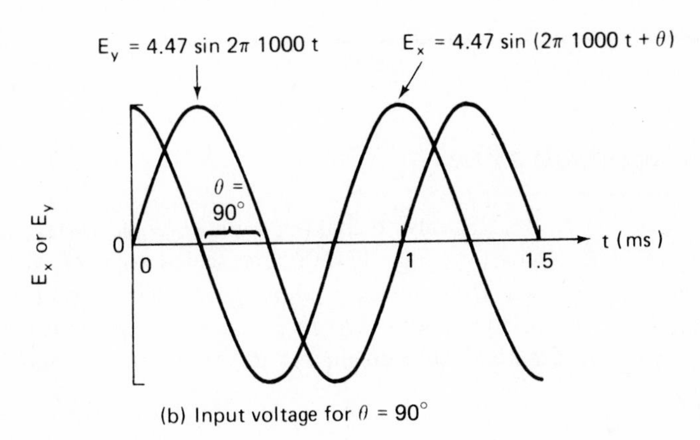

(b) Input voltage for θ = 90°

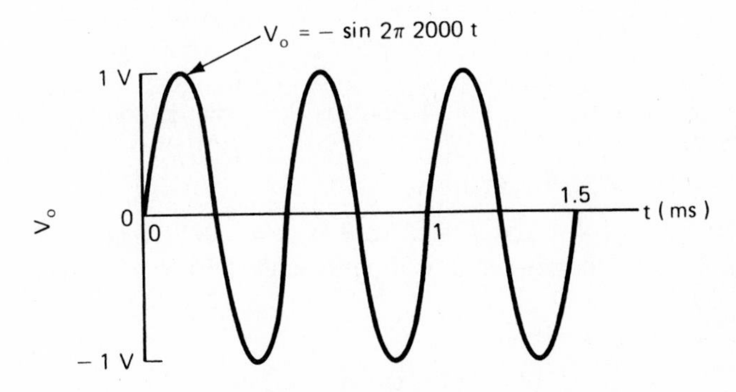

(c) Output voltage for θ = 90°, dc term is 0 V

FIGURE 12-6 Multiplier used to measure the phase-angle difference between two equal frequencies.

Example 12-5

In Fig. 12-5, $E_{xp} = E_{yp} = 5$ V and the dc component of V_o is 1.25 V from Eq. (12-4b). Prove that there is 0° phase angle between E_x and E_y (since they are the same voltage).

Solution From Eq. (12-4b),

$$\frac{20 \times 1.25}{5 \times 5} = \cos \theta = \frac{25}{25} = 1$$

Since $\cos 0° = 1$, $\theta = 0°$.

12-4.2 Phase-Angle Meter

Equation (12-4b) points the way to making a phase-angle meter. Assume that the peak values of E_x and E_y in Fig. 12-6(a) are scaled to 4.47 V by amplifiers or voltage dividers. Then a dc voltmeter is connected as shown in Fig. 12-6(a) to measure just the dc voltage component. The meter face then can be calibrated directly in degrees. The procedure is developed further in Examples 12-6 and 12-7.

Example 12-6

The average value of V_o in Fig. 12-6(c) is 0, so the dc component of $V_o = 0$ V. Calculate θ.

Solution From Eq. (12-4b), $\cos \theta = 0$, so $\theta = \pm 90°$. Note that the dc component of V_o cannot distinguish between a lead phase angle (+) or a lag phase angle (−).

Example 12-7

Calculate $V_{o\,dc}$ for phase angles of (a) $\theta = \pm 30°$; (b) $\theta = \pm 45°$; (c) $\theta = \pm 60°$.

Solution From a trigonometric table, obtain the $\cos \theta$ and apply Eq. (12-4a).

θ (deg)	$\cos\theta$	$V_{o\,dc}$ (V)
±30	0.866	0.866
±45	0.707	0.707
±60	0.500	0.500
± 0	1.000	1.000
±90	0.000	0.000

(The last two rows of this table come from Examples 12-5 and 12-6.)

The 0- to 1-V voltmeter scale can now be calibrated in degrees, 0 V for a 90° phase angle and 1.0 V for 0° phase angle. At 0.866 V, $\theta = 30°$, and so forth. The phase angle meter does not indicate whether θ is a leading or lagging phase angle but only the phase difference between E_x and E_y.

12-4.3 Phase Angles Greater than ±90°

The cosine of phase angles greater than +90° or −90° is a negative value. Therefore, V_o will be negative. This extends the capability of the phase angle meter in Example 12-7.

Example 12-8

Calculate $V_{o\,dc}$ for phase angles of (a) $\theta = \pm 90°$; (b) $\theta = \pm 120°$; (c) $\theta = \pm 135°$; (d) $\theta = \pm 150°$; (e) $\theta = \pm 180°$.

Solution Using Eq. (12-4a) and tabulating results, we have

θ	±90°	±120°	±135°	±150°	±180°
$V_{o\,dc}$	0 V	−0.5 V	−0.70 V	−0.866 V	−1 V

From the results of Examples 12-7 and 12-8, a ±1-V voltmeter can be calibrated to read from 0 to ±180°.

12-5 INTRODUCTION TO AMPLITUDE MODULATION

12-5.1 Need for Amplitude Modulation

Low-frequency audio or data signals cannot be transmitted from antennas of reasonable size. Audio signals can be transmitted by changing or *modulating* some characteristic of a higher-frequency *carrier* wave. If the amplitude of the carrier wave is changed in proportion to the audio signal, the process is called *amplitude modulation* (AM). Changing the frequency or the phase angle of the carrier wave results in *frequency modulation* (FM) and *phase-angle modulation* (PM), respectively.

Of course, the original audio signal must eventually be recovered by a process called *demodulation* or detection. The remainder of this section concentrates on using the multiplier for amplitude modulation. ("Modulate" is from the ancient Greek language meaning to "change." Curiously, it is the Latin prefix "de" that converts the meaning to "change back.")

12-5.2 Defining Amplitude Modulation

The introduction to amplitude modulation begins with the amplifier in Fig. 12-7(a). The input voltage E_c is amplified by a constant gain A. Amplifier output V_o is the product gain of A and E_c. Now suppose that the amplifier's gain is varied. This concept is represented by an arrow through A in Fig. 12-7(b). Assume that A is varied from 0 to a maximum and back to 0 as shown in Fig. 12-7(b) by the plot of A versus t. This means that the amplifier multiplies the input voltage E_c by a different value (gain) over a period of time. V_o is now the amplitude of input E_c varied or multiplied by an amplitude of A. This process is an example of amplitude modulation, and the output voltage V_o is called the *amplitude modulated signal*. Therefore, to obtain an amplitude-modulated signal (V_o), the amplitude of a high-frequency carrier signal (E_c) is varied by an intelligence or data signal A.

12-5.3 The Multiplier Used as a Modulator

From Section 12-5.2 and Fig. 12-7(b), V_o equals E_c multiplied by A. Therefore, amplitude modulation is a *multiplication process*. As shown in Fig. 12-7(c), E_c is applied to a multiplier's x input. E_m [having the same shape as A in Fig. 12-7(b)] is applied to the multiplier's y input. E_c is multiplied by a voltage that varies from 0 through a maximum and back to 0. So V_o has the same envelope as E_m. The multiplier can be considered a *voltage-controlled gain device* as well as an amplitude modulator. The waveshape shown is that of a *balanced modulator*. The reason for this name will be given in Section 12-6.3.

Note carefully in Fig. 12-7(c) that V_o is not a sine wave; that is, the peak values of successive half-cycles are different. This principle is used in Section 12-10 to show how a *frequency-shifter* (*heterodyne*) circuit works. But first, we examine amplitude modulation in greater detail.

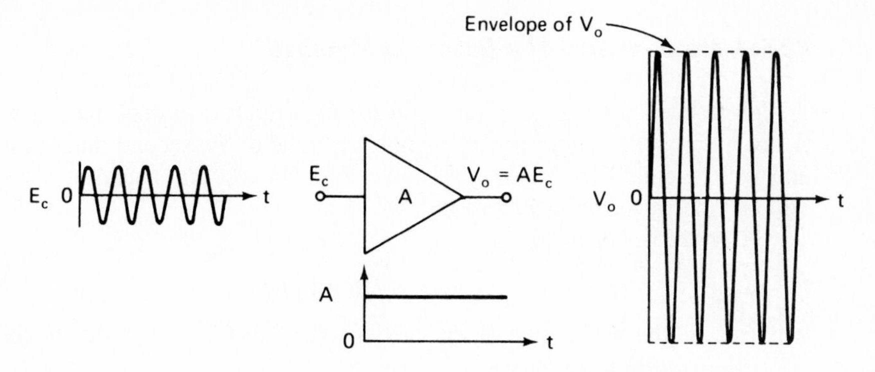

(a) Input E_c is amplified by constant gain A to give output $V_o = AE_c$

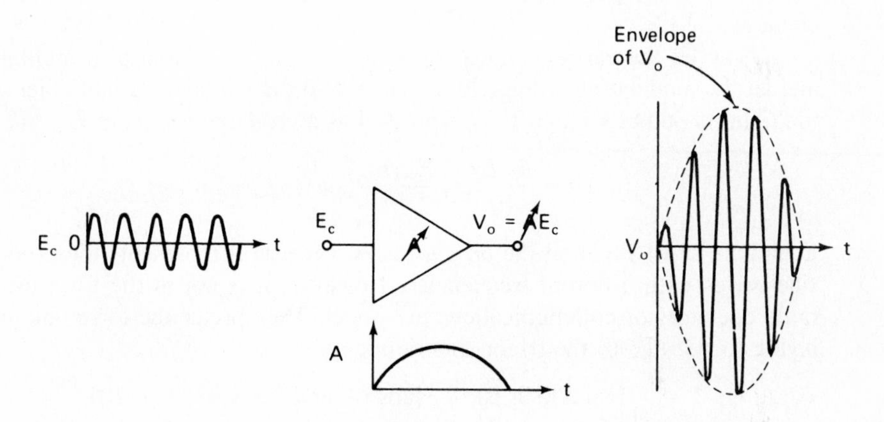

(b) If amplifier gain A is varied with time, the envelope of V_o is varied with time

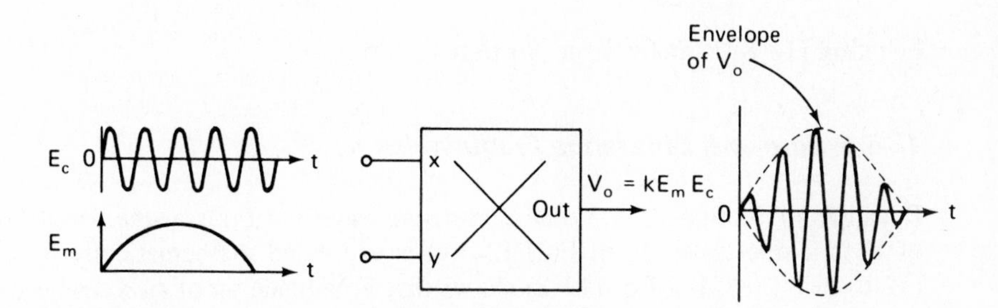

(c) If E_m varies as A in part (b), then V_o has the same general shape as in part (b)

FIGURE 12-7 Introduction to modulation.

12-5.4 Mathematics of a Balanced Modulator

A high-frequency sinusoidal *carrier wave* E_c is applied to one input of a multiplier. A lower-frequency audio or data signal is applied to the second input of a modulator and will be called the *modulating wave,* E_m. For test and analysis, both E_c and E_m will be sine waves described as follows.

Carrier wave, E_c:

$$E_c = E_{cp} \sin 2\pi f_c t \tag{12-5a}$$

where E_{cp} is the peak value of the carrier wave and f_c is the carrier frequency.

Modulating wave, E_m:

$$E_m = E_{mp} \sin 2\pi f_m t \tag{12-5b}$$

where E_{mp} is the peak value of the modulating wave and f_m is the modulating frequency.

Now let the carrier voltage E_c be applied to the x input of a multiplier as E_x, and let the modulating voltage E_m be applied to the y input of a multiplier as E_y. The multiplier's output voltage V_o is expressed as a *product term* from Eq. (12-1b) as

$$V_o = \frac{E_m E_c}{10} = \frac{E_{mp} E_{cp}}{10}(\sin 2\pi f_m t)(\sin 2\pi f_c t) \tag{12-6}$$

Equation (12-6) is called the *product term,* because it represents the product of two sine waves with different frequencies. However, it is not in the form used by ham radio operators or communications personnel. They prefer the form obtained by applying to Eq. (12-6) the trigonometric identity

$$(\sin A)(\sin B) = \tfrac{1}{2}[\cos (A - B) - \cos (A + B)] \tag{12-7}$$

Substituting Eq. (12-7) into Eq. (12-6), where $A = E_c$ and $B = E_m$, we have

$$V_o = \frac{E_{mp} E_{cp}}{20} \cos 2\pi (f_c - f_m)t - \frac{E_{mp} E_{cp}}{20} \cos 2\pi (f_c + f_m)t \tag{12-8}$$

Equation (12-8) is analyzed in Section 12-5.5.

12-5.5 Sum and Difference Frequencies

Recall from Section 12-5.3 that E_c is a sine wave and E_m is a sine wave, but no part of V_o is a sine wave. V_o in Fig. 12-7(c) is expressed mathematically by either Eq. (12-6) or (12-8). But Eq. (12-8) shows that V_o is made up of *two* cosine waves with frequencies *different* from either E_m or E_c. They are the *sum frequency* $f_c + f_m$ and the *difference frequency* $f_c - f_m$. The sum and difference frequencies are evaluated in Example 12-9.

Example 12-9

In Fig. 12-8, carrier signal E_c has a peak voltage of $E_{cp} = 5$ V and a frequency of $f_c = 10{,}000$ Hz. The modulating signal E_m has a peak voltage of $E_m = 5$ V and a frequency of $f_m = 1000$ Hz. Calculate the peak voltage and frequency of (a) the sum frequency; (b) the difference frequency.

Solution From Eq. (12-8), the peak value of both sum and difference voltages is

$$\frac{E_{mp}E_{cp}}{20} = \frac{5\text{ V} \times 5\text{ V}}{20} = 1.25\text{ V}$$

The sum frequency is $f_c + f_m = 10{,}000\text{ Hz} + 1000\text{ Hz} = 11{,}000$ Hz; the difference frequency is $f_c - f_m = 10{,}000\text{ Hz} - 1000\text{ Hz} = 9000$ Hz. Thus V_o is made up of the difference of two cosine waves:

$$V_o = 1.25 \cos 2\pi 9000t - 1.25 \cos 2\pi 11{,}000t$$

This result can be verified by connecting a wave or spectrum analyzer to the multiplier's output; a 1.25-V deflection occurs at 11,000 Hz and at 9000 Hz. The original input signals of 1 kHz and 10 kHz do *not* exist at the output.

An oscilloscope can be used to show input and output voltages of the multiplier of Example 12-9. The product term for V_o is found from Eq. (12-6):

$$V_o = 2.5\text{ V}\underbrace{(\sin 2\pi 10{,}000t)}_{E_c}\underbrace{(\sin 2\pi 1000t)}_{E_m}$$

$$= 2.5 \times E_c \times E_m$$

V_o is shown with E_m in the top drawing and with E_c in the bottom drawing of Fig. 12-8. Observe that E_m and E_c have peak voltages of 5 V. The peak value of V_o is 2.5 V. Note that the upper and lower envelopes of V_o are *not* the same shape as E_m. Therefore, we *cannot* rectify and filter V_o to recover E_m. This characteristic distinguishes the balance modulator.

12-5.6 Side Frequencies and Sidebands

Another way of displaying the output of a modulator is by a graph showing the peak amplitude as a vertical line for each frequency. The resulting *frequency spectrum* is shown in Fig. 12-9(a). The sum and difference frequencies in V_o are called *upper* and *lower side* frequencies because they are above and below the carrier frequency on the graph. When more than one modulating signal is applied to the modulator

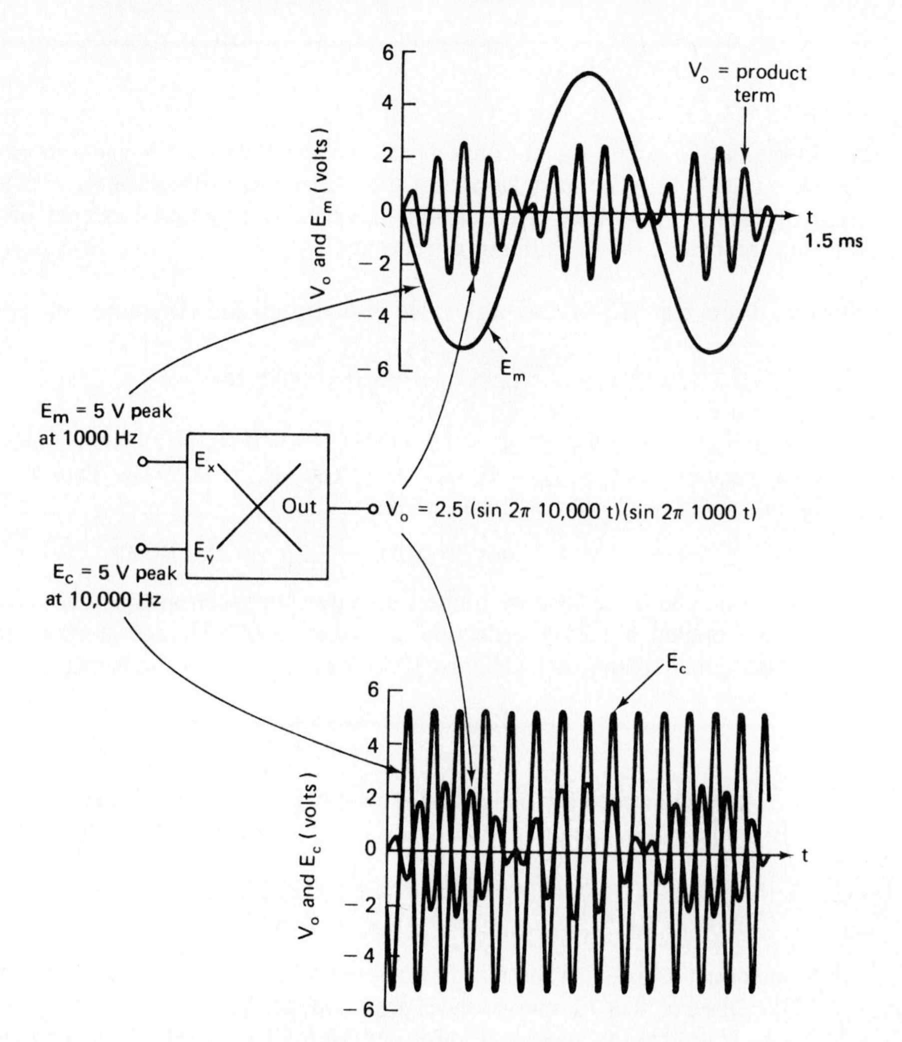

FIGURE 12-8 The multiplier as a balanced modulator.

(y input) input in Fig. 12-8, each generates a sum and difference frequency in the output. Thus, there will be two side frequencies for each y input frequency, placed symmetrically on either side of the carrier. If the expected range of modulating frequencies is known, the resulting range of side frequencies can be predicted. For example, if the modulating frequencies range between 1 and 4 kHz, the lower side frequencies fall in a band between (10 − 4) kHz = 6 kHz and (10 − 1) kHz = 9 kHz. The band between 6 and 9 kHz is called the *lower sideband*. For this same example, the *upper sideband* ranges from (10 + 1) kHz = 11 kHz to (10 + 4) kHz = 14 kHz. Both upper and lower sidebands are shown in Fig. 12-9(b).

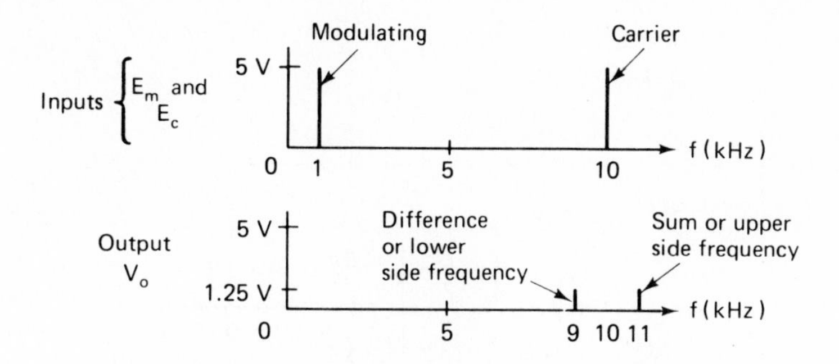

(a) Frequency spectrum for f_c = 10 kHz and f_m = 1 kHz in Example 12-8

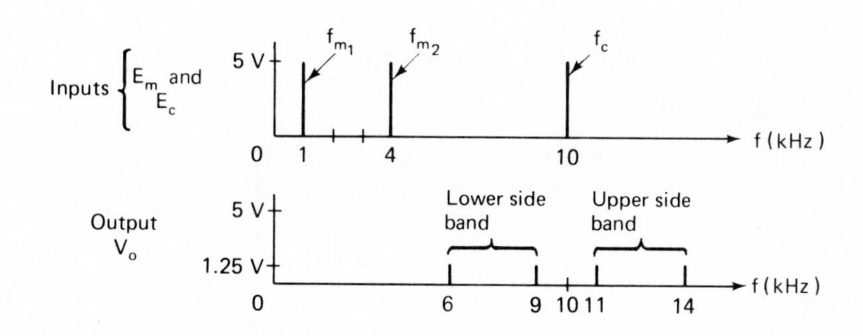

(b) Frequency spectrum for f_c = 10 kHz and f_{m_1} = 1 kHz, f_{m_2} = 4 kHz

FIGURE 12-9 Frequency spectrum for a balanced modulator.

12-6 STANDARD AMPLITUDE MODULATION

12-6.1 Amplitude Modulator Circuit

The circuits of Section 12-5 multiplied the carrier and modulating signals to generate a balanced output that is expressed either as (1) a product term, or (2) a sum and difference frequency. The term "balanced modulator" originated in the days of vacuum tube–transformer technology. It was very difficult to "balance out" the carrier. With today's modern multipliers, the absence of carrier E_c in the output is a zero-cost bonus. The classical or standard amplitude modulator (AM) adds the carrier term to the output. The AM car radio uses standard AM. One way of adding the carrier term to generate a standard AM output is shown in Fig. 12-10(a). The modulating signal is fed into one input of an adder. A dc voltage equal to the peak value of

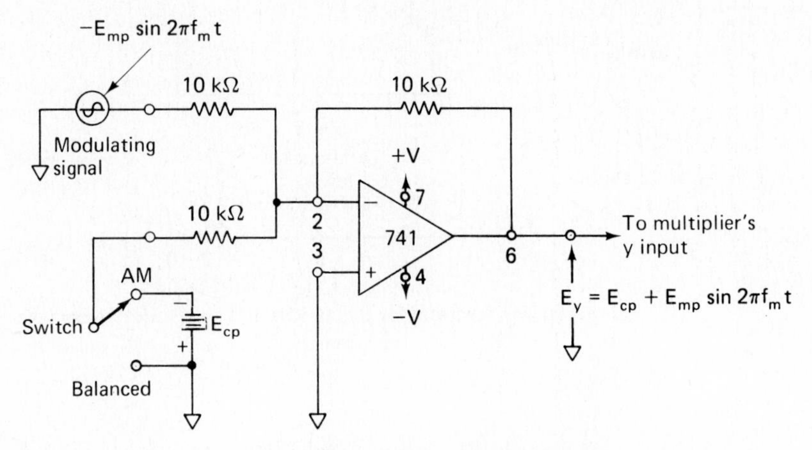

(a) Adder circuit to add carrier signal

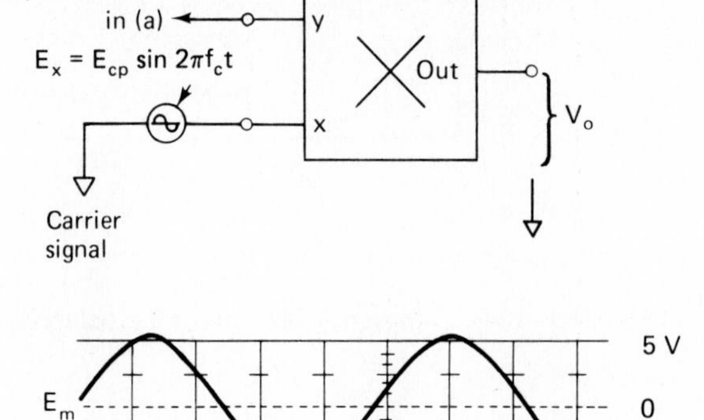

(b) Multiplier as a modulator

FIGURE 12-10 Circuit to demonstrate amplitude modulation or balanced modulation (see also Fig. 12-12).

the carrier voltage E_{cp} is fed into the other input. The output of the adder is then fed into the y input of a multiplier, as shown in Fig. 12-10(b). The carrier signal is fed into the x input. The multiplier multiplies E_x by E_y, and its output voltage is the standard AM voltage given by either of the following equations:

$$V_o = \begin{cases} \dfrac{E_{cp}^2}{10} \sin 2\pi f_c t & \text{(carrier term)} \\ \quad + & \\ \dfrac{E_{cp} E_{mp}}{10} (\sin 2\pi f_c t)(\sin 2\pi f_m t) & \text{(product term)} \end{cases} \tag{12-9}$$

or

$$V_o = \begin{cases} \dfrac{E_{cp}^2}{10} \sin 2\pi f_c t & \text{(carrier term)} \\ \quad + & \\ \dfrac{E_{cp} E_{mp}}{20} \cos 2\pi (f_c - f_m) t & \text{(lower side frequency)} \\ \quad - & \\ \dfrac{E_{cp} E_{mp}}{20} \cos 2\pi (f_c + f_m) t & \text{(upper side frequency)} \end{cases} \tag{12-10}$$

The output voltage V_o is shown in Fig. 12-10(b). The voltage levels are worked out in the following example.

Example 12-10

In Fig. 12-10, $E_{cp} = E_{mp} = 5$ V. The carrier frequency $f_c = 10$ kHz, and the modulating frequency is $f_m = 1$ kHz. Evaluate the peak amplitudes of the output carrier and product terms.

Solution From Eq. (12-9), the carrier term peak voltage is

$$\frac{(5\text{ V})(5\text{ V})}{10} = 2.5\text{ V}$$

The product term peak voltage is

$$\frac{(5\text{ V})(5\text{ V})}{10} = 2.5\text{ V}$$

The side frequencies peak voltages are

$$\frac{(5\text{ V})(5\text{ V})}{20} = 1.25\text{ V}$$

The waveshape of V_o is shown in Fig. 12-10(b). Observe that both upper and lower envelopes of V_o are the same shape as E_m. This is characteristic of a standard amplitude modulator, AM, *not* of the balanced modulator. It allows easy recovery of audio signal E_m by a half-wave rectifier and suitable filter capacitor.

12-6.2 Frequency Spectrum of a Standard AM Modulator

The signal frequencies present in V_o for the standard AM output of Fig. 12-10 are found from Eq. (12-10). Using the voltage values in Example 12-10, we have

$$\text{Carrier term} = 2.5\text{-V peak at } 10{,}000\text{ Hz}$$

$$\text{Lower side frequency} = 1.25\text{-V peak at } 9000\text{ Hz}$$

$$\text{Upper side frequency} = 1.25\text{-V peak at } 11{,}000\text{ Hz}$$

These frequencies are plotted in Fig. 12-11 and should be compared with the balanced modulator of Fig. 12-9.

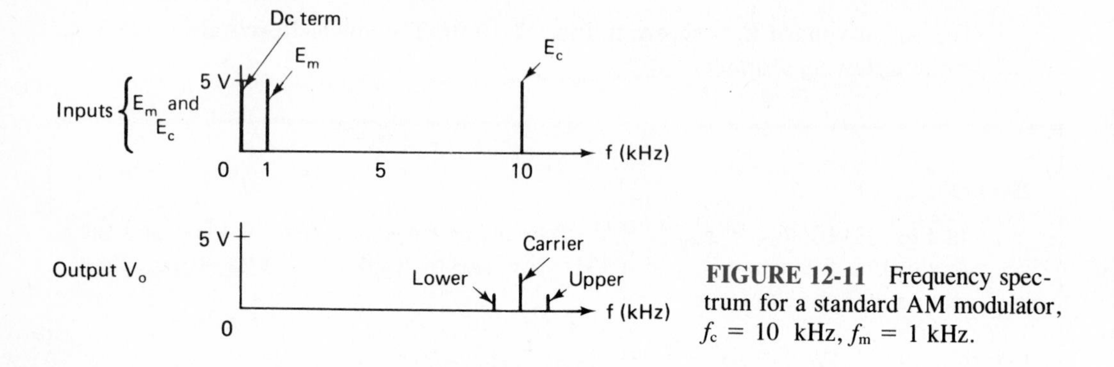

FIGURE 12-11 Frequency spectrum for a standard AM modulator, $f_c = 10$ kHz, $f_m = 1$ kHz.

12-6.3 Comparison of Standard AM Modulators and Balanced Modulators

If the switch in Fig. 12-10(a) is positioned to AM, V_o will contain three frequencies, f_c, $f_c + f_m$, and $f_c - f_m$, carrier plus sum and difference frequencies. Observe that the envelopes of V_o have the same shape as the intelligence signal E_m. This observation can be used to recover E_m from the AM signal, as stated in Example 12-10. Note that if there is no signal frequency, the station still transmits carrier f_c. Radio receivers use this fact to activate signal-strength meters, tuning lights, and automatic volume control (AVC).

If the switch in Fig. 12-10(a) is positioned to "Balanced," V_o will contain only the product term with only two frequencies $f_c + f_m$ and $f_c - f_m$. The envelope of V_o does *not* follow E_m. Since V_o does *not* contain f_c, this type of modulation is called *balanced modulation* in the sense that the carrier has been balanced out. It is also called *suppressed carrier modulation,* since the carrier is suppressed in the output. If no modulating frequency is present, the radio station does not transmit. This is a good system for clandestine operation. For a comparison of balanced and standard AM modulation, both outputs are shown together in Fig. 12-12.

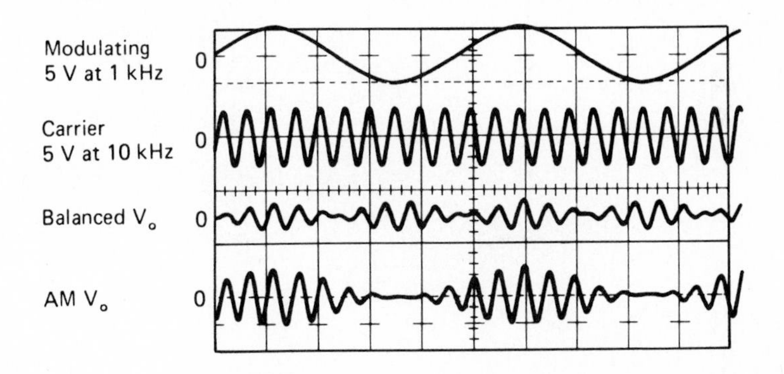

FIGURE 12-12 Comparison of balanced modulation and standard AM from Fig. 12-10.

12-7 DEMODULATING AN AM VOLTAGE

Demodulation, or *detection,* is the process of recovering a modulating signal E_m from the modulated output voltage V_o. To explain how this is accomplished, the AM modulated wave is applied to the y input of a multiplier as shown in Fig. 12-13. Each y input frequency is multiplied by the x input carrier frequency and generates a sum and difference frequency as shown in Fig. 12-13(b). Since only the 1-kHz frequency is the modulating signal, use a low-pass filter to extract E_m. Thus the demodulator is simply a multiplier with the carrier frequency applied to one input, and the AM signal to be demodulated is fed into the other input. The multiplier's output is fed into a low-pass filter whose output is the original modulating data signal E_m. Thus a multiplier plus a low-pass filter and carrier signal equals a demodulator.

V_o from Figure 12-10(b)
9, 10, and 11 kHz
E_{x2} = 10,000 Hz at
2.5 V peak
y
x
Out
V_{o2}
Low-pass filter
E_m

(a) Multiplier used as a demodulator

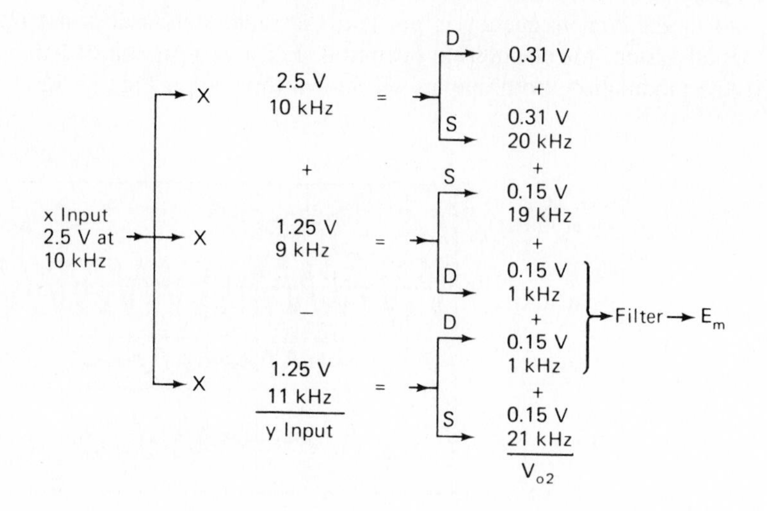

(b) Frequency and peak amplitude of signal components at x-input, y-input, multiplier output, and filter

FIGURE 12-13 The demodulator is a multiplier plus a low-pass filter.

Waveshapes at inputs and outputs of both the AM modulator and demodulator are shown in Fig. 12-14. Note the unusual shape of V_{o2} because it contains six components [detailed in Fig. 12-13(b)].

12-8 DEMODULATING A BALANCED MODULATOR VOLTAGE

Modulating signal E_m is recovered from a balanced modulator by means of the same technique employed in Fig. 12-13 and Section 12-7. The only difference is due to the absent carrier frequency of 10 kHz at the demodulator's y input. This missing 10 kHz also eliminates both the dc and 20-kHz term in V_{o2}. The circuit arrangement of Fig. 12-15 was built to demonstrate the demodulating technique and show the result-

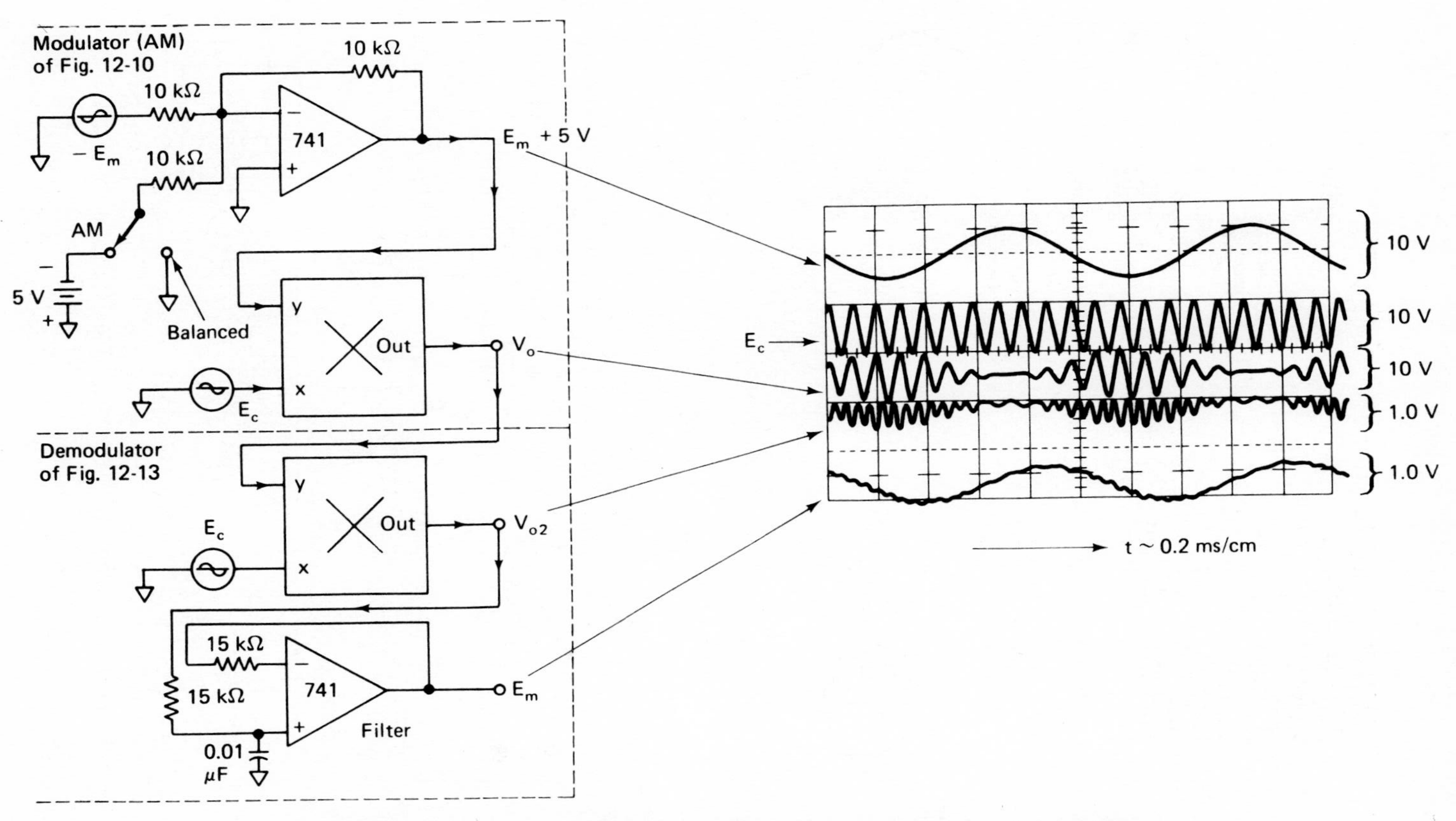

FIGURE 12-14 Voltage waveforms in an amplitude modulator and demodulator (f_c = 10 kHz, f_m = 1kHz).

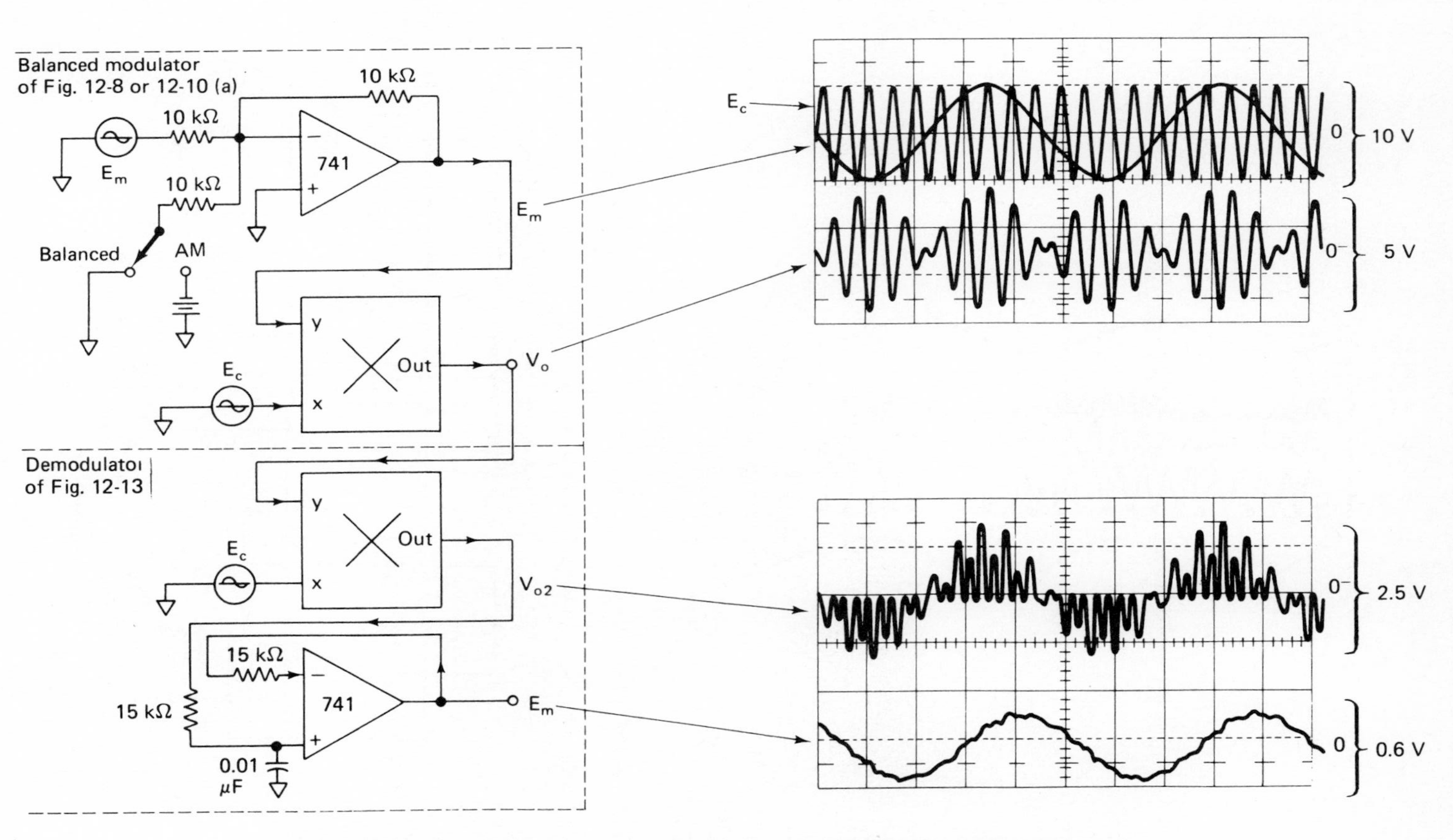

FIGURE 12-15 Demonstration of balanced modulator and demodulator with waveshapes.

ing waveshapes. The demodulated E_m is not a pure sine wave, because only a simple filter was used. If f_c is increased to 100 kHz, E_m will be closer to being a pure sine wave. The carrier's frequency fed into the demodulator should be *exactly* equal to the carrier frequency driving the modulator.

12-9 SINGLE-SIDEBAND MODULATION AND DEMODULATION

In the balanced modulator of Figs. 12-8 and 12-9, we could add a high-pass filter (see Chapter 11) to the modulator's output. If the filter removed all the lower side frequencies, the output is *single sideband* (SSB). If the filter only attenuated the lower side frequencies (to leave a *vestige* of the lower sideband), we would have a *vestigial* sideband modulator.

Assume that only one modulating frequency f_m is applied to our single-sideband modulator together with carrier f_c. Its output would be a single upper side frequency $f_c + f_m$. To demodulate this signal and recover f_m, all we have to do is connect the SSB signal $f_c + f_m$ to one multiplier input and f_c to the other input. According to the principles set forth in Section 12-5.4, the demodulator's output would have a sum frequency of $(f_c + f_m) + f_c$ and a difference frequency of $(f_c + f_m) - f_c = f_m$. A low-pass filter would recover the modulating signal f_m and easily eliminate the high-frequency signal, whose frequency is $2f_c + f_m$.

12-10 FREQUENCY SHIFTING

In radio communication circuits, it is often necessary to shift a carrier frequency f_c with its accompanying side frequencies down to a lower intermediate frequency f_{IF}. This shift of each frequency is accomplished with the multiplier connections of Fig. 12-16(a). The modulated carrier signals are applied to the *y* input. A local oscillator is adjusted to a frequency f_o equal to the sum of the carrier and desired intermediate frequency and applied to the *x* input. The frequencies present in the output of the multiplier are calculated in the following example.

Example 12-11

In Fig. 12-16(a) amplitudes and frequencies of an AM modulated wave are present at each input as follows:

y input:

Peak amplitude (V)	Frequency (kHz)
1	$(f_c + f_m) = 1005$
4	$f_c = 1000$
1	$(f_c - f_m = 995$

where f_c is the broadcasting station's carrier frequency and $(f_c + f_m)$ and $(f_c - f_m)$ are the upper and lower side frequencies due to a 5-kHz modulating frequency, and 1000-kHz carrier frequency.

x input: The local oscillator is set for a 5-V peak sine wave at 1445 kHz, because the desired intermediate frequency is 455 kHz. Find the peak value and frequency of each signal component in the output of the multiplier.

Solution From Eq. (12-10), the peak amplitude of each y input frequency is multiplied by the peak amplitude of the local oscillator frequency. This product is multiplied by $\frac{1}{20}$ ($\frac{1}{10}$ for the scale factor $\times$ $\frac{1}{2}$ from the trigonometric identity) to obtain the peak amplitude of the resulting sum and difference frequencies at the multiplier's output. The results are tabulated in Fig. 12-16.

All frequencies present in the multiplier's output are plotted on the frequency spectrum of Fig. 12-16(c). A low-pass filter or band-pass filter is used to pass only the three lower intermediate frequencies of 450, 455, and 460 kHz. The upper intermediate frequencies of 2450, 2455, and 2460 kHz may be used if desired, but they are usually filtered out.

We conclude from Example 12-11 that each frequency present at the y input is shifted down and up to new intermediate frequencies. The lower set of intermediate frequencies can be extracted by a filter. Thus, the information contained in the carrier f_c has been preserved and shifted to another subcarrier or intermediate frequency. The process of frequency shifting is also called *heterodyne*. The heterodyne

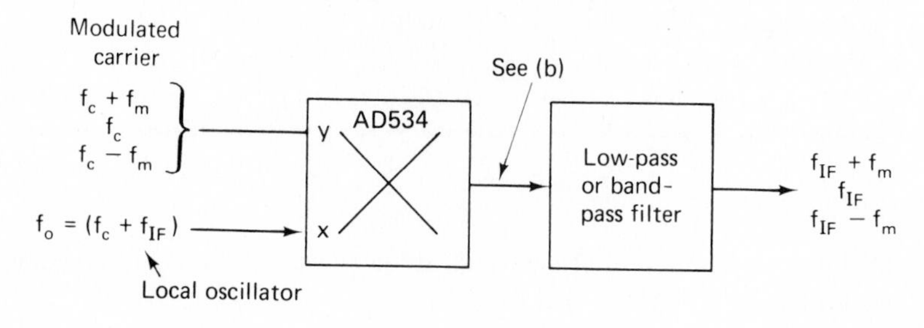

(a) Circuit for a frequency shifter

FIGURE 12-16 The multiplier as a frequency shifter.

Frequency at y input (kHz)	Multiplier output: Peak (V)	Multiplier output: Frequency (kHz)
1005	$\frac{1 \times 5}{20} = 0.25$	1455 + 1005 = 2460 1455 − 1005 = 450
1000	$\frac{4 \times 5}{20} = 1.0$	1455 + 1000 = 2455 1455 − 1000 = 455
995	$\frac{1 \times 5}{20} = 0.25$	1455 + 995 = 2450 1455 − 995 = 460

(b) Frequencies present in multiplier output

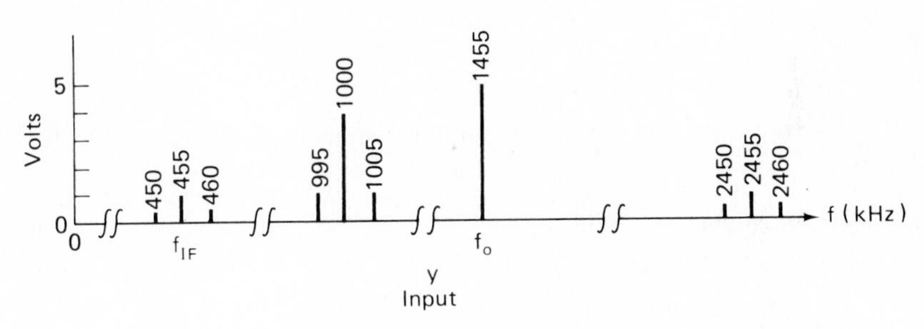

(c) y Input frequencies are shifted to the intermediate frequency

FIGURE 12-16 *(cont.)*

principle will be used in Section 12-13 to construct a universal AM receiver that will demodulate standard AM, and balance modulator and single-sideband signals.

12-11 ANALOG DIVIDER

An analog divider will give the ratio of two signals or provide gain control. It is constructed as shown in Fig. 12-17 by inserting a multiplier in the feedback loop of an op amp. Since the op amp's (−) input draws negligible current, the current I is equal in the equal resistors R. Therefore, the output voltage of the multiplier V_m is equal in magnitude but opposite in polarity (with respect to ground) to E_Z or

$$E_Z = -V_m \tag{12-11a}$$

But V_m is also equal to one-tenth (scale factor) of the product of input E_x and output of the op amp V_o. Substituting for V_m yields

$$E_Z = \frac{V_o E_x}{10} \tag{12-11b}$$

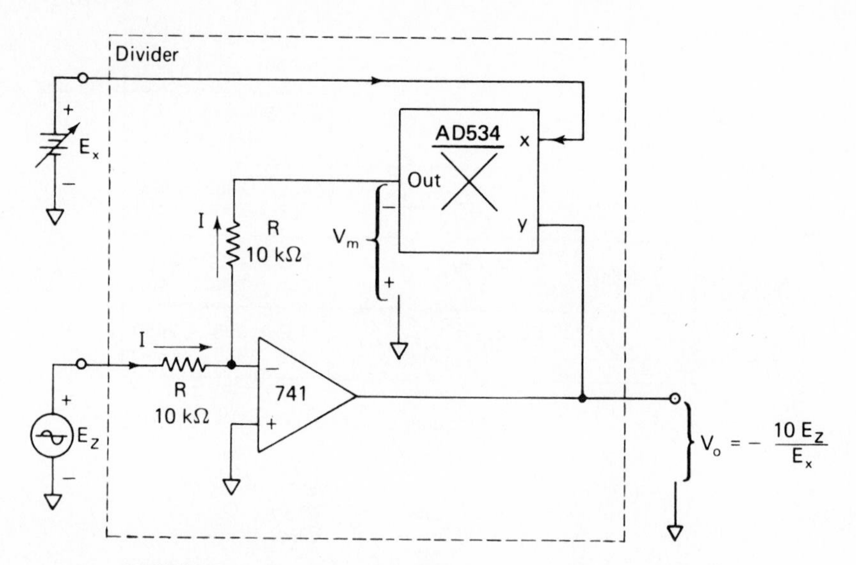

FIGURE 12-17 Division with an op amp and a multiplier.

Solving for V_o, we obtain

$$V_o = \frac{10E_Z}{E_x} \qquad (12\text{-}11c)$$

Equation (12-11c) shows that the divider's output V_o is proportional to the ratio of inputs E_Z and E_x. E_x should never be allowed to go to 0 V or to a negative voltage, because the op amp will saturate. E_Z can be either positive, negative, or 0 V. Note that the divider can be viewed as a voltage gain $10/E_x$ acting on E_Z. So if E_x is changed, the gain will change. This voltage control of the gain is useful in automatic gain-control circuits.

12-12 FINDING SQUARE ROOTS

A divider can be made to find square roots by connecting both inputs of the multiplier to the output of the op amp (see Fig. 12-18). Equation (12-11a) also pertains to Fig. 12-18. But now V_m is one-tenth (scale factor) of $V_o \times V_o$ or

$$-E_Z = V_m = \frac{V_o^2}{10} \qquad (12\text{-}12a)$$

Solving for V_o (eliminate $\sqrt{-1}$) yields

$$V_o = \sqrt{10|E_Z|} \qquad (12\text{-}12b)$$

Equation (12-12b) states that V_o equals the square root of 10 times the *magnitude* of E_Z. E_Z must be a negative voltage, or else the op amp saturates. The range of E_Z is

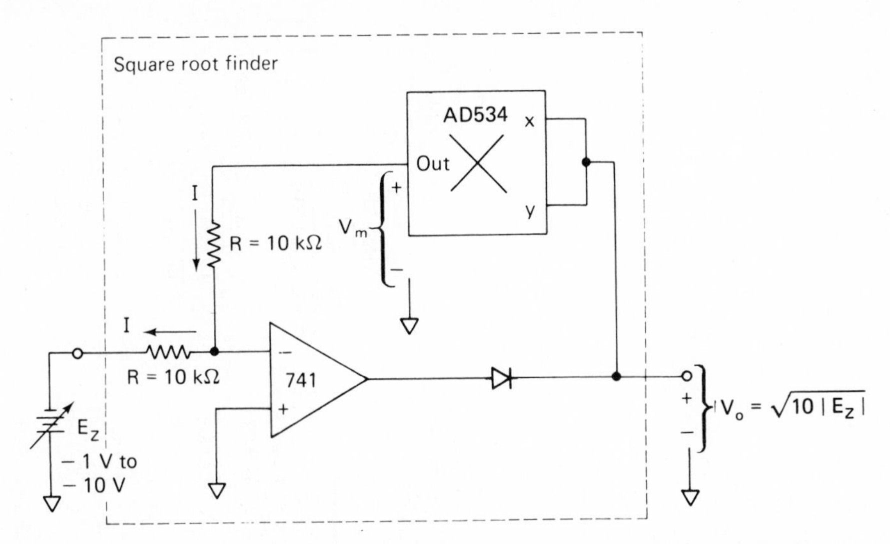

FIGURE 12-18 Square rooting with an op amp and a multiplier.

between −1 and −10 V. Voltages smaller than −1 V will cause inaccuracies. The diode prevents (−) saturation for positive E_Z. If E_Z has positive values, reverse the diode.

12-13 UNIVERSAL AMPLITUDE MODULATION RECEIVER

12-13.1 Tuning and Mixing

The ordinary automobile or household AM radio can receive only standard AM signals that occupy the AM broadcast band, about 500 to 1500 kHz. This type of radio receiver cannot extract the audio or data signals from single-sideband (CB) or suppressed carrier transmission.

Figure 12-19 shows a receiver that will receive any type of AM transmission, carrier plus sidebands, sidebands without carrier, or a single sideband (either upper or lower). To understand its operation, assume that a station is transmitting a 5-kHz audio signal that modulates a 1005-kHz carrier wave. The station transmits a standard AM frequency spectrum of the 1005-kHz carrier and both lower and upper side frequencies of 1000 and 1010 kHz [see Fig. 12-19(a)].

In Fig. 12-19(b) the receiver's *tuner* is tuned to select this one station's 10-kHz band of frequencies out of the entire broadcast band of frequencies that are present on the receiver's antenna. A *local oscillator* in the radio is designed to produce a signal that *tracks* the tuner and is always 455 kHz higher than the tuner frequency. The oscillator and tuner output frequencies are multiplied by the *IF mixer*. The IF mixer acts as a frequency shifter to shift the incoming radio-frequency carrier down to an *intermediate-frequency* (IF) carrier of 455 kHz.

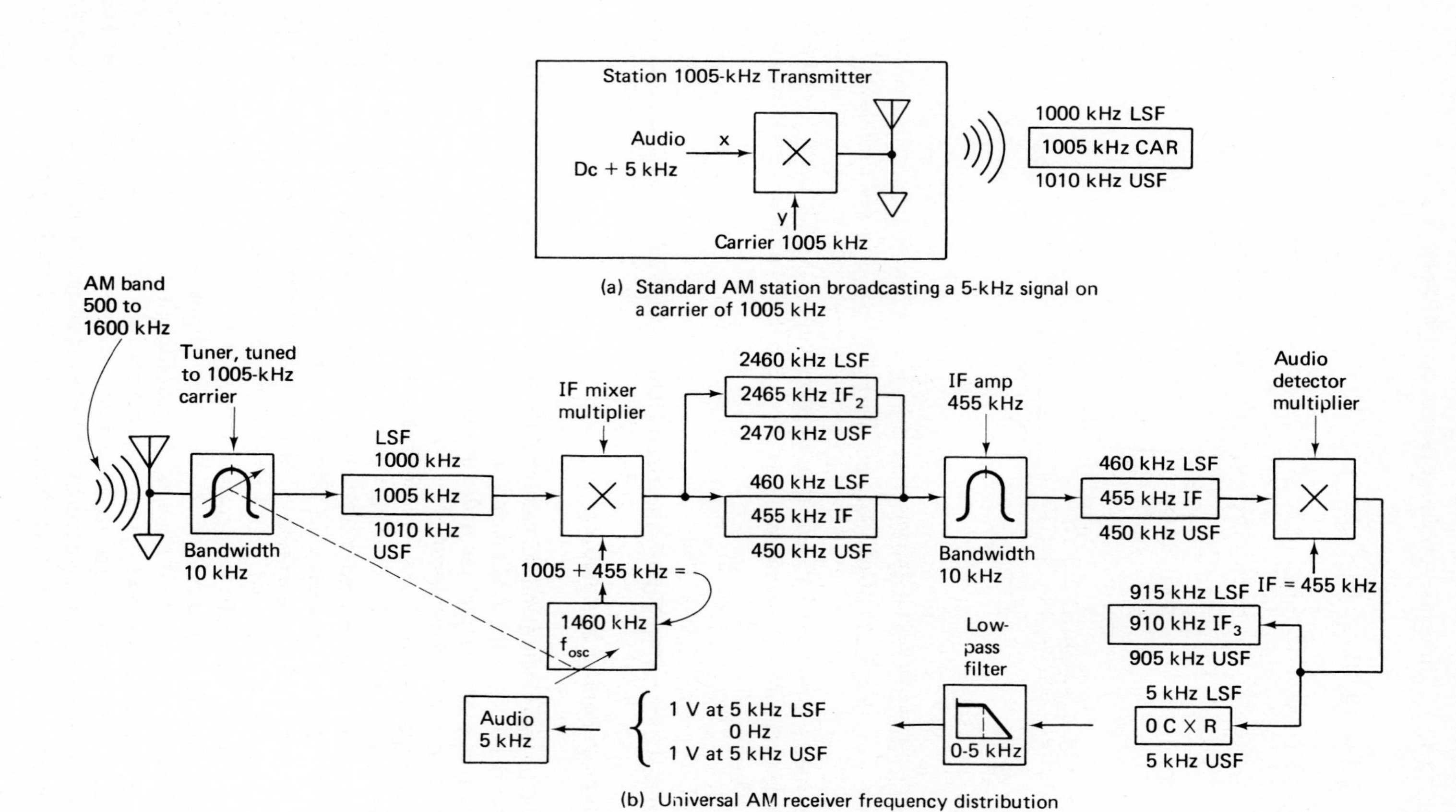

FIGURE 12-19 The superheterodyne receiver in (b) can demodulate or detect audio signals from the standard AM transmission in (a) and also single-sideband or suppressed carrier AM.

12-13.2 Intermediate-Frequency Amplifier

The output of the IF mixer contains both sum frequencies (2465-kHz IF_2 carrier) and difference frequencies (455-kHz IF carrier). Only the difference frequencies are amplified by the tuned high-gain *IF amplifier*. This first frequency shift (heterodyning) is performed so that most of the signal amplification is done by a single narrowband tuned-IF amplifier that usually has three stages of gain. *Any* station carrier that is selected by the tuner is shifted by the local oscillator and mixer multiplier down to the IF frequency for amplification. This frequency downshift scheme is used because it is much easier to build a reliable narrowband IF amplifier (10-kHz bandwidth centered on a 455-kHz carrier) than it is to build an amplifier that can provide equal amplification of, *and select* 10-kHz bandwidths over, the *entire* AM broadcast band.

12-13.3 Detection Process

The output of the IF amplifier is multiplied by the IF frequency in the *audio detector multiplier*. The term *detection* means that we are going to *detect* or *demodulate* the audio signal from the 455-kHz IF carrier. The audio detector shifts the incoming IF carrier and side frequencies up and down as sum and difference frequencies. Only the difference frequencies are transmitted through the low-pass filter in Fig. 12-19(b). The astute reader will note that the *low-pass filter* output frequencies are *not* labeled (+) 5 kHz for the upper side frequency and (−) 5 kHz for the lower side frequency. If you work the mathematics out using sine waves for audio, carrier, local oscillator, and IF signals, it turns out that both audio 5-kHz signals are in phase (as negative cosine waves). The output of the low-pass filter is applied to an audio amplifier and finally to a speaker.

12-13.4 Universal AM Receiver

Why will this receiver do what most other receivers cannot do? The previous sections dealt with a standard AM transmission carrier plus *both* upper and lower side frequencies. Suppose that you eliminated the 1005-kHz carrier at the transmitter in Fig. 12-19(a). Note how the carrier is identified by enclosure in a rectangle as it progresses through the receiver in Fig. 12-19(b). If no carrier enters the receiver, but only the side frequencies, both audio signals (USF and LSF) will still enter the audio amplifier. Thus this receiver can recover audio information from either standard AM or balanced AM modulation. An AM car radio will *not* recover the audio signal from balanced AM transmissions.

Next suppose that you eliminated (by filters) the carrier *and* upper side frequency from the transmitter in Fig. 12-19(a); only the lower side frequency of 1000 kHz would be broadcast. The entire lower sideband would occupy 1000 to 1005 kHz. This is *single-sideband transmission*. At the receiver, the tuner would select 1000 kHz (to 1005 kHz). The IF amplifier would output 460 kHz (to 455 kHz). Finally, the low-pass filter would output 5 kHz (to 0 kHz); thus this receiver can also receive single-sideband transmission. The versatility of this type of receiver is inherent in the design. It requires no switches to activate circuit changes for different types of AM modulation.

LABORATORY EXERCISES

One of the easiest analog multipliers to use is the AD534 internally trimmed, precision IC multiplier. It is available with excellent data sheets and supporting application notes from Analog Devices, Inc. No external trimpots are required. A lower-cost alternative is the AD533 of Fig. 12-3. Laboratory experience can be gained by using the AD533 or AD534 to recreate the waveshapes in the frequency doubler of Fig. 12-5.

You will need two signal or function generators to make the amplitude modulators of Figs. 12-8 and 12-10 and the systems of Figs. 12-14 and 12-15. Stay with a 1-kHz modulating signal and a 10-kHz signal for the carrier. Always keep E_m on the *A* input of a CRO and trigger the CRO from the positive zero crossing of E_m. With patience, you can manually synchronize the 10-kHz carrier with the 1-kHz signal.

An interesting demonstration can be performed with the circuit of Fig. 12-14. Tune a standard AM receiver to an empty frequency of about 600 kHz. Connect about 10 ft of wire to the output of the modulator. Adjust the frequency of the carrier oscillator to 600 kHz. Tune the radio until you pick up the 1-kHz signal. The range is only about 20 ft and therefore should not violate an FCC regulation.

PROBLEMS

12-1. Find V_o in Fig. 12-1 for the following combination of inputs: **(a)** $x = 5$ V, $y = 5$ V; **(b)** $x = -5$ V, $y = 5$ V; **(c)** $x = 5$ V, $y = -5$ V; **(d)** $x = -5$ V, $y = -5$ V.

12-2. State the operating point quadrant for each combination in Problem 12-1 [see Fig. 12-2(a)].

12-3. What is the name of the procedure used to make $V_o = 0$ when both x and y inputs are at 0 V?

12-4. Find V_o in Fig. 12-4 if $E_i = -3$ V.

12-5. The peak value of E_i in Fig. 12-5 is 8 V, and its frequency is 400 Hz. Evaluate the output's **(a)** dc terms; **(b)** ac term.

12-6. In Fig. 12-6, $E_{xp} = 10$ V, $E_{yp} = 10$ V, and $\theta = 30°$. Find V_o.

12-7. Repeat Problem 12-6 for $\theta = -30°$.

12-8. In the balanced modulator of Fig. 12-8, E_x is a 5-kHz sine wave at 8 V peak and E_y is a 3-kHz sine wave at 5 V peak. Find the peak voltage of each frequency in the output.

12-9. In Fig. 12-8, the carrier frequency is 15 kHz. The modulating frequencies range between 1 and 2 kHz. Find the upper and lower side bands.

12-10. The switch is on AM in Fig. 12-10. The modulating frequency is 10 kHz at 5 V peak. The carrier is 100 kHz at 8 V peak. Identify the peak value and each frequency contained in the output.

12-11. If the switch is thrown to "Balanced" in Problem 12-10, what changes result in the output?

12-12. The x input of Fig. 12-13 is three sine waves of 5 V at 20 kHz, 2.0 V at 21 kHz, and 2.0 V at 19 kHz. The y input is 5 V at 20 kHz. What are the output signal frequency components?

12-13. It is desired to shift a 550-kHz signal to a 455-kHz intermediate frequency. What frequency should be generated by the local oscillator?

12-14. $E_x = 10$ V and $E_Z = -1$ V in Fig. 12-17. Find V_o.

CHAPTER 13

Integrated-Circuit Timers

LEARNING OBJECTIVES

Upon completion of this chapter on integrated-circuit timers, you will be able to:

- Name three operating states of a 555 timer and tell how they are controlled by the trigger and threshold terminals.
- Draw circuits that produce a time delay or an initializing pulse upon application of power.
- Connect the 555 to make an oscillator for any desired frequency.
- Use 555 oscillators to make a tone-burst oscillator or voltage-controlled frequency shifter.
- Explain the operation of a 555 when it is wired to perform as a one-shot or monostable multivibrator.
- Use the 555 one-shot as a touch switch, frequency divider, or missing pulse detector.

- Describe the operation of an XR2240 programmable timer/counter.
- Connect the XR2240 as a long-interval timer, free-running oscillator, binary pattern generator, or frequency synthesizer.
- Build a switch programmable timer.

13-0 INTRODUCTION

Applications such as oscillators, pulse generators, ramp or square-wave generators, one-shot multivibrators, burglar alarms, and voltage monitors all require a circuit capable of producing timing intervals. The most popular integrated-circuit timer is the 555, first introduced by Signetics Corporation (see Appendix 4). Similar to general purpose op amps, the 555 is reliable, easy to use in a variety of applications, and low in cost. The 555 can also operate from supply voltages of +5 V to +18 V, making it compatible with both TTL (transistor-transistor logic) circuits and op amp circuits. The 555 timer can be considered a functional block that contains two comparators, two transistors, three equal resistors, a flip-flop, and an output stage. These are shown in Fig. 13-1.

Besides the 555 timer, there are also available counter timers such as Exar's XR—2240. The 2240 contains a 555 timer plus a programmable binary counter in a single 16-pin package. A single 555 has a maximum timing range of approximately 15 min. Counter timers have a maximum timing range of days. The timing range of both can be extended to months or even years by cascading. Our study of timers will begin with the 555 and its applications and then proceed to the counter timers.

13-1 OPERATING MODES OF THE 555 TIMER

The 555 IC timer has two modes of operation, either as an astable (free-running) multivibrator or as a monostable (one-shot) multivibrator. Free-running operation of the 555 is shown in Fig. 13-2(a). The output voltage switches from a high to a low state and back again. The time the output is either high or low is determined by a resistor–capacitor network connected externally to the 555 timer (see Section 13-2). The value of the high output voltage is slightly less than V_{CC}. The value of the output voltage in the low state is approximately 0.1 V.

When the timer is operated as a one-shot multivibrator, the output voltage is low until a negative-going trigger pulse is applied to the timer; then the output switches high. The time the output is high is determined by a resistor and capacitor connected to the IC timer. At the end of the timing interval, the output returns to the low state. Monostable operation is examined further in Sections 13-5 and 13-6. To understand how a 555 timer operates, a brief description of each terminal is given in Section 13-2.

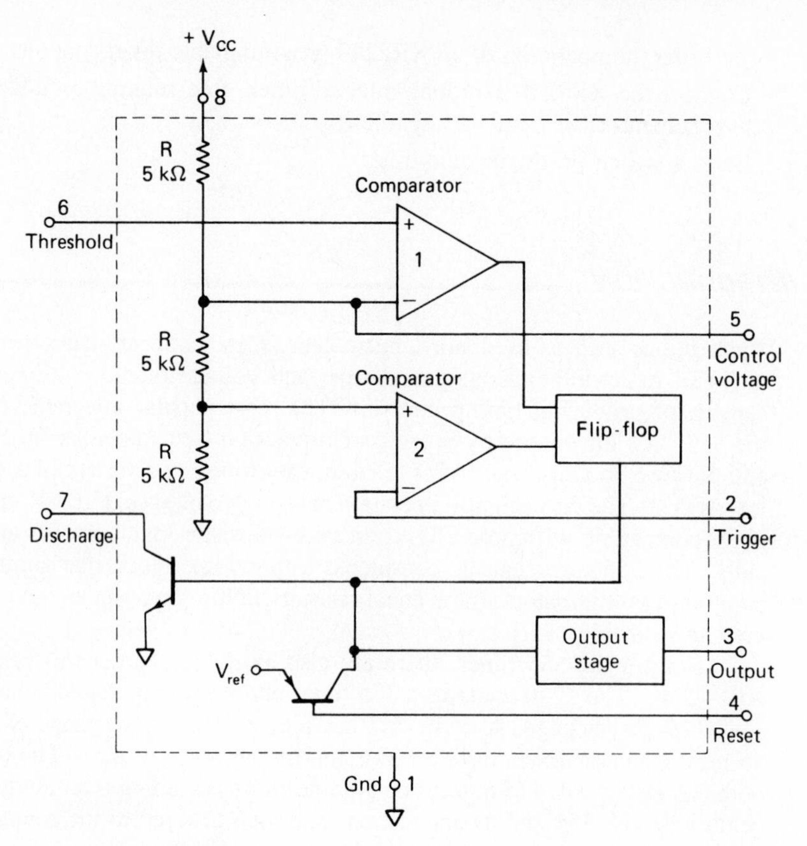

FIGURE 13-1 A 555 integrated-circuit timer.

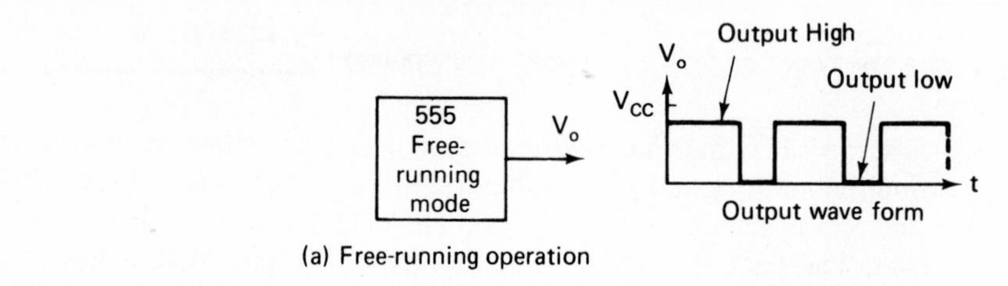

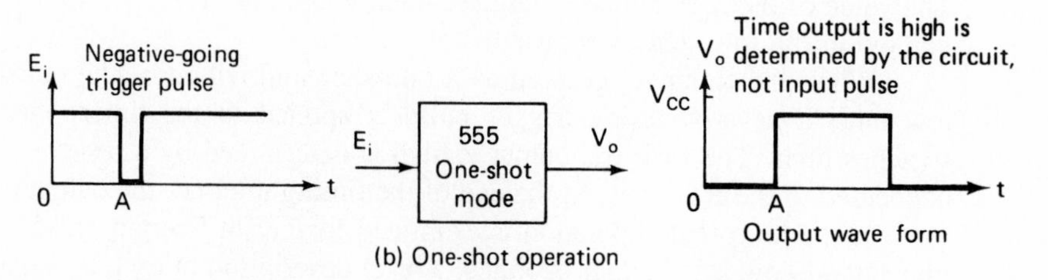

FIGURE 13-2 Operating modes of a 555 timer.

13-2 TERMINALS OF THE 555

13-2.1 Packaging and Power Supply Terminals

The 555 timer is available in two package styles, TO-99 and DIP, as shown in Fig. 13-3(a) and Appendix 4. Pin 1 is the common, or ground terminal, and pin 8 is the

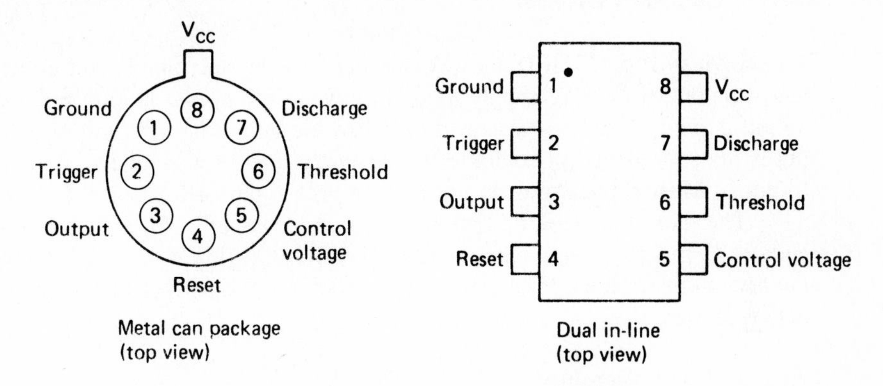

(a) 555 pin connections and package styles

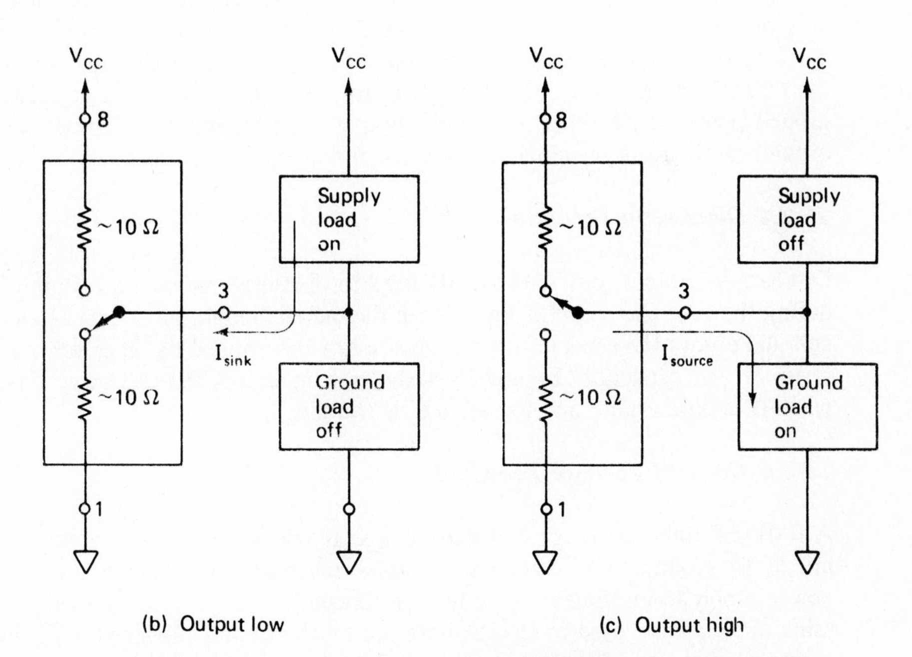

(b) Output low (c) Output high

FIGURE 13-3 The 555 timer output operation and package terminals. Either a grounded or a supply load can be connected, although usually not simultaneously.

positive voltage supply terminal V_{CC}. V_{CC} can be any voltage between +5 V and +18 V. Thus the 555 can be powered by existing digital logic supplies (+5 V), linear IC supplies (+15 V), and automobile or dry cell batteries. Internal circuitry requires about 0.7 mA per supply volt (10 mA for V_{CC} = +15) to set up internal bias currents. Maximum power dissipation for the package is 600 mW.

13-2.2 Output Terminal

As shown in Fig. 13-3(b) and (c), the output terminal, pin 3, can either source or sink current. A *floating* supply load is *on* when the *output* is *low,* and *off* when the output is *high.* A *grounded* load is *on* when the *output* is *high,* and *off* when the output is *low*. In normal operation either a supply load or a grounded load is connected to pin 3. Most applications do not require both types of loads at the same time.

The maximum sink or source current is technically 200 mA, but more realistically is 40 mA. The high output voltage [Fig. 13-3(c)] is about 0.5 V below V_{CC}, and the low output voltage [Fig. 13-3(b)] is about 0.1 V above ground, for load currents below 25 mA.

13-2.3 Reset Terminal

The reset terminal, pin 4, allows the 555 to be disabled and override command signals on the trigger input. When not used, the reset terminal should be wired to $+V_{CC}$. If the reset terminal is grounded or its potential reduced below 0.4 V, both the output terminal, pin 3, and the discharge terminal, pin 7, are at approximately ground potential. In other words, the output is held low. If the output was high, a ground on the reset terminal immediately forces the output low.

13-2.4 Discharge Terminal

Discharge terminal, pin 7, is usually used to discharge an external timing capacitor during the time the output is low. When the output is high, pin 7 acts as an open circuit and allows the capacitor to charge at a rate determined by an external resistor or resistors and capacitor. Figure 13-4 shows a model of the discharge terminal for when C is discharging and for when C is charging.

13-2.5 Control Voltage Terminal

A 0.01-μF filter capacitor is usually connected from the control voltage terminal, pin 5, to ground. The capacitor bypasses noise and/or ripple voltages from the power supply to minimize their effect on threshold voltage. The control voltage terminal may also be used to change both the threshold and trigger voltage levels. For example, connecting a 10-kΩ resistor between pins 5 and 8 changes threshold voltage to 0.5 V_{CC} and the trigger voltage to 0.25 V_{CC}. An external voltage applied

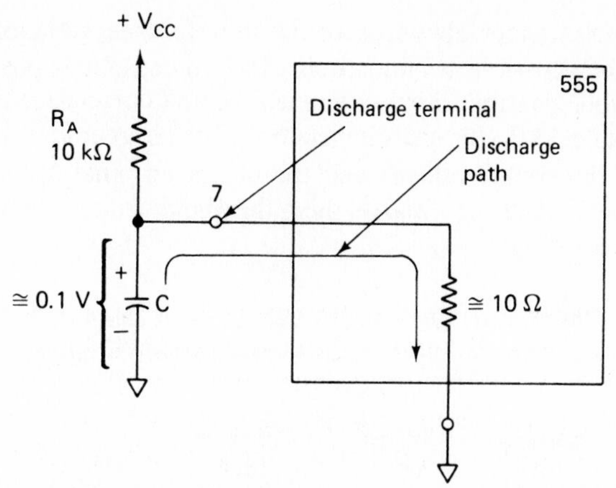

(a) Model of the discharge terminal when the output is low, and capacitor is discharging

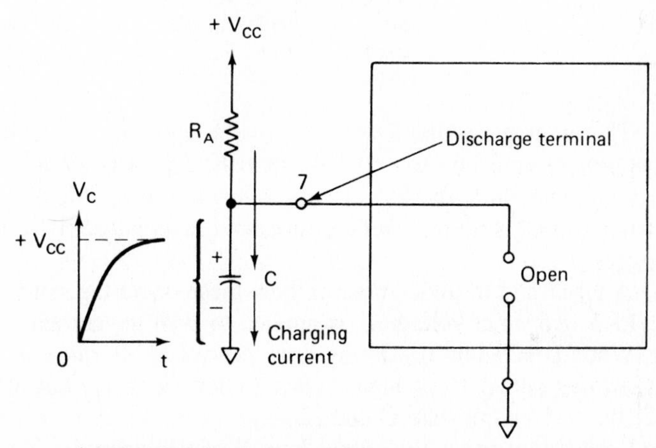

(b) Model of the discharge terminal when the output is high, and capacitor is charging

FIGURE 13-4 Operation of discharge terminal.

to pin 5 will change both threshold and trigger voltages and can also be used to modulate the output waveform.

13-2.6 Trigger and Threshold Terminals

The 555 has two possible operating states and one memory state. They are determined by *both* the *trigger* input, pin 2, and the *threshold* input, pin 6. The trigger input is compared by comparator 1 in Fig. 13-1, with a lower threshold voltage V_{LT} that is equal to $V_{CC}/3$. The threshold input is compared by comparator 2 with a higher threshold voltage V_{UT} that is equal to $2V_{CC}/3$. Each input has two possible voltage levels, either above or below its reference voltage. Thus with two inputs there are four possible combinations that will cause four possible operating states.

The four possible input combinations and corresponding states of the 555 are given in Table 13-1. In operating state A, *both* trigger and threshold are *below* their respective threshold voltages and the output terminal (pin 3) is *high*. In operating state D, *both* inputs are *above* their threshold voltages and the output terminal is *low*.

TABLE 13-1 OPERATING STATES OF A 555 TIMER: $V_{UT} = 2V_{CC}/3$, $V_{LT} = V_{CC}/3$; HIGH $\simeq V_{CC}$, LOW OR GROUND $\simeq$ 0 V

Operating state	Trigger pin 2	Threshold pin 6	State of terminals: Output 3	State of terminals: Discharge 7
A	Below V_{LT}	Below V_{UT}	High	Open
B	Below V_{LT}	Above V_{UT}	High	Open
C	Above V_{LT}	Below V_{UT}	Remembers last state	Remembers last state
D	Above V_{LT}	Above V_{UT}	Low	Ground

The observation that low inputs give a high output, and high inputs give a low output, might lead you to conclude that the 555 acts as an inverter. However, as shown in Table 13-1, the 555 also has a *memory* state. Memory state C occurs when the trigger input is *above, and* the threshold input is *below* their respective reference voltages.

A visual aid in understanding how these operating states occur is presented in Fig. 13-5. An input voltage E_i is applied to *both* trigger and threshold input terminals. When E_i is below V_{LT} during time intervals A–B and E–F, state A operation results, so that output V_{o3} is high. When E_i lies above V_{LT} but below V_{UT}, within time B–C, the 555 enters state C and remembers its last A state. When E_i exceeds V_{UT}, state D operation sends the output low. When E_i drops between V_{UT} and V_{LT} during time D–E, the 555 remembers the last D state and its output stays low. Finally, when E_i drops below V_{LT} during time E–F, the A state sends the output high.

By plotting output V_{o3} against E_i in Fig. 13-5, we see a hysteresis characteristic. Recall from Chapter 4 that a *hysteresis loop* means that the circuit has *memory*. This also means that if the inputs are in one of the memory states, you cannot tell what state the output is now in, unless you know the previous state. Two *power-on* applications will now be given to show how to analyze circuit operation from Table 13-1.

13-2.7 Power-on Time Delays

There are two types of timing events that may be required during a power-on application. You may wish to apply power to one part of a system and wait for a short interval before starting some other part of a system. For example, you may need to re-

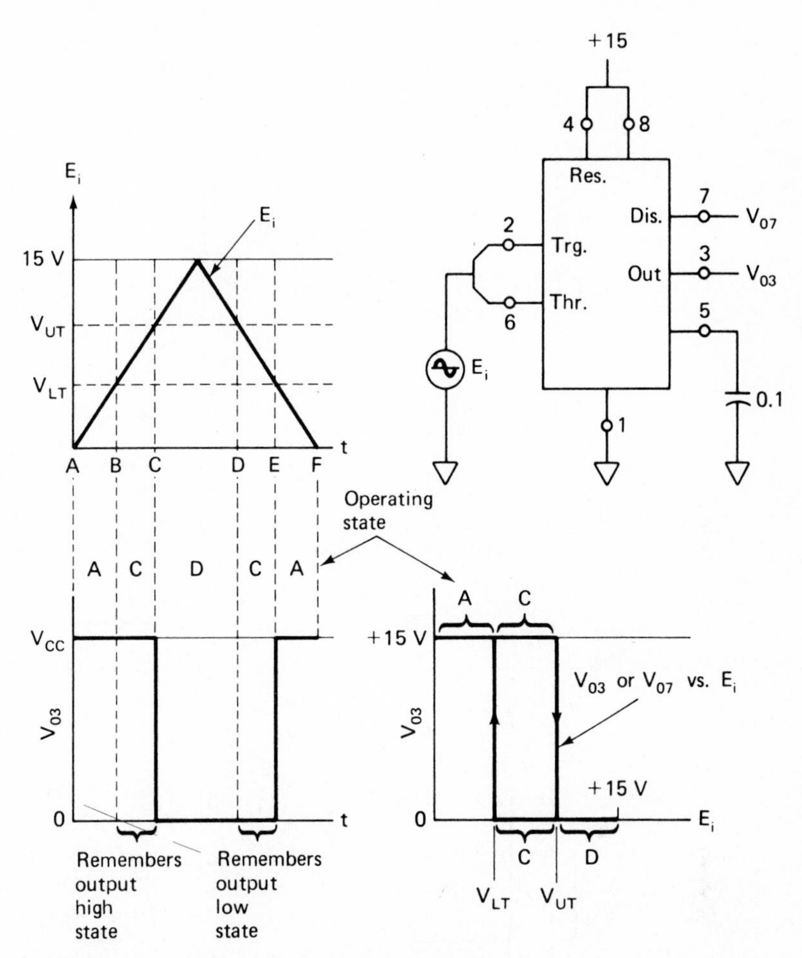

FIGURE 13-5 Three of the four operating states of a 555 timer are shown by a test circuit to measure E_i and V_{o3} versus time and V_{o3} versus E_i.

set all counters to zero before starting a personal computer at the beginning of a business day. A circuit that solves this problem is shown in Fig. 13-6(a). When the power switch is thrown to on at $t = 0$, the initial capacitor voltage is zero. Therefore, both pins 2 and 6 are above their respective thresholds and the output stays low in operating state D. As capacitor C charges, threshold drops below V_{UT} while trigger is still above V_{LT}, forcing the 555 into memory state C. Finally, both trigger and threshold drop just below V_{LT}, where the 555 enters state A and forces the output high at time T.

The net result is that an output from pin 3 of the 555 is delayed for a time interval T after the switch closure at $t = 0$. The time delay is found from $T = 1.1\, R_A C$.

By interchanging R_A and C, a time delay with a high output can be generated.

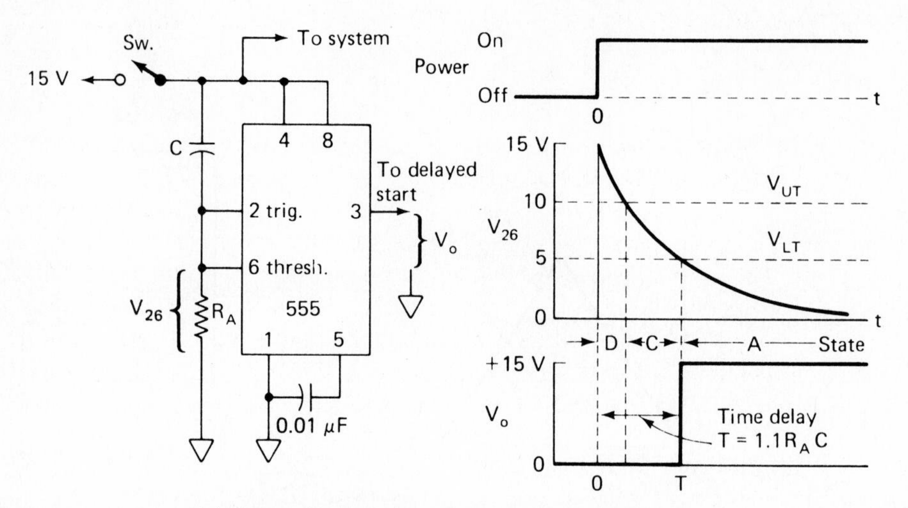

(a) Output V_o does not go high until a time interval T elapses after application of power at t = 0

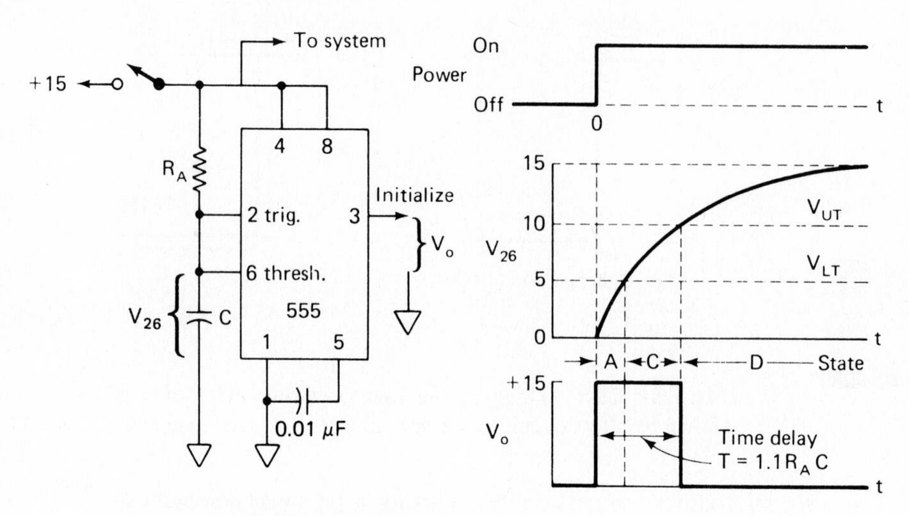

(b) Output V_o goes high for a time interval T after power is applied

FIGURE 13-6 Power-on time-delay applications are analyzed by reference to Table 13-1.

In the circuit of Fig. 13-6(b), power is applied to a system when the switch is closed. The 555's output goes high for a period of time T and then goes low. T is found from Eq. (13-9). This type of startup pulse is typically used to reset counters and initialize computer sequences after a power failure. It also can allow time for an operator to exit after an alarm system has been turned on before arming the system.

13-3 FREE-RUNNING OR ASTABLE OPERATION

13-3.1 Circuit Operation

The 555 is connected as a free-running multivibrator in Fig. 13-7(a). Refer to the waveshapes in Fig. 13-7(b) to follow the circuit's operation. At time A both pins 2 and 6 go just below $V_{LT} = \frac{1}{3}V_{CC}$ and output pin 3 goes high (state A). Pin 7 also becomes an open, so capacitor C charges through $R_A + R_B$. During output high time A–B, the 555 is in memory state C, remembering the previous A state. When V_C goes just above $V_{UT} = \frac{2}{3}V_{CC}$ at time B, the 555 enters state D and sends the output low. Pin 7 also goes low and capacitor C discharges through resistor R_B. During output low time B–C, the 555 is in memory state C, remembering the previous state D. When V_C drops just below V_{LT}, the sequence repeats.

13-3.2 Frequency of Oscillation

The output stays high during the time interval that C charges from $\frac{1}{3}V_{CC}$ to $\frac{2}{3}V_{CC}$ as shown in Fig. 13-7(b). This time interval is given by

$$t_{high} = 0.695(R_A + R_B)C \tag{13-1}$$

The output is low during the time interval that C discharges from $\frac{2}{3}V_{CC}$ to $\frac{1}{3}V_{CC}$ and is

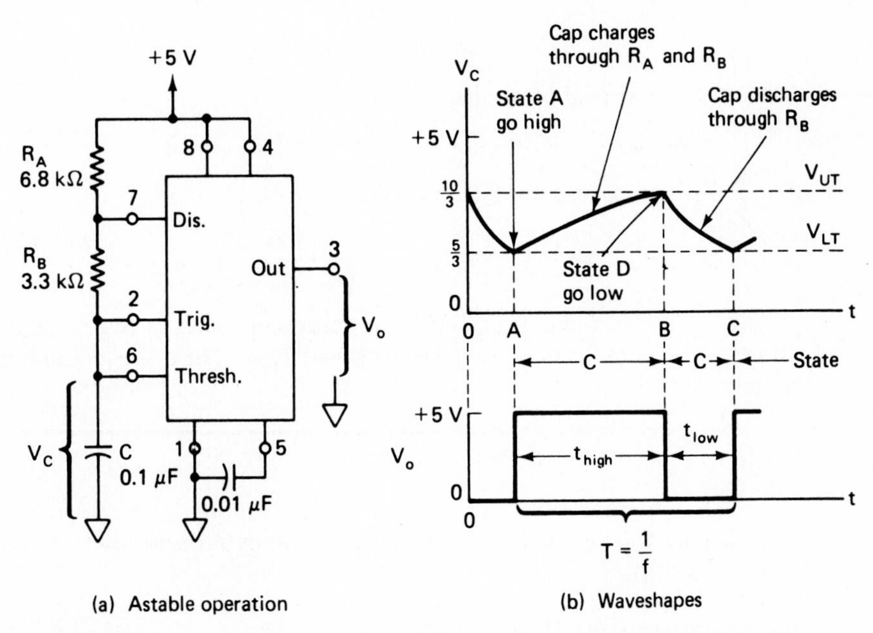

(a) Astable operation

(b) Waveshapes

FIGURE 13-7 Waveshapes are shown in (b) for the free-running astable multivibrator in (a). Frequency of operation is determined by the resistance and capacitance values in (c).

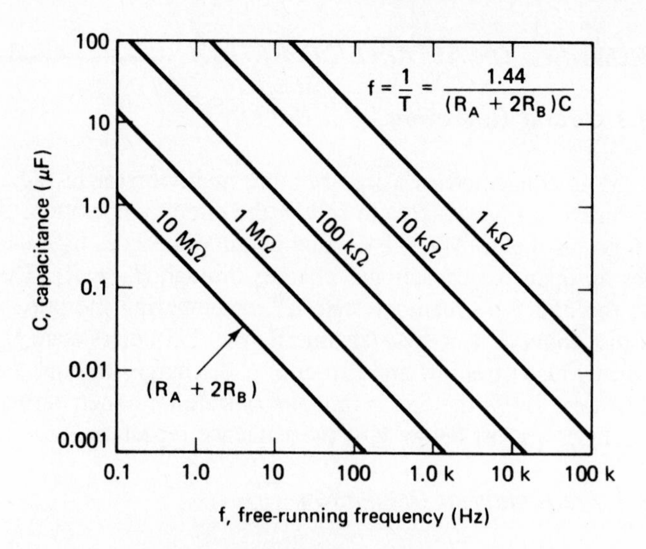

(c) Frequency dependence on R_A, R_B, and C

FIGURE 13-7 *(cont.)*

given by

$$t_{low} = 0.695 R_B C \tag{13-2}$$

Thus the total period of oscillation T is

$$T = t_{high} + t_{low} = 0.695(R_A + 2R_B)C \tag{13-3}$$

The free-running frequency of oscillation f is

$$f = \frac{1}{T} = \frac{1.44}{(R_A + 2R_B)C} \tag{13-4}$$

Figure 13-7(c) is a plot of Eq. (13-4) for different values of $(R_A + 2R_B)$ and quickly shows what combinations of resistance and capacitance are needed to design an astable multivibrator.

Example 13-1

Calculate (a) t_{high}, (b) t_{low}, and (c) the free-running frequency for the timer circuit of Fig. 13-7(a).

Solution (a) By Eq. (13-1),

$$t_{high} = 0.695(6.8\ \text{k}\Omega + 3.3\ \text{k}\Omega)(0.1\ \mu\text{F}) = 0.7\ \text{ms}$$

(b) By Eq. (13-2),

$$t_{low} = 0.695(3.3\ \text{k}\Omega)(0.1\ \mu\text{F}) = 0.23\ \text{ms}$$

(c) By Eq. (13-4),

$$f = \frac{1.44}{[6.8\ \text{k}\Omega + (2)(3.3\ \text{k}\Omega)][0.1\ \mu\text{F}]} = 1.07\ \text{kHz}$$

The answer to part (c) agrees with results obtainable from Fig. 13-7(c).

13-3.3 Duty Cycle

The ratio of time when the output is low t_{low} to the total period T is called the *duty cycle D*. In equation form,*

$$D = \frac{t_{low}}{T} = \frac{R_B}{R_A + 2R_B} \tag{13-5}$$

Example 13-2

Calculate the duty cycle for the values given in Fig. 13-7(a).

Solution By Eq. (13-5),

$$D = \frac{3.3\ \text{k}\Omega}{6.8\ \text{k}\Omega + 2(3.3\ \text{k}\Omega)} = 0.25$$

This checks with Fig. 13-6(b) which shows that the timer's output is low for approximately 25% of the total period T. Equation (13-5) shows that it is impossible to obtain a duty cycle of $\frac{1}{2}$ or 50%. As presented, the circuit of Fig. 13-7(a) is not capable of producing a square wave. The only way D in Eq. (13-5) can equal $\frac{1}{2}$ is for R_A to equal 0. Then there would be a short circuit between V_{CC} and pin 7. However, R_A must be large enough so that when the discharge transistor is "on," current through it is limited to 0.2 A. Thus the minimum value of R_A in ohms is given by

$$\text{minimum } R_A \simeq \frac{V_{CC}}{0.2\ \text{A}} \tag{13-6}$$

In practice, keep R_A equal to or greater than 1 kΩ.

*The original literature published by Signetics (the maker of the 555) defined duty cycle as shown. We go along with the originator in this chapter. In most other texts and papers, duty cycle is expressed, in percent, as the ratio of high time to period.

The next section presents an inexpensive way of extending the duty cycle.

13-3.4 Extending the Duty Cycle

The duty cycle for the circuit of Fig. 13-7(a) can never be equal to or greater than 50%, as discussed in Section 13-3.3. By connecting a diode in parallel with R_B in Fig. 13-8(a), a duty cycle of 50% or greater can be obtained. Now the capacitor charges through R_A and the diode, but discharges through R_B. The times for the output waveform are

$$t_{high} = 0.695 R_A C \tag{13-7a}$$

$$t_{low} = 0.695 R_B C \tag{13-7b}$$

$$T = 0.695(R_A + R_B)C \tag{13-7c}$$

Equations (13-7a) and (13-7b) show that if $R_A = R_B$, then the duty cycle is 50%, as shown in Fig. 13-8(b) and (c).

13-4 APPLICATIONS OF THE 555 AS AN ASTABLE MULTIVIBRATOR

13-4.1 Tone-Burst Oscillator

With the switch in Fig. 13-9 set to the "continuous" position, the B 555 timer functions as a free-running multivibrator. The frequency can be varied from about 1.3 kHz to 14 kHz by the 10-kΩ potentiometer. If the potentiometer is replaced by a thermistor or photoconductive cell, the oscillating frequency will be proportional to temperature or light intensity respectively.

The A 555 timer oscillates at a slower frequency. The 1-MΩ potentiometer sets the lowest frequency at about 1.5 Hz. Lower frequencies are possible by replacing the 1-μF capacitor with a larger value. When the connecting switch is thrown to the "burst" position, output pin 3 of the *A* timer alternately places a ground or high voltage on reset pin 4 of the B 555 timer. When pin 4 of the *B* timer is grounded, it cannot oscillate, and when ungrounded the timer oscillates. This causes the *B* timer to oscillate in bursts. The output of the tone-burst generator is V_o and is taken from pin 3 of timer *B*. V_o can drive either an audio amplifier or a stepdown transformer directly to a speaker. The 556 IC timer contains two 555 timers in a single 14-pin dual-in-line package. Thus the tone-burst generator can be made with one 556.

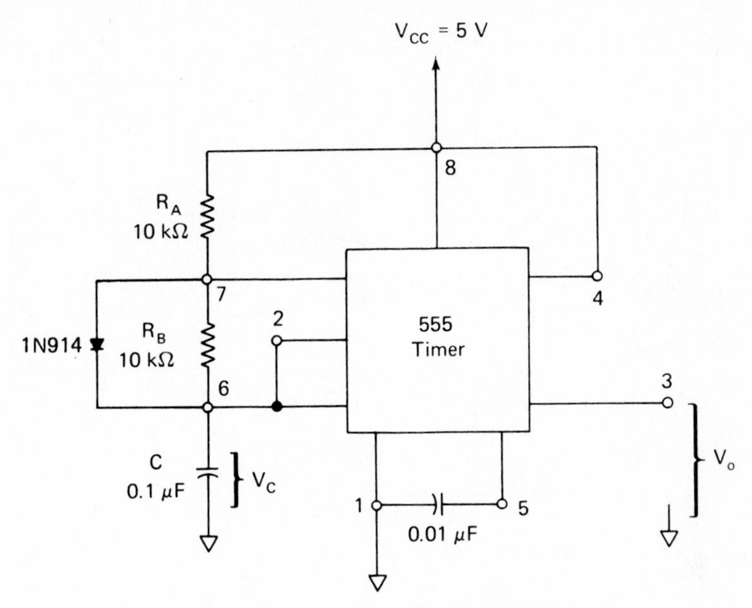

(a) Timer circuit to produce a 50% duty cycle

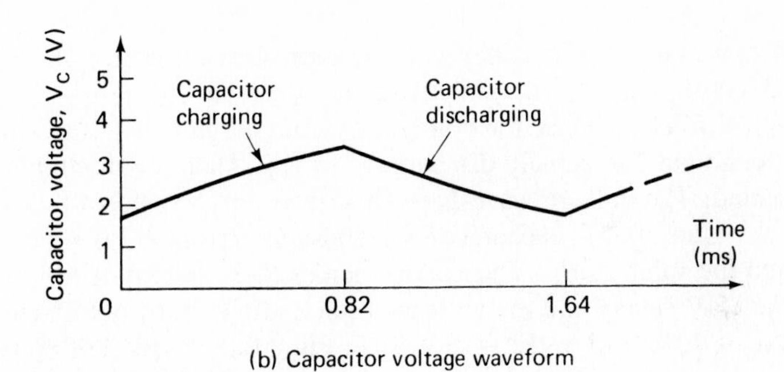

(b) Capacitor voltage waveform

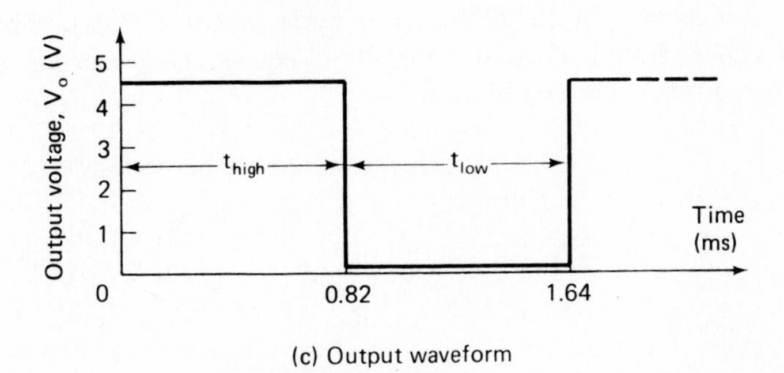

(c) Output waveform

FIGURE 13-8 Connecting a diode across R_B to produce duty cycles of 50%.

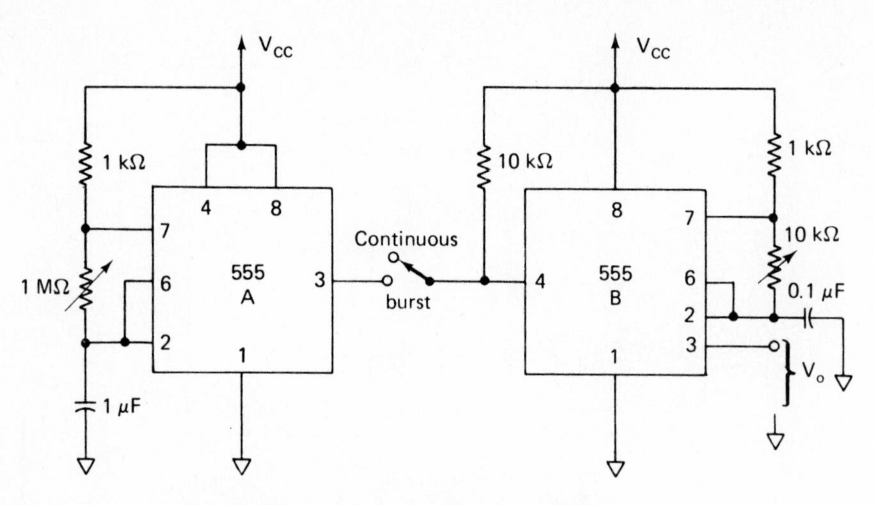

FIGURE 13-9 Tone-burst oscillator.

13-4.2 Voltage-Controlled Frequency Shifter

A low-cost, low-frequency voltage-controlled frequency shifter is presented in Fig. 13-10(a). Since the 555 timer is powered by $V_{CC} = 5$ V, $V_{UT} = 5\text{ V}(\frac{2}{3})$ and $V_{LT} = 5\text{ V}(\frac{1}{3})$. Capacitor voltage V_C will charge to V_{UT}, at which time the 555 will ground pin 7 to rapidly discharge C to V_{LT}. Then the discharge path for C is disconnected. The voltage waveshape described for C is shown in Fig. 13-10(c).

Capacitor C is charged by a constant current I. I is set by the voltage across R_E and the value of R_E. The voltage across R_E is determined by the difference between the 15-V supply and the voltage at pin 2 ($10\text{ V} - E$) of voltage follower 741B. Thus $V_{RE} = 15 - (10 - E) = 5\text{ V} + E$. Pin 2 follows pin 3 of op amp B, or V_{oA}. The inverting adder, 741A, has an output voltage of $V_{oA} = 10 - E$.

Charge lost by C for each cycle equals $C(\Delta V_C)$, where $\Delta V_C = 5\text{ V}(\frac{2}{3}) - 5\text{ V}(\frac{1}{3}) = 5\text{ V}/3$. Charge stored by C equals charge current I times period T (the charge time). For equilibrium

$$\text{charge stored} = \text{charge lost} \tag{13-8a}$$

$$IT = C\,\Delta V \tag{13-8b}$$

$$\frac{5\text{ V} - E}{R_E}T = C\frac{5\text{ V}}{3} \tag{13-8c}$$

Since period $T = 1/f_{out}$, we can rewrite Eq. (13-8c) as

$$f_{out} = \text{center frequency } f_c + \text{shift frequency } \Delta f \tag{13-8d}$$

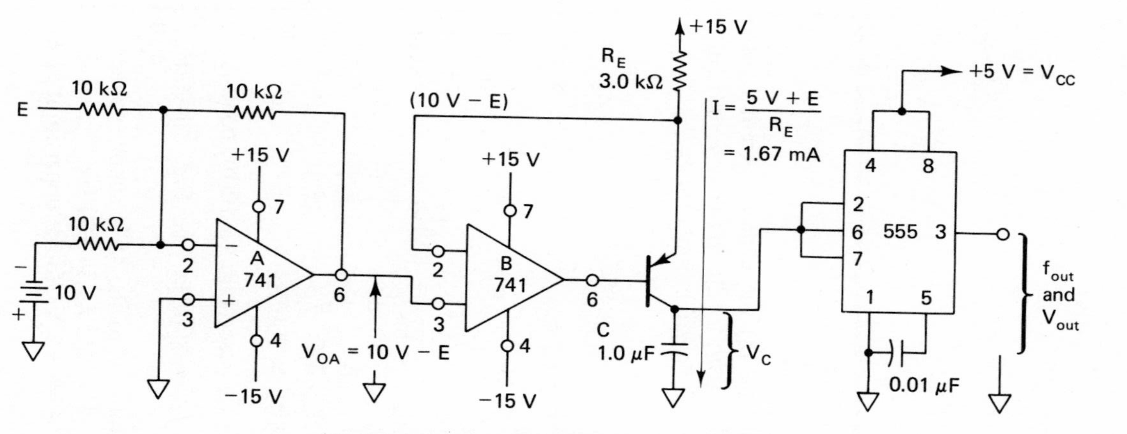

(a) Circuit for voltage-controlled frequency shifter

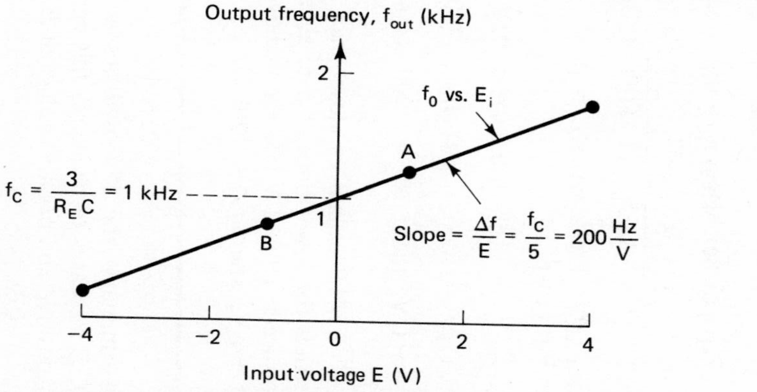

(b) Output–input characteristic of the frequency shifter: when E = 0, the output voltage is a 10-μs negative spike with a frequency of 1 kHz; output frequency f_{out} increases by 200 Hz for every positive-going 1 V of E. If E goes negative, f_{out} decreases at the same rate

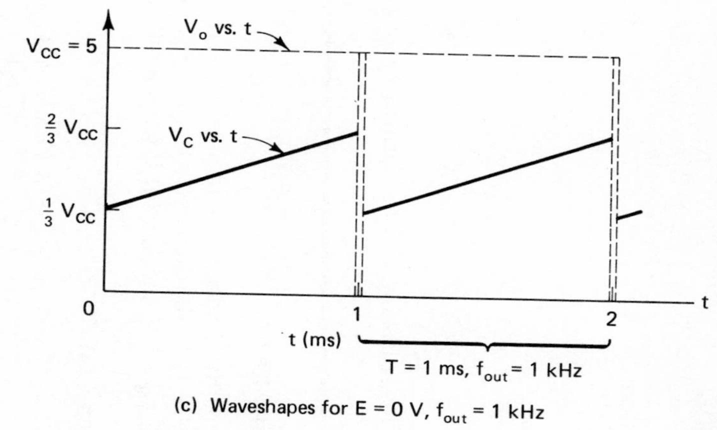

(c) Waveshapes for E = 0 V, f_{out} = 1 kHz

FIGURE 13-10 Voltage-controlled frequency shifter.

where

$$f_c = \frac{3}{R_E C} \text{ when } E = 0 \text{ V} \tag{13-8e}$$

and

$$\Delta f = 0.2 f_c E \tag{13-8f}$$

This reasoning will be clarified by an example.

Example 13-3

For the frequency shifter of Fig. 13-10 calculate (a) the charge current I for $E = 0$ V; (b) the center frequency f_c when $E = 0$ V; (c) the frequency shift for $E = \pm 1$ V and f_{out}. (d) Point out the positive and negative limits for E_i.

Solution (a)

$$I = \frac{5\text{ V} - E}{R_E} = \frac{5\text{ V}}{3\text{ k}\Omega} = 1.67\text{ mA}$$

(b) From Eq. (13-8e), when $E = 0$,

$$f_c = \frac{3}{R_E C} = \frac{3}{(3\text{ k}\Omega)(1 \times 10^{-6}\text{ F})} = \frac{3}{3\text{ ms}} = 1\text{ kHz}$$

Therefore, $f_{out} = 1$ kHz when $E = 0$.
(c) From Eq. (13-8f),

$$\Delta f = 0.2 f_c E = 0.2(1000)1 = 200\text{ Hz}$$

$f_{out} = f_c + \Delta f = 1000 + 200$ Hz $= 1200$ Hz [see Fig. 13-10(b), point A]

For $E = -1$ V, then, $\Delta f = -200$ Hz and $f_{out} = 800$ Hz [see Fig. 13-10(b), point B].
(d) Pin 2 of 741B cannot go closer than about 1 V to the 15-V supply voltage. This restricts the lower limit of E to about -4 V and f_{out} to 200 Hz. V_{CE} of the transistor needs about 2 V headroom above $V_{UT} = 3.3$ V. So the upper limit of E is about $+4$ V, where $f_{out} = 1800$ Hz.

In summary, the 555 oscillates at a center frequency f_c determined by Eq. (13-8e). E increases or decreases this center frequency by an amount of $0.2\,f_c$ per volt for positive and negative values of E, respectively.

13-5 ONE-SHOT OR MONOSTABLE OPERATION

13-5.1 Introduction

Not all applications require a continuous repetitive wave such as that obtained from a free-running multivibrator. Many applications need to operate only for a specified length of time. These circuits require a one-shot or monostable multivibrator. Figure 13-11(a) is a circuit diagram using the 555 for monostable operation. When a negative-going pulse is applied to pin 2, the output goes high and terminal 7 removes a short circuit from capacitor C. The voltage across C rises from 0 volts at a rate determined by R_A and C. When capacitor voltage reaches $\frac{2}{3}V_{CC}$, comparator 1 in Fig. 13-1 causes the output to switch from high to low. The input and output voltage waveforms are shown in Fig. 13-11(a). The output is high for a time given by

$$t_{high} = 1.1R_AC \tag{13-9}$$

Figure 13-11(b) is a plot of Eq. (13-9) and quickly shows the wide range of output pulses that are obtainable and the required values of R_A and C. Figure 13-11(a) gives the idea of a one-shot. In practice we must add more parts to make a workable circuit (see Section 13-5.2).

Example 13-4

If $R_A = 9.1\ \text{k}\Omega$, find C for an output pulse duration of 1 ms.

Solution Rearrange Eq. (13-9):

$$C = \frac{t_{high}}{1.1R_A} = \frac{1 \times 10^{-3}\ \text{s}}{1.1(9.1 \times 10^3)\ \Omega} = 0.1\ \mu\text{F}$$

This answer checks with that obtainable at point B in Fig. 13-11(b). For the 555 timer to trigger properly in this type of operation, the width of the trigger pulse must be less than t_{high} and a trigger input pulse network is needed so that the output does not switch on the positive-going edge of the trigger pulse (point P).

13-5.2 Input Pulse Circuit

Figure 13-12 shows the multivibrator wired for monostable operation in contrast with Fig. 13-11(a). R_i, C_i, and diode D are needed to generate a single output pulse for one input pulse. Resistor R_A and capacitor C determine the time that the output is high, as given by Eq. (13-9). Resistor R_i is connected between V_{CC} and pin 2 to ensure that the output is normally low. C_i is charged to $(V_{CC} - E_i)$ until the negative

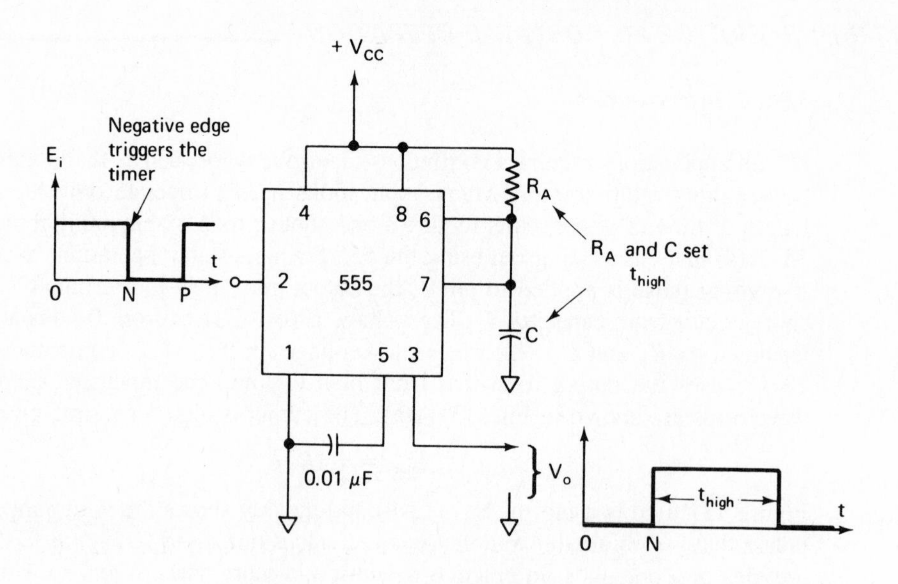

(a) 555 timer wired for monostable operation. Peak value of E_i must be greater than or equal to $\frac{2}{3}$ V_{CC}

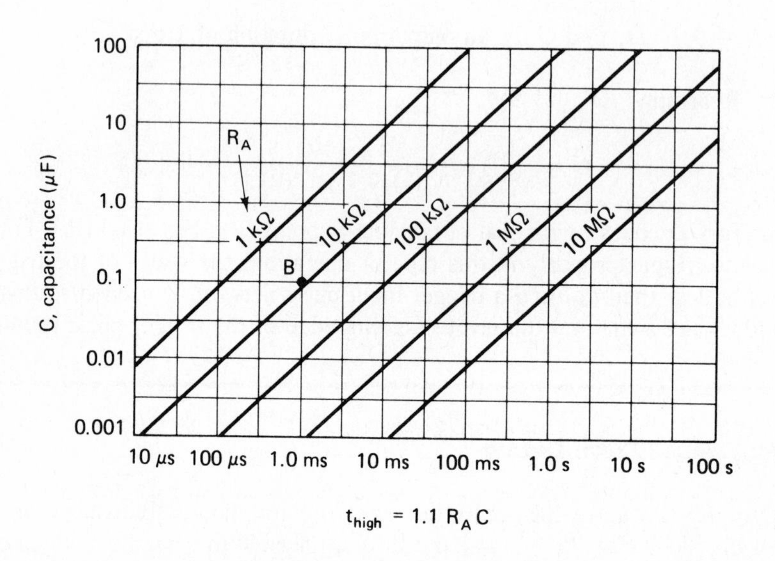

(b) Design aid to determine output pulse duration

FIGURE 13-11 Monostable operation.

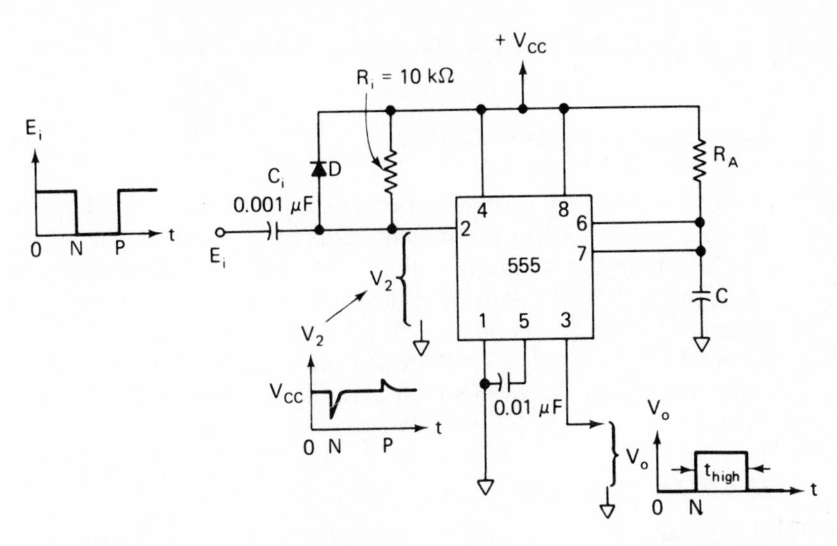

FIGURE 13-12 For satisfactory monostable operation, the input pulse network of R_i, C_i, and D is needed.

trigger pulse occurs. The time constant of R_i and C_i should be small with respect to the output timing interval t_{high}. Diode D prevents the 555 timer from triggering on positive-going edges of E_i. Waveforms for the input pulse, E_i, the pulse at pin 2, V_2, and the output pulse, V_o, are all shown in Fig. 13-12.

Example 13-5

(a) If $R_A = 10\ \text{k}\Omega$ and $C = 0.2\ \mu\text{F}$ in Fig. 13-12, find t_{high}. (b) What is the time constant of R_i and C_i in Fig. 13-12?

Solution (a) By Eq. (13-19),

$$t_{high} = 1.1(10 \times 10^3)(0.2 \times 10^{-6}) = 2.2\ \text{ms}$$

(b) Time constant $= R_i C_i = (10 \times 10^3)(0.001 \times 10^{-6}) = 0.01$ ms.

Just as with astable operation, reset terminal pin 4 is normally tied to supply voltage, V_{CC}. If pin 4 is grounded at any time, the timing cycle is stopped. When the reset terminal is grounded, both output pin 3 and discharge terminal 7 go to ground potential. Thus the output goes low and any charge accumulated by the timing capacitor C is removed. As long as the reset terminal is grounded, these conditions remain.

13-6 APPLICATIONS OF THE 555 AS A ONE-SHOT MULTIVIBRATOR

13-6.1 Water-Level Fill Control

In Fig. 13-13(a), the start switch is closed and the output of the 555 is low. When the start switch is opened, the output goes high to actuate the pump. The output is high for a time interval given by Eq. (13-9). Upon completion of the timing interval, the output of the 555 returns to its low state, turning the pump off. The height of the water level is set by the timing interval which is set by R_A and C. In the event of a potential overflow, the overfill switch must place a ground on reset pin 4, which causes the timer's output to go low and stops the pump.

13-6.2 Touch Switch

The 555 is wired as a one-shot multivibrator in Fig. 13-13(b) to perform as a touch switch. A 22-MΩ resistor to pin 2 holds the 555 in its idle state. If you scuff your feet to build up a static charge, the 555 will produce a single shot output pulse when you touch the finger plate. If the electrical noise level is high (due, for example, to fluorescent lights) the 555 may oscillate when you touch the finger plate. Reliable and consistent triggering will occur if a thumb is placed on a ground plate and fingers of the same hand tap the finger plate. An isolated power supply or batteries should be used for safety.

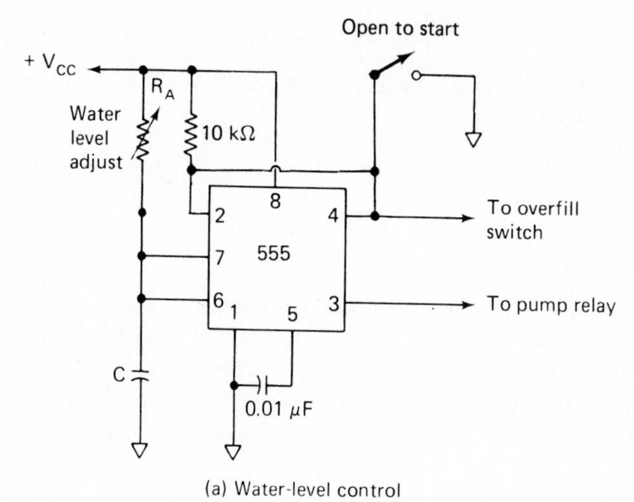

FIGURE 13-13 Basic one-shot application of the 555.

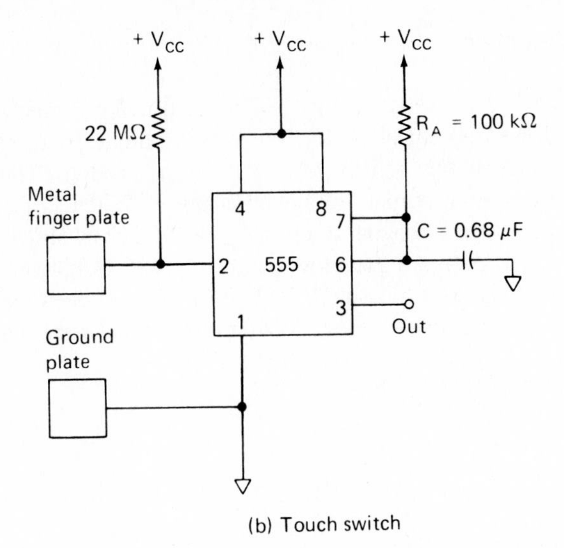

FIGURE 13-13 *(cont.)*

13-6.3 Frequency Divider

Figure 13-12 also can be used as a frequency divider if the timing interval is adjusted to be longer than the period of the input signal E_i. For example, suppose that the frequency of E_i is 1 kHz, so that its period is 1 ms. If $R_A = 10\ \text{k}\Omega$ and $C = 0.1\ \mu\text{F}$, the timing interval given by Eq. (13-9) is $t_{high} = 1.1$ ms. Therefore, the one shot will be triggered by the first negative-going pulse of E_i, but the output will still be high when the second negative-going pulse occurs. The one-shot will, however, be re-triggered on the third negative-going pulse. In this example, the one-shot triggers on every other pulse of E_i, so there is only one output for every two input pulses; thus E_i is divided by 2.

Example 13-6

(a) Calculate the timing interval in Fig. 13-12 if $R_A = 10\ \text{k}\Omega$ and $C = 0.1\ \mu\text{F}$.
(b) What value of R_A should be installed to divide a 1-kHz input signal by 3?

Solution (a) By Eq. (13-9), $t_{high} = 1.1(10 \times 10^3)(0.1 \times 10^{-6}) = 1.1$ ms. (b) t_{high} should exceed two periods of E_i, or 2 ms, and be less than three periods (3 ms). Choose $t_{high} = 2.2$ ms; then $2.2\text{ ms} = 1.1 R_A \times 0.1 \times 10^{-6}\text{ F}$; $R_A = 20\ \text{k}\Omega$.

13-6.4 Missing Pulse Detector

Transistor Q is added to the 555 one-shot in Fig. 13-14(a) to make a missing pulse detector. When E_i is at ground potential (0 V), the emitter diode of transistor Q clamps capacitor voltage V_C to a few tenths of a volt above ground. The 555 is forced into its idle state with a high output voltage V_o at pin 3. When E_i goes high, the transistor cuts off and capacitor C begins to charge. This action is shown by the waveshapes in Fig. 13-14(b). If E_i again goes low before the 555 completes its timing cycle, the voltage across C is reset to about 0 V. If, however, E_i does *not* go low before the 555 completes its timing cycle, the 555 enters its normal state and output

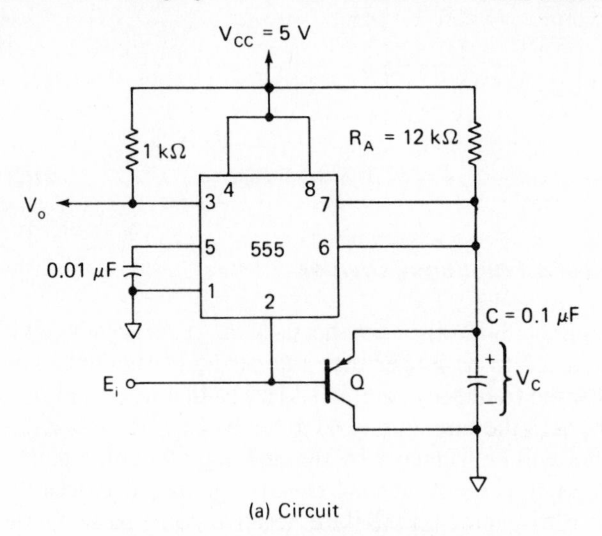

(a) Circuit

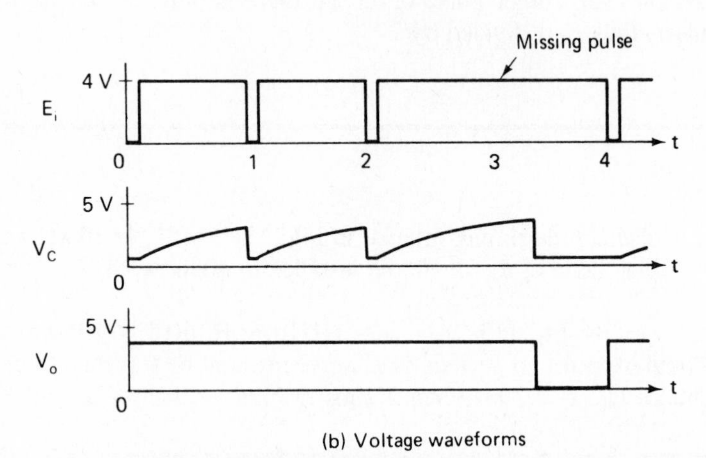

(b) Voltage waveforms

FIGURE 13-14 Missing pulse detector.

V_o goes low. This is exactly what happens if the $R_A C$ timing interval is slightly longer than the period of E_i and E_i suddenly misses a pulse. This type of circuit can detect a missing heartbeat. If E_i pulses are generated from a rotating wheel, this circuit tells when the wheel speed drops below a predetermined value. Thus the missing pulse detector circuit also performs speed control and measurement.

13-7 INTRODUCTION TO COUNTER TIMERS

When a timer circuit is connected as an oscillator and is used to drive a counter, the resultant circuit is *a counter timer*. Typically, the counter has many separate output terminals. One output terminal gives one pulse for each period T of the oscillator. A second output terminal gives one output pulse for every two periods ($2T$) of the oscillator. A third output terminal gives one output pulse for every four oscillator periods ($4T$), and so on, depending on the design of the counter. Thus each output terminal is rated in terms of the basic oscillator period T.

Some counters are designed so that their outputs can be connected together. The resultant output pulse is the *sum* of the individual output pulses. For example, if the first, second, and third output terminals are wired together, the result is one output pulse for every $1T + 2T + 4T = 7T$ oscillator periods. A counter with this capability is said to be *programmable,* because the user can program the counter to give one output pulse for any combination of timer outputs. One such programmable timer/counter is Exar's XR 2240. This integrated-circuit device is representative of the timer/counter family and some of its features will be studied next.

13-8 THE XR 2240 PROGRAMMABLE TIMER/COUNTER

13-8.1 Circuit Description

As shown in Fig. 13-15, the XR 2240 consists of one modified 555 timer, one 8-bit binary counter, and a control circuit. They are all contained in a single 16-pin dual-in-line package.

A positive-going pulse applied to *trigger* input 11 starts the 555 time base oscillator. A positive-going pulse on *reset* pin 10 stops the 555 time base oscillator. The threshold voltage for both trigger and reset terminals is about +1.4 V.

The time base period T for one cycle of the 555 oscillator is set by an external RC network connected to the *timing* pin 13. T is calculated from

$$T = RC \tag{13-10}$$

where R is in ohms, C in farads, and T in seconds. R can range from 1 kΩ to 10 MΩ and C from 0.05 to 1000 μF. Thus the period of the 555 can range from microseconds to hours.

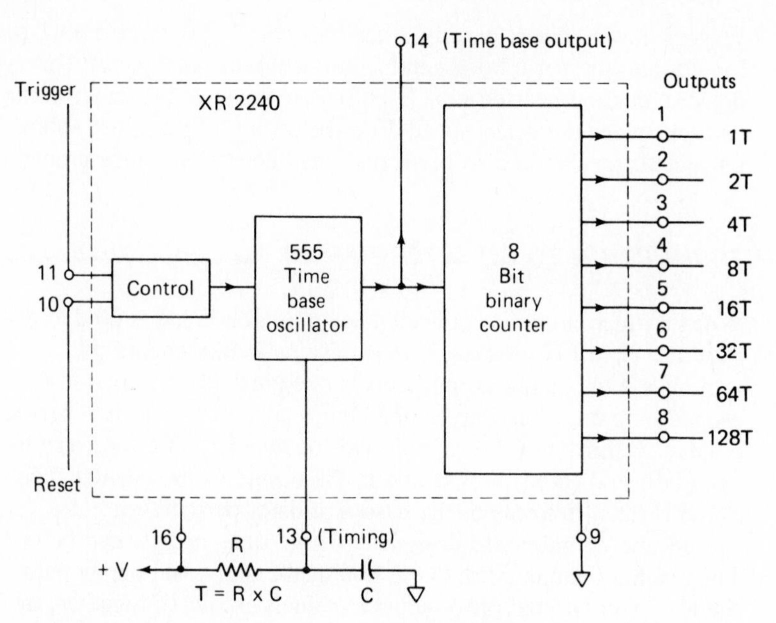

FIGURE 13-15 Block diagram of the XR 2240 programmable timer/counter.

The output of the 555 time base oscillator is available for measurement at pin 14 and also drives the 8-bit binary counter. Operation of the counter is discussed in Section 13-8.2.

13-8.2 Counter Operation

A simplified schematic of the 8-bit binary counter is shown in Fig. 13-16. Output of the 555 time base oscillator is shown as a switch. One side of the switch is connected to ground while the other side is wired to a 20-kΩ pull-up resistor. A regulated positive voltage is available at pin 15. Each negative-going edge from the 555 steps the 8-bit counter up by one count.

Normally, the 2240 is in its *reset* position. That is, all 8 output pins (pins 1 to 8) act like open circuits, as shown by the output switch models in Fig. 13-16. Pull-up resistors (10 kΩ) should be installed, as shown, to those terminals that are going to be used. Outputs 1 and 4 will then be high in the reset condition.

When the 2240 is triggered (pulse applied to pin 11), all output switches of the counter are closed by the control circuit and outputs 1 to 8 go low. Thus the counter begins its count with all outputs essentially grounded. At the end of every time base period, the 555 steps the counter once. The counter's T switch on terminal 1 opens after the first time base period (output 1 goes high) and closes after the second time base period. This counting action of the timer is shown in Fig. 13-16(b).

Output pin 2 is labeled $2T$ in Fig. 13-16(a). It is seen from Fig. 13-16(b) that

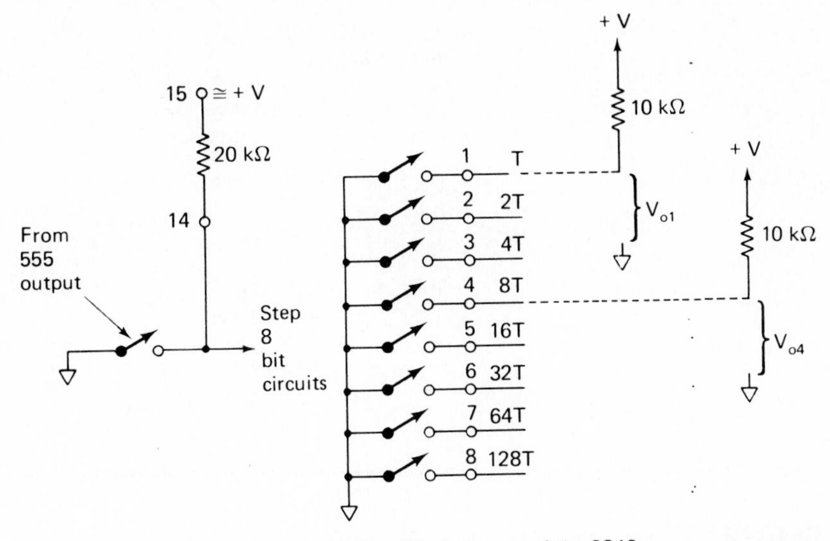

(a) Simplified outputs of the 2240

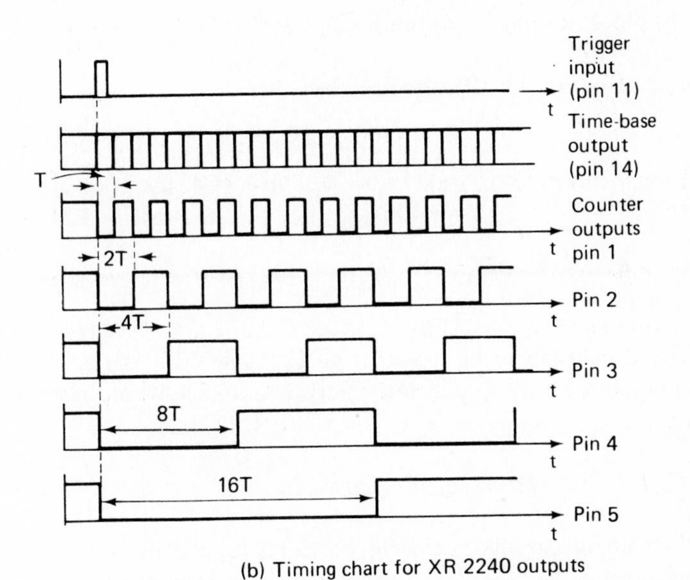

(b) Timing chart for XR 2240 outputs

FIGURE 13-16 Counter operation.

the output on pin 2 has stayed low for two time base periods ($2T$). Thus the second output stays low for twice the time interval of the first output. This conclusion may be generalized to all outputs of the binary counter; that is, each output stays low for twice the time interval of the preceding output. Time intervals for pins 1 to 5 are shown in Fig. 13-16(b) and are given for all outputs in Table 13-2.

TABLE 13-2 OUTPUT TERMINAL TIME CHART

Terminal number	Time output stays low after trigger pulse
1	T
2	$2T$
3	$4T$
4	$8T$
5	$16T$
6	$32T$
7	$64T$
8	$128T$

Example 13-7

After triggering, how long will the following output terminals stay low? (a) pin 3; (b) pin 4; (c) pin 7; (d) pin 8. $R = 100\ \text{k}\Omega$ and $C = 0.01\ \mu\text{F}$.

Solution By Eq. (13-10), the time base period is

$$T = (100 \times 10^3)(0.01 \times 10^{-6}) = 1\ \text{ms}$$

From Table 13-2, (a) $t_{\text{low}} = 4(1\ \text{ms}) = 4\ \text{ms}$; (b) $t_{\text{low}} = 8(1\ \text{ms}) = 8\ \text{ms}$; (c) $t_{\text{low}} = 64(1\ \text{ms}) = 64\ \text{ms}$; (d) $t_{\text{low}} = 128(1\ \text{ms}) = 128\ \text{ms}$.

The conclusion to be drawn from Example 13-7 is that after triggering, there are eight pulses of different time intervals available from the counter timer.

13-8.3 Programming the Outputs

The output circuits are designed to be used either individually or wired together, which is called *wire-or*. The term *wire-or* indicates that two or more output terminals can be jumpered together with a common wire (output bus) to a single pull-up resistor, as shown in Fig. 13-17(a). The resultant timing cycle for V_o is found by redrawing the individual timing of pins 4 and 5 in Fig. 13-17(b). Here we see that as long as either pin 4 *or* pin 5 is low, V_o will be low. Only when both outputs go high (output switches open) will the output go high. Thus the timing cycle for the output bus is found simply by calculating the sum T_{sum} of the individual outputs.

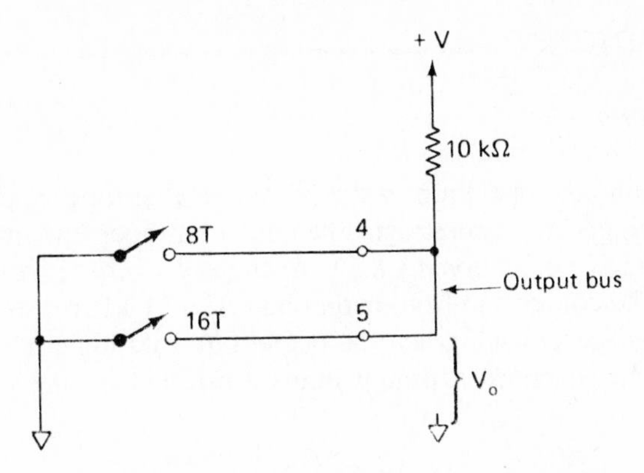

(a) Pins 4 and 5 are wired together to program 24T

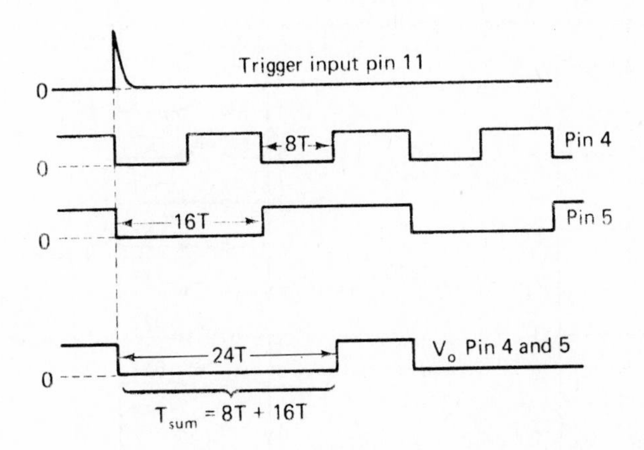

(b) Common bus V_o stays low as long as either pin 4 or pin 5 stays low

FIGURE 13-17 Programming the outputs.

Example 13-8

Calculate the timing cycle for (a) Fig. 13-17(a); (b) a circuit where pins 3, 6, and 7 are jumpered to a common bus. Let $T = 1$ s.

Solution (a) $T_{sum} = 8T + 16T = 24T = 24 \times 1 \text{ s} = 24 \text{ s}$; (b) $T_{sum} = 4T + 32T + 64T = 100T = 100 \times 1 \text{ s} = 100 \text{ s}$.

By using switches instead of jumper wires, T_{sum} can be easily changed or *programmed* for any desired timing cycle from T to $255T$.

13-9 TIMER/COUNTER APPLICATIONS

13-9.1 Timing Applications

The 2240 is wired for monostable operation in the programmable timer application of Fig. 13-18. When the trigger input goes high, the output bus goes low for a timing cycle period equal to T_{sum} (see Section 13-8.3). At the end of the timing cycle, the output bus goes high. The connection from output bus via a 51-kΩ resistor to reset pin 10 forces the timer to reset itself when the output bus goes high. Thus after each trigger pulse, the 2240 generates a timing interval selected by the program switches.

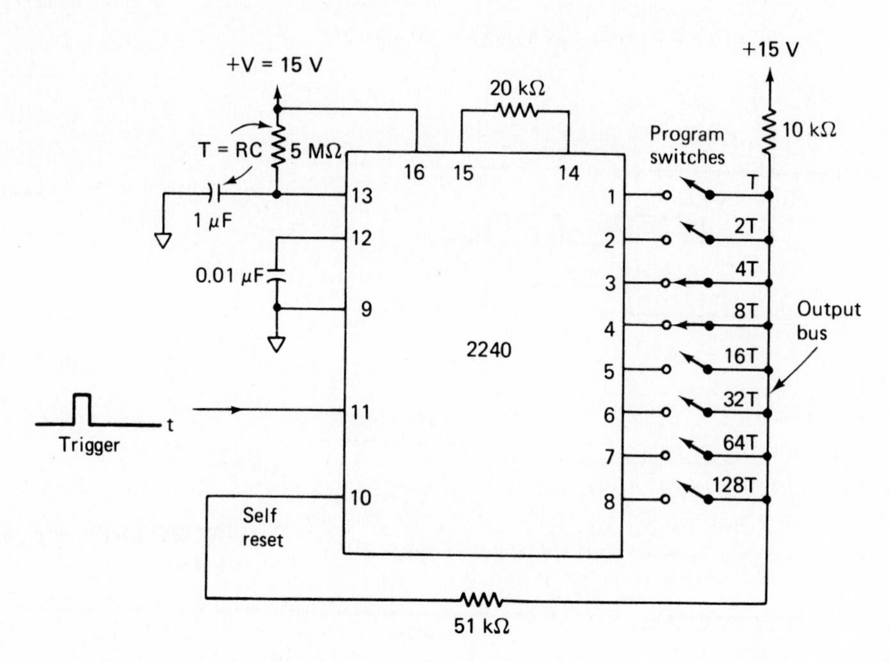

FIGURE 13-18 Programmable timer 5 s to 21 min 15 s in 5-s intervals.

Example 13-9

In Fig. 13-18, $C = 1.0\ \mu F$ and $R = 5\ M\Omega$ to establish a time base period given by Eq. (13-10) to be 5 s. What is (a) the timing cycle for switch positions shown in Fig. 13-18; (b) the minimum programmable timing cycle; (c) the maximum programmable timing cycle?

Solution (a) $T_{sum} = 4T + 8T = 12T = 12 \times 5\text{ s} = 60\text{ s} = 1\text{ min}$; (b) minimum timing cycle is $1T = 5$ s; (c) with all program switches closed.

$$T_{sum} = T + 2T + 4T + 8T + 16T + 32T + 64T + 128T = 255T$$

$$255T = 255 \times 5\text{ s} = 1275\text{ s} = 21\text{ min } 15\text{ s}$$

13-9.2 Free-Running Oscillator, Synchronized Outputs

The 2240 operates as a free-running oscillator in the circuit of Fig. 13-19. The reset terminal is grounded so that the 2240 will stay in its timing cycle once it is started. When power is applied, R_R and C_R couple a positive-going pulse into trigger input 11 to start the internal time base oscillator running.

Each output is wired through an external control switch to an individual pull-up resistor. A square-wave output voltage is available at each counter output. Their frequencies have a binary relationship. That is, the frequency available at each pin is one-half the frequency present at the preceding pin. The waveshapes are identical to those in Fig. 13-16(b). Observe that the *period* of the f_1 frequency at pin 1 is twice the time base period rating T or $2(T)$. Thus $f_1 = 1/2T$. At pin 4, the period is $2(8T)$ and $f_4 = 1/16T$.

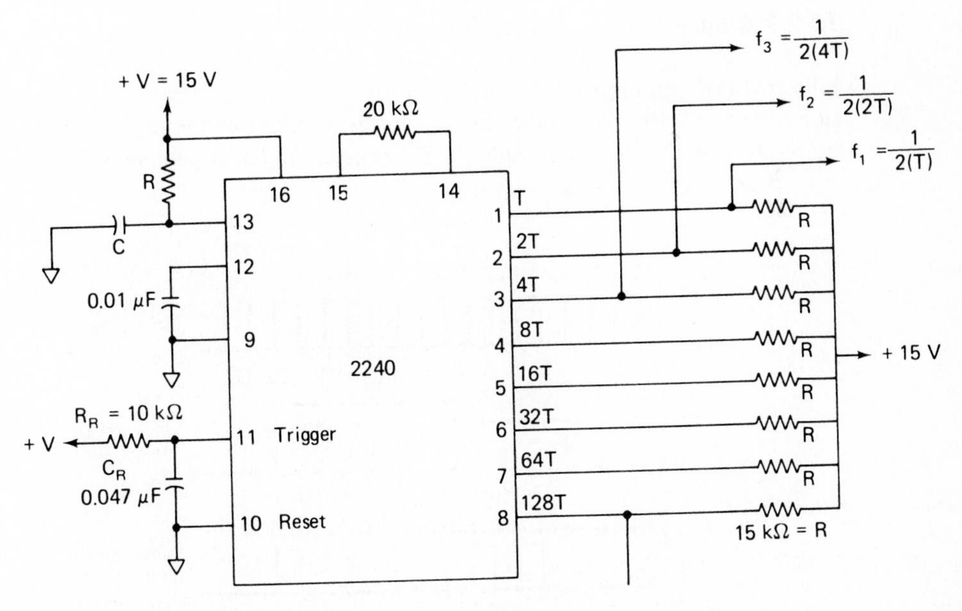

FIGURE 13-19 Free-running oscillator with synchronized outputs.

Example 13-10

In Fig. 13-19, $T = 2.5$ ms; what frequencies are present at (a) output 1; (b) output 2; (c) output 3; (d) output 4?

Solution Tabulating calculations, we obtain:

Pin number	Time base rating	Period	Frequency (Hz)
1	T	$2T = 5$ ms	200
2	$2T$	$4T = 10$ ms	100
3	$4T$	$8T = 20$ ms	50
4	$8T$	$16T = 40$ ms	25

The connections to pins 10 and 11 may be removed to allow the oscillator to be started with a positive-going trigger pulse at pin 11. To stop oscillation, apply a positive-going pulse to reset pin 10.

13-9.3 Binary Pattern Signal Generator

Pulse patterns similar to those shown in Fig. 13-20 are generated by a modified version of Fig. 13-19. The modification requires the eight output resistors to be replaced by program switches and a single 10-kΩ resistor similar to that shown in Fig. 13-18.

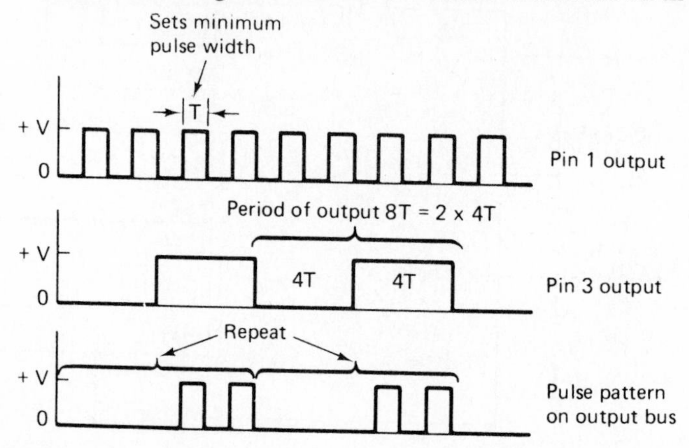

FIGURE 13-20 Binary pattern signal generator with outputs T and 4T connected to output bus.

Also eliminate the 51-kΩ resistor between the output bus and the self-reset terminal.

The output is a train of pulses that depends on which program switches are closed. The period of the pulse pattern is set by the highest program switch that is closed, and the pulse width is set by the lowest program switch that is closed. For example, if the $4T$ (pin 3) and $1T$ (pin 1) switches are closed, the pulse pattern is repeated every $2 \times 4T = 8T$ seconds (see Fig. 13-20). The minimum pulse width is $1T$. To determine the actual pulse pattern, refer to the timing chart in Fig. 13-16(b). If switches $1T$ and $4T$ are closed, there is an output pulse only when there are high output pulses from each line. The repeating pulse patterns are shown in Fig. 13-20.

13-9.4 Frequency Synthesizer

The output bus in Fig. 13-21(a) is capable of generating any one of 255 related frequencies. Each frequency is selected by closing the desired program switches to program a particular frequency at output V_o.

To understand circuit operation, assume that the output bus goes high. This will drive reset pin 10 high and couple a positive-going pulse into trigger pin 11. The reset terminal going positive resets the 2240 (all outputs low). The positive pulse on pin 10 retriggers the 2240 time base oscillator, to begin generation of a time period that depends on which program switches are closed. For example, assume that switches T and $4T$ are closed in Fig. 13-21(a). The timing for these switches is shown in Fig. 13-21(b). The output bus stays low for $4T$ from pin 3 plus $1T$ from pin 1 before going high (to initiate a reset-retrigger sequence noted above). The time period and frequency of the output signal V_o is thus expressed by

$$\text{period} = T_{\text{sum}} + T \tag{13-11a}$$

and

$$f = \frac{1}{\text{period}} \tag{13-11b}$$

where T_{sum} is found by adding the time base rating for each output terminal connected to the output bus.

Example 13-11

Find the output frequency for Fig. 13-21(a).

Solution From Eq. (13-11a), $T_{\text{sum}} = 1T + 4T = 5T$, and period $= (T_{\text{sum}} + T) = 6T = 6 \times 1\text{ ms} = 6\text{ ms}$. By Eq. (13-11b),

$$f = \frac{1}{6 \times 10^{-3}\text{ s}} = 166\text{ Hz}$$

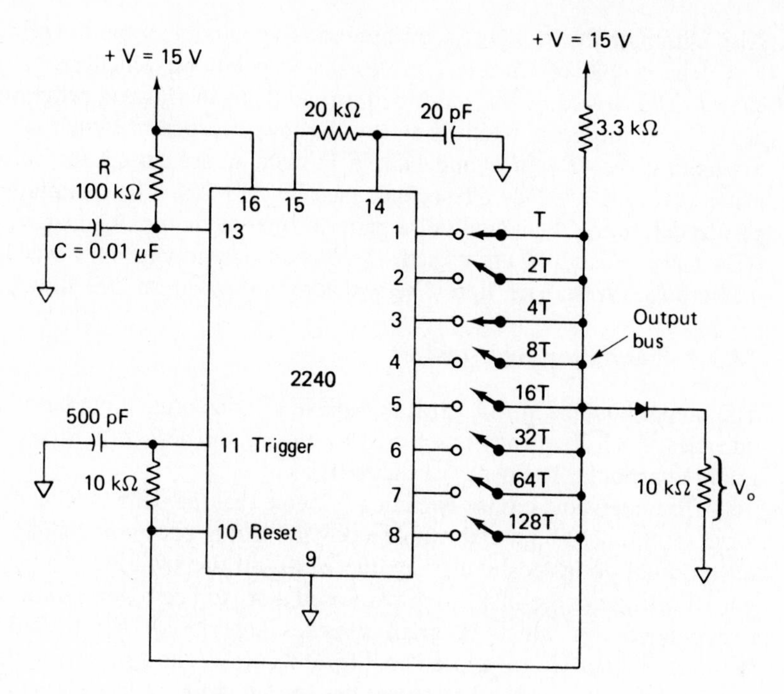

(a) Frequency synthesizer connections

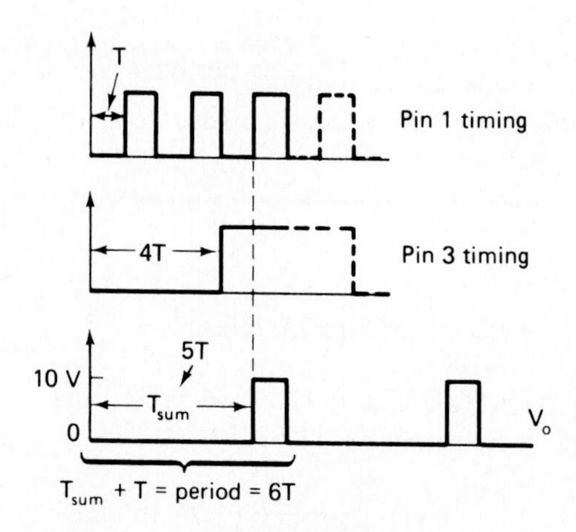

(b) Output voltage for program switches 1 and 4 closed

FIGURE 13-21 Frequency synthesizer, $T = 1$ ms, $f = 166$ Hz.

13-10 SWITCH PROGRAMMABLE TIMER

13-10.1 Timing Intervals

We close this chapter with a useful programmable timer. In the version of Fig. 13-22, the basic timing interval is adjusted for $T = 5.0$ s by timing capacitor C and timing resistor R. A 16-pin, eight-circuit DIP switch is used to select the desired time interval, as shown in Example 13-9. Close switch 1 for a 5-s timer. Closing SW2 adds 10 s, SW3 adds 20 s and so on. With all switches closed, the maximum time interval is $2^8 \times 5$ s = 1280 s or 21 min 20 s. The basic timing interval can be changed by picking new values for R and/or C.

13-10.2 Circuit Operation

Start switch S_s is opened momentarily, driving the trigger pin high to start the timer. All outputs of the XR2240 go low. This low is extended by any closed select switch(es) S_1 through S_8, to (−) in of the 301 comparator. The 301's output goes high to light LED D_1 and turn on transistor Q_1. The on-transistor energizes the relay, thus switching the contacts from the NC position to the NO position. This completes the power circuit to turn *on* any appliance for the timing interval.

To turn *off* an appliance during the timing interval, simply move the ac common wire from terminal A of the relay to terminal C. The supply voltage must be regulated to obtain repeatable results.

LABORATORY EXERCISES

13-1. The first 555 lab experience should begin with the circuit of Fig. 13-5. Use a manually adjustable power supply (or a 10-kΩ pot across pins 8 and 1) to vary E_i. Connect CRO A input to E_i and CRO B input to V_o (dc coupled). Plot E_i and V_o vs. time; dial an *x-y* plot to see the hysteresis loop. Once you learn to recognize the three possible states of the 555 (high, low, memory) you can analyze or design rather complex 555 circuits without confusion.

13-2. The multivibrator circuit of Fig. 13-7 and the one-shot of Fig. 13-12 are two basic circuit types that should be explored. The chirp oscillator or tone burst circuit of Fig. 13-9 teaches operation of the reset pin, pin 4. As an experiment of a terrible sound, connect a 10-kΩ pot as a volume control from pin 3 of 555B to ground. Connect a 0.1-μF capacitor from the wiper arm to an audio amplifier. Note that this circuit could drive yourself or others from the room.

PROBLEMS

13-1. What are the operating modes of the 555 timer?

13-2. In Fig. 13-6(a), $R_A = R_B = 10$ kΩ, $C = 0.1$ μF. Find **(a)** t_{high}; **(b)** t_{low}; **(c)** frequency of oscillation.

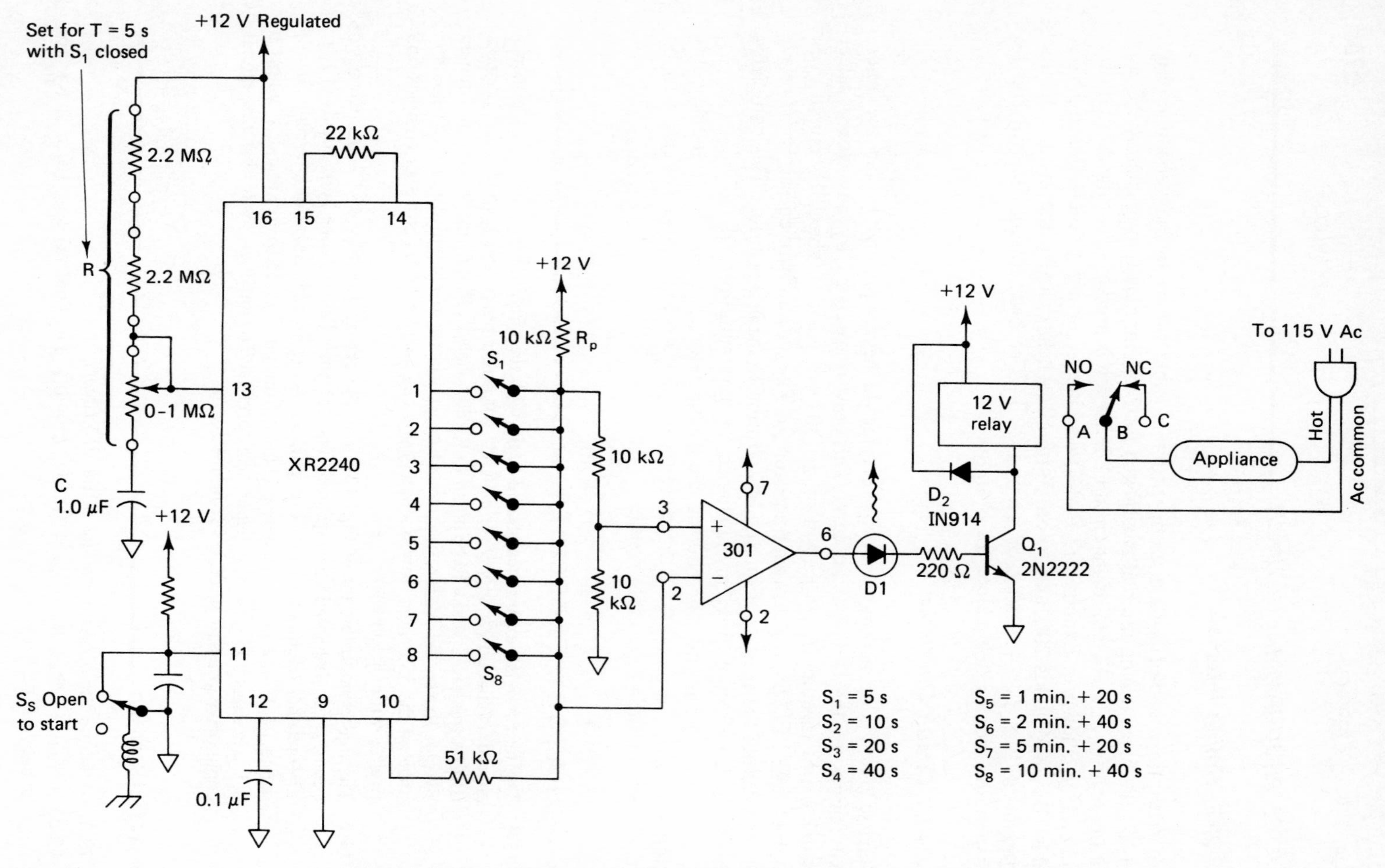

FIGURE 13-22 Switch programmable timer. A basic time interval is set to $T = 5$ s by R and C. Time select switches S_1 through S_8 determine how long the appliance load will be turned on after momentary-open start switch S_S is depressed. $S_1 = 5$ s. S_2 adds 10 s. The remaining switches extend the time interval as shown above.

13-3. Using the graph of Fig. 13-7, estimate the free-running frequency of oscillation f if $(R_A + 2R_B) = 1\ \text{M}\Omega$ and $C = 0.02\ \mu\text{F}$.

13-4. What is the duty cycle in Problem 13-2?

13-5. In Example 13-1, R_A and R_B are increased by a factor of 10 to 68 kΩ and 33 kΩ. Find the new frequency of oscillation.

13-6. In Fig. 13-8, R_A and R_B are each reduced to 5 kΩ. What is the effect on **(a)** the duty cycle; **(b)** the period T of the output?

13-7. In Fig. 13-9, at what value should the 10-kΩ resistor be set for a 2-kHz output from the B 555?

13-8. Capacitor C is changed to 0.1 μF in Fig. 13-10. Calculate **(a)** the center frequency f_c when $E = 0$ V; **(b)** the frequency shift for $E = \pm 2$ V.

13-9. In Fig. 13-11(a), $R_A = 100\ \text{k}\Omega$ and $C = 0.1\ \mu\text{F}$. Find t_{high}.

13-10. R_A is changed to 20 kΩ in Example 13-5. Find t_{high}.

13-11. In Example 13-6(b), what value of R_A is required to divide a 1-kHz signal by 2?

13-12. Refer to Example 13-7. How long will the following output terminals stay low? **(a)** pin 1; **(b)** pin 2; **(c)** pin 5; **(d)** pin 6.

13-13. In Fig. 13-17(a), T is set for 1 ms and pins 2, 4, 6, and 8 are connected to the output bus. Find the timing interval.

13-14. In Problem 13-13, the odd-numbered pins, 1, 3, 5, and 7, are connected to the output bus. Find the timing interval.

13-15. In Example 13-9, C is changed to 0.1 μF and R to 500 kΩ. Find **(a)** the time base period; **(b)** the timing cycle for switch positions shown in Fig. 13-18; **(c)** the maximum timing cycle.

13-16. In Example 13-10, what frequencies are present at pins **(a)** 5; **(b)** 6; **(c)** 7; **(d)** 8?

13-17. In Fig. 13-21, only switches to pins 1, 2, 3, and 4 are closed. Find the output frequency.

CHAPTER 14

Digital-to-Analog and Analog-to-Digital Converters

LEARNING OBJECTIVES

Upon completion of this chapter on digital-to-analog and analog-to-digital converters, you will be able to:

- Write the general output–input equations of a DAC or ADC and calculate their outputs for any given input.
- Draw an *R*–2*R* resistance ladder network, calculate all its currents and explain how it is used to convert an analog input voltage to a digital output
- Distinguish between a DAC and an MDAC.
- Tell what features must be present to make a DAC or ADC compatible with a microprocessor.
- Explain how a microprocessor selects only one DAC out of all peripheral devices and sends data to it.

- Explain how a microprocessor selects only one ADC out of all peripherals and reads its data.
- Name the three most common types of ADCs and tell how each one operates.
- Calculate the maximum sine-wave frequency that can be digitized to an accuracy of $\pm\frac{1}{2}$ LSB by an ADC or sample-and-hold amplifier.
- Dynamically test a microprocessor-compatible DAC, the AD558.
- Operate a microprocessor-compatible ADC, the AD670, *without* a microprocessor.

14-0 INTRODUCTION

Real-world processes produce analog signals which vary continuously. The rate may be very slow, such as a change in temperature variations, or very fast, such as in an audio system. Analog processes are best described by decimal numbers and letters of the alphabet. Microprocessors and computers, however, use binary patterns to represent numbers, letters, or symbols.

It is not easy to store, manipulate, compare, calculate, or retrieve data with accuracy using analog technology. However, computers can perform these tasks quickly and on an almost unlimited mass of data with precision, using digital techniques. Thus the need for *converters* to interface between the analog and digital worlds emerged. Analog-to-digital converters, ADCs, allow the analog world to communicate with computers. Computers communicate with people and physical processes via digital-to-analog converters, DACs. Our study of this intercommunication between the analog and digital world begins by developing the output–input equation first for a DAC and then for an ADC.

14-1 DAC CHARACTERISTICS

We ask three questions whose answers describe the most important characteristics of a DAC. First, how many different output values can be provided by the DAC? Second, by how much will the analog output voltage *change* in response to a digital input word *change* of one LSB? (The answer to both of these questions is presented in the next section under the topic *resolution*.) Third, what is the output–input equation of the DAC that allows you to predict the output voltage if you know the digital input word?

14-1.1 Resolution

Both circuit symbol and output–input characteristics of a 4-bit DAC are shown in Fig. 14-1. There are four digital inputs, indicating a 4-bit DAC. Each digital input requires an electrical signal representing either a logic 1 or a logic 0. D_0 is the *least*

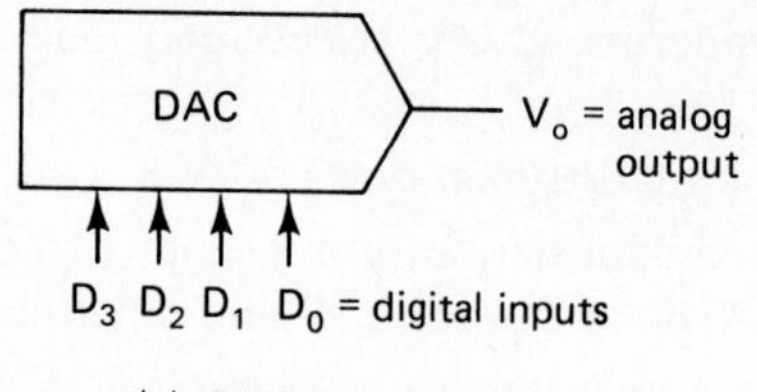

(a) DAC circuit symbol

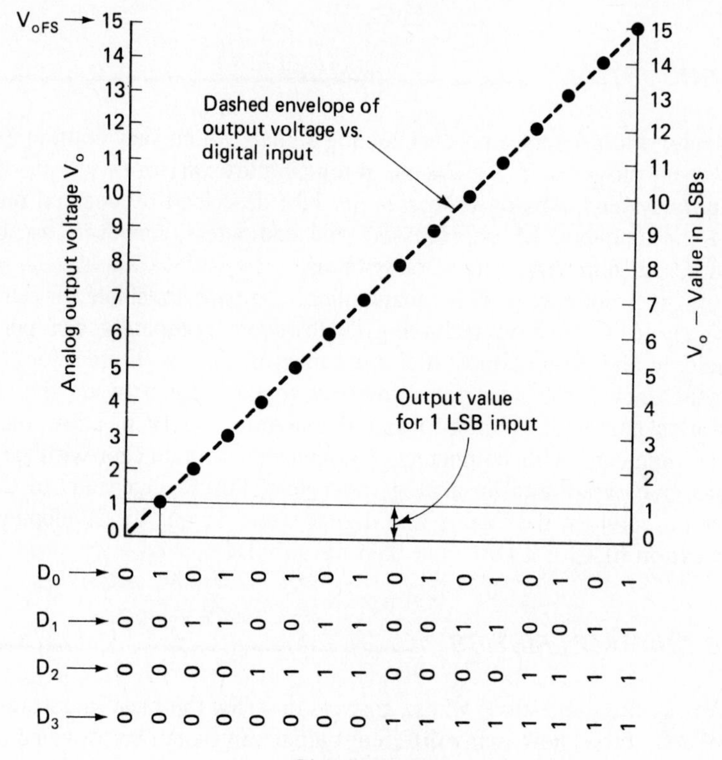

(b) Plot of analog output voltage vs. digital input code for a 4-bit DAC

FIGURE 14-1 Circuit symbol in (a) and output–input characteristic in (b).

significant bit, LSB. D_3 is the *most significant bit, MSB*. In Fig. 14-1(b), analog output voltage V_o is plotted against all 16 possible digital input words. V_o is also shown with its value in LSBs.

Resolution is defined in two ways:

1. Resolution is the number of different analog output values that can be provided

by a DAC. For an n-bit DAC,

$$\text{resolution} = 2^n \tag{14-1a}$$

2. Resolution is also defined as the ratio of a *change* in output voltage resulting from a *change* of 1LSB at the digital inputs.

To calculate resolution for this definition, you need two pieces of information from the data sheets; full-scale output voltage V_{oFS}, and the number of digital inputs, n. V_{oFS} is defined as the voltage resulting when all digital inputs are 1's. Therefore, resolution is calculated from

$$\text{resolution} = \frac{V_{oFS}}{2^n - 1} \tag{14-1b}$$

In Fig. 14-1(b) there are $n = 4$ digital inputs. Therefore, V_o will have $2^4 = 16$ different output values, 0 through 15. Note that $V_{oFS} = 15$ V when the digital input word is 1111. The decimal value of binary 1111 is $(2^4 - 1) = 16 - 1 = 15$. Thus resolution equals 15 V/15 = 1 V/LSB.

Example 14-1

An 8-bit DAC has an output voltage range of 0 – 2.55 V. Define its resolution in two ways.

Solution (a) From Eq. (14-1a),

$$\text{resolution} = 2^n = 2^8 = 256$$

The output voltage can have 256 different values (including zero).
(b) From Eq. (14-1b),

$$\text{resolution} = \frac{V_{oFS}}{2^n - 1} = \frac{2.55\text{ V}}{2^8 - 1} = \frac{2.55\text{ V}}{255} = \frac{10\text{ mV}}{1\text{ LSB}}$$

An input change of 1 LSB causes the output to change by 10 mV.

Now we have the basics for determining the output–input equation.

14-1.2 Output–Input Equation

For a DAC, a transfer function or output–input equation answers the question: What is the change in analog output voltage resulting from a digital input word? This output–input equation is obtained by multiplying the resolution by the change in LSBs. In equation form,

$$V_o = \text{resolution} \times D \tag{14-2}$$

where V_o is the analog output voltage, resolution is given by Eq. (14-1b), and D is the *decimal value of the digital input*.

Example 14-2

Suppose that the digital input word for a 4-bit DAC changes from 0000 to 0110. Calculate the DAC's final output voltage.

Solution The decimal value of 0110 is 6. This value represents D, the digital input word. From previous discussion when $n = 4$, $V_{oFS} = 15$. Applying Eq. (14-1b) yields

$$\text{resolution} = \frac{15}{2^4 - 1} = 1 \text{ V/LSB}$$

Now using Eq. (14-2), we obtain

$$V_o = (1 \text{ V/LSB}) \times 6 \text{ LSB} = 6 \text{ V}$$

Example 14-3

An 8-bit DAC has a resolution of 10 mV/LSB. Find (a) V_{oFS} and (b) V_o when the digital input code is 10000000.

Solution (a) V_{oFS} occurs when the input digital word is 11111111. Binary 11111111 contains 255 LSBs and has a decimal value of 255. Thus $D = 255$, and applying Eq. (14-2) yields

$$V_{oFS} = \text{resolution} \times D = \frac{10 \text{ mV}}{\text{LSB}} \times 255 \text{ LSB} = 2.55 \text{ V}$$

(b) The equivalent decimal value of 10000000 is 128. Thus $D = 128$ and

$$V_o = \text{resolution} \times D = \frac{10 \text{ mV}}{\text{LSB}} \times 128 \text{ LSB} = 1.28 \text{ V}$$

14-2 ADC CHARACTERISTICS

14-2.1 Output–Input Equation

The digital output of an ideal 4-bit ADC is plotted against analog input voltage in Fig. 14-2(b). Similar to DACs, the resolution of an ADC is defined in two ways. First, it is the maximum number of digital output codes. This expression of ADC

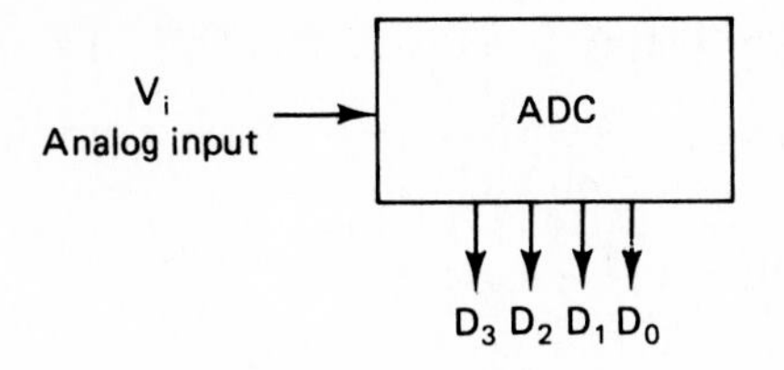

(a) Symbol for a 4-bit ADC

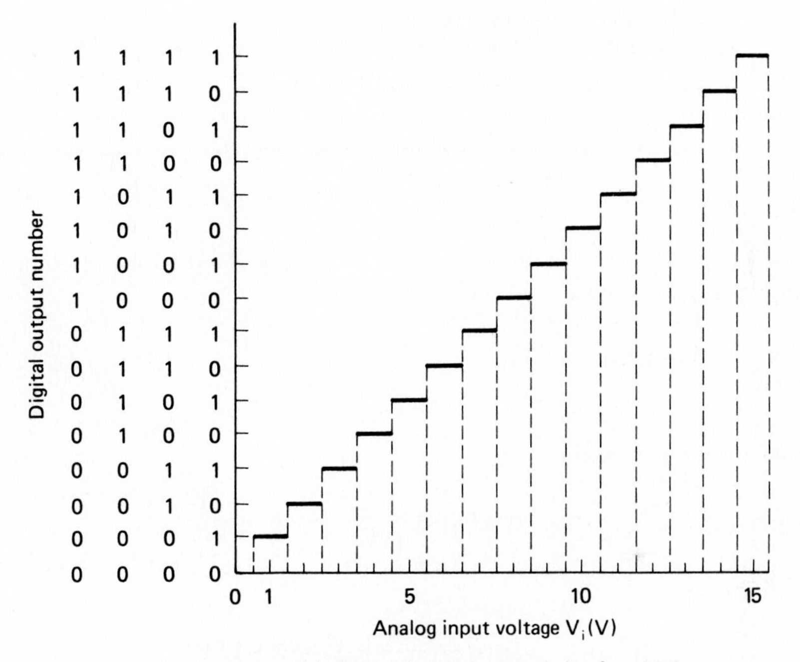

(b) Output-input characteristic of an ADC

FIGURE 14-2 Circuit symbol and output–input characteristic for a 4-bit DAC.

resolution is the same as for the DAC and is repeated here:

$$\text{resolution} = 2^n \tag{14-3a}$$

Resolution is also defined as the ratio of a change in value of input voltage, V_i, needed to change the digital output by 1 LSB. If you know the value of full-scale input voltage, V_{iFS}, required to cause a digital output of all 1's, resolution can be calculated from

$$\text{resolution} = \frac{V_{iFS}}{2^n - 1} \tag{14-3b}$$

In simplest form, the output–input equation of an ADC is given by

$$\text{digital output code} = \text{binary equivalent of } D \tag{14-4}$$

where D equals decimal value of the digital output, or D equals the number of LSBs in the digital output and D is found from

$$D = \frac{V_i}{\text{resolution}} \tag{14-5}$$

For example, refer to Fig. 14-2(b), where $n = 4$ and $V_{iFS} = 15$ V. Resolution $= 15 \text{ V}/(2^4 - 1) = 1$ V/LSB. If $V_i = 5$ V, then $D = 5 \text{ V}/(1 \text{ V/LSB}) =$ 5 LSB. The digital code for $D = 5$ is 0101.

Example 14-4

An 8-bit ADC outputs all 1's when $V_i = 2.55$ V. Find its (a) resolution and (b) digital output when $V_i = 1.28$ V.

Solution (a) From Eq. (14-3a),

$$\text{resolution} = 2^8 = 256$$

and from Eq. (14-3b),

$$\text{resolution} = \frac{2.55 \text{ V}}{2^8 - 1} = \frac{10 \text{ mV}}{\text{LSB}}$$

(b) From Eq. (14-5),

$$D = \frac{1.28 \text{ V}}{10 \text{ mV/LSB}} = 128 \text{ LSBs}$$

From Eq. (14-4),

$$\text{digital output code} = \text{binary equivalent of } 128 = 10000000$$

14-2.2 Quantization Error

Figure 14-2(b) shows that the binary output is 0101 for all values of V_i between 4.5 and 5.5 V. There is an unavoidable uncertainty about the exact value of V_i when the output is 0101. This uncertainty is specified as *quantization error*. Its value is $\pm\frac{1}{2}$ LSB. Increasing the number of bits results in a finer resolution and a smaller quantization error.

Example 14-5

What is the quantization error for the ADC of Example 14-4?

Solution In Example 14-4 resolution was found to be 10 mV per LSB. The quantization error is $\pm\frac{1}{2}$ LSB or ± 5 mV.

14-3 DIGITAL-TO-ANALOG CONVERSION PROCESS

14-3.1 Block Diagram

The block diagram for a basic DAC is drawn in Fig. 14-3. Reference voltage V_{ref} is connected to a resistance network. A digital input code, via control circuitry, flips switches (one for each bit), connected to the resistance network. The output of the resistance network is in the form of a current. This current may then be converted to a voltage. Both current and voltage outputs are analog representations of the digital input code.

The actual digital-to-analog conversion takes place within the resistance network. Accordingly, we begin our study of DAC circuitry by looking next at the standard resistance network. It is called an R–$2R$ ladder network.

14-3.2 R–2R Ladder Network

A 4-bit R–$2R$ ladder network is drawn in Fig. 14-4. Each digital input controls the position of its corresponding current switch. A current switch steers its ladder current either into a real ground (position 0) or a virtual ground (position 1). Thus the wiper of each switch is always at ground potential, so that the rung currents are constant except for the brief transition time of each switch.

In Fig. 14-4, rail currents flow horizontally; rung currents flow down through the bit switches. The rail current I_1 enters node 0, where it sees a resistance R_0. R_0 is the equivalent resistance of a $2R$ resistor via switch D_0 to ground in parallel with a $2R$ terminate resistor. Thus $R_0 = 2R \parallel 2R = R$. As rail current I_1 leaves node 1, it sees R in series with $R_0 = R$ or $2R$. If we work our way back from the terminate end to the voltage source, the value of resistance "looking" into a node is R. As shown in Fig. 14-4, $R_3 = R_2 = R_1 = R_0 = R$. R is called the *characteristic resistance* of the ladder network. In other words, V_{ref} sees the entire ladder network as a single resistor equal to R.

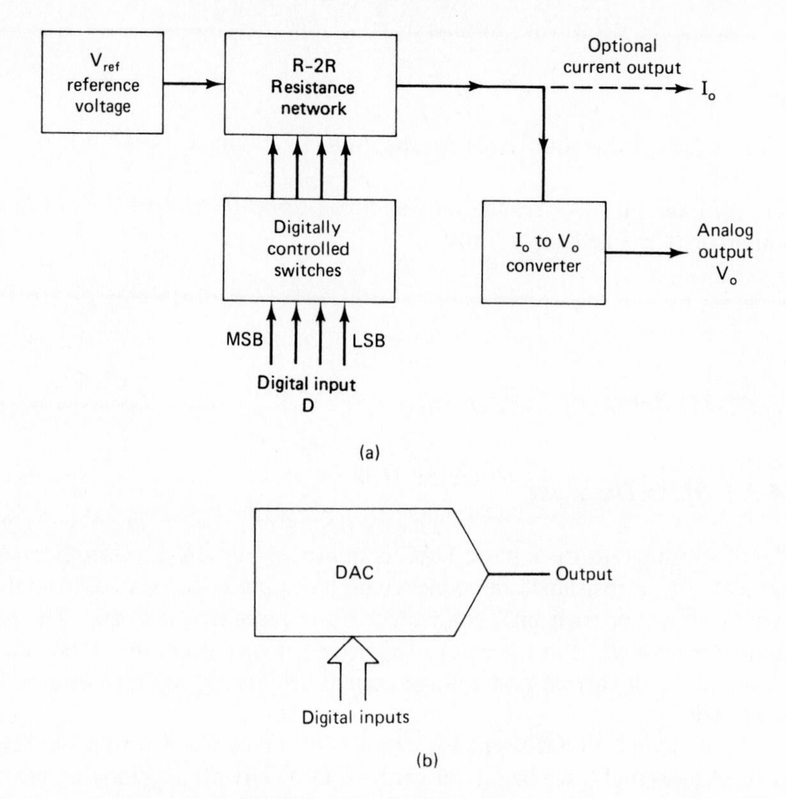

FIGURE 14-3 Block diagram and circuit symbol for a basic DAC.

14-3.3 Ladder Currents

Since V_{ref} sees the ladder network as a resistance R, rail current I_{ref} is

$$I_{ref} = \frac{V_{ref}}{R} \tag{14-6}$$

The current pattern of the R–$2R$ network in Fig. 14-4 is analyzed as follows: The current I_{ref} splits into two equal parts at node 3. The rung current $I_3 = I_{ref}/2$ and the rail current $I_3 = I_{ref}/2$. Each rail current divides equally again at each node as it proceeds down the ladder. Rung currents are evaluated from

$$\begin{aligned} I_3 &= \frac{I_{ref}}{2} & \qquad I_2 &= \frac{I_3}{2} = \frac{I_{ref}}{4} \\ I_1 &= \frac{I_2}{2} = \frac{I_{ref}}{8} & \qquad I_0 &= \frac{I_1}{2} = \frac{I_{ref}}{16} \end{aligned} \tag{14-7}$$

I_0 is the current controlled by the LSB switch.

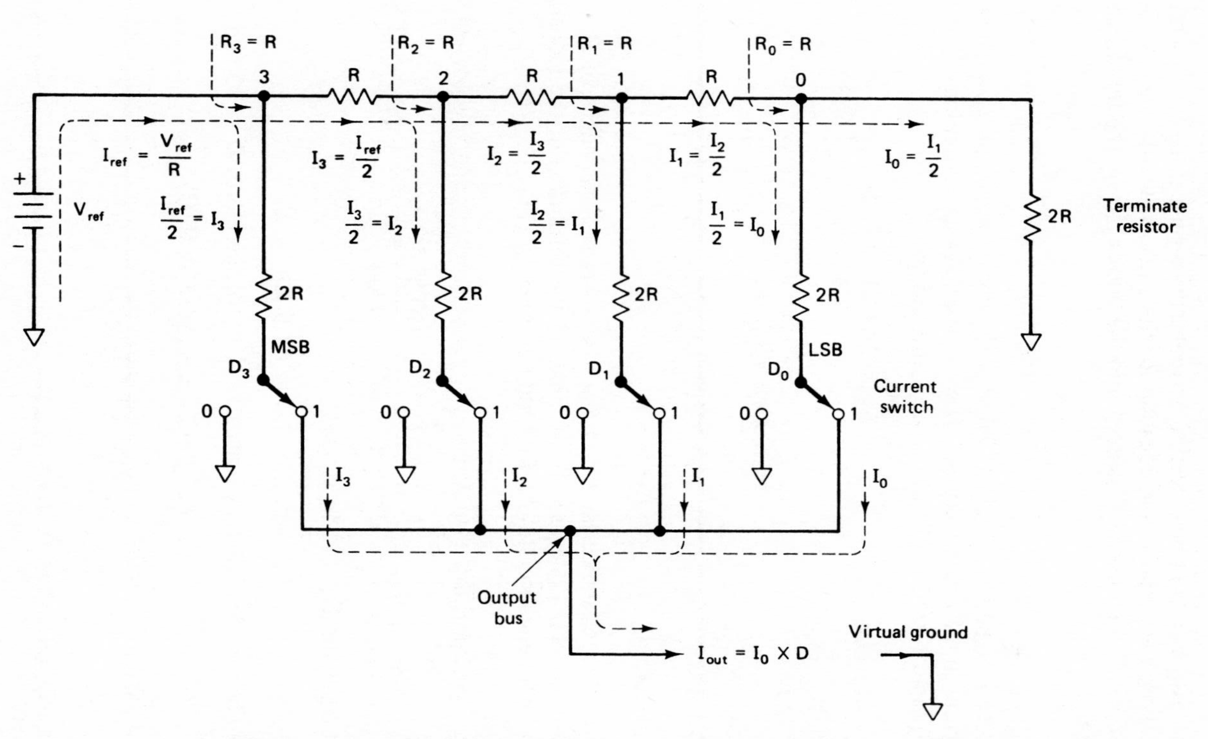

FIGURE 14-4 This R–$2R$ ladder converts the digital input code into an analog output current I_{out}.

14-3.4 Ladder Equation

The output current bus receives current from a rung if the bit switch is in position 1. To write the output–input equation for the ladder network, we observe that I_{out} is the sum of all rung currents steered into the output bus by the bit switches. In equation form,

$$I_{out} = I_0 \times D \tag{14-8}$$

where D equals the decimal value of the digital input and I_0 is the smallest value of current in the ladder network. If we define I_0 as the resolution of the ladder, the output–input Equation (14-8) can be expressed as

$$I_{out} = \text{resolution} \times D \tag{14-9}$$

where

$$\text{resolution} = I_0 = \frac{I_{ref}}{2^n} = \frac{1}{2^n} \times \frac{V_{ref}}{R} \tag{14-10}$$

Example 14-6

The 4-bit resistance ladder of Fig. 14-4 has resistor values of $R = 10\ \text{k}\Omega$ and $2R = 20\ \text{k}\Omega$. V_{ref} equals 10 V. Find (a) the resolution of the ladder; (b) its output–input equation; (c) I_{out} for a digital input of 1111.

Solution (a) From Eq. (14-10),

$$\text{resolution} = \frac{1}{2^n} \times \frac{V_{ref}}{R} = \frac{1}{2^4} \times \frac{10\text{ V}}{10\text{ k}\Omega} = \frac{1}{16} \times 1\text{ mA} = 62.5\ \mu\text{A}$$

(b) Applying Eq. (14-9) yields

$$I_{out} = 62.5\ \mu\text{A} \times D$$

(c) The decimal value of binary 1111 is 15; therefore, $D = 15$ and

$$I_{out} = 62.5\ \mu\text{A} \times 15 = 0.9375\ \mu\text{A}$$

14-4 VOLTAGE OUTPUT DACs

As shown in Fig. 14-5, the output ladder current can be converted into a voltage by adding an op amp and a feedback resistor. Output voltage V_o is given by

$$V_o = -I_{out} R_F \tag{14-11a}$$

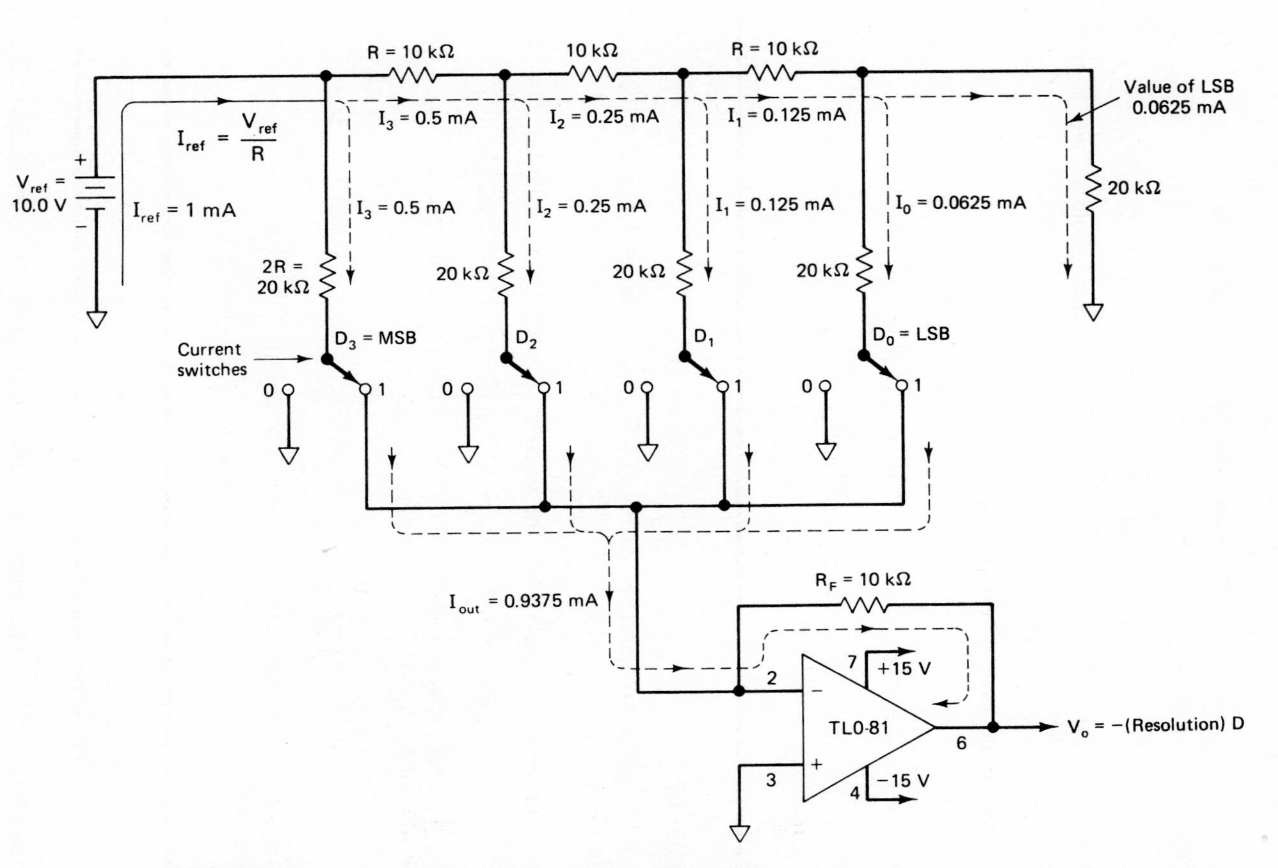

FIGURE 14-5 I_{out} of the R–$2R$ ladder is converted into a voltage by the op amp and feedback resistor R_F.

Substituting for I_{out} from Eq. (14-9) gives

$$V_o = -(\text{current resolution} \times D) \times R_F \tag{14-11b}$$

Rewrite Eq. (14-11b) as

$$V_o = -(\text{current resolution} \times R_F) \times D \tag{14-11c}$$

The coefficient of D is the voltage resolution or simply resolution and is given as

$$\text{resolution} = I_0 R_F \tag{14-11d}$$

and V_o can be written very simply as

$$V_o = -\text{resolution} \times D \tag{14-11e}$$

In terms of the actual hardware, V_o is expressed as

$$V_o = -\left(\frac{V_{ref}}{R} \times \frac{1}{2^n} R_F\right) \times D \tag{14-12}$$

Example 14-7

For the voltage output DAC of Fig. 14-5, find (a) its resolution and (b) V_o when the digital input is 1111.

Solution (a) From Example 14-6, the value of $I_0 = 62.5\ \mu\text{A}$. From Eq. (14-11d),

$$\text{resolution} = I_o R_F = 62.5\ \mu\text{A} \times 10\ \text{k}\Omega = 0.625\ \text{V}$$

A 1-bit input change causes a 0.625-V output voltage change.
(b) Calculate the DAC's performance equation from Eq. (14-12).

$$V_o = -\left(\frac{10\ \text{V}}{10\ \text{k}\Omega} \times \frac{1}{2^4} \times 10\ \text{k}\Omega\right) \times D = -0.625\ \text{V} \times D$$

For a digital input of 1111, $D = 15$. Therefore,

$$V_o = -0.625\ \text{V} \times 15 = -9.375\ \text{V}$$

14-5 MULTIPLYING DAC

Equation (14-12) can be rewritten to show how an MDAC or multiplying DAC operates.

$$V_o = (\text{constant}) \times V_{ref} \times D \tag{14-13}$$

where

$$\text{constant} = -\frac{R_F}{2^n R}$$

Equation (14-13) shows that V_o is the *product* of two input signals, V_{ref} and D, and *both* signals can be variables. One example for MDAC use is the volume control of an audio signal by a microprocessor.

Suppose that V_{ref} is an audio signal that varies from 0 to 10 V in Fig. 14-5. From Example 14-7, if $D = 0001$, V_o would vary from 0 to 0.625 V. If the digital input is 1000, $D = 8$ and the volume of V_o would increase, from 0 to 8×0.625 V = 5 V. Maximum volume of 15×0.625 V = 9.375 V will occur when the digital input word is 1111, $D = 15$. Thus the MDAC performs as a digitally operated volume control.

14-6 8-BIT DIGITAL-TO-ANALOG CONVERTER: THE DAC-08

The DAC-08 is a low-cost, fast MDAC, housed in a 16-pin DIP. Its operating principles are examined by reference to the task performed by each of its terminals in Fig. 14-6.

14-6.1 Power Supply Terminals

Pins 13 and 3 are the positive and negative supply terminals, respectively, and can have any value from ±4.5 to ±18 V. They should be bypassed with 0.1-μF capacitors, as shown in Fig. 14-6a.

14-6.2 Reference (Multiplying) Terminal

Flexibility of the DAC-08 is enhanced by having *two* rather than one reference input. Pins 14 and 15 allow positive or negative reference voltages, respectively. A positive reference voltage input is shown in Fig. 14-6a.

The user can adjust the DAC-08's input ladder current I_{ref} quite easily from 4 μA to 4 mA with a typical value of 2 mA.

$$I_{ref} = \frac{V_{ref}}{R_{ref}} \tag{14-14}$$

14-6.3 Digital Input Terminals

Pins 5 through 12 identify the digital input terminals. Pin 5 is the most significant bit (MSB), D_7. Pin 12 is the LSB, D_0. The terminals are TTL or CMOS compatible. Logic input "0" is 0.8 V or less. Logic "1" is 2.0 V or more, regardless of the power

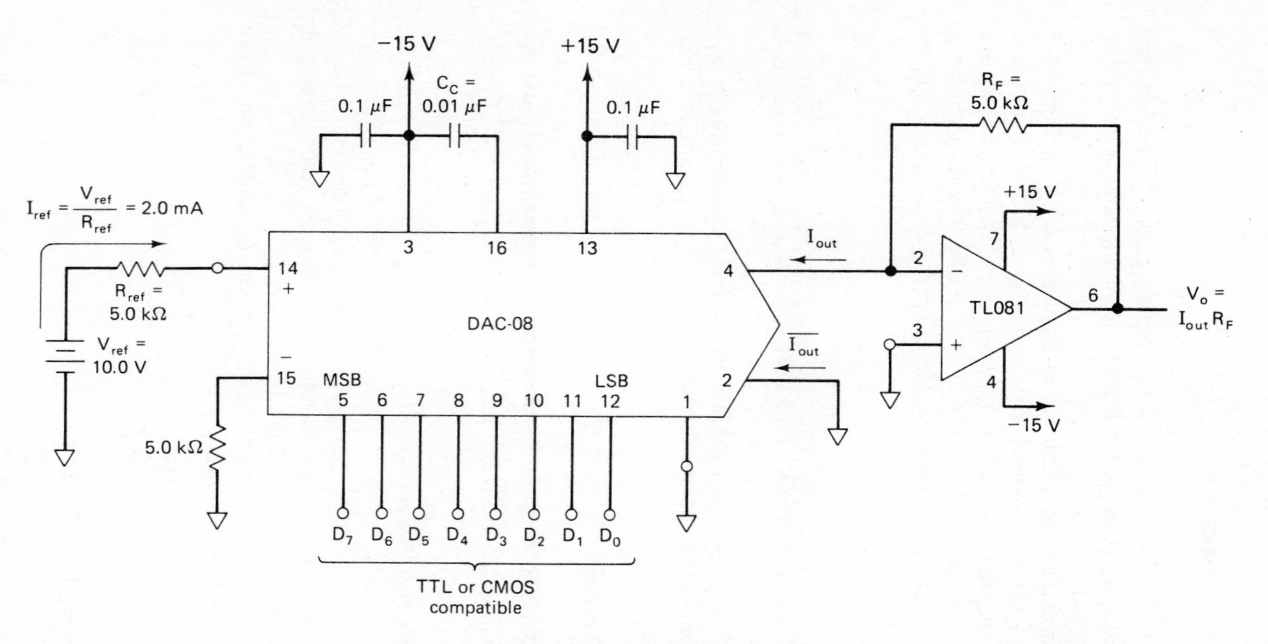

(a) DAC-08 wired for positive output voltages

	Digital inputs								Analog output	
	D_7	D_6	D_5	D_4	D_3	D_2	D_1	D_0	I_{out}	V_o
LSB	0	0	0	0	0	0	0	1	7.812 μA	39 mV
Half-scale	1	0	0	0	0	0	0	0	1.000 mA	5.0 V
Full-scale	1	1	1	1	1	1	1	1	1.992 mA	9.96 V

(b) Summary of Examples 14-7 and 14-8

FIGURE 14-6 An 8-bit DAC-08 is wired for unipolar output voltage in (a). I_{out} has values given in (b) for three digital input words. The op amp converts I_{out} to voltage V_o.

supply voltages. Usually, pin 1, V_{LC}, is grounded. However, it can be used to adjust the *logic input threshold* voltage V_{TH} according to $V_{TH} = V_{LC} + 1.4$ V. These digital inputs control eight internal current switches.

14-6.4 Analog Output Currents

Two current output terminals are provided in Fig. 14-6 to increase the DAC-08's versatility. Pin 4 conducts output current I_{out} and pin 2 conducts its complement, $\overline{I_{out}}$. If an internal switch is positioned to "1," its ladder rung current flows in the I_{out} bus. If positioned to "0," ladder rung current flows in the $\overline{I_{out}}$ bus.

The current value of 1 LSB (resolution) is found from

$$\text{resolution} = (\text{value of 1 LSB}) = \frac{V_{ref}}{R_{ref}} \times \frac{1}{2^n} \qquad (14\text{-}15a)$$

I_{out} is calculated from

$$I_{out} = (\text{value of 1 LSB}) \times D \qquad (14\text{-}15b)$$

where D is the decimal value of the digital input word. The full-scale output current in the pin 4 output bus occurs when the digital input is 11111111, so that $D = 255$. Let's define this current as I_{FS}, where

$$I_{FS} = (\text{value of 1 LSB}) \times 255 \qquad (14\text{-}16a)$$

The sum of all ladder rung currents in the DAC-08 equals I_{FS}. Since this sum always divides between the I_{out} and $\overline{I_{out}}$, the value of $\overline{I_{out}}$ is given by

$$\overline{I_{out}} = I_{FS} - I_{out} \qquad (14\text{-}16b)$$

Example 14-8

Calculate (a) the ladder input current I_{ref} of the DAC-08 in Fig. 14-6; (b) the current value of 1 LSB.

Solution (a) From Eq. (14-14),

$$I_{ref} = \frac{10 \text{ V}}{5 \text{ k}\Omega} = 2 \text{ mA}$$

(b) From Eq. (14-15a),

$$\text{current value of 1 LSB or resolution} = \frac{10 \text{ V}}{5 \text{ k}\Omega} \times \frac{1}{2^8} = 7.812\ \mu\text{A}$$

Example 14-9

For the DAC-08 circuit in Fig. 14-6, find the values of I_{out} and $\overline{I_{out}}$ when the digital input words are (a) 00000001; (b) 10000000; (c) 11111111.

Solution Example 14-8 showed that current outut resolution is 7.812 μA/bit. From Eq. (14-16a), evaluate I_{FS}.

$$I_{FS} = (\text{resolution})255 = 7.812\ \mu\text{A} \times 255 = 1.992\ \text{mA}$$

The value of D is 1 for (a), 128 for (b), and 255 for (c). I_{out} can now be found from Eq. (14-15b):

(a) $I_{out} = 7.812\ \mu\text{A} \times 1 = 7.812\ \mu\text{A}$ for 00000001 input
(b) $I_{out} = 7.812\ \mu\text{A} \times 128 = 1.000\ \text{mA}$ for 10000000 input
(c) $I_{out} = 7.812\ \mu\text{A} \times 255 = 1.992\ \text{mA}$ for 11111111 input

From Eq. (14-16b),

(a) $\overline{I_{out}} = 1.992\ \text{mA} - 7.812\ \mu\text{A} = 1.984\ \text{mA}$
(b) $\overline{I_{out}} = 1.992\ \text{mA} - 1.0\ \text{mA} = 0.992\ \text{mA}$
(c) $\overline{I_{out}} = 1.992\ \text{mA} - 1.992\ \text{mA} = 0$

The results of Examples 14-8 and 14-9 are tabulated in Fig. 14-6(b).

14-6.5 Unipolar Output Voltage

In Fig. 14-6(a), the DAC-08's current output I_{out} is converted to an output voltage V_o by an external op amp and resistor R_F. This voltage output has a resolution of

$$\text{resolution} = \frac{V_{ref}}{R_{ref}} \times R_F \times \frac{1}{2^n} \tag{14-17a}$$

and V_o is given by

$$V_o = \text{resolution} \times D = I_{out} R_F \tag{14-17b}$$

Example 14-10

For the DAC-08 circuit of Fig. 14-6(a), find V_o for digital inputs of (a) 00000001; (b) 11111111.

Solution From Eq. (14-17a),

$$\text{resolution} = (10\ \text{V})\frac{5\ \text{k}\Omega}{5\ \text{k}\Omega} \times \frac{1}{256} = 39.0\ \text{mV/bit}$$

(a) From Eq. (14-17b) with the value of $D = 1$,

$$V_o = 39.0\ \text{mV} \times 1 = 39.0\ \text{mV for 00000001 input}$$

(b) The value of $D = 255$. From Eq. (14-17b),

$$V_o = 39.0\ \text{mV} \times 255 = 9.961\ \text{V for 11111111 input}$$

14-6.6 Bipolar Analog Output Voltage

The versatility of the DAC-08 is shown by wiring it to give a bipolar analog output voltage in response to a digital input word [Fig. 14-7(a)]. The op amp and two resistors convert the *difference* between I_{out} and $\overline{I_{out}}$ into a voltage V_o:

$$V_o = (I_{out} - \overline{I_{out}})R_F \qquad (14\text{-}18)$$

I_{out} drives V_o positive and $\overline{I_{out}}$ drives V_o negative. If the digital input word increases by 1 bit, I_{out} increases by 1 LSB. However, $\overline{I_{out}}$ must therefore *decrease* by 1 LSB. Therefore, the differential output current changes by 2 LSB; thus we would expect the bipolar output voltage span to be twice that of a unipolar output (Section 14-6.5).

V_{ref} has been increased slightly in Fig. 14-7(a) so that I_{ref} increases to 2.048 mA [Eq. (14-14)]. This increases the current value of 1 LSB to an even 8 μA [Eq. (14-15a)]. We show how the output voltage responds to digital inputs by an example.

Example 14-11

For the circuit of Fig. 14-7(a), calculate V_o for digital inputs of (a) 00000000; (b) 01111111; (c) 10000000; (d) 11111111.

Solution The current value of 1 LSB equals 8 μA. From Eq. (14-16a), I_{FS} = (8 μA)255 = 2.040 mA.

(a) From Eq. (14-15b), I_{out} = (8 μA) × 0 = 0. Then from Eq. (14-16b), $\overline{I_{out}}$ = 2.040 m − 0 = 2.04 mA. Find V_o from Eq. (14-18):

$$V_o = (0 - 2.04 \text{ mA})(5 \text{ k}\Omega) = -10.20 \text{ V}$$

The values of I_{out}, $\overline{I_{out}}$, and V_o are calculated for (b), (c), and (d) and are summarized in Fig. 14-7(b).

Note that full-scale negative output voltage of −10.20 V occurs for an all-zero digital input. All ones give a positive full-scale output of plus 10.20 V. Note also that V_o never goes to precisely zero volts. When I_{out} is less than $\overline{I_{out}}$ by 8 μA (01111111), V_o equals −40 mV. Since this is the closest that V_o approaches to 0 V from negative full-scale, V_o = −40 mV is called negative zero.

Many DACs must operate under the control of a microprocessor or computer. We therefore present next a microprocessor-compatible DAC.

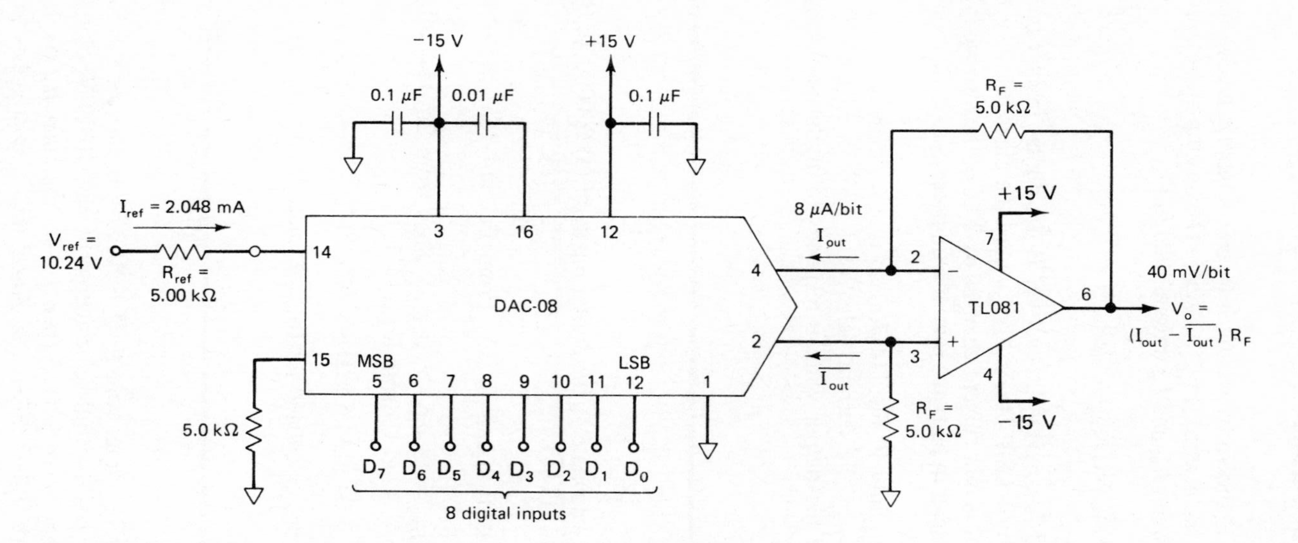

(a) DAC-08 wired for bipolar output voltage

	Digital inputs								Analog outputs		
	D_7	D_6	D_5	D_4	D_3	D_2	D_1	D_0	I_{out} (mA)	$\overline{I_{out}}$ (mA)	V_o (V)
Negative full scale	0	0	0	0	0	0	0	0	0	2.040	−10.20
Negative zero	0	1	1	1	1	1	1	1	1.016	1.024	−0.040
Positive zero	1	0	0	0	0	0	0	0	1.024	1.016	0.040
Positive full scale	1	1	1	1	1	1	1	1	2.040	0	10.20

(b) Tabulated solutions for Example 14-9

FIGURE 14-7 The op amp converts complementary output currents of the DAC-08 into bipolar output voltages. The op amp is wired as a differential current-to-voltage converter.

14-7 MICROPROCESSOR COMPATIBILITY

14-7.1 Interfacing Principles

The programmer views the location of a DAC's register(s) or any other peripheral chip's registers as an address in the total memory space. DACs have "write only" registers. This means that a DAC has a register that the microprocessor can send binary digits to via the data bus. An ADC's registers are "read only" registers. These devices have a register(s) whose contents can be "read" by the microprocessor via the data bus. Both DACs and ADCs have logic that permits selection via the address bus.

14-7.2 Memory Buffer Registers

"Read only" or "write only" registers have two operating states: *transparent* and *latching*. An idle register always remembers (latches) the last digital word written into it and the register can be disconnected from the data bus. More specifically, the interface between data bus and register can be in the high-Z state, or essential open circuit.

When a register is *transparent*, it is connected to the data bus. For example, an 8-bit register in a DAC would allow its 8 data bits D_7 through D_0 to "read" the logic 1's and/or 0's present on each corresponding wire of the data bus placed there by the microprocessor (see Fig. 14-8).

How does the microprocessor tell one ADC or one DAC, out of all other peripherals or memory addresses, that it is selected? This question is answered next.

14-7.3 The Selection Process

Every DAC or ADC register has an address just like any memory location of the microprocessor. To write to one particular DAC, the microprocessor places the address of that DAC on the address bus (see Fig. 14-8). One output of a local decoder goes low to enable a chip select, $\overline{CS}$, terminal on the selected DAC. The DAC's digital input buffer registers do not yet become transparent. The DAC is only partially selected.

To fully select the DAC, the microprocessor places a low on a line to the *chip enable,* $\overline{CE}$, terminal. This line is controlled by the microprocessor's read/write line, which may also be referred to as MRW, MEMW, or R/$\overline{W}$. When both the $\overline{CS}$ and $\overline{CE}$ terminals are low, only one DAC can communicate with the microprocessor. Its internal register becomes transparent and accepts data from the data bus. The DAC immediately converts this digital data to an analog output voltage, V_o. This completes a "write only" operation. When either $\overline{CS}$ or $\overline{CE}$ go high, the DAC's register enters the latching state and remembers the last data written into it. Thus V_o is held at the analog equivalent voltage.

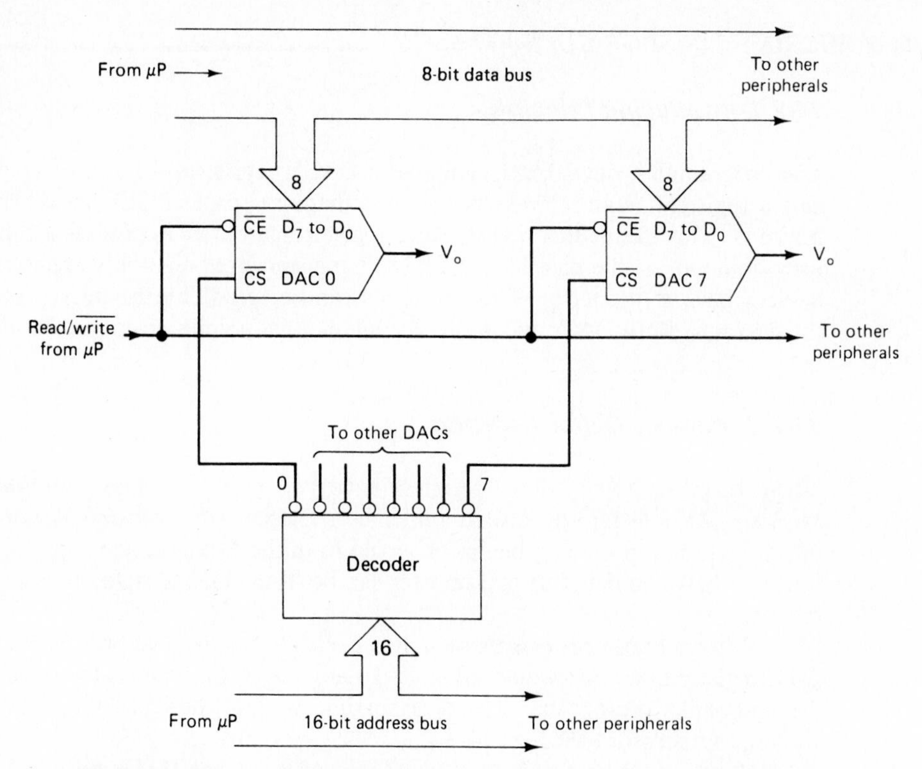

FIGURE 14-8 To select one DAC, the microprocessor places its address on the address bus. One ouput of the decoder goes low in response to its corresponding address code and enables the chip select terminal, $\overline{CS}$, of the DAC selected.

The selection process for an ADC is similar. That is, both $\overline{CS}$ and $\overline{CE}$ lines of the desired ADC are brought low. Now the ADC's output register is in the transparent mode and its contents can be "read" by the microprocessor via the data bus.

14-8 AD558 MICROPROCESSOR-COMPATIBLE DAC

14-8.1 Introduction

An example of a complete 8-bit microprocessor-compatible D/A converter is introduced in Fig. 14-9. The AD558 can operate either continuously or it can be controlled by a microprocessor. It is complete with an on-board precision reference voltage, latching digital inputs, and select terminals. It also contains an op amp to give an analog output voltage that is pin-programmable for output ranges of 0 to 2.56 V or 0 to 10.0 V. Operation of the AD558 is studied by analyzing those tasks performed by its terminals.

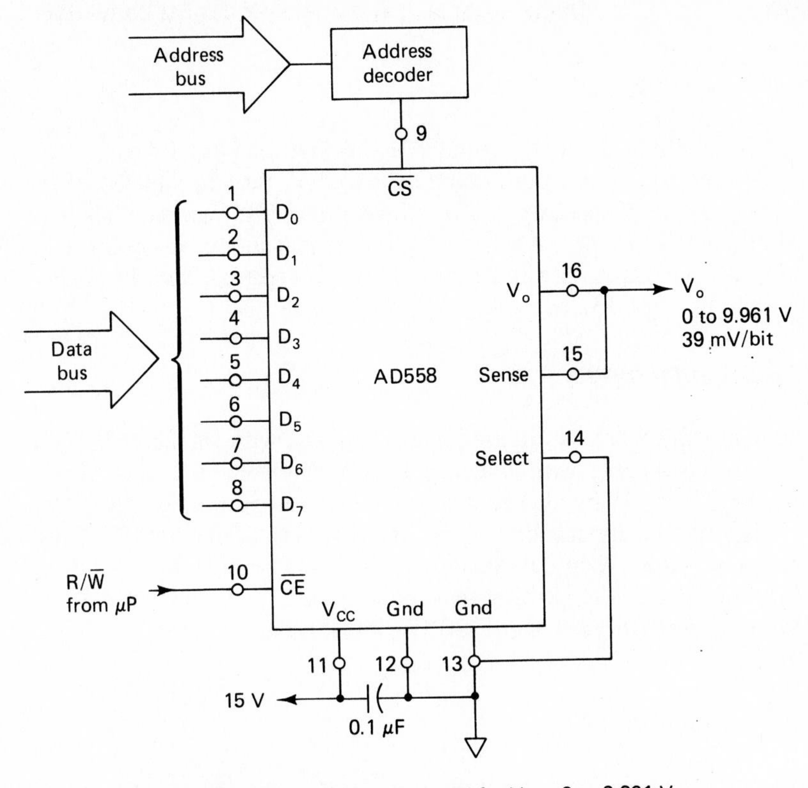

(a) AD558 terminals. Pin-programmed for V_o = 0 to 9.961 V

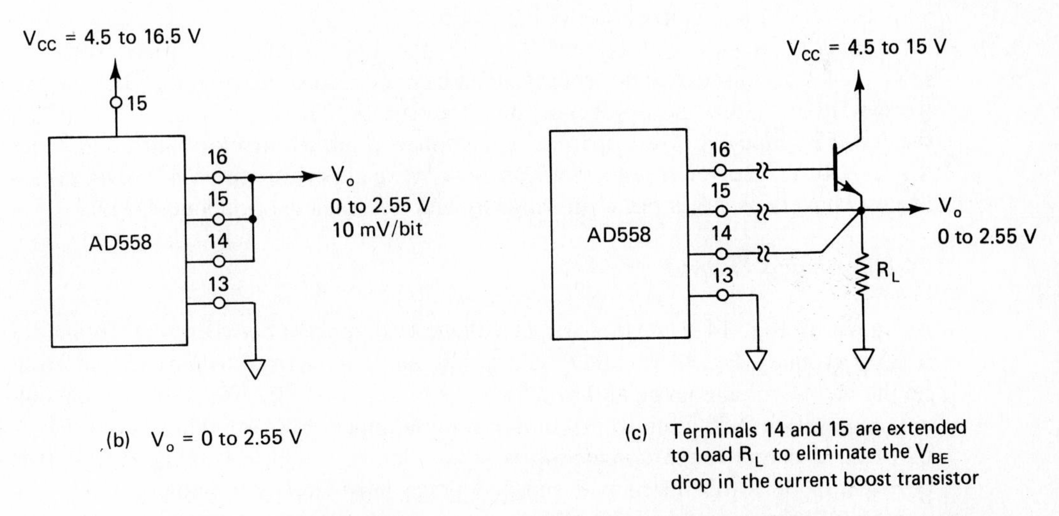

(b) V_o = 0 to 2.55 V

(c) Terminals 14 and 15 are extended to load R_L to eliminate the V_{BE} drop in the current boost transistor

FIGURE 14-9 The AD588 is an 8-bit microprocessor-compatible D/A converter whose pinouts are shown in (a). It may be pin-programmed for 0 to 2.55 V as shown in (b). Terminal 15 and gain select terminal 14 may be extended to a load for current boost as shown in (c).

14-8.2 Power Supply

Pin 11 is the power supply terminal V_{CC} in Fig. 14-9(a). It requires a minimum of +4.5 V and has a maximum rating of +16.5 V. Pins 12 and 13 are the digital and analog grounds, respectively. This allows the user to maintain *separate* analog and digital grounds throughout a system, joining them at *only* one point. Usually, pins 12 and 13 are wired together and a 0.1-μF bypass capacitor *must* be connected between V_{CC} and pin 12 or 13.

14-8.3 Digital Inputs

Pins 1 through 8 are the digital inputs D_0 to D_7, with D_0 the LSB and D_7 the MSB. They are compatible with standard TTL or *low*-voltage CMOS. Logic 1 is 2.0 V minimum for a "1" bit. Logic 0 is 0.8 V maximum for a "0" bit.

The digital input pins connect the data bus to the AD558's internal *memory latching register,* when the AD558 is selected. This condition is called *transparent.* When unselected, the latching register is essentially disconnected from the data bus and remembers the last word written into the latching register. This condition is called "latching."

14-8.4 Logic Circuitry

The microprocessor executes a *write* command over the address bus via an address decoder and write line to the AD558's logic control pins 9 and 10. They are called chip select ($\overline{CS}$) and chip enable ($\overline{CE}$), respectively.

If a "1" is present on either $\overline{CS}$ or $\overline{CE}$, the digital inputs are in the "latching" mode. They are disconnected from the data bus. The input latches remember the last word written by the microprocessor over the data bus. If both $\overline{CS}$ and $\overline{CE}$ are "0," the AD558's inputs are "transparent" and connect the input memory latch register to the data bus. The microprocessor can now write data into the DAC. Digital-to-analog conversion takes place immediately and is completed in about 200 ns.

14-8.5 Analog Output

As shown in Fig. 14-9, analog output voltage (V_o) appears between pins 16 and 13 (analog ground). Pin 14 is called "select" (V_o gain). It is wired to pins 15 and 16 to set the output voltage range at 0 to 2.56 V, as in Fig. 14-9(b). The actual analog output range is 0 to 2.55 V or 10 mV/bit for a digital input of 00000000 to 11111111. A 0- to 10-V output range connection is shown in Fig. 14-9a. Actual range is 0 to 9.961 V or 38.9 mV/bit (power supply voltage must *exceed* maximum V_o by 2 V minimum). Sense terminal 15 allows remote load-voltage sensing to eliminate effects of *IR* drops in long leads to the load. It can also be used for current boost as in Fig. 14-9(c).

14-8.6 Dynamic Test Circuit

A single AD558 can be tested dynamically *without* a microprocessor by the stand-alone, low-parts-count test circuit of Fig. 14-10. Pins 9 and 10 of the AD558 are grounded. This connects the AD558's input register (transparent) to an 8-bit synchronous counter that simulates a data bus.

The test circuit consists of three ICs. One 555 timer is wired as a 1-kHz clock. It steps an 8-bit synchronous binary counter made from two CD4029s. The counter's outputs are wired to the digital inputs of the DAC. A CRO is connected (dc-coupled) to display V_o. It should be externally triggered from the negative edge of the MSB (pin 8 of the AD558 or pin 2 of the right CD4029 in Fig. 14-10). The analog voltage waveshape appearing at V_o will resemble a staircase.

Each clock pulse steps the counter up by one count and increases V_o by 10 mV. Thus the risers will equal 10 mV and the staircase will have 256 treads from 0 to 2.55 V. The tread on each step will occupy about 1 ms. Thus one staircase waveshape is generated every 256 ms. Any glitch, nonlinearity, or other abnormality will be quite apparent. The capacitor C_T can be changed to give faster or slower clock frequencies, as indicated in Fig. 14-10. Any visible glitches can be minimized but *not* eliminated from V_o. However, a *sample-and-hold* or *follow-and-hold amplifier* can be connected to V_o. It waits until the glitch settles down, samples V_o, and holds this correct value. The principle of sample-and-hold is presented in Section 14-15.2.

14-9 INTEGRATING ADC

14-9.1 Types of ADCs

General characteristics of ADCs were introduced in Section 14-2. There are three standard types classified according to their conversion times. The *slow integrating ADC* typically requires 300 ms to perform a conversion. It is the best choice for measuring slowly varying dc voltages. The faster *successive approximation ADC's* conversion time is a few microseconds and can digitize audio signals. Fastest of all are the more costly *flash* converters, which can digitize video signals.

14-9.2 Principles of Operation

The block diagram of a typical dual-slope integrating A/D converter such as Intersil's 7106/7107 is shown in Fig. 14-11. An on-board divide-by-4 counter drives the control logic at a rate of 12 kHz. This frequency is set by the user, via external timing resistor R_T and C_T. It must be a multiple of the local line frequency (50 or 60 Hz) to render the ADC immune to line-frequency noise.

The control logic unit activates a complex network of logic circuits and analog

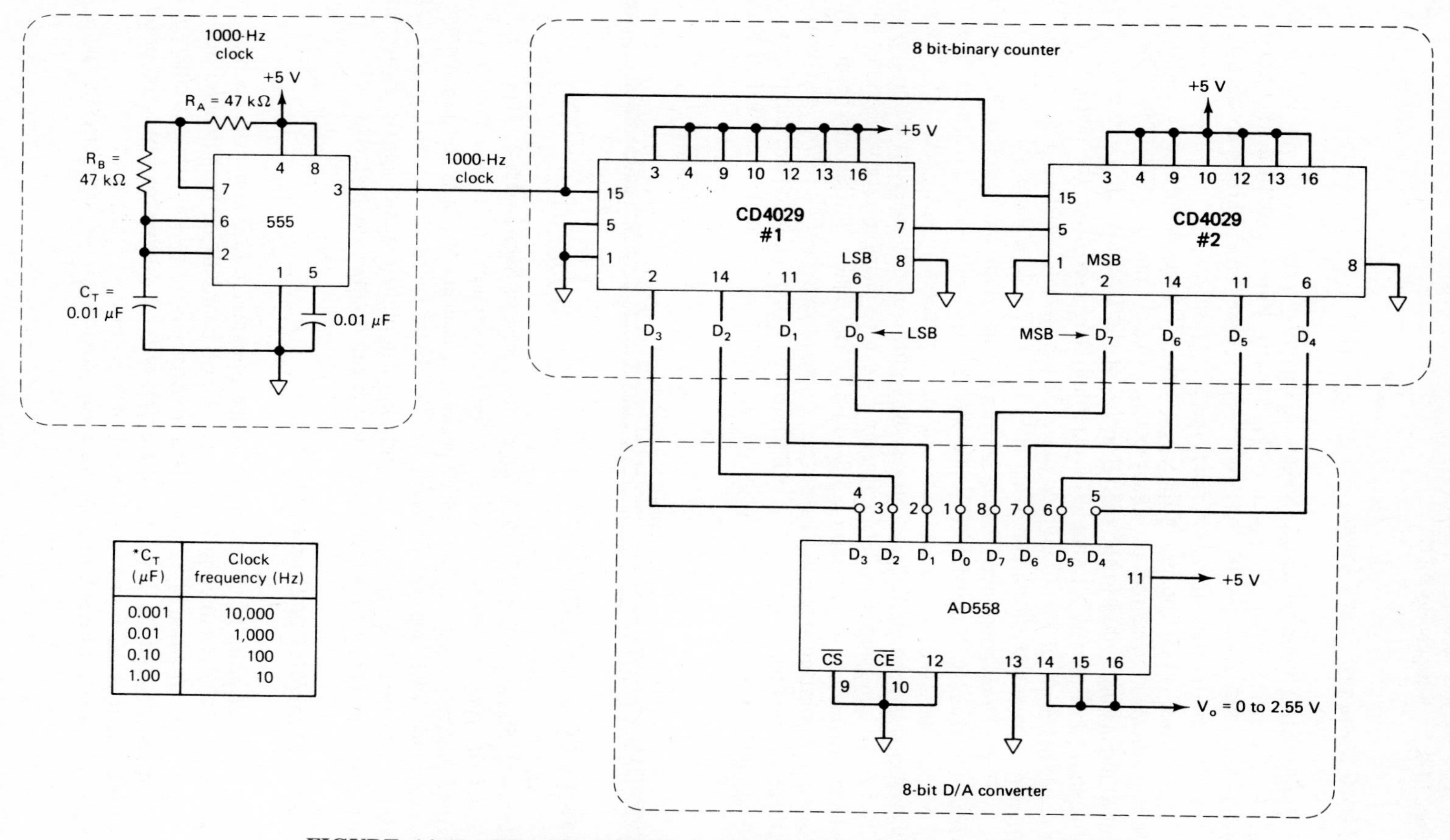

*C_T (μF)	Clock frequency (Hz)
0.001	10,000
0.01	1,000
0.10	100
1.00	10

FIGURE 14-10 The 555 clock drives an 8-bit binary counter made from two CD4049 ICs. The outputs count in binary from 00000000 to 11111111 and then repeat. The digital count is converted by the DAC into an analog voltage that resembles a staircase.

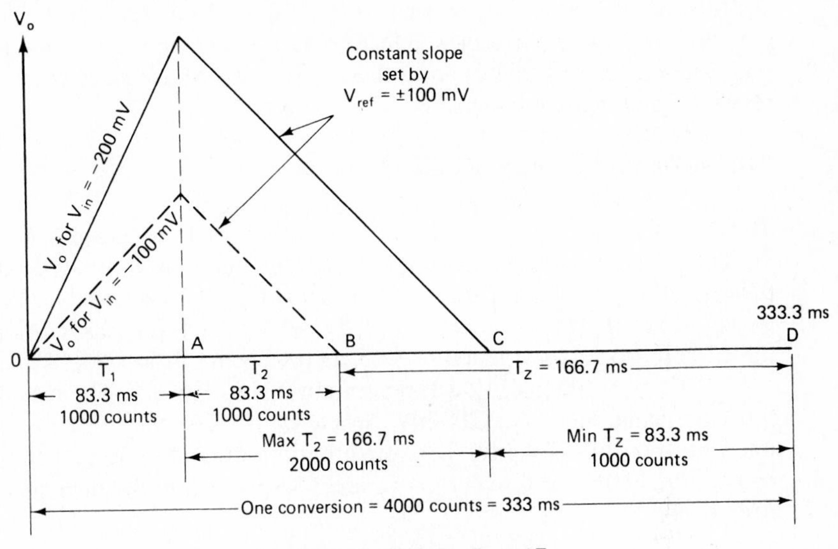

(a) Timing for phases T_1, T_2, and T_Z

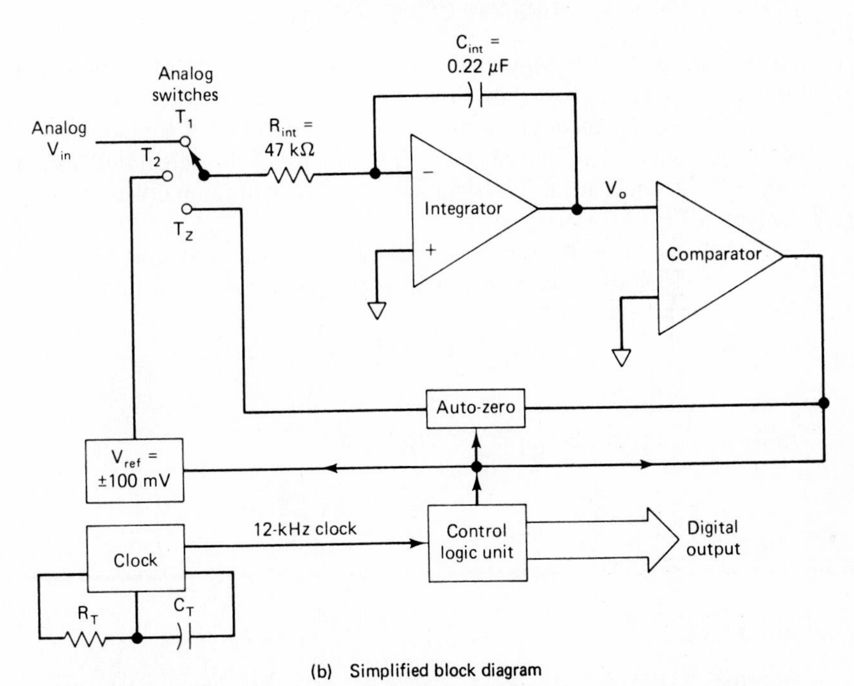

(b) Simplified block diagram

FIGURE 14-11 (a) Timing diagram of a typical dual-slope integrating ADC; (b) simplified block diagram of a dual-slope integrating ADC. In (a), one A/D conversion takes place in three phases: signal integrating phase T_1, reference T_2, and auto-zero T_z.

switches to convert analog input voltage V_{in} into a digital output. The analog-to-digital conversion is performed in three phases and requires about one-third of a second. These operating phases are called *signal integrate phase* T_1, *reference integrate phase* T_2, and *auto-zero phase* T_z. These will be discussed in sequence.

14-9.3 Signal Integrate Phase, T_1

The control logic unit of Fig. 14-11(b) connects V_{in} to an integrator to begin phase T_1. The integrator or ramp generator's output V_o ramps up or down depending on the polarity of V_{in} and at a rate set by V_{in}, R_{int}, and C_{int}. If V_{in} is negative, V_o ramps up, as shown in Fig. 14-11(a). Time T_1 is set by the logic unit for 1000 clock pulses. Since the 12-kHz clock has a period of 83.3 μs per count, T_1 lasts 83.33 ms.

If $V_{in} = -100$ mV, V_o will ramp from 0 V to 833 mV. The maximum allowed full-scale value of V_{in} is ± 200 mV. When $V_{in} = -200$ mV, V_o will rise to a maximum of 1666 mV. Clearly, V_o is directly proportional to V_{in}. At the end of 1000 counts, the logic unit disconnects V_{in} and connects V_{ref} to the integrator. This action ends T_1 and begins T_2.

14-9.4 Reference Integrate Phase, T_2

During phase T_1, the logic unit determined the polarity of V_{in} and charged a reference capacitor, C_{ref} (not shown), to a reference voltage $V_{ref} = 100$ mV. At the beginning of phase T_2, the logic unit connects C_{ref} to the integrator so that V_{ref} has a polarity opposite to V_{in}. Consequently, V_{ref} will ramp the integrator back toward zero. Since V_{ref} is constant, the integrator's output V_o will ramp down at a *constant* rate, as shown in Fig. 14-11(a).

When V_o reaches zero, a comparator tells the logic unit to terminate phase T_2 and begin the next auto-zero phase. T_2 is thus proportional to V_o and consequently, V_{in}. The exact relationship is

$$T_2 = T_1 \frac{V_{in}}{V_{ref}} \tag{14-19a}$$

Since $T_1 = 83.33$ ms and $V_{ref} = 100$ mV,

$$T_2 = \left(0.833 \frac{\text{ms}}{\text{mV}}\right) V_{in} \tag{14-19b}$$

Example 14-12

For the ADC of Fig. 14-11, calculate T_2 if (a) $V_{in} = \pm 100$ mV; (b) $V_{in} = \pm 200$ mV.

Solution (a) From Eq. (14-19b),

$$T_2 = \left(0.833 \frac{\text{ms}}{\text{mV}}\right)(100 \text{ mV}) = 83.33 \text{ ms}$$

(b)
$$T_2 = \left(0.833 \frac{\text{ms}}{\text{mV}}\right)(200 \text{ mV}) = 166.6 \text{ ms.}$$

14-9.5 The Conversion

The actual conversion of analog voltage V_{in} into a digital count occurs during T_2 as follows. The control unit connects the clock to an internal binary-coded-decimal counter at the beginning of phase T_2. The clock is disconnected from the counter at the end of T_2. Thus the counter's content becomes the digital output. This digital output is set by T_2 and the clock frequency:

$$\text{digital output} = \left(\frac{\text{counts}}{\text{second}}\right)T_2 \tag{14-20a}$$

but T_2 is set by V_{in} from Eq. (14-19a) and therefore

$$\text{digital output} = \left(\frac{\text{counts}}{\text{second}}\right)(T_1)\left(\frac{V_{in}}{V_{ref}}\right) \tag{14-20b}$$

Since clock frequency is 12 kHz for the 7106/7107 ADC, $T_1 = 83.33$ ms, and $V_{ref} = 100$ mV, the output–input equation is

$$\text{digital output} = \left(12{,}000 \frac{\text{counts}}{\text{second}}\right)\left(\frac{83.33 \text{ ms}}{100 \text{ mV}}\right)V_{in}$$

or

$$\text{digital output} = \left(10 \frac{\text{counts}}{\text{mV}}\right)V_{in} \tag{14-20c}$$

The counter's output is connected to an appropriate $3\frac{1}{2}$-digit display.

Example 14-13

V_{in} equals +100 mV in the ADC of Fig. 14-11. Find the digital output.

Solution From Eq. (14-20c),

$$\text{digital output} = \left(10 \frac{\text{counts}}{\text{mV}}\right)(100 \text{ mV}) = 1000 \text{ counts}$$

Example 14-13 shows the need for some human engineering. The display reads 1000, but it *means* that V_{in} equals 100 mV. *You* must wire in a decimal point to display 100.0 and paste a "mV" sign beside the display.

14-9.6 Auto-Zero

The block diagram of Fig. 14-11(b) contains a section labeled "auto-zero." During the third and final phase of conversion, T_z, the logic unit activates several analog switches and connects an auto-zero capacitor C_{AZ} (not shown).

The auto-zero capacitor is connected across the integrating capacitor, C_{int}, and any input offset voltages of both integrating and comparator op amps. C_{AZ} charges to a voltage approximately equal to the average error voltage due to C_{int} and the offset voltages. During the following phases T_1 and T_2, the error voltage stored on C_{AZ} is connected to cancel any error voltage on C_{ref}. Thus the ADC is automatically zeroed for every conversion.

14-9.7 Summary

Refer to the timing diagram in Fig. 14-11(a). The logic unit allocates 4000 counts for one conversion. At 83.33 μs per count, the conversion takes 333 ms. The control unit always allocates 1000 counts or 83.3 ms to phase T_1.

The number of counts required for T_2 depends on V_{in}. Zero counts are used for V_{in} = 0 V and a maximum of 2000 counts or 166.7 ms are used when V_{in} is at its maximum limit of $\pm$200 mV.

T_2 and T_z always share a total of 3000 counts for a total of 250 ms. For V_{in} 0 V, T_2 = 0 counts and T_z = 3000 counts. For V_{in} = $\pm$200 mV, T_2 = 2000 counts and T_z = 1000 counts.

Intersil markets a complete $3\frac{1}{2}$-digit digital voltmeter kit. The kit contains a 40-pin dual-slope integrating A/D converter (7106 or 7107), all necessary parts, printed circuit board, and instructions. The instructions make it easy to make, easy to use, and forms an excellent tutorial on integrating ADCs.

14-10 SUCCESSIVE APPROXIMATION ADC

The block diagram of a successive approximation register (ADC) is shown in Fig. 14-12. It consists of a DAC, a comparator, and a *successive approximation register* (SAR). One terminal is required for analog input voltage V_{in}. The digital output is available in either serial or parallel form. A minimum of three control terminals are required. *Start conversion* initiates an A/D conversion sequence and *end of conversion* tells when the conversion is completed. An external clock terminal sets the time to complete each conversion.

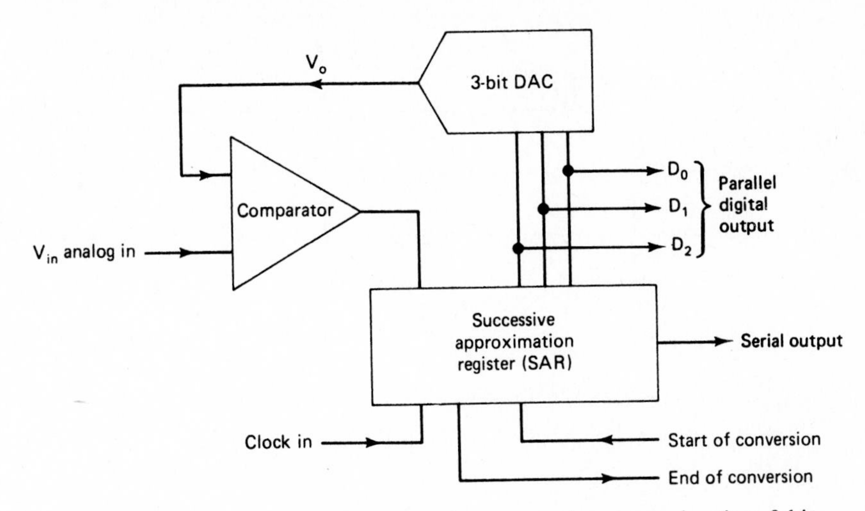

FIGURE 14-12 Block diagram of a successive approximation 3-bit ADC.

14-10.1 Circuit Operation

Refer to Fig. 14-12. An input *start conversion* command initiates one analog-to-digital conversion cycle. The successive approximation register (SAR) connects a sequence of digital numbers, one number for each bit to the inputs of a DAC. This process is explained in Section 14-3.

The DAC converts each digital number into an analog output V_o. Analog input voltage, V_{in}, is compared to V_o by a comparator. The comparator tells the SAR whether V_{in} is greater or less than DAC output V_o, once for each bit. For a 3-bit output, three comparisons would be made.

Comparisons are made beginning with the MSB and ending with the LSB, as will be explained. At the end of the LSB comparison, the SAR sends an end-of-conversion signal. The digital equivalent of V_{in} is now present at the SAR's digital output.

14-10.2 Successive Approximation Analogy

Suppose that you had 1-, 2-, and 4-lb weights (SAR) plus a balance scale (comparator and DAC). Think of the 1-lb weight as 1 LSB and the most significant 4-lb weight as 4-LSB. Refer to Figs. 14-12 and 14-13. V_{in} corresponds to an unknown weight.

Let us convert V_{in} = 6.5 V to a digital output (unknown weight = 6.5 lb). You would place the unknown weight on one platform of the balance, the 4-lb weight on the other, and compare if the unknown weight (V_{in}) exceeded the 4-lb weight. The SAR uses one clock pulse to apply the MSB 100 to the DAC in Fig. 14-13. Its out-

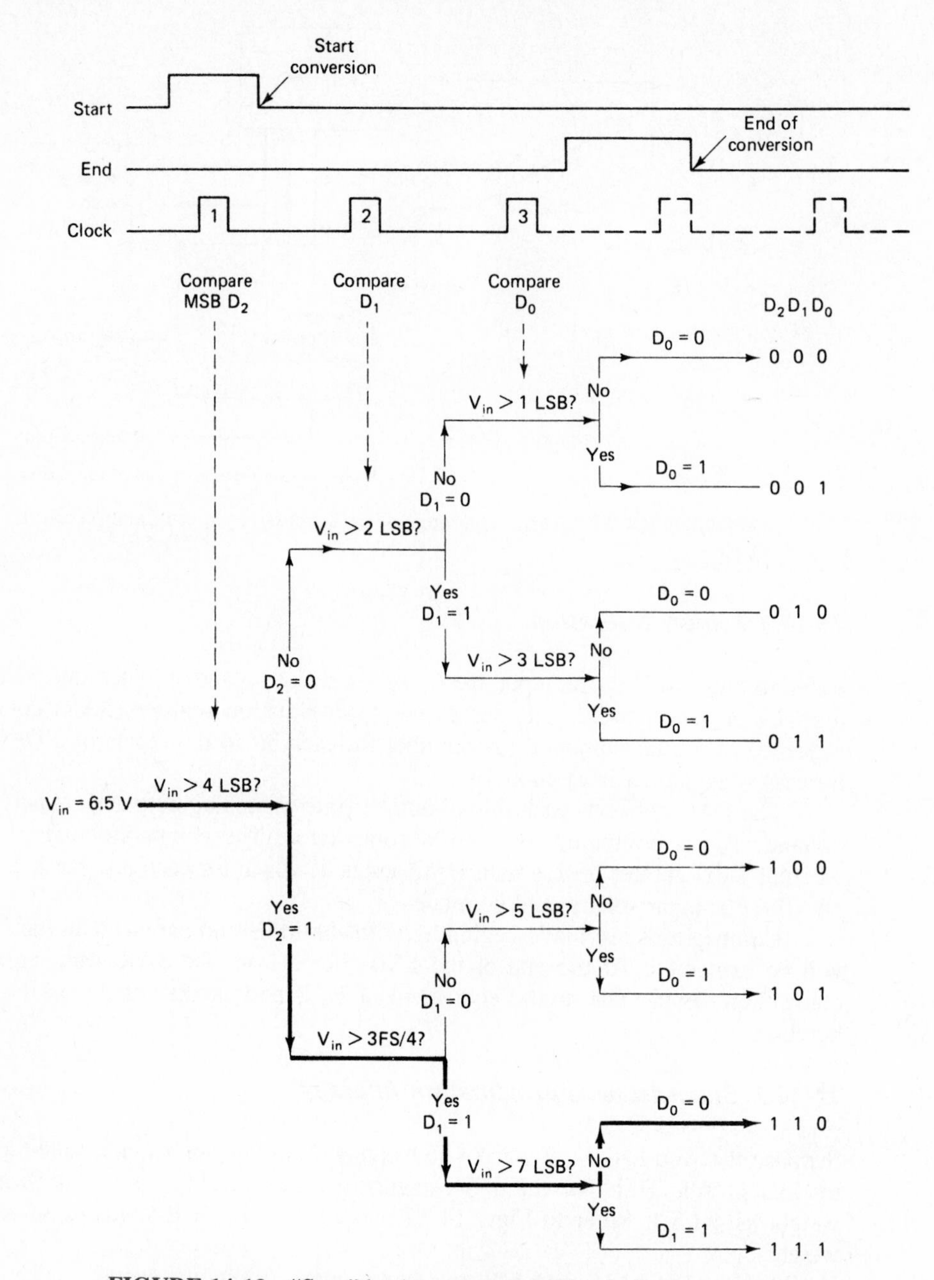

FIGURE 14-13 "Start" begins operation of this 3-bit successive approximation register. Beginning with the MSB, the weight of each bit is compared with V_{in} by the comparator of Fig. 14-12. If V_{in} is greater, the SAR's output is set to 1, or to 0 if V_{in} is smaller. Dark lines show the conversion for $V_{in} = 6.5$ V.

put, $V_o = 4$ V, is compared with V_{in}. The MSB (D_2) is set to 1 if $V_{in} > V_o$. This is analogous to you leaving the 4-lb weight on the scale.

The SAR then applies 110 (add a 2-lb weight) to the DAC, D_1 is set to 1 since $V_{in} = 6.5$ V is greater than $V_o = 6$ V. Finally, the SAR applies 111 to the DAC (add 1 lb). Since $V_{in} = 6.5$ V is less than 7 V, D_0 is set to zero (1-lb weight removed).

14-10.3 Conversion Time

Figure 14-13 shows that one clock pulse is required for the SAR to compare each bit. However, an additional clock pulse is usually required to reset the SAR prior to performing a conversion. The time for one analog-to-digital conversion must depend on both the clock's period T and number of bits n. The relationship is

$$T_C = T(n + 1) \tag{14-21}$$

Example 14-14

An 8-bit successive approximation ADC is driven by a 1-MHz clock. Find its conversion time.

Solution The time for one clock pulse is 1 μs. From Eq. (14-21),

$$T_C = 1\ \mu\text{s}(8 + 1) = 9\ \mu\text{s}$$

14-11 ADCs FOR MICROPROCESSORS

The microprocessor "views" a peripheral ADC simply as a "read only" address in the microprocessor's memory map. Refer to Fig. 14-14. The ADC must have a tri-state *memory buffer register* (MBR). In the idle state, the MBR will contain a digital code resulting from the ADC's *last* conversion. Also, the MBR will be disconnected from the data bus.

The microprocessor uses the address bus and decoders to select one ADC out of all the others by bringing its *chip select* terminal low. This process is similar to that shown in Fig. 14-8. A low on the $\overline{\textit{chip select}}$ terminal in Fig. 14-14 tells the ADC that a command is coming to its read/$\overline{\text{write}}$ terminal. If read/$\overline{\text{write}}$ is brought low by the microprocessor, the ADC converts V_{in} into a digital code and loads or writes it into its own MBR. When read/$\overline{\text{write}}$ is high *and* chip select is low, the ADC's memory buffer register is connected (transparent) to the data bus.

It is important to look at this operation from the microprocessor's viewpoint. A *read* command means that the microprocessor is going to read data stored in the ADC's memory buffer register. The ADC's digital tri-state outputs must go from high-Z (high impedance) to transparent and connect the digital word to the data bus.

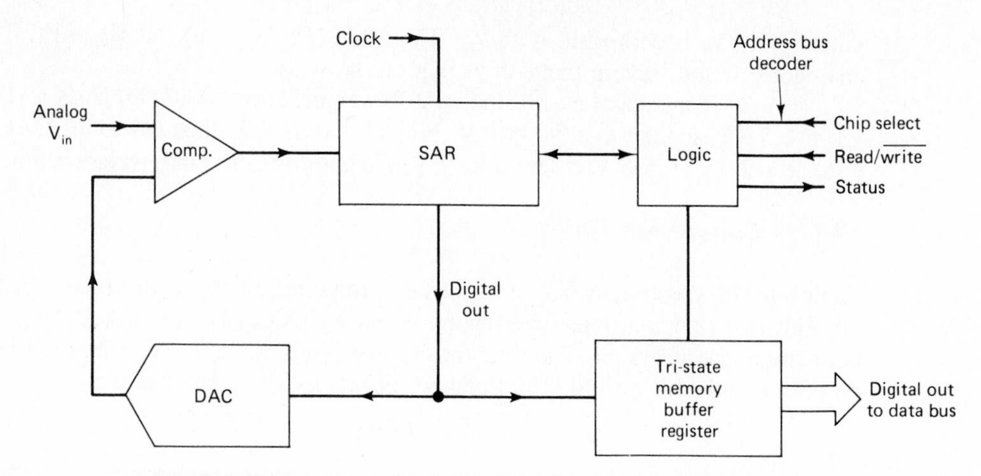

FIGURE 14-14 To be compatible with microprocessors, the ADC of Fig. 14-12 requires selection logic and a memory buffer register.

A *write command is actually a start conversion* command to the ADC. The microprocessor thus tells the ADC: (1) perform a conversion; (2) store (and *write*) it in your memory; and (3) don't tell me the result until I want to *read* it.

Finally, the microprocessor-compatible ADC must tell the microprocessor via its *status* terminal when a conversion is in progress; status goes high. If a conversion is completed, status goes low to signal the microprocessor that data is valid and ready for reading. We select Analog Devices AD670 to learn how all the foregoing features are available in a single 20-pin integrated circuit.

14-12 AD670 MICROPROCESSOR-COMPATIBLE ADC

The AD670 is an 8-bit microprocessor-compatible successive approximation analog-to-digital converter. The 20-pin package of Fig. 14-15 contains all the features described in Section 14-10 and Fig. 14-14. In addition, it contains an on-board clock, voltage reference, and instrumentation amplifier, and needs only a single 5-V supply. To understand how the AD670 operates, we examine the tasks performed by each of its terminals and associated circuit blocks.

14-12.1 Analog Input Voltage Terminals

Four analog input terminals are pins 16, 17, 18, and 19 in Fig. 14-15. They are inputs to an instrumentation amplifier configured to handle unipolar or bipolar analog input voltages. They are also pin-programmable to make it easy for the user to select resolution. Figure 14-15(a) shows operation for an analog input of 0 to 2.55 V,

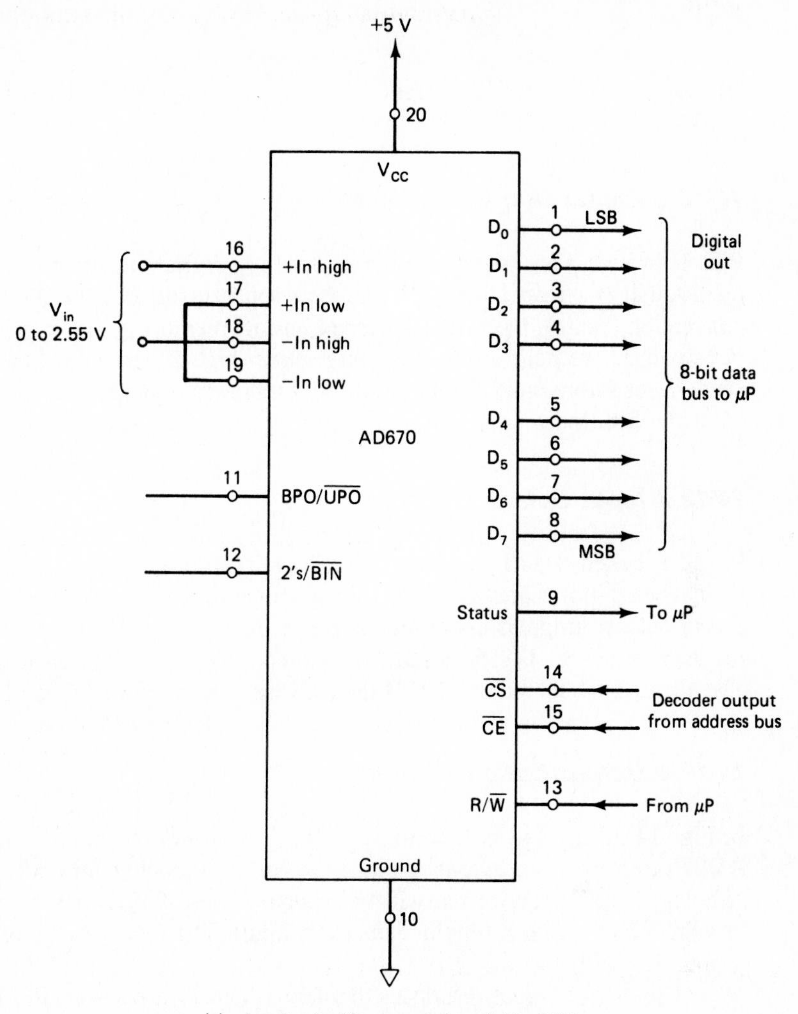

(a) Inputs wired for a resolution of 10 mV/bit

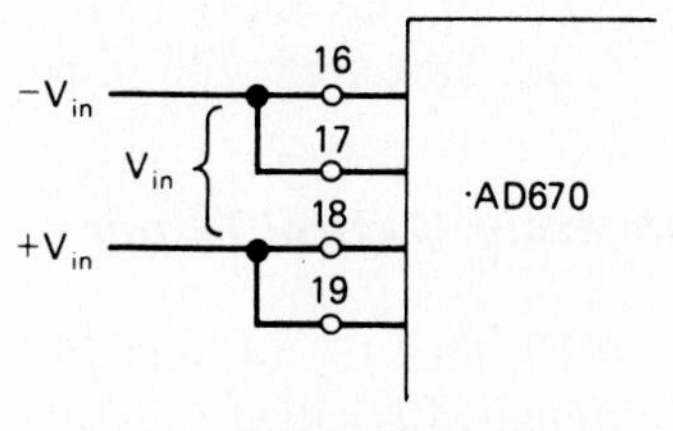

(b) Inputs wired for 1 mV/bit

FIGURE 14-15 (a) AD670 ADC pin connections. Full-scale analog input voltages are 0 to 2.55 V or 0 to ±1.28 V and 0 to 255 mV or 0 to ±128 mV in (b).

resolution = 10 mV/LSB. Figure 14-15(b) shows operation for 0 to 255 mV or 1 mV/LSB.

14-12.2 Digital Output Terminals

Pins 1 through 8 are tristate, buffered, latching digital outputs for the data bus digits, D_0 through D_7, respectively. When a microprocessor tells the AD670 to perform a conversion (write), the result is latched into its memory buffer register. Tri-state output switches are held in the *high-impedance* (high-Z) state until the microprocessor sends a read command. Thus the ADC's memory register is normally disconnected from the data bus.

14-12.3 Input Option Terminal

Pin 11 is called BPO/$\overline{\text{UPO}}$ and allows the microprocessor to tell the AD670 whether to accept a bipolar analog input voltage range or a unipolar input range. A low on pin 11 selects unipolar operation. A range of 0 to 2.55 V or 0 to 255 mV is set by the user as in Fig. 14-15(a) and (b). A high sent to pin 11 selects bipolar operation. The V_{in} range is then ±1.28 V [Fig. 14-15(a)] or ±128 mV [Fig. 14-15(b)].

14-12.4 Output Option Terminal

In Fig. 14-15, pin 12 is labeled "2's/$\overline{\text{BIN}}$." It allows the microprocessor to tell the AD670 to present an *output format* in either 2's-complement code, or binary code. A binary output code format will be *straight binary* if V_{in} is unipolar (pin 11 = low) or *offset binary* if V_{in} is bipolar (pin 11 = high). The four possible options are shown in Fig. 14-16(a).

The digital output responses to analog input V_{in} are shown in Fig. 14-16(b) and (c). V_{in} is the *differential* input voltage and is defined by

$$V_{in} = (+V_{in}) - (-V_{in}) \tag{14-22}$$

where $+V_{in}$ and $-V_{in}$ are measured with respect to ground.

14-12.5 Microprocessor Control Terminals

As shown in Fig. 14-15, pins 13, 14, and 15 are used by a microprocessor to control the AD670. Terminal 14 is called *chip select* ($\overline{\text{CS}}$) and terminal 15 is called *chip enable* ($\overline{\text{CE}}$). Pin 13 is called read/$\overline{\text{write}}$ (R/$\overline{\text{W}}$).

If $\overline{\text{CS}}$, $\overline{\text{CE}}$, and R/$\overline{\text{W}}$ are all brought low, the ADC converts continuously. It performs one conversion every 10 μs or less. The result of each conversion is latched into the output buffer register. However, the digital output code is *not* connected to the data bus because the outputs are high-impedance. This condition is

Pin 11 BPO/$\overline{UPO}$	Input range	Pin 12 2's/$\overline{BIN}$	Output format
0	Unipolar	0	Straight binary
1	Bipolar	0	Offset binary
0	Unipolar	1	2's complement
1	Bipolar	1	2's complement

(a)

Differential V_{in}	Unipolar/straight binary, pin 11 = 0, 12 = 0
0	0000 0000
1 mV	0000 0001
128 mV	1000 0000
255 mV	1111 1111

(b)

Differential V_{in}	Bipolar/offset binary, pin 11 = 1, 12 = 0	Bipolar/2's complement, pin 11 = 1, 12 = 1
−128 mV	0000 0000	1000 0000
−1 mV	0111 1111	1111 1111
0	1000 0000	0000 0000
1 mV	1000 0001	0000 0001
127 mV	1111 1111	0111 1111

(c)

FIGURE 14-16 Input ranges unipolar or bipolar, and output formats are determined by pins 11 and 12 in (a). Output codes are given for unipolar inputs in (b) and for bipolar inputs in (c). (a) Input range and output format are controlled by pins 12 and 11, respectively; (b) digital output code for unipolar V_{in} inputs wired as in Fig. 14-15a; (c) digital output codes for bipolar V_{in} inputs wired as in Fig. 14-15b.

called a *write* and *convert* command. That is, the microprocessor tells the AD670 to write converted data into its own buffer register. If $\overline{CS}$ or R/$\overline{W}$ or $\overline{CE}$ is high, the AD670 is unselected (high impedance) and retains the last conversion in its register.

Status terminal, pin 9, stays high during a conversion. When a conversion is completed, pin 9 outputs a low to tell the microprocessor that data is valid in the AD670's buffer register. To read data out of the AD670, the microprocessor brings R/$\overline{W}$ high while status and $\overline{CS}$ *and* $\overline{CE}$ are low. This is a *read* command from the microprocessor.

The AD670's buffer becomes transparent and connects the eight digital outputs (D_7 through D_0) to the data bus. Data will remain on the bus until the AD670 is disconnected by bringing $\overline{CS}$ high, or $\overline{CE}$ high, or R/$\overline{W}$ low.

Summary

1. A low on $\overline{CE}$ and $\overline{CS}$ selects the AD670. What happens next depends on R/$\overline{W}$.
2. If R/$\overline{W}$ is low (for at least 0.3 μs), a conversion is performed and the result is written into the buffer register. Outputs are high-impedance. The conversion requires 10 μs.

3. If $R/\overline{W}$ is high, the last conversion is stored in the buffer and the outputs are transparent. No further conversions are performed. The contents of the register can now be read by the microprocessor via the data bus.
4. Status tells the microprocessor what is going on within the AD670. Status = high means that conversion is being performed. Status = low tells the microprocessor that data are valid. The microprocessor is free to read the selected AD670's data by placing a high to $R/\overline{W}$.

14-13 TESTING THE AD670

Figure 14-17 shows how to wire an AD670 to perform continuous conversions *without* a microprocessor. This circuit can be used as a laboratory exercise to gain experience operating ADCs. Each data output, D_0 to D_7, is connected to an inverter, resistor, and LED. These components simulate a data bus. An LED lights to signify that a logic 1 is present on its associated data bus wire.

Pins 14 and 15 are wired so that $\overline{CS}$ and $\overline{CE}$ are low. This causes continuous conversion. The 555 timer drives $R/\overline{W}$ low for 5 μs to simulate a write command. $R/\overline{W}$ thus returns high before a conversion is completed in the 10-μs conversion time. At the end of 10 μs, the high on $R/\overline{W}$ simulates a read command and data are displayed on the LEDs. If R_T = 1.5 MΩ, the AD670 makes one conversion and one readout 1000 times per second. Reduce R_T to 120 kΩ for convert/reads of 10,000 times per second.

14-14 FLASH CONVERTERS

14-14.1 Principles of Operation

Fastest of all A/D converters is the *flash* converter, shown in Fig. 14-18(a). A reference voltage and resistor divider network establishes a resolution of 1 V/LSB. Analog input voltage V_{in} is applied to the + inputs of all comparators. Their outputs drive an 8-line-to-3-line priority encoder. The encoder logic outputs a binary code that represents the analog input.

For example, suppose that V_{in} = 5.0 V. The outputs of comparators 1 through 5 would go high and 6 through 8 would go low. As shown in Fig. 14-18(b), the digital output would be 101.

14-14.2 Conversion Time

The conversion time of the flash converter is limited only by response time of comparators and logic gates. They can digitize video or radar signals. The flash converter's high speed becomes more expensive as resolution is increased. Figure 14-18

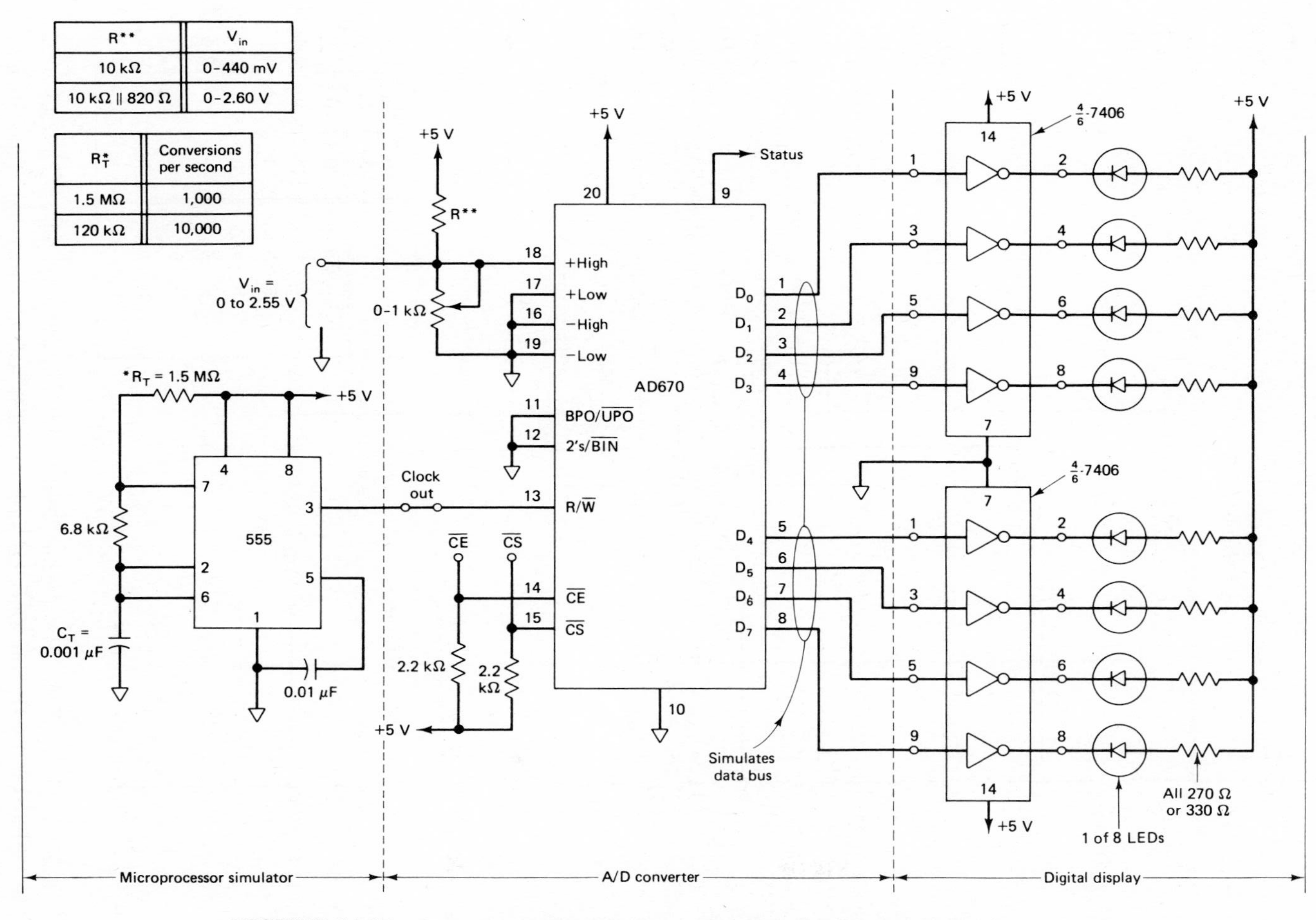

R**	V_{in}
10 kΩ	0–440 mV
10 kΩ ‖ 820 Ω	0–2.60 V

R_T^*	Conversions per second
1.5 MΩ	1,000
120 kΩ	10,000

FIGURE 14-17 Operation of the AD670 can be studied without the need for a microprocessor. Pins 14 and 15 can be grounded to simulate a microprocessor selection via an address bus. The 555 timer simulates continuous read/$\overline{write}$ commands from a microprocessor.

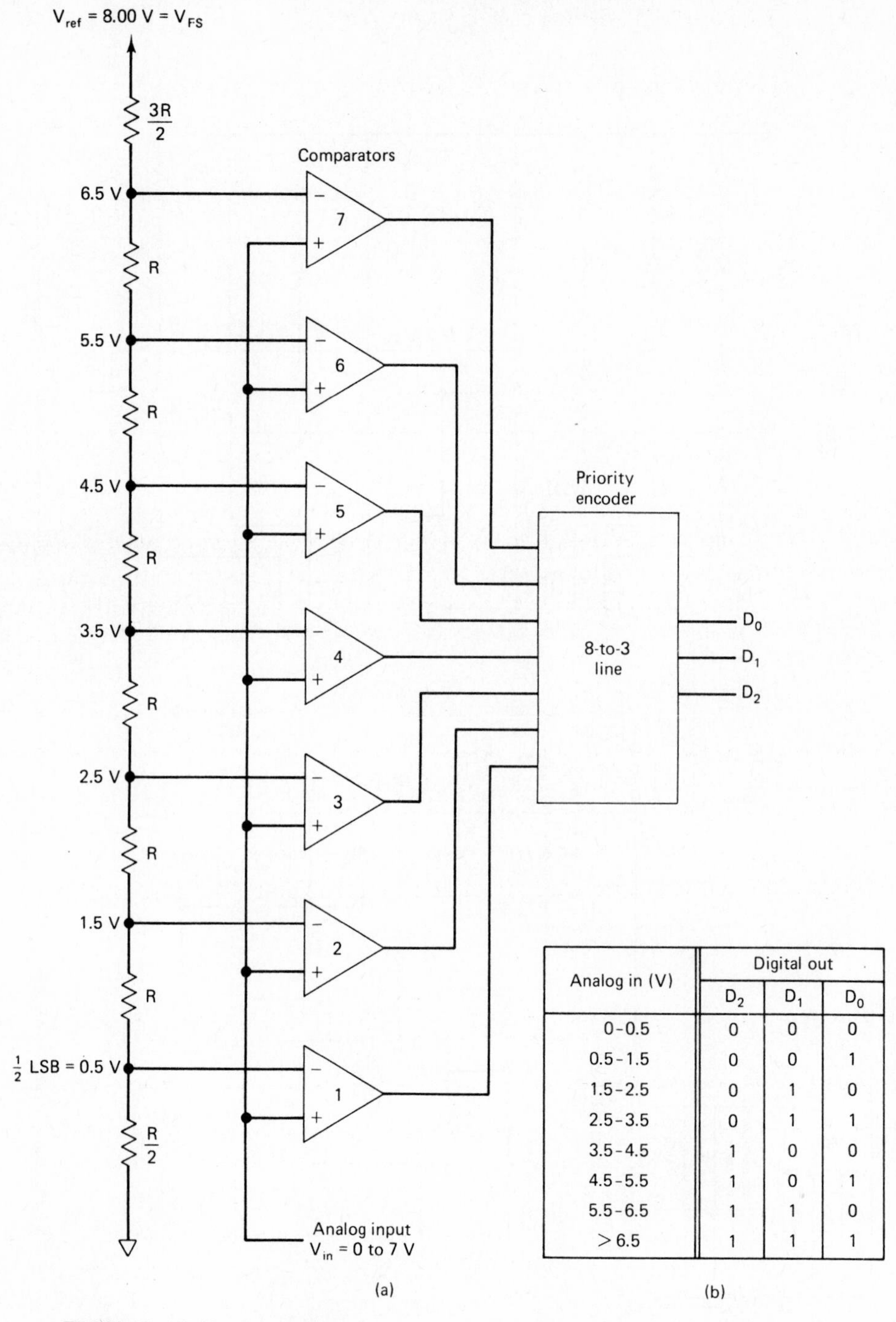

Analog in (V)	Digital out		
	D_2	D_1	D_0
0–0.5	0	0	0
0.5–1.5	0	0	1
1.5–2.5	0	1	0
2.5–3.5	0	1	1
3.5–4.5	1	0	0
4.5–5.5	1	0	1
5.5–6.5	1	1	0
>6.5	1	1	1

FIGURE 14-18 (a) Three-bit flash (parallel) A/D converter; (b) output versus input.

shows that the flash converter requires seven comparators (or $2^3 - 1$) to perform a 3-bit conversion. The number of comparators required for n-bit resolution is

$$\text{number of comparators} = 2^n - 1 \qquad (14\text{-}23)$$

For example, an 8-bit flash converter requires ($2^8 - 1$) or 255 comparators. Encoder logic would be more complex requiring a 256-line-to-8-line priority encoder.

14-15 FREQUENCY RESPONSE OF ADCs

14-15.1 Aperture Error

During conversion time, T_C, the analog input voltage must not change by more than $\pm\frac{1}{2}$ LSB (total 1 LSB), or the conversion will be incorrect. This type of inaccuracy is called *aperture error*. The rate of change of V_{in} with respect to time is called *slew rate*. If V_{in} is a sine wave, its slew rate is maximum at its zero crossings. The sine wave's slew rate is determined by both its peak voltage and frequency.

For an A/D converter, the maximum frequency for a sine wave V_{in} to be digitized within an accuracy of $\pm\frac{1}{2}$ LSB is

$$f_{max} \simeq \frac{1}{2\pi(T_C)2^n} \qquad (14\text{-}24)$$

Example 14-15

The AD670 is an 8-bit ADC with a conversion time of 10 μs. Find the maximum frequency of an input sine wave that can be digitized without aperture error.

Solution From Eq. (14-24),

$$f_{max} \simeq \frac{1}{2\pi(2^8)10\ \mu\text{s}} = \frac{1}{2\pi(256)10 \times 10^{-6}\ \text{s}} = 62\ \text{Hz}$$

Example 14-15 shows that the frequency response of even a fast ADC is surprisingly low. For a 10-bit integrating ADC with a conversion time of $\frac{1}{3}$ s, the highest sine frequency is about 0.5 mHz, or 1 cycle per 2000 s.

Summary. An 8-bit converter with a 10-μs conversion time can theoretically perform $[1/(10\ \mu\text{s})]\mu\text{s} = 100{,}000$ conversions per second. Yet the highest frequency sine wave that can be converted without slew-rate limiting is about 62 cycles per second. To raise the frequency response, we must add another circuit block, the sample-and-hold or follower-and-hold amplifier.

14-15.2 Sample-and-Hold Amplifier

The sample-and-hold (S/H) or follow-and-hold amplifier of Fig. 14-19 is made from two op amps, a hold capacitor (C_H), and a high-speed analog switch. This amplifier is connected between an analog input signal and the input to an ADC.

When the S/H amplifier is in the *sample* mode, the switch is closed and hold capacitor (C_H) voltage *follows* V_{in}. A *hold* command opens the switch and C_H retains a charge equal to V_{in} at the moment of switching. The S/H amplifier thus acts to hold V_{in} (stored on C_H) constant, while the ADC performs a conversion.

Conversion time of the ADC no longer limits frequency response. Instead, the limited is the *aperture time* of the S/H amplifier, which can be made much less than the conversion time. Aperture time is the time elapsed between a hold command and a switch opening. If the hold command is advanced by a time equal to the aperture time, C_H will hold the desired sample of V_{in}. Then the only remaining error is *aperture time uncertainty,* the switch jitter variation for each hold command.

Commercial S/H amplifiers have aperture time uncertainties lower than 50 ns. An example shows the improvement in frequency response due to an added S/H amplifier.

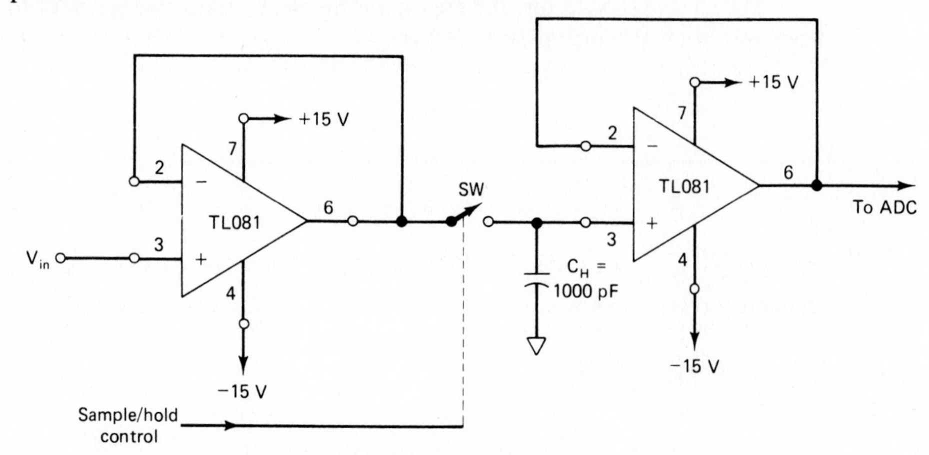

FIGURE 14-19 Sample-and-hold amplifier.

Example 14-16

A S/H amplifier with an aperture time uncertainty of 50 ns is connected to an 8-bit ADC. Find the highest-frequency sine wave that can be digitized within an error of 1 LSB.

Solution Replace conversion time by aperture uncertainty time in Eq. (14-24):

$$f_{max} \simeq \frac{1}{2\pi(2^8)50 \times 10^{-9}\text{ s}} = 12.4\text{ kHz}$$

LABORATORY EXERCISES

14-1. *DAC-08 Digital-to-Analog Converter*. The least expensive DAC circuit is that of Fig. 14-6. Use an OP-07 or TL081 op amp to minimize errors due to bias currents and input offset voltage. Ground all digital inputs with jumpers and check that V_o is close to ground potential. Change the jumper on LSB pin 12 to +5V for a digital input of 00000001. Check that V_o increases by about 39 mV. To measure resolution, jumper all inputs to +5 V for an input code of 11111111. Measure V_{oiFS}. Calculate resolution from Eq. (14-1b). The presence of glitches (Section 14-8.6) cannot be seen by these static tests. Use the dynamic test presented next.

14-2. *AD558 Digital-to-Analog Converter*. To dynamically test a DAC, use the circuit of Fig. 14-10. Measure resolution as follows: **(a)** Wire only the AD558 circuit of Fig. 14-10 and ground all the digital inputs. Measure V_o. **(b)** Remove the input grounds and jumper them to +5 V. Measure V_{oiFS}. **(c)** Calculate resolution from Eq. (14-1b). Remove the input jumpers.

Counter setup. Wire only the 555 circuit. Use an oscilloscope to measure the clock frequency at pin 3. Wire the two CD4029 (or any other 8-bit synchronous counter) to the 555. Use an oscilloscope to verify the square-wave frequency at pin 2 (D_7) of CD4029 #2. This pin outputs the most significant bit, MSB. The frequency at pin 2 is 1/256 of the clock frequency.

Staircase measurement. Wire the counter's outputs to the AD558's inputs as in Fig. 14-10. Connect a dual-trace oscilloscope (dc coupled) to display staircase generator as follows.

(a) Connect the external trigger of the oscilloscope to MSB pin 8 of the AD558. Set to trigger external on the negative edge.

(b) Connect channel 1 of the oscilloscope to measure V_o at pin 16 of the AD558. Zero the trace at the bottom of the oscilloscope screen. The vertical amplifier is set for 0.5 V/div dc-coupled. Time base = 5 ms/div.

(c) Connect channel 2 of the oscilloscope to pin 8 of the AD558 to monitor the MSB. Set the vertical sensitivity for 2 V/div and dc-coupled.

(d) Sketch the waveshape for V_o and also the MSB.

Increase the vertical amplifier gain of channel 1 until you can see the voltage increments for V_o. This value should correspond to 10 mV and is the resolution of change in V_o to a change of 1 bit. Save the AD558 circuit if you wish to operate it together with an ADC in part 14-4 of the laboratory exercise.

14-3. *Operating an AD670 without a Microprocessor*. Refer to the circuit of Fig. 14-17. Wire only the 555 circuit. Connect a CRO to measure the clock signal at pin 3 (time base = 5 μs/div). Measure the high time of the signal and low time. *Note:* When pin 3 goes low, you complete the selection process and initiate a conversion.

Digital displays. Wire up the 7406 inverter and LED circuitry. Jumper 5 V to each inverter input to see if the associated LED goes on. With *no* connections to the inputs, measure their open-circuit voltage. This value will tell you later when the AD670 outputs are in the high-impedance state.

Testing the AD670. Wire the AD670 to both 555 timers and display as in Fig. 14-17. Adjust V_{in} to 2.60 V.

1. Jumper a ground to both $\overline{\text{CS}}$ and $\overline{\text{CE}}$ to make the output buffer registers transparent. All LEDs should be lit. Remove the ground from either $\overline{\text{CS}}$ or $\overline{\text{CE}}$ to latch the

AD670. Ground AD670 pin 18 to make $V_{in} = 0$. The LEDs should remain lighted.

2. Replace the ground on $\overline{CS}$ or $\overline{CE}$ with V_{in} still zero. The AD670 converts and the LEDs should go off. Remove the ground from V_{in} and adjust V_{in} until the LEDs indicate 10000000. Measure V_{in} and divide it by 128 to obtain resolution in mV/bit.

Conversion time. Keep V_{in} set for an AD670 output of 10000000, with $\overline{CS}$ and $\overline{CE}$ grounded. Connect a dual-trace oscilloscope (with both channels set for 5 V/div and a time base = 5 μs/div) as follows:

1. Connect channel A to AD670's pin 13. On this signal's negative edge, a conversion begins.
2. Connect channel B to the status pin, pin 9. Measure the high time of status line. This is the conversion time.
3. Move channel B to monitor the MSB pin 8 and also D_4 at pin 5. Sketch all waveshapes and indicate when (a) AD670 outputs are in the high-impedance state; (b) AD670 outputs are transparent; (c) when a conversion begins; and (d) when a conversion ends.

14-4. *Communicating from an AD670 ADC to an AD558 DAC*. Remove the digital display circuitry from the AD670 of Fig. 14-17. Keep the 555 and AD670 circuitry. Set $V_{in} = 1.28$ V. Connect the digital outputs of the AD670 to the corresponding digital inputs of the AD558 (circuit saved from part 2). Basically, we are connecting the AD670 in Fig. 14-17 to the AD558 of Fig. 14-10. V_o of the DAC should equal (reasonable) V_{in} of the ADC. Remove the voltage-divider network at pin 18 of the AD670 (Fig. 14-17). For V_{in} connect a 0 to 2 V sine wave. Monitor V_{in} and V_o with a dual-trace oscilloscope to see that $V_{in} = V_o$.

PROBLEMS

Digital-to-Analog Converters

14-1. Give two definitions for a DAC's resolution.

14-2. Equation (14-2) is the output–input equation for a DAC. How do you evaluate its D term?

14-3. A 10-bit DAC has a resolution of 1 mV/bit. Find **(a)** the number of possible output voltages; **(b)** V_{oFS}.

14-4. What is the quantization error of the DAC in Problem 14-3?

14-5. Refer to the R–$2R$ ladder network of Fig. 14-5. Let $V_{ref} = 10$ V, $R_F = 5$ kΩ, $R = 5$ kΩ, and $2R = 10$ kΩ. Find **(a)** characteristic ladder resistance; **(b)** I_o; **(c)** voltage resolution; **(d)** output–input equation; **(e)** V_{oFS}.

14-6. An 8-bit DAC has a resolution of 5 mV/bit. Find **(a)** its output–input equation, **(b)** V_{oFS}; **(c)** V_o when the input is 10000000.

14-7. What is the voltage at pins 2 and 3 of the op amp in Fig. 14-7(a) when the digital inputs are 10000000? Also calculate the voltage across the feedback resistor to get V_o.

14-8. A basic DAC consists of a reference voltage, ladder network, current switches, and op amp. Name two additional features required to make the DAC microprocessor compatible.

14-9. Can the AD558 and DAC-08 be used as multiplying DACs?

14-10. These questions refer to the AD558. **(a)** Name the terminals that allow this DAC to be selected. **(b)** Describe the digital input register's latching mode and **(c)** transparent mode of operation.

14-11. **(a)** What is the output–input equation for an AD558?
(b) Find V_o for an input code of 10000000.

Analog-to-Digital Converters

14-12. Name three types of ADCs and indicate their relative conversion speeds (slow or fast).

14-13. $V_{in} = 50$ mV in the integrating ADC circuit of Fig. 14-11.
(a) What is the duration for integrating phase T_1 and the value of V_o?
(b) What is the name of phase T_2, the value of V_{ref}, and the duration of T_2?
(c) Find the circuit output.

14-14. Name the three components of a successive approximation 8-bit ADC.

14-15. A microprocessor issues a write command to an ADC. Does the ADC send data to the microprocessor or perform a conversion?

14-16. An input voltage with a range of 0 to 2.55 V is applied to pin 16 of an AD670 and pin 18 is grounded. Which other input pins should be jumpered or grounded to select this range?

14-17. How do you pin-program the AD670 for straight binary output?

14-18. How does a microprocessor tell an AD670 to **(a)** perform a conversion; **(b)** place the result on the data bus? **(c)** How does the microprocessor know when the AD670 has finished a conversion and its data are valid?

14-19. **(a)** What is the conversion time for an AD670?
(b) How many conversions can it perform per second?
(c) What is the maximum sine wave frequency that it can convert without adding a sample-and-hold amplifier?

14-20. If the sample-and-hold amplifier of an 8-bit ADC has an aperture uncertainty time of 10 ns, what maximum sine wave frequency can it convert within $\pm\frac{1}{2}$ LSB?

14-21. How many comparators are required to make an 8-bit flash converter?

CHAPTER 15

Power Supplies

LEARNING OBJECTIVES

Upon completion of this chapter on power supplies, you will be able to:

- Draw the schematic for a full-wave bridge (FWB) rectifier unregulated power supply.
- Identify the components of an FWB, tell what each component does in the circuit.
- Design an FWB rectifier; choose the specifications for the transformer, diodes, and capacitor; purchase these components from standard stock; build the rectifier; test it; and document its performance.
- Measure the percent regulation and percent ripple, draw the load voltage waveshapes at no load or full load, and plot the regulation curve for an FWB unregulated power supply.
- Design or analyze a bipolar or two-value unregulated power supply.
- Explain the need for voltage regulators.

- Connect an IC voltage regulator to an unregulated FWB rectifier circuit to make a voltage-regulated power supply.
- Design, build, and test a $\pm 15\text{-}V$ regulated power supply for analog ICs.
- Build a regulated 5-V supply for TTL logic.
- Connect an LM317 to an unregulated supply to obtain a laboratory-type voltage regulator that can be adjusted precisely to a required voltage.

15-0 INTRODUCTION

Most electronic devices require dc voltages to operate. Batteries are useful in low-power or portable devices, but operating time is limited unless the batteries are recharged or replaced. The most readily available source of power is the 60-Hz 110-V ac wall outlet. The circuit that converts this ac voltage to a dc voltage is called a *dc power supply*.

The most economical dc power supply is some type of rectifier circuit. Unfortunately, some ac ripple voltage rides on the dc voltage, so the rectifier circuit does not deliver pure dc. An equally undesirable characteristic is a reduction in dc voltage as more load current is drawn from the supply. Since dc voltage is *not* regulated (that is, constant with changing load current), this type of power supply is classified as *unregulated*. Unregulated power supplies are introduced in Sections 15-1 and 15-2. It is necessary to know their limitations before such limitations can be minimized or overcome by adding regulation. It is also necessary to build an unregulated supply before you connect a voltage regulator to it.

Without a good regulated voltage supply, none of the circuits in this text (or any other text for that matter) will work. Therefore, this chapter shows the simplest way to analyze or design power supplies for linear or digital ICs.

It is possible to make a good voltage regulator with an op amp plus a zener diode, resistors, and a few transistors. However, it is wiser to use a modern integrated circuit voltage regulator. The types of superb regulators are so vast there is no problem in finding one that will suit your needs.

We will present an op amp regulator to illustrate the workings of a few of the features within an IC regulator. Then we will proceed to a representative sampling of some of the widely used IC voltage regulators. But we begin with the unregulated supply.

15-1 INTRODUCTION TO THE UNREGULATED POWER SUPPLY

15-1.1 Power Transformer

A transformer is required for reducing the nominally 115-V ac wall outlet voltage to the lower ac value required by transistors, ICs, and other electronic devices. Transformer voltages are given in terms of rms values. In Fig. 15-1, the transformer is

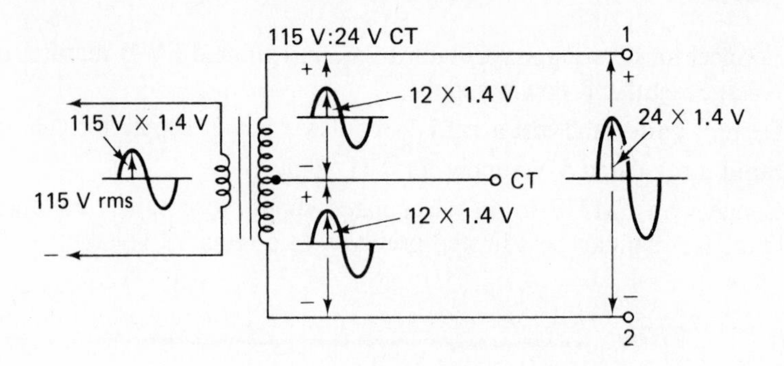

(a) Peak voltages for the positive half-cycle

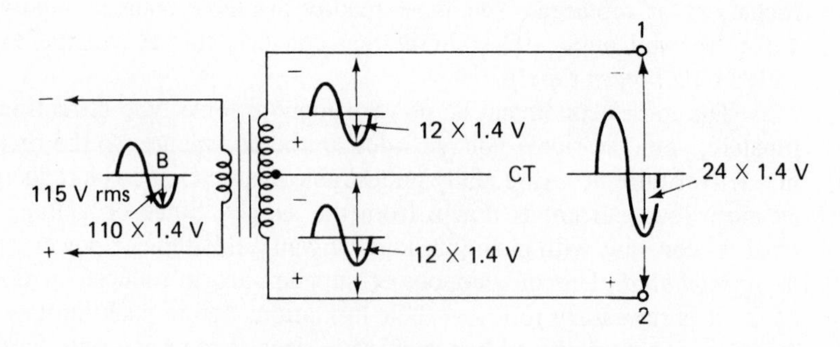

(b) Peak voltages for the negative half-cycle

FIGURE 15-1 Power transformer 115 V/24 VCT.

rated as 115 to 24 V center tap. With the 115-V rms connected to the primary, 24 V rms is developed between secondary terminals 1 and 2. A third lead, brought out from the center of the secondary, is called a center tap, CT. Between terminals CT and 1 or CT and 2, the rms voltage is 12 V.

An oscilloscope would show the sinusoidal voltages shown in Fig. 15-1. The maximum instantaneous voltage E_m is related to the rms value E_{rms} by

$$E_m = 1.4(E_{rms}) \tag{15-1}$$

In Fig. 15-1(a), voltage polarities are shown for the positive primary half-cycle; those for the negative half-cycle are shown in Fig. 15-1(b).

Example 15-1

Find E_m in Fig. 15-1 between terminals 1 and 2.

Solution By Eq. (15-1), $E_m = 1.4(24 \text{ V}) = 34 \text{ V}$.

15-1.2 Rectifier Diodes

The next step in building a dc power supply is to convert the lower secondary ac voltage of the transformer to a pulsated dc voltage. This is accomplished by silicon diodes.

In Fig. 15-2(a), four diodes are arranged in a diamond configuration called a full-wave bridge rectifier. They are connected to terminals 1 and 2 in the transformer of Fig. 15-1. When terminal 1 is positive with respect to terminal 2, diodes D_1 and D_2 conduct. When terminal 2 is positive with respect to terminal 1, diodes D_3 and D_4 conduct. The result is a pulsating dc voltage between the output terminals.

15-1.3 Positive versus Negative Supplies

Note that the bridge has two input terminals labeled ac. The output terminals are labeled (+) and (−), respectively. Also note that the output dc voltage *cannot* as yet

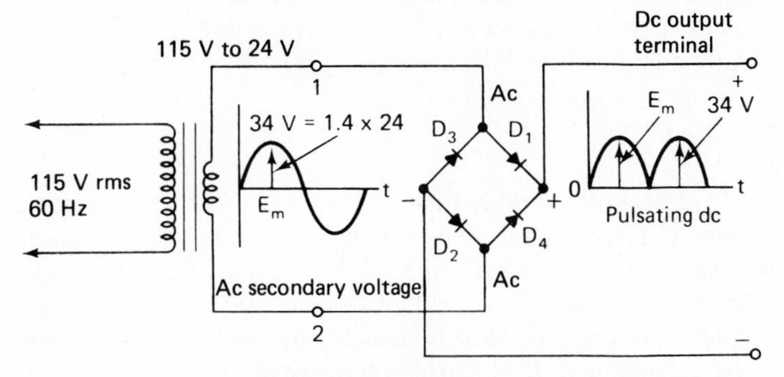

(a) Transformer and four diodes reduce 162 V peak ac primary voltage to 34 V peak pulsating dc

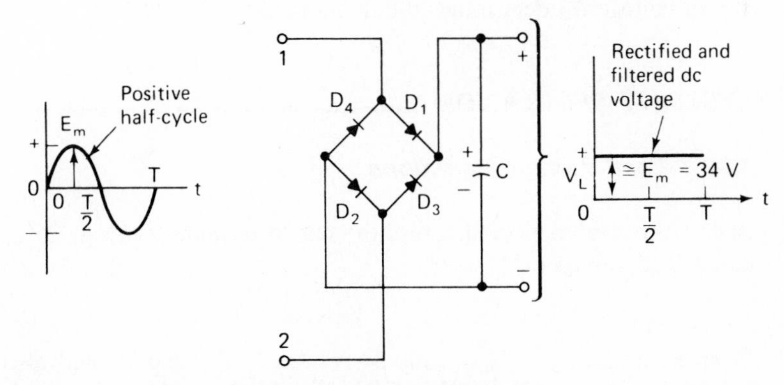

(b) Capacitor C filters the pulsating dc in (a) to give a dc load voltage

FIGURE 15-2 Transformer plus rectifier diodes plus filter capacitor equals unregulated power supply.

be designated positive or negative. It is a *floating* supply. If you want a "positive" supply, you must "earth ground" the negative terminal.

The third (green) wire of the line cord extends earth ground from the "U"-shaped terminal of the wall outlet usually to the metal chassis. This connection is to protect the user. Simply extend the green wire terminal to the negative terminal and call this terminal *power supply common.* All voltage measurements are with respect to power supply common and it is designated in a schematic by a ground symbol. To make a negative supply, simply earth ground the positive terminal of the bridge.

15-1.4 Filter Capacitor

The pulsating dc voltage in Fig. 15-2(a) is not pure dc, so a filter capacitor is placed across the dc output terminals of the bridge rectifier [see Fig. 15-2(b)]. This capacitor smooths out the dc pulsations and gives an almost pure dc output load voltage, V_L. V_L is the unregulated voltage that supplies power to the load. The filter capacitor is typically a large electrolytic capacitor, 500 μF or more.

15-1.5 Load

In Fig. 15-2(b), nothing other than the filter capacitor is connected across the dc output terminals. The unregulated power supply is said to have no load. This means that the *no-load current,* or 0-load current, I_L, is drawn from the output terminals. Usually, the maximum expected load current, or full-load current, to be furnished by the supply is known. The load is modeled by resistor R_L as shown in Fig. 15-3(a). As stated in Section 15-0, the load voltage changes as the load current changes in an unregulated power supply. The manner in which this occurs is examined next. But the key idea to power supply analysis now becomes clear. The peak value of secondary ac voltage E_m determines the dc no-load voltage of V_L.

15-2 DC VOLTAGE REGULATION

15-2.1 Load Voltage Variations

A dc voltmeter connected across the output terminals in Fig. 15-2(b) measures the dc no-load voltage of V_L, or

$$V_{\text{dc no load}} = E_m \tag{15-2}$$

From Example 15-1, $V_{\text{dc no load}}$ is 34 V. An oscilloscope would also show the same value with no ac ripple voltage, as in Fig. 15-3(b). Now suppose that a load R_L was connected to draw a full-load dc current of $I_L = 1$ A, as in Fig. 15-3(a). An oscilloscope now shows that the load voltage V_L has a lower *average,* or dc value V_{dc}. Moreover, the load voltage has an ac ripple component, ΔV_o, superimposed on the

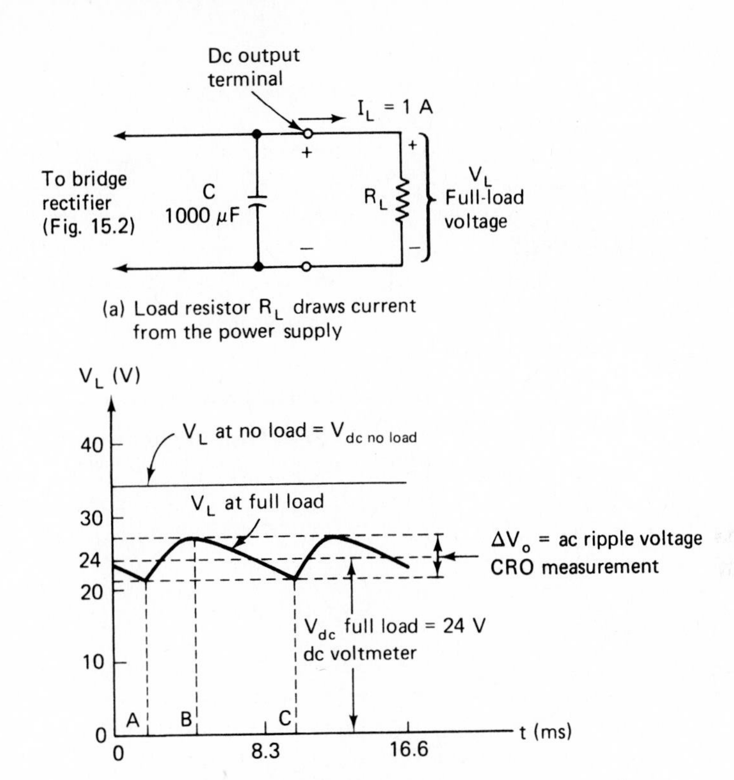

(a) Load resistor R_L draws current from the power supply

(b) Load voltage changes from 34 V at no load to 24 V plus ripple at full load

FIGURE 15-3 Variation of dc load voltage and ac ripple voltage from no-load current to full-load current.

dc value. The average value measured by a dc voltmeter is 24 V and is called V_{dc} full load. The peak-to-peak ripple voltage is called ΔV_o and measures 5 V in Fig. 15-3(b).

There are two conclusions to be drawn from Fig. 15-3(b). First, the dc load voltage goes *down* as dc load current goes *up*; how much the load voltage drops can be estimated by a technique explained in Section 15-2.2. Second, the ac ripple voltage increases from 0 V at no-load current to a large value at full-load current. As a matter of fact, the ac ripple voltage increases directly with an increase in load current. The amount of ripple voltage can also be estimated, by a technique explained in Section 15-3.

15-2.2 DC Voltage Regulation Curve

In the unregulated power supply circuit Fig. 15-4(a), the load R_L is varied so that we can record corresponding values of dc load current and dc load voltage. The dc meters respond only to the *average* (dc) load current or voltage. If corresponding values

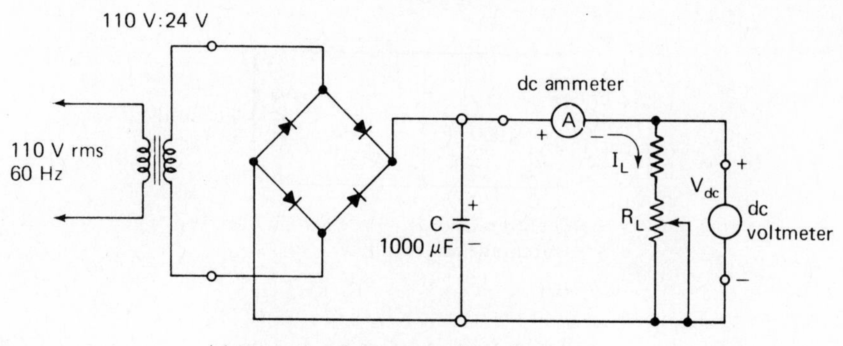

(a) Unregulated power supply performance measured with dc ammeter and voltmeter

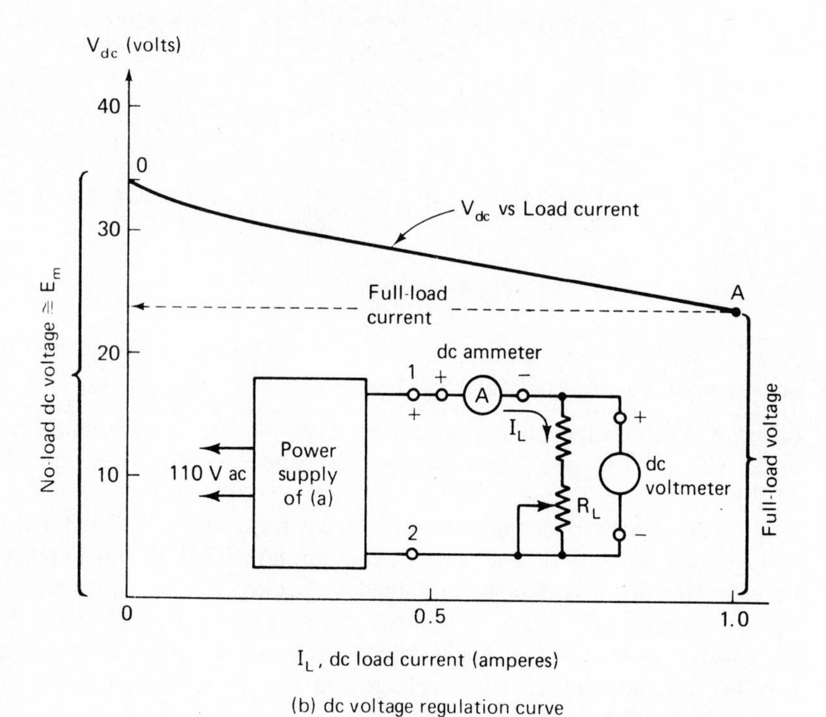

(b) dc voltage regulation curve

FIGURE 15-4 DC load voltage varies with load current in (a) as shown by the voltage regulation curve in (b).

of current and voltage are plotted, the result is the *dc voltage regulation curve* of Fig. 15-4(b). For example, point 0 represents the no-load condition, $I_L = 0$ and $V_{\text{dc no load}} = E_m = 34$ V. Point *A* represents the full-load condition, $I_L = 1$ A and $V_{\text{dc full load}} = 24$ V.

15-2.3 DC Model of a Power Supply

Figure 15-5(a) shows results obtained when making measurements of (a) no-load voltage with no-load current and (b) full-load voltage at full-load current. As you draw *more* dc current from the supply, something inside the supply causes an *increasing* internal voltage drop to leave less voltage available across the load. The simplest way to account for this behavior is to blame it on an *internal* or *output resistance* R_o. Accordingly, this dc behavior can be described by

$$V_{\text{dcNL}} = V_{\text{dcFL}} + I_{\text{LFL}} R_o \tag{15-3a}$$

or

$$V_{\text{dcFL}} = \frac{R_L}{R_o + R_L}(V_{\text{dcNL}}) \tag{15-3b}$$

We use the data of both the no-load and full-load measurements in Fig. 15-5 and the voltage regulation curve of Fig. 15-4 to measure R_o.

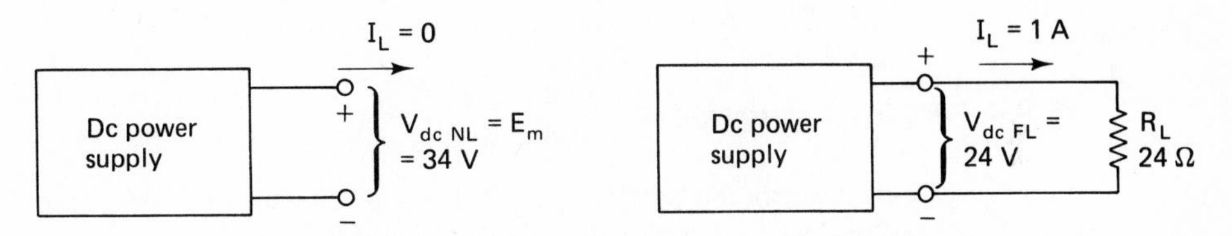

(a) No-load and full-load dc voltages of a power supply

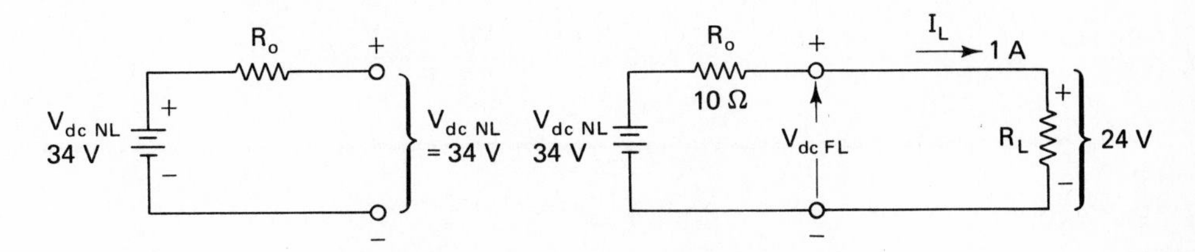

(b) Dc circuit model to explain the measurements made to plot the dc voltage regulator curve of Fig. 15.4

FIGURE 15-5 A dc model is developed to explain the dc voltage measurements for an unregulated power supply. $V_{\text{dcNL}} = V_{\text{dcFL}} + I_L R_o$.

Example 15-2

(a) Calculate output resistance R_o from the no-load and full-load measurements of Figs. 15-4 and 15-5. (b) Predict dc output voltage at half-load where $I_L = 0.5$ A.

Solution (a) At no load, $I_L = 0$ and $V_{dcNL} = 34$ V; at full load, $I_L = 1$ A and $V_{dcFL} = 24$ V. From Eq. (15-3a),

$$R_o = \frac{V_{dcNL} - V_{dcFL}}{I_{LFL}} = \frac{(34 - 24)\text{ V}}{1\text{ A}} = 10\ \Omega$$

(b) Use Eq. (15-3a) again, but modify it for $I_L = 0.5$ A.

$$V_{dc} = V_{dcNL} - I_L R_o = 34\text{ V} - (0.5\text{ A})(10\ \Omega)$$
$$= 34\text{ V} - 5\text{ V} = 29\text{ V}$$

Note that the 5 V appears *internally* within the power supply to cause 5 V × 0.5 A = 2.5 W of heat.

R_o models the net effect of internal losses within the power supply. These losses occur because of the transformer, diodes, capacitor, and even wires going to the wall outlet. It is pointless to track down the contributions of each one. We are only interested in their net effect R_o.

15-2.4 Percent Regulation

Another way to describe dc performance is by a specification called percent regulation. You measure the supply's no-load voltage and full-load voltage. Percent regulation is then calculated from

$$\%\text{ regulation} = \frac{V_{dcNL} - V_{dcFL}}{V_{dcFL}} \times 100 \qquad (15\text{-}4)$$

Example 15-3

Find percent regulation for the dc power supply data in Figs. 15-4 and 15-5.

Solution From the data, $V_{dcNL} = 34$ V and $V_{dcFL} = 24$ V. From Eq. (15-4),

$$\%\text{ regulation} = \frac{(34 - 24)\text{ V}}{24\text{ V}} \times 100 = 41.7\%$$

Percent regulation tells you by *what percent the full-load voltage will rise* when you remove the load.

15-3 AC RIPPLE VOLTAGE

15-3.1 Predicting AC Ripple Voltage

Figure 15-6(b) shows how to measure both ac and dc performance of a power supply. Dc measurements (average values) are made with dc meters. Their measured dc values, I_L and V_{dc}, are summarized and plotted as the (dc) voltage regulation curve of Fig. 15-4(b).

The peak-to-peak ac ripple voltage ΔV_o is centered on V_{dc}. ΔV_o can be estimated from

$$\Delta V_o \simeq \frac{I_L}{200C} \tag{15-5a}$$

where ΔV_o is in volts, I_L in amperes, and C is the size of the filter capacitor in farads. If load voltage V_L is measured with an ordinary ac voltmeter, it will indicate the rms value of the ripple voltage V_{rms}. A coupling capacitor within the meter eliminates the dc component. V_{rms} is related to ΔV_o by the approximation

$$\Delta V_o = 3.5\ V_{rms} \tag{15-5b}$$

We need one other characteristic of the power supply that will be used later in the chapter. It tells us how to design a power supply for a voltage regulator. It is called minimum instantaneous load voltage and it occurs at full load. As seen in Fig. 15-6(b),

$$\text{minimum } V_L = V_{dcFL} - \frac{\Delta V_o}{2} \tag{15-5c}$$

These principles will be illustrated by an example.

Example 15-4

A full-wave bridge rectifier has (1) a full-load current of 1 A, (2) a full-load voltage of 24 V, and (3) a filter capacitor of 1000 μF. Calculate (a) the peak-to-peak and rms value of ripple voltage at full load; (b) minimum instantaneous output voltage.

Solution (a) From Eq. (15-5a),

$$\Delta V_o = \frac{1\text{ A}}{(200)(1000 \times 10^{-6}\text{ F})} = 5\text{ V}$$

From Eq. (15-5b),

$$V_{\text{rms}} = \frac{\Delta V_o}{3.5} = \frac{5\text{ V}}{3.5} = 1.43\text{ V}$$

(b) From Eq. (15-5c),

$$\text{minimum } V_{\text{L}} = V_{\text{dcFL}} - \frac{\Delta V_o}{2} = 24\text{ V} - \frac{5\text{ V}}{2} = 21.5\text{ V}$$

15-3.2 Ripple Voltage Frequency and Percent Ripple

The ripple voltage frequency of a full-wave rectifier with capacitor filter is 120 Hz, or twice the line voltage frequency. This is because the capacitor must charge and discharge twice for each cycle of line voltage. As shown in Fig. 15-6(a), the period for each cycle of ripple voltage is 8.3 ms.

The ac performance of a power supply can also be specified by a single percentage number. You measure the worst-case ripple voltage V_{rms}. This occurs at full-load current [see Fig. 15-6(a)]. Measure V_{dcFL} and calculate *percent ripple* from

$$\%\text{ ripple} = \frac{V_{\text{rms}}\text{ at full load}}{V_{\text{dcFL}}} \times 100 \tag{15-6}$$

Example 15-5

Calculate the percent ripple for the power supply specified in Example 15-4.

Solution From Example 15-4, $V_{\text{dcFL}} = 24$ V and $V_{\text{rms}} = 1.43$ V. From Eq. (15-6),

$$\%\text{ ripple} = \frac{1.43}{24} \times 100 = 6\%$$

15-3.3 Controlling Ripple Voltage

Equation (15-5a) tells us that ΔV_o depends directly on load current. Therefore, for the same-value filter capacitor, if the load current doubles, the ripple voltage doubles. Equation (15-5a) also shows that ΔV_o is inversely proportional to C. If C dou-

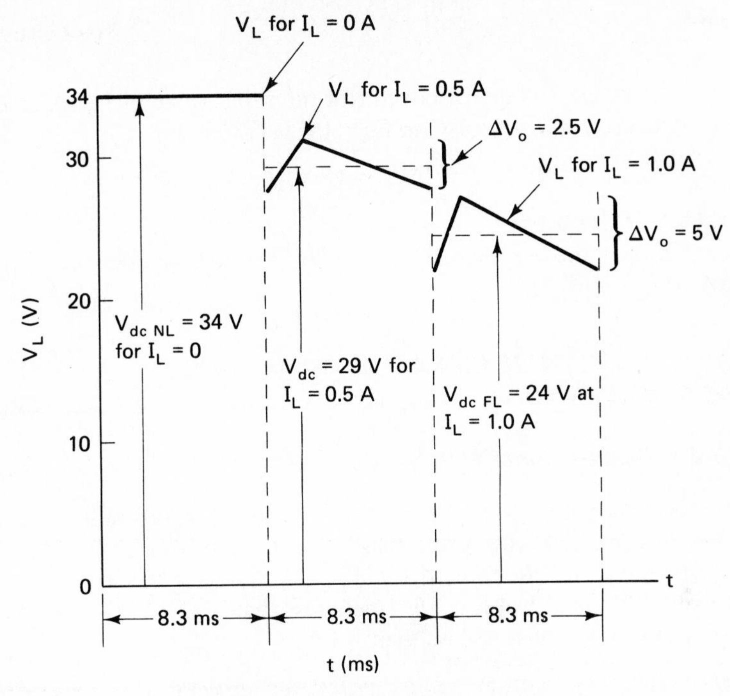

(a) A dc-coupled CRO shows both ac and dc components of the load voltage V_L

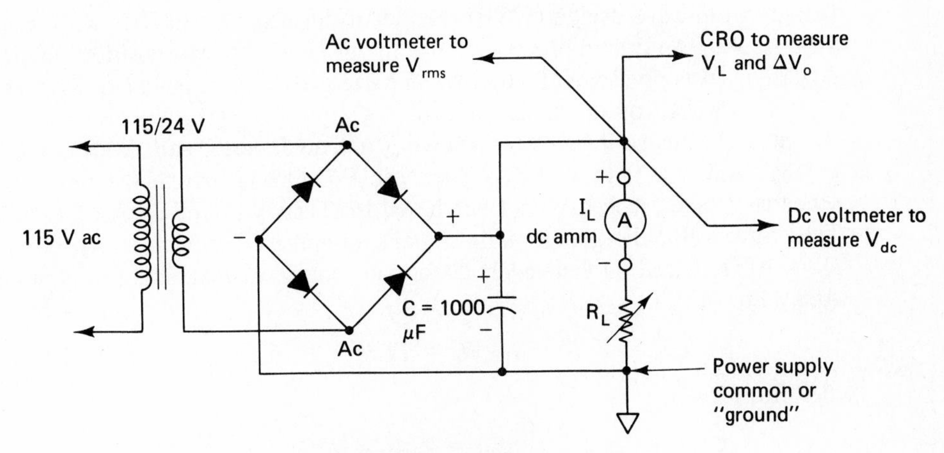

(b) Test circuit to measure ac and dc performance of an unregulated power supply

FIGURE 15-6 This test circuit setup allows simultaneous measurement of both ac and dc performance of an unregulated power supply. A direct-coupled CRO measures instantaneous load voltage V_L and peak-to-peak ripple voltage L_o. Dc meters measure average (dc) load voltage V_{dc} and current I_L. An ac voltmeter measures the rms value of the ripple voltage V_{rms}.

bles, the ripple voltage is halved (for the same value of I_L). A nice rule of thumb is to substitute 1000 μF for C in Eq. (15-5a) to obtain

$$\Delta V_o = (5 \text{ V})(I_L \text{ in amperes}) \tag{15-7}$$

when $C = 1000\ \mu$F.

If $I_L = 0.5$ A, then $\Delta V_o = 2.5$ V. Double C to 2000 μF if you want to reduce ΔV_o to $2.5\text{ V}/2 = 1.25$ V.

15-4 DESIGN PROCEDURE FOR A FULL-WAVE BRIDGE UNREGULATED SUPPLY

15-4.1 Design Specification, General

The dc requirement for a power supply is specified typically as "I need 12 volts at 1 ampere." We now can define these values as (1) $V_{dcFL} = 12$ V and (2) $I_{LFL} = 1$ A. (Incidentally, R_L for full load is $12\text{ V}/1\text{ A} = 12\ \Omega$.) The ac requirement can be specified in many ways, (1) % ripple less than 5%, (2) V_L minimum = 11 V, or (3) ΔV_o at full load must be less than 2.1 V.

Design Example 15-6

Design a full-wave bridge (FWB) rectifier to furnish 12 V at 1 A with less than 10% ripple. The design procedures use the principle of superposition. (a) Do the dc design. This gives you the transformer size and diode ratings. (b) Do the ac design. This gives you the capacitor size.

(a) **Dc design procedure showing a transformer and diodes.** Use Eqs. (15-3a) and (15-2) to find E_m. Then use Eq. (15-1) to rate the transformer's secondary. Since you have not purchased the parts yet, you cannot measure R_o. Therefore, as the designer, *you* must make an *informed estimate* of R_o. Guess that $R_o = 10\ \Omega$. Based on your *guess* (informed estimate), the rest is straightforward. Apply Eq. (15-3a),

$$V_{dcNL} = V_{dcFL} + I_{LFL}R_o = 12\text{ V} + (1\text{ A})(10\ \Omega) = 22\text{ V}$$

then Eq. (15-2),

$$E_m = V_{dcNL} = 22\text{ V}$$

and finally, Eq. (15-1),

$$E_{rms} = \frac{E_m}{1.4} = 22\text{ V}/1.4 = 15.5\text{ V}$$

Your problem appears to worsen because you cannot buy a 115/15.5 V transformer from stock. So we learn the first basic lesson of power supply design. *You usually cannot get the dc voltage you want*. Even if you could get 12 V at

1 A, V_L would be 22 V at 0 A. The voltage regulator was invented to solve this problem. However, your company stocks 115 V/12.6 VCT transformers with two secondaries. Connect one 12.6 V winding in series aiding with a center tap and proper terminal of the other winding (6.3 V) to obtain $E_s = 6.3\text{ V} + 12.6\text{ V} = 18.9\text{ V}$.

(You learned also that this is no way to power digital, linear, or automobile electronics, but it will give a fine opportunity to *analyze* a design later, in Example 15-7.) All that remains is to choose the transformers's ac current rating as follows: For a FWB rectifier, choose $I_{secrms} = 1.8 I_{LFL}$. Therefore, choose $I_{secondary} = 2$ A. *Choosing the bridge diodes.* Diode selection is easy. They are rated by average current, I_{av}, and peak inverse voltage, PIV. A prudent design guide is given as follows:

1. Choose a diode with $I_{av} \geq I_{LFL} = 1$ A.
2. Choose a diode with a PIV rating greater than $V_{dcNL} + 20\%\ V_{dcNL}$.

(b) **AC design procedure: choosing filter capacitor.** The original specifications required less than 10% ripple at $V_{dcFL} = 12$ V. Even knowing that we *may not* realize this dc value, we proceed on the original specification. (You may get lucky and find an R_o of 15 Ω for the 115 V/18 V transformer that you purchase.)

First, calculate ΔV_o as follows. Apply Eq. (15-6),

$$V_{rms} = \frac{(\%\text{ ripple})V_{dcNL}}{100} = \frac{10(12\text{ V})}{100} = 1.2\text{ V}$$

then Eq. (15-5b),

$$\Delta V_o = 3.5(1.2\text{ V}) = 4.2\text{ V}$$

and finally, Eq. (15-5a),

$$C = \frac{I_L}{200\ \Delta V_o} = \frac{1\text{ A}}{(200)(4.2\text{ V})} = 1190\ \mu\text{F}$$

Choose C greater than 1190 μF or $C = 1500$ μF. If necessary, you can construct C from three 500-μF caps in parallel.

Note: Electrolytic capacitors also have a voltage rating called "working volts dc" (WVDC). The filter capacitor should be rated the same as a diode:

$$\text{WVDC} = V_{dcNL} + 20\%\ V_{dcNL}$$

Design Summary We end up with an FWB rectifier that has the following parts:

1. Transformer: 115 V/18 V at 2.0 A
2. Four diodes: each 1 A at PIV = 25 V (or more)
3. Three capacitors: 500 μF connected in parallel, WVDC = 25 V (or more)

We do not have a satisfactory design, but that is often the nature of practical solutions for unregulated power supplies. However, let us use this design to analyze and predict performance of an FWB rectifier.

Analysis Example 15-7: FWB Unregulated Supply

Given the power supply design schematic of Fig. 15-7(a), (a) predict its dc performance by plotting a dc regulation curve and calculating percent regulation. *Assume*: $R_o = 7\ \Omega$, $I_{LFL} = 1.0$ A; (b) plot the no-load and full-load instantaneous voltage V_L that you would expect to see on a CRO. Also calculate percent ripple.

Solution (a) Since $E_{rms} = 18\ V_{rms}$, calculate E_m and V_{dcNL} from Eqs. (15-1) and (15-2):

$$V_{dcNL} = E_m = 1.4E_{rms} = 25.5\text{ V}$$

Calculate V_{dcFL} from Eq. (15-3a):

$$V_{dcFL} = V_{dcNL} - I_{LFL}R_o = 25.5\text{ V} - (1\text{ A})(7\ \Omega) = 18.5\text{ V}$$

and percent regulation from Eq. (15-4):

$$\%\text{ regulation} = \frac{V_{dcNL} - V_{dcFL}}{V_{dcFL}} \times 100 = \frac{(25.5 - 18.5)\text{ V}}{18.5} \times 100 \simeq 38\%$$

(b) Calculate ΔV_o from Eq. (15-5a) and V_{rms} from Eq. (15-5b):

$$\Delta V_o \simeq \frac{I_L}{200C} = \frac{1\text{ A}}{200(1000 \times 10^{-6}\text{ F})} = 5\text{ V}$$

$$V_{rms} = \frac{\Delta V_o}{3.5} = \frac{5\text{ V}}{3.5} = 1.43\text{ V}$$

Finally, calculate percent ripple from Eq. (15-6):

$$\%\text{ ripple} = \frac{V_{rms}}{V_{dcFL}} 100 = \frac{1.43}{18.5} \times 100 = 7.7\%$$

Dc performance of the unregulated power supply is depicted in Fig. 15-7(b). Ac performance is summarized by the time plots at both no load and full load in Fig. 15-7(c). Note that $\Delta V_o = 0$ V at no-load current. Also, ΔV_o is (approximately) centered on V_{dcFL}.

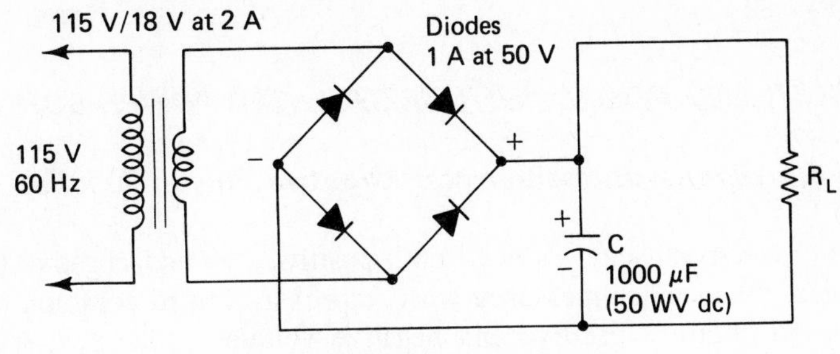

(a) Circuit for analysis example 14.6

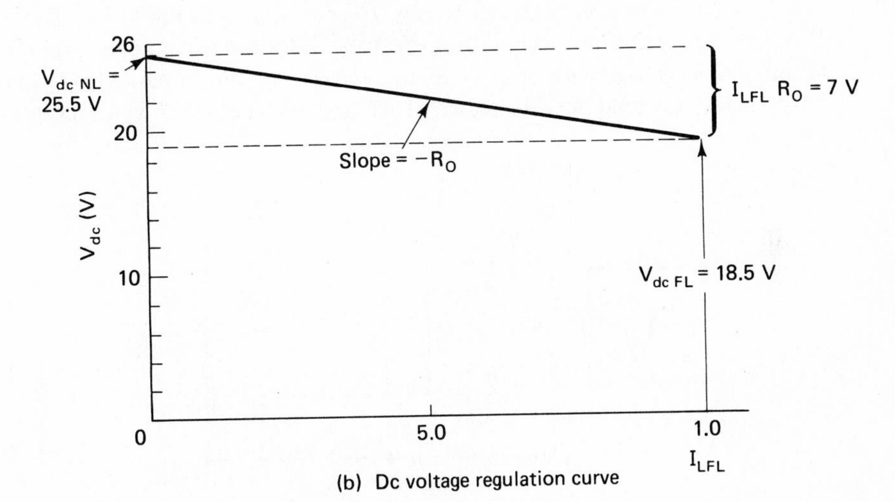

(b) Dc voltage regulation curve

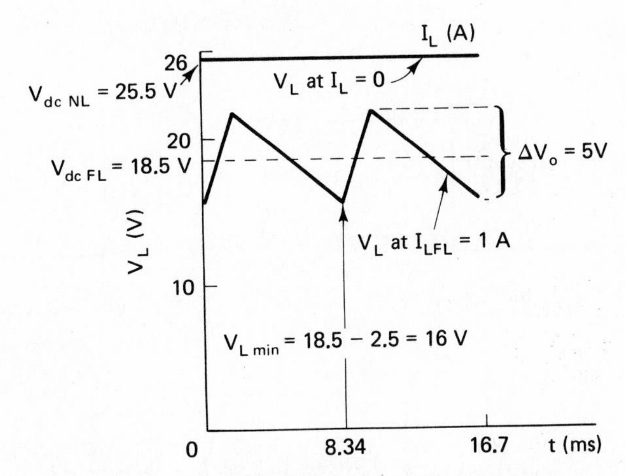

(c) No-load and full-load instantaneous voltage

FIGURE 15-7 The predicted dc performance for the filtered FWB rectifier in (a) is shown in (b) and its ac performance in (c).

15-5 BIPOLAR AND TWO-VALUE UNREGULATED POWER SUPPLIES

15-5.1 Bipolar or Positive and Negative Power Supplies

Many electronic devices need both positive (+) and negative (−) supply voltages. These voltages are measured with respect to a third common (or grounded) terminal. To obtain a positive and negative voltage, either two secondary transformer windings or one center-tapped secondary winding is needed.

A transformer rated at 115 V : 24 V CT is shown in Fig. 15-8. Diodes D_1 and D_2 make terminal 1 positive with respect to center tap CT. Diodes D_3 and D_4 make terminal 2 negative with respect to the center tap. From Eq. (15-1) and Section 15-2.1, both no-load dc voltages are 1.41 × 12 V rms = 17 V. Capacitors $C+$ and

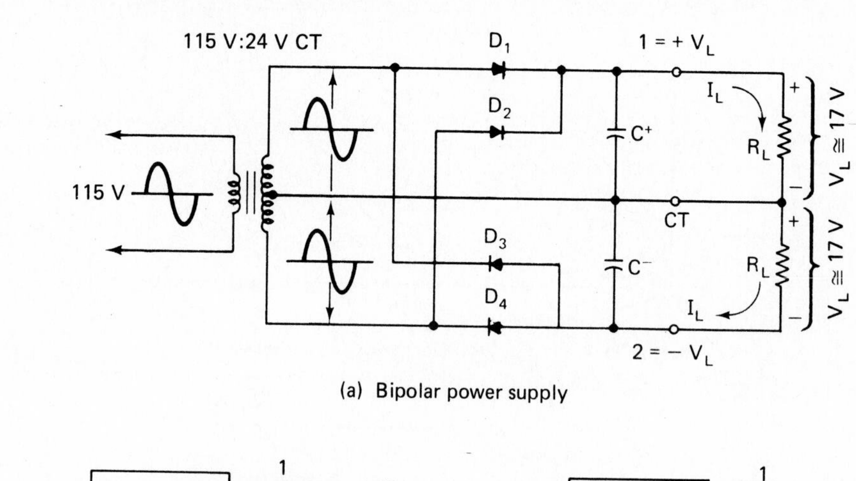

(a) Bipolar power supply

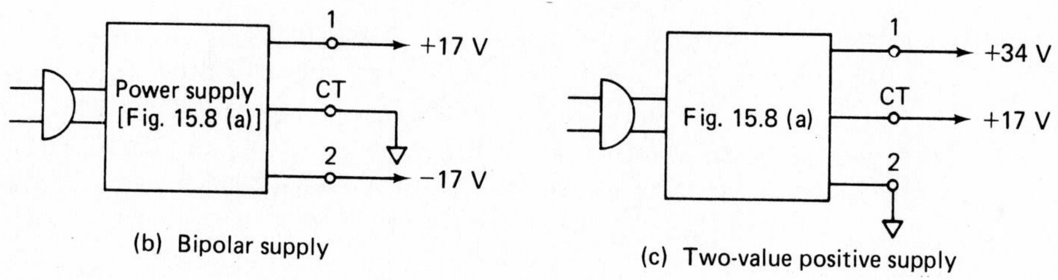

(b) Bipolar supply

(c) Two-value positive supply

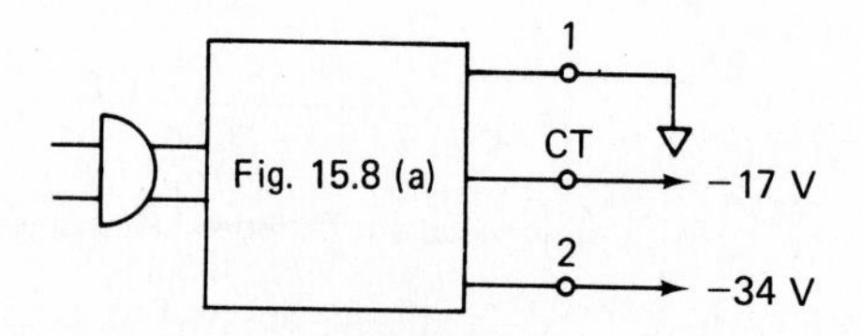

(d) Two-value negative supply

FIGURE 15-8 Bipolar and two-value power supplies.

$C-$, respectively, filter the positive and negative supply voltages. As shown in Sections 15-4 and 15-5, the ac ripple voltage and dc voltage regulation may be predicted for both no-load and full-load voltages.

15-5.2 Two-Value Power Supplies

If the center tap of the power supply of Fig. 15-8 is grounded, we have a *bipolar* power supply. It is shown schematically in Fig. 15-8(b). If terminal 2 is grounded as in Fig. 15-8(c), we have a two-value positive supply. Finally, by grounding terminal 1 in Fig. 15-8(d), we get a two-value negative power supply. This indicates the versatility of the center-tapped transformer.

15-6 NEED FOR VOLTAGE REGULATION

Previous sections have shown that the unregulated power supply has two undesirable characteristics: the dc voltage decreases and the ac ripple voltage increases as load current increases. Both disadvantages can be minimized by adding a *voltage-regulator section* to the *unregulated supply* as in Fig. 15-9. The resulting power supply is classified as a *voltage-regulated supply*.

15-7 THE HISTORY OF LINEAR VOLTAGE REGULATORS

15-7.1 The First Generation

An excellent dc voltage regulator can be built from an op amp, zener diode, two resistors or one potentiometer, and one or more transistors. In 1968, Fairchild Semiconductor Division integrated all of these components (plus others) into a single IC and called it the μA723 monolithic voltage regulator. Because of its flexibility, it has survived to the present day. It does, however, require a number of support components, has minimal internal protection circuitry, and requires the user to add boost transistors for more current capability and a resistor for limiting short-circuit current.

The race was on to make a three-terminal fixed-voltage regulator. National Semiconductor won with the LM309, in a close finish with Fairchild's 7800 series. The LM309 and μA7805 have three terminals. To use one, all you have to do is connect an unregulated supply between its input and common terminals. Then connect a load between the output and common and the design is complete. (Connect a decoupling capacitor across both input terminals and output terminals to improve performance.) These devices have internal protection circuitry that will be discussed later.

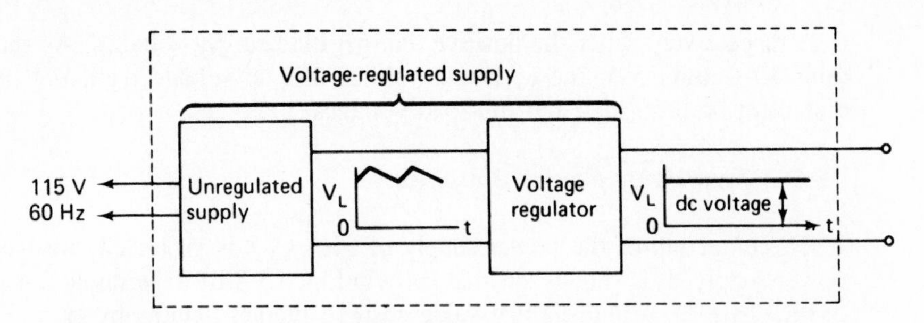

FIGURE 15-9 Unregulated supply plus a voltage regulator gives a voltage-regulated power supply.

15-7.2 The Second Generation

The success of the +5-V regulators changed the philosophy of many system designers. There was no need to have a central regulator supplying current to each circuit board in the system and suffering the large I^2R loss. Now each printed circuit card could have its own on-board local regulator. The local regulator also protected its ICs against line voltage transients.

The +5-V regulator's success spawned an array of three-terminal regulators of 6, 8, 9, 12, 15, 18, and 24 V and their negative counterparts. The geniuses who made these devices finally did what Thomas Edison swore could never be done. They invented a device that you could think of as a dc transformer. Now, if you needed a 15-V regulator to furnish 1 A, you simply bought one.

15-7.3 The Third Generation

Linear IC regulators were so popular that they created serious problems for original equipment manufacturers (OEMs). How do you stock all these sizes, and how do you make enough to suit the growing number of voltage requirements?

The LM117 was the first successful superior-performance adjustable positive IC voltage regulator. It was followed by the LM137 adjustable negative regulator. We will present only a few of the bewildering array of linear IC regulators. Space does not permit presentation of the switching regulators.

15-8 LINEAR IC VOLTAGE REGULATORS

15-8.1 Classification

Linear IC voltage regulators are classified by four characteristics:

1. *Polarity:* negative, positive, or dual tracking.
2. *Terminal count:* three-terminal or multiterminal.

3. *Fixed or adjustable output voltage:* standard fixed voltages are ±5, ±12, and ±15 V. Adjustable range is typically 1.2 to 37 V or −1.2 to −37 V.
4. *Output current:* Typical output current capabilities are 0.1, 0.2, 0.25, 0.5, 1.5, and 3 A and the new 5 and 10 A.

15-8.2 Common Characteristics

The instantaneous voltage at the input of an IC regulator must always exceed the dc output voltage by a value that is typically equal to 0.5 to 3 V. This requirement is called *minimum instantaneous input–output voltage, dropout voltage,* or simply *headroom.* As shown in Fig. 15-10(a), the LM340-15 voltage regulator has an output voltage of 15 V at a load of 1 A.

Suppose that the unregulated power supply that feeds the regulator has a 1000-μF capacitor and thus a ripple voltage of $\Delta V_o = 5$ V. As shown in Fig. 15-10(a), you need a minimum input voltage of

$$V_{\text{Lmin}} = V_{o\,\text{reg}} + \text{headroom} \tag{15-8}$$

or

$$V_{\text{Lmin}} = 15\text{ V} + 3\text{ V} = 18\text{ V}$$

This means that V_{dcFL} must be 20.5 V at the very least [see Eq. (15-5c)]. Although you might be tempted to make V_{dcFL} high to give plenty of headroom, you must remember that the worst regulator heat power is $I_{\text{LFL}}(V_{\text{dcFL}} - V_{o\,\text{reg}})$. So there is your trade-off. A higher V_{dcFL} wastes more heat in the regulator.

15-8.3 Self-Protection Circuits

The internal circuitry of these devices senses the load current. If the load current exceeds a specified value, the output current is automatically limited until the overload is removed. They also measure both their input-output difference voltage and load current to be sure that no disallowed combination occurs. If it does, the regulator shuts down. This feature is called *safe area protection.*

Finally, these regulators even measure their own temperature to see if you heat-sinked them properly. If the internal die temperature exceeds 150 to 175°C, they shut down. If you remove the fault, the regulator goes back to work.

15-8.4 External Protection

Despite the well-designed internal protection circuitry, regulators can still be damaged by misuse, sabotage, or certain failures of external circuits. The measures you can take to safeguard against these eventualities are given in the data sheets of a particular regulator.

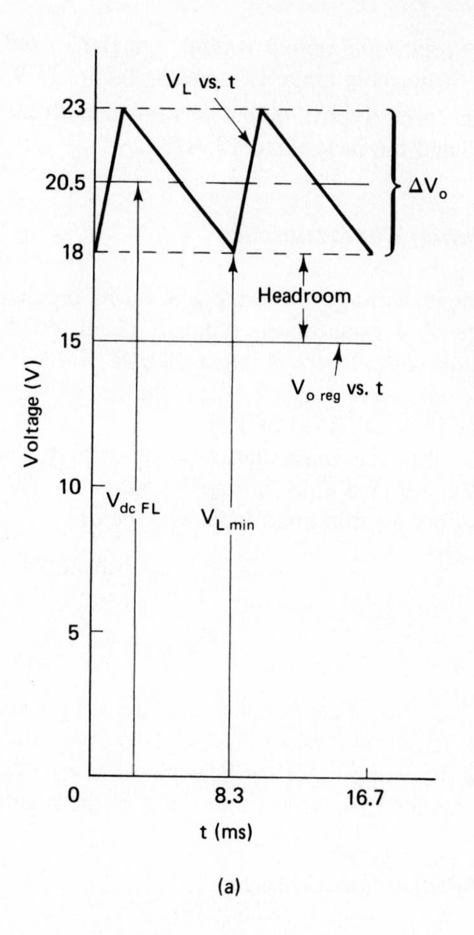

(a)

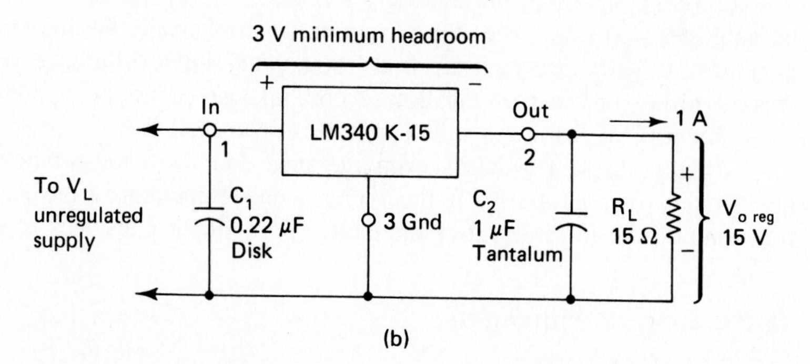

(b)

FIGURE 15-10 All voltage regulators need approximately 1 to 3 V between input and output terminals to ensure operation of the internal circuitry.

15-8.5 Ripple Reduction

Manufacturers of linear IC regulators specify their ac performance by a parameter called *ripple rejection*. It is the ratio of the peak-to-peak input ripple voltage ΔV_{ounreg} to the peak-to-peak output ripple voltage ΔV_{oreg}. It is typically 60 dB or more. That is a reduction in ripple voltage of at least 1000 : 1. For example, if 5 V of ripple is at the regulator's input, less than 5 mV appear across the load. We now turn our attention to specific applications for IC regulators.

15-9 POWER SUPPLY FOR LOGIC CIRCUITS

15-9.1 The Regulator Circuit

A +5 V digital power supply for TTL logic or certain microprocessors is shown in Fig. 15-11. The K package of the LM340-05 is a steel TO-3 case and should be heat sinked for a case-to-ambient thermal resistance of a 6°C/W or less. This means that you should use a 0.002-in.-thick insulating mica washer with a good thermal joint compound (Wakefield Thermalloy Thermocote) between the TO-3 case and its heat sink (or use the chassis as a heat sink).

The LM340K-05 can furnish up to 1.5 A. It has internal current limit at 2.1 A for pulse operation. It also has safe area protection that protects its output transistor. It has thermal shutdown protection at a junction temperature of 150°C to prevent burnout. The added diode protects the regulator against short circuits occurring at its input terminals.

15-9.2 The Unregulated Supply

V_{Lmin} for the unregulated FWB supply should allow 3 V of headroom and should be greater than (3 + 5) V = 8 V (see Fig. 15-11). Choose a 12.6-V transformer at (1.8 × 1 A) at a 2-A current rating. From Sections 15-2 and 15-3: (1) V_{dcNL} =

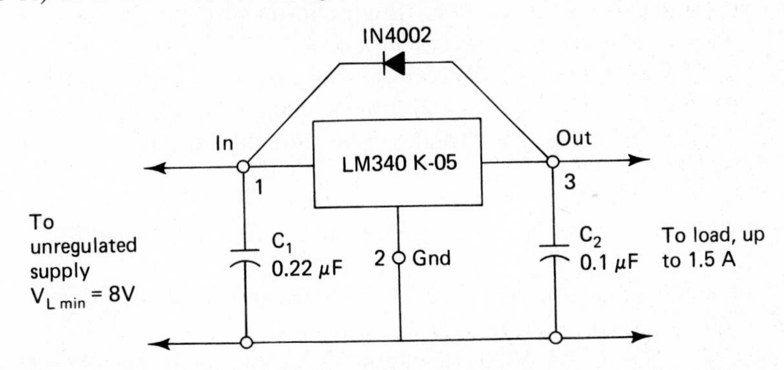

FIGURE 15-11 A TTL digital logic regulated power supply (5 V at 1 A.)

12.6 V × 1.4 = 17.8 V. (2) Assume that $R_o = 6\ \Omega$. If $I_{LFL} = 1$ A, then $V_{dcFL} = 17.8\text{ V} - 6\ \Omega(1\text{ A}) = 11.8$ V. (3) Pick a 1000-μF filter capacitor to give a $\Delta V_o = 5$ V. Thus, V_{Lmin} should be at least 11.8 V − (5 V/2) = 9.3 V, leaving some margin. Pick the capacitor WVDC ≥ 25 V and diodes with a 25-V PIV rating. The diode current rating should exceed 1 A.

15-10 ±15-V POWER SUPPLIES FOR LINEAR APPLICATION

15-10.1 High-Current ±15-V Regulator

Figure 15-12(a) presents a bipolar ±15 V supply that can furnish 1 A from either (+) or (−) terminal. The LM340K-15 is a +15-V regulator with load current capability up to 1.5 A. To use it as stand-alone +15-V supply, (1) remove the diodes, R_2, C_N, and the LM320-15; and (2) replace R_1 with a short circuit.

The LM320K-15 is a −15-V regulator with current capability up to 1.5 A. Both regulators have current limit, safe area, and thermal shutdown protection. They should be heat-sinked as directed in Section 15-10.2.

Resistor R_1 is needed to ensure that the positive regulator starts up when the negative regulator has a heavy load. R_2 offsets the effect on ±15-V regulation caused by adding R_1.

15-10.2 Low-Current ±15-V Regulator

Since one op amp rarely draws more than 5 mA, you need only a ±100-mA supply to power well over 20 op amps. For this reason an inexpensive low-power supply is shown in Fig. 15-12(b). The LM325H is a dual-tracking ±15-V supply in a 10-pin metal can package that can furnish ±100 mA. It has internal current limiting and thermal overload protection. (Buy a clip-on heat sink, or epoxy about a 2 in. by 2 in. piece of aluminum to the top surface.)

Notice that the LM325 has two excellent voltage regulators packed into a single IC. The output capacitors provide energy storage to improve transient response. The input capacitors are needed if the unregulated supply is more than 4 in. from the LM325.

15-10.3 Unregulated Supply for the ±15-V Regulators

The unregulated supply required is shown in Fig. 15-8. Select:

1. $C^+ = C^- = 1000\ \mu$F minimum for both high- and low-current supplies, with WVDC = 30 V.

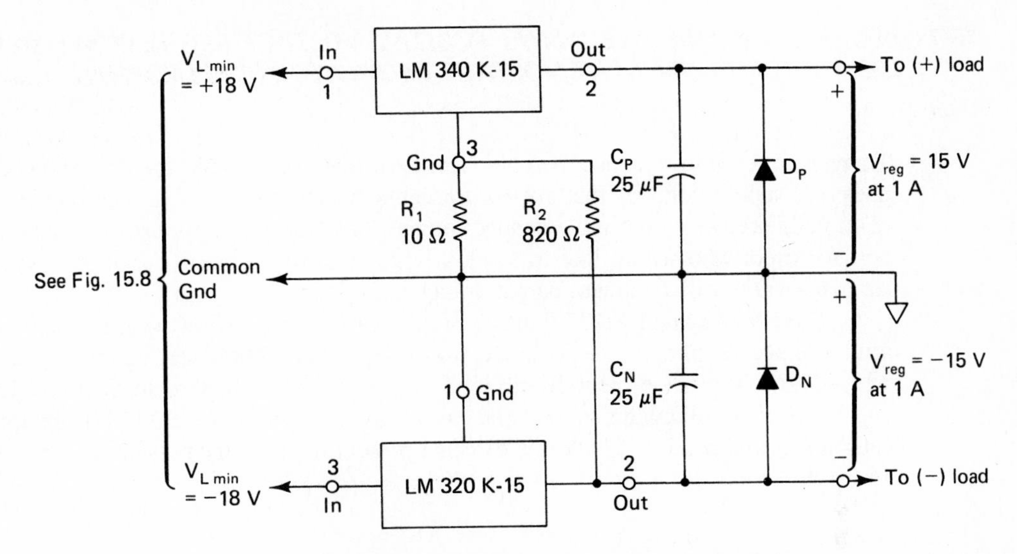

(a) Regulated ±15 V power supply for current up to ±1 A; the unregulated supply would be similar to that of Fig. 15.8; requires two separate regulators

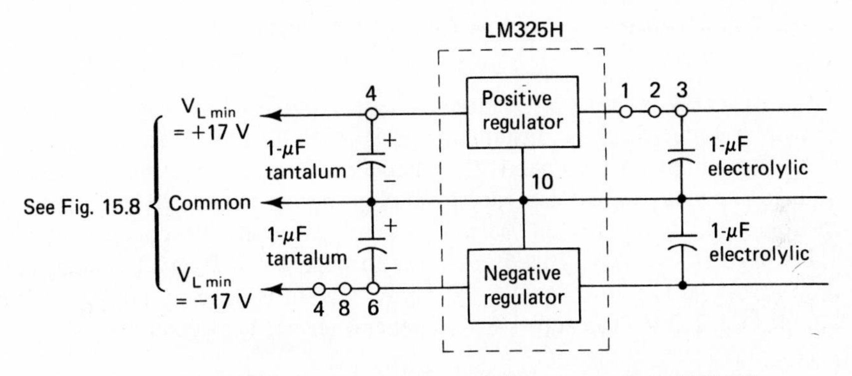

(b) Regulated ±15-V power supply for currents up to ±100 mA

FIGURE 15-12 Two selections are presented for ±15-V supplies for op amps. The heavy-current 1 A version in (a) required two ICs. The modest-current version in (b) needs but a single IC.

2. For the ±1-A supply, select a transformer of 115 V/36 VCT at 2 A. For the ±100-mA supply, select a 115 V/30 VCT at 0.2 A.
3. Diodes should be rated for $I_{av} \geq 1.0$ A for the high-current supply and ≥0.1 A for the low-current supply. PIV ratings for both should exceed 30 V (50 V is a standard size).

15-11 ADJUSTABLE THREE-TERMINAL POSITIVE VOLTAGE REGULATOR (THE LM317HV) AND NEGATIVE VOLTAGE REGULATOR (THE LM337HV)

There is a need for (1) regulated load voltages that are variable for laboratory supplies, (2) supply voltages that are *not* available as standard fixed-voltage regulators, (3) a very precisely adjustable supply voltage, or (4) providing a price-break lower cost for users who would like to stock a large quantity of one IC regulator type to furnish a variety of regulated output voltages.

The LM117 and LM137 families of adjustable three-terminal positive and negative voltage regulators, respectively, were developed. They are superb regulators with all the internal protection circuitry listed for the regulators in Sections 15-9 and 15-10. Since they are so versatile, they will encounter a variety of hostile applications, so it is prudent to add the external protection circuitry presented in Section 15-12.4.

15-12 LOAD VOLTAGE ADJUSTMENT

15-12.1 Adjusting the Positive Regulated Output Voltage

The LM317HV adjustable positive voltage regulator has only three terminals as shown in Fig. 15-13(a). Installation is simple, as shown in Fig. 15-13(b). The LM317 maintains a nominal 1.25 V between its output and adjust terminals. This voltage is called V_{ref} and can vary from chip to chip from 1.20 to 1.30 V. A 240-Ω resistor, R_1, is connected between these terminals to conduct a current of 1.2 V/240 Ω = 5 mA. This 5 mA flows through R_2. If R_2 is adjustable, the voltage drop across it, V_{R2}, will equal $R_2 \times 5$ mA. Output voltage of the regulator is set by V_{R2} plus the 1.2-V drop across R_1. In general terms, V_o is given by

$$V_o = \frac{1.2\text{ V}}{R_1}(R_1 + R_2) \qquad (15\text{-}9a)$$

Normally, $R_1 = 240\ \Omega$. Thus any desired value of regulated output voltage is set by trimming R_2 to a value determined from

$$V_o = 1.2\text{ V} + (5\text{ mA})(R_2) \qquad (15\text{-}9b)$$

For example, if you need a 5-V supply for TTL logic, make $R_2 = 760\ \Omega$. If you need a +15-V supply for an op amp or CMOS, make $R_2 = 2760\ \Omega$. $R_2 = 2160\ \Omega$ will give you 12-V car voltage. Make R_2 a 3-kΩ pot and you can adjust V_o to any voltage between 1.2 V (D-cell battery) and 16.2 V.

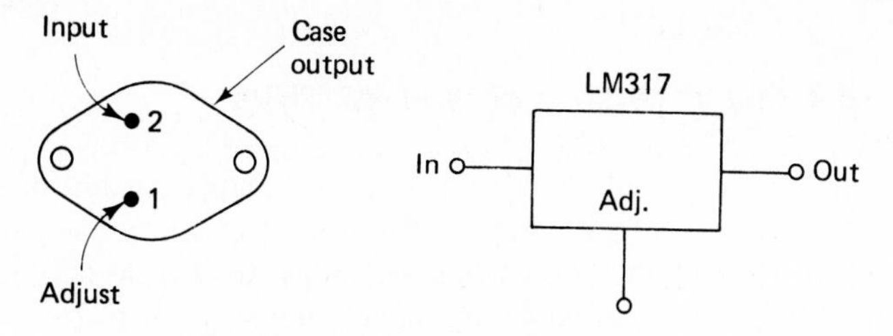

(a) LM317K connection diagram for TO-3 steel case and circuit schematic

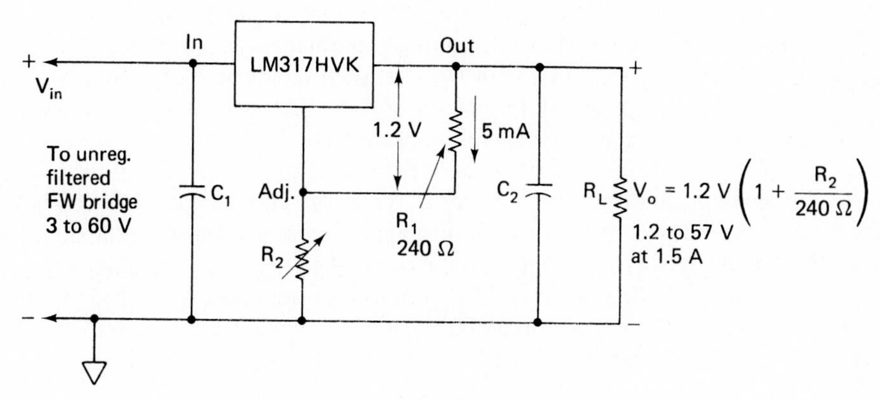

(b) Connecting the LM317HVK to act as an adjustable positive voltage regulator

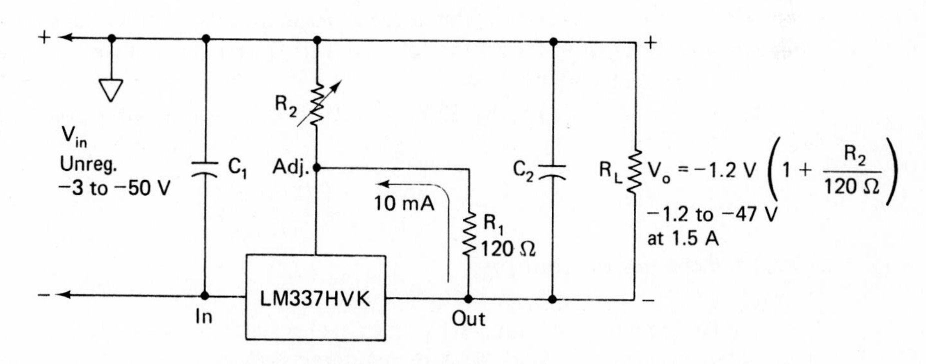

(c) Connecting the LM337HVK to act as an adjustable negative regulator

FIGURE 15-13 Adjustable three-terminal positive (LM317) and negative (LM337) IC regulators are easy to use.

15-12.2 Characteristics of the LM317HVK

The LM317HVK will provide a regulated output current of up to 1.5 A, provided that it is not subjected to a power dissipation of more than about 15 W (TO-3 case). This means it should be electrically isolated from, and fastened to, a large heat sink such as the metal chassis of the power supply. A 5-in. by 5-in. piece of aluminum chassis stock also makes an adequate heat sink (see Section 15-10.1).

The LM317 requires a minimum "dropout" voltage of 3 V across its input and output terminals or it will drop out of regulation. Thus the upper limit of V_o is 3 V below the minimum input voltage from the unregulated supply.

It is good practice to connect bypass capacitors C_1 and C_2 (1-μF tantalum) as shown in Fig. 15-13(b). C_1 minimizes problems caused by long leads between the rectifier and LM317. C_2 improves transient response. Any ripple voltage from the rectifier will be reduced by a factor of over 1000 if R_2 is bypassed by a 1-μF tantalum capacitor or 10-μF aluminum electrolytic capacitor.

The LM317HVK protects itself against overheating, too much internal power dissipation, and too much current. When the chip temperature reaches 175°C, the 317 shuts down. If the product of output current and input-to-output voltage exceeds 15 to 20 W, or if currents greater than about 1.5 A are required, the LM317 also shuts down. When the overload condition is removed, the LM317 simply resumes operation. All of these protection features are made possible by the remarkable internal circuitry of the LM317.

15-12.3 Adjustable Negative-Voltage Regulator

An adjustable three-terminal *negative*-voltage regulator is also available [see the LM337HVK in Fig. 15-13(c)]. The negative regulator operates on the same principle as the positive regulator except that R_1 is a 120-Ω resistor and the maximum input voltage is reduced to 50 V.

V_o is given by Eq. (15-9b). If $R_1 = 120\ \Omega$, then V_o depends upon R_2 according to

$$V_o = 1.25\text{ V} + (10\text{ mA})R_2 \tag{15-10}$$

15-12.4 External Protection

It is standard practice to connect C_1 and C_2 to a regulator (see Fig. 15-14) for reasons stated in Section 15-12.2. Any regulator should have diode D_1 to protect it against input shorts; otherwise, load capacitance can pump current back into its output and destroy it.

Capacitor C_3 is added to greatly improve ac ripple voltage rejection. However, if a short circuit occurs across the regulator's output, C_3 will try to pump current back into the adjust terminal. Diode D_2 steers this current instead into the short circuit.

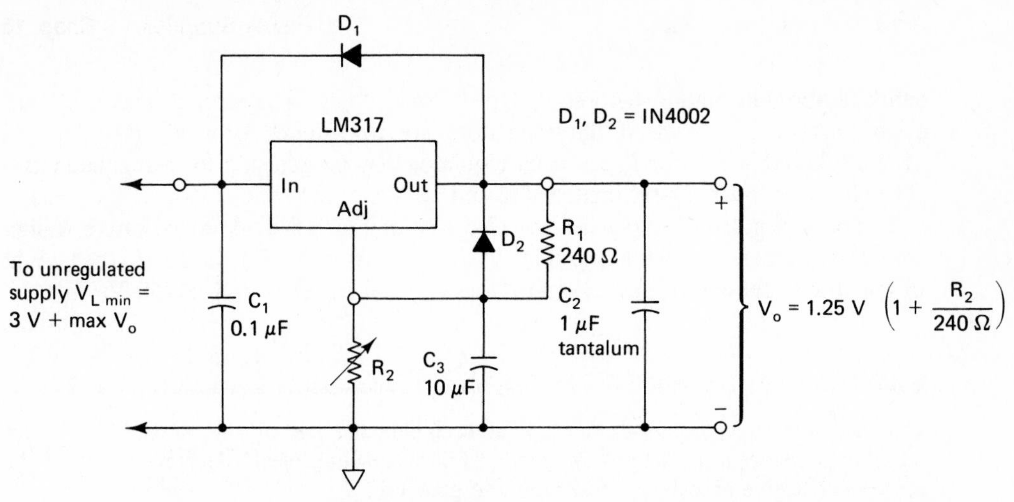

FIGURE 15-14 Variable-voltage positive regulated supply with external protection. D_1 protects the regulator from input short circuits. D_2 protects the regulator from output short circuits.

15-13 ADJUSTABLE LABORATORY-TYPE VOLTAGE REGULATOR

A standard LM317K and LM337K IC positive and negative regulator, respectively, are interconnected with support components in Fig. 15-15. They form an indepen-

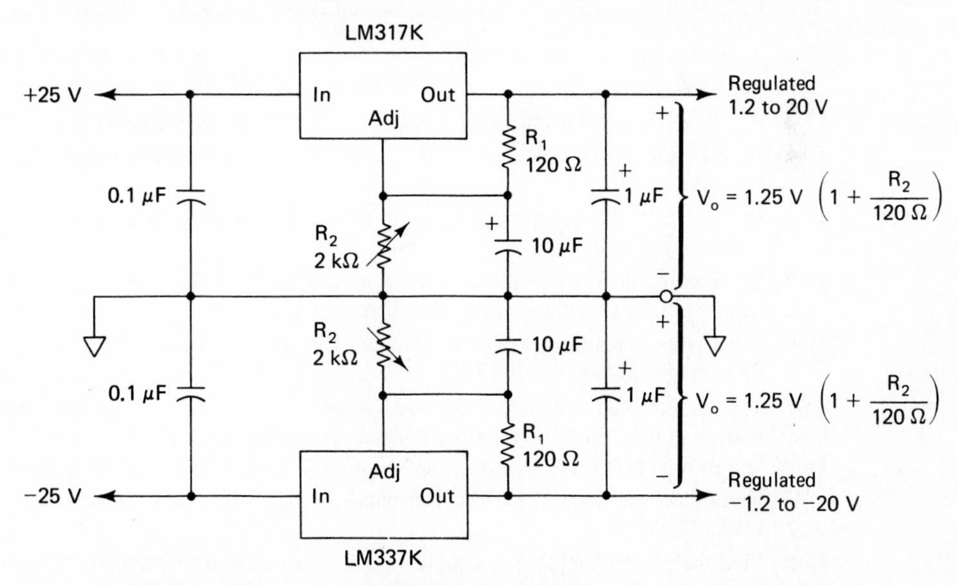

FIGURE 15-15 Adjustable bipolar laboratory-type voltage regulator. Positive and negative outputs can be adjusted independently for any voltage between 1.2 and 20 V.

dently adjustable bipolar laboratory-type power supply. The steel K packages will easily furnish 1 A each if the regulators are heat-sinked properly (see Section 15-9.1). Variable resistor R_2 for each regulator may be adjusted for a regulated output voltage between approximately 1.2 and 20 V.

The unregulated supply has the circuitry of Fig. 15-8. A conservative design would select (1) a transformer, 115 V/50 VCT at 2 A, (2) diodes $I_{av} > 1$ A at PIV $\geq$ 50 V (IN 4002), and (3) capacitors of 1000 μF at WVDC $\geq$ 50 V.

LABORATORY EXERCISE

The regulators shown in this chapter have all been built and used for student experiments, for new circuit development, and for research and development projects. They have withstood a remarkable degree of abuse, misuse, and also good use.

These circuits should *not* be constructed on ordinary breadboards. They should be constructed on a printed circuit board or wirewrap board. The IC regulators must be properly heatsinked.

It is easier to heat sink the less expensive TO-220 plastic package. (They have a package code letter T: for example, 317T and 337T.) They are also easier to attach to a pc board and perform almost as well as the steel packages. If you can get only about 300 mA from your regulator, tighten the heatsink mounting screws after checking that the thermal compound has been used.

PROBLEMS

15-1. A transformer is rated at 115 to 28 V rms at 1 A. What is the peak secondary voltage?

15-2. A 115/28 V transformer is used in Fig. 15-4(a). Find V_{dc} at no load.

15-3. As dc load current decreases, what happens to **(a)** dc load voltage; **(b)** ac ripple voltage?

15-4. In Fig. 15-4, the transformer rating is 115 to 28 V at 1 A. What is V_{dc} at a full-load current of $I_L = 0.5$ A?

15-5. Dc measurements of a power supply give $V_{dcNL} = 17.8$ V and $V_{dcFL} = 13.8$ V at $I_{LFL} = 0.5$ A. Calculate: **(a)** R_o; **(b)** percent regulation.

15-6. What voltage readings would be obtained with an ac voltmeter for peak-to-peak ripple voltages of **(a)** 1 V; **(b)** 3 V?

15-7. The dc full-load voltage of a power supply is 28 V and the peak-to-peak ripple voltage is 6 V. Find the minimum instantaneous load voltage.

15-8. A 110 V/28 V CT transformer is installed in Fig. 15-8. What no-load dc voltage would be measured **(a)** between terminals 1 and 2; **(b)** from 1 to CT; **(c)** from 2 to CT?

15-9. Find the percent ripple in Example 15-7 if C is changed to 2000 μF.

15-10. Design an FWB unregulated power supply to output +15 V at 1 A, with less than 5% ripple. Assume that $R_o = 8\ \Omega$. Choose the transformer from available ratings of

115/12 V, 115/18 V, and 115/24 V all at 2 A. Available capacitors are 500 μF and 1000 μF at WVDC = 50 V.

15-11. Given 115/25.2 V at a 3-A transformer and C = 1000 μF in an FWB rectifier. Assume R_o = 6 Ω and I_{LFL} = 1 A. Calculate **(a)** V_{dcNL}; **(b)** V_{dcFL}; **(c)** percent regulation; **(d)** ΔV_o; **(e)** ripple; **(f)** V_{Lmin}.

15-12. An unregulated power supply has V_{dcNL} = 18 V, V_{dcFL} = 10 V at 1 A, ΔV_o = 5 V, and C = 1000 μF. Calculate ΔV_o if you **(a)** double C to 2000 μF, or **(b)** reduce I to 0.5 A, or **(c)** both reduce I to 0.5 A and double C to 2000 μF.

15-13. An ac voltmeter indicates a value of 1.71 V_{rms} across the FWB supply, and a dc voltmeter indicates 12 V_{dc}. Draw the expected value of V_L that would be seen on a dc coupled CRO for a time interval of 16.7 ms (see Fig. 15-7).

15-14. If I_L measures 1 A in Problem 15-13, find **(a)** the value of C; **(b)** the value of R_L.

15-15. You need a regulated output voltage of 24.0 V. If R_1 = 240 Ω in Fig. 15-13(a), find the required value of R_2.

15-16. Assume that the regulator of Problem 15-15 delivers a load current of 1.0 A. If its overage dc input voltage is 30 V, show that the regulator must dissipate 6 W.

15-17. Find V_o if R_2 is short-circuited in **(a)** Fig. 15-13(a); **(b)** Fig. 15-13(b).

15-18. Adjustable resistor R_2 = 0 to 2500 Ω in Fig. 15-13(a). Find the upper and lower limits of V_o as R_2 is adjusted from 2500 Ω to 0 Ω.

15-19. Suppose that a 1200-Ω low-stop resistor, and a 2500-Ω pot are connected in series in place of the single resistor R_2 in Fig. 15-13(a). Find the upper and lower limits of V_o as the pot is adjusted from 2500 Ω to 0 Ω.

15-20. If the dropout voltage of an LM317 is 3 V, what is the minimum instantaneous input voltage for the regulator circuit of Problem 15-18?

15-21. Calculate the required values for R_2 in Fig. 15-15 to give outputs of ±12 V.

Answers to Selected Odd-Numbered Problems

CHAPTER 1

1-1. Mathematical operation
1-3. Part identification number
1-5. Package style
1-7. Pin 8
1-9. **(a)** The common connection to the positive and negative supplies or a ground symbol. **(b)** Make all voltage measurements with respect to power supply common.

CHAPTER 2

2-5. **(a)** $V_o = 0$ V **(b)** $I_{os} = 25$ mA (typical)

2-7.

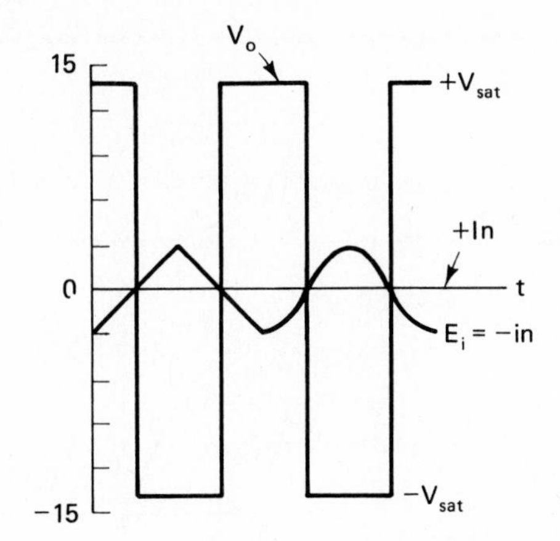

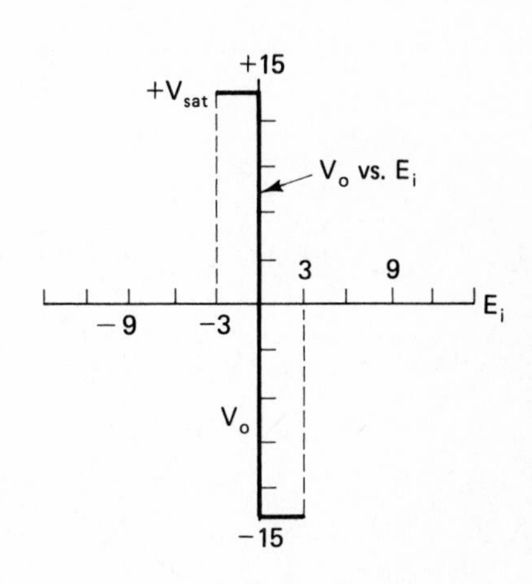

2-9 Problem	E_i applied to	zero-crossing type:
2-7:	(−) input	inverting
2-8:	(+) input	noninverting

2-11. **(a)** If E_i is above V_{ref}, V_o will be above 0 V at $+V_{sat}$. **(b)** If signal is wired directly (or via a resistor) to the (+) input, the circuit is noninverting.
2-13. **(a)** Arbitrarily choose first a 0–5 V adjustment. **(b)** We now need a divider of 5 V/50 mV = 100 to 1. **(c)** Pick an isolating resistor equal to or greater than 10 × the 5-kΩ pot. Pick 100 kΩ.

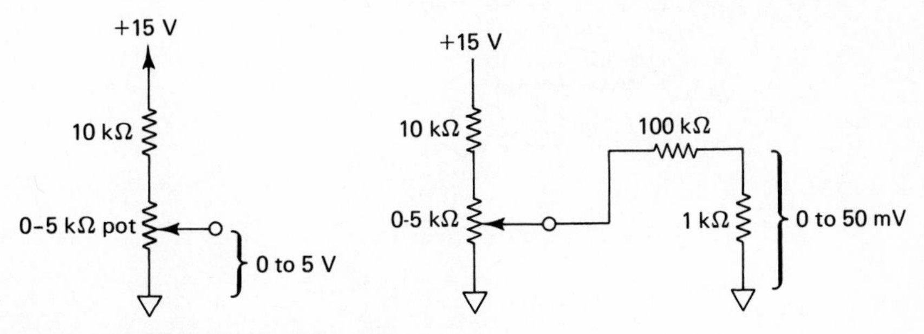

Therefore, the divider resistor is 100 kΩ ÷ 100 or 1 kΩ.

2-15. V_{temp} (V)	T_{hi} (ms)
0	0
5	10
10	20

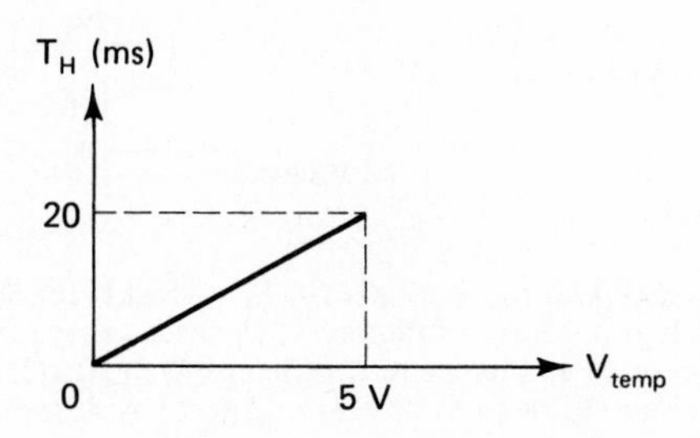

CHAPTER 3

3-1. Negative
3-3. If a circuit has negative feedback and V_o is not in saturation, **(a)** $E_d = 0$ V and **(b)** op amp inputs draw negligible signal current.
3-5. **(a)** $V_o = -10$ V, op amp I_o sinks -1 mA **(b)** $V_o = 4$ V, I_o sources 0.4 mA

3-7.

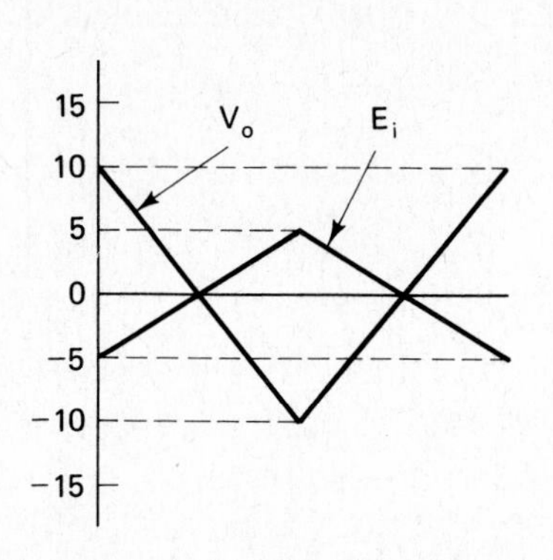

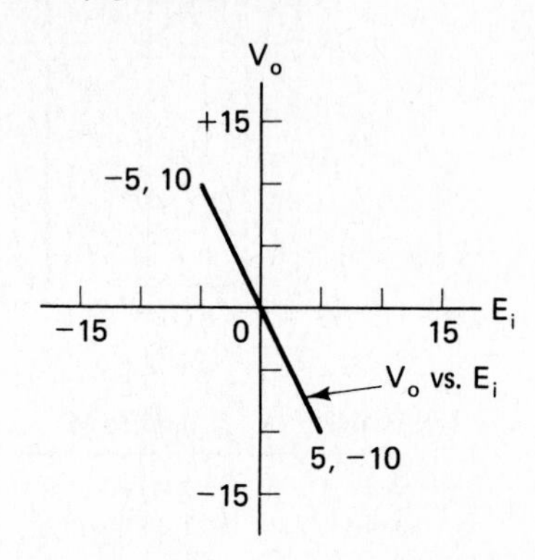

3-9. Noninverting amplifier: **(a)** $V_o = 10$ V; **(b)** $V_o = -4$ V. In comparison, the gain magnitudes are equal.
3-11. **(a)** $R_i = 10\ \text{k}\Omega$ **(b)** $R_f = 50\ \text{k}\Omega$

3-13.

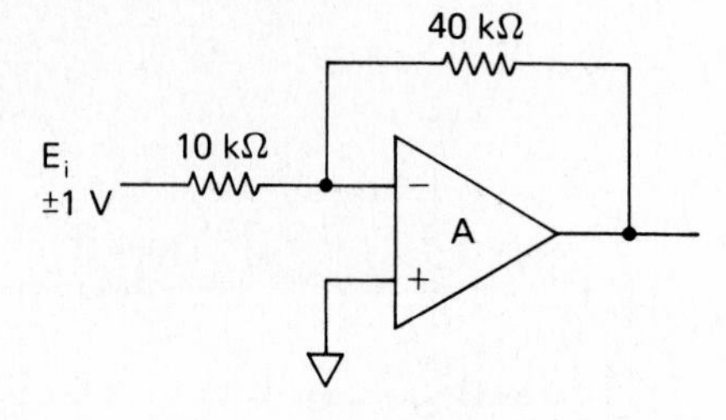

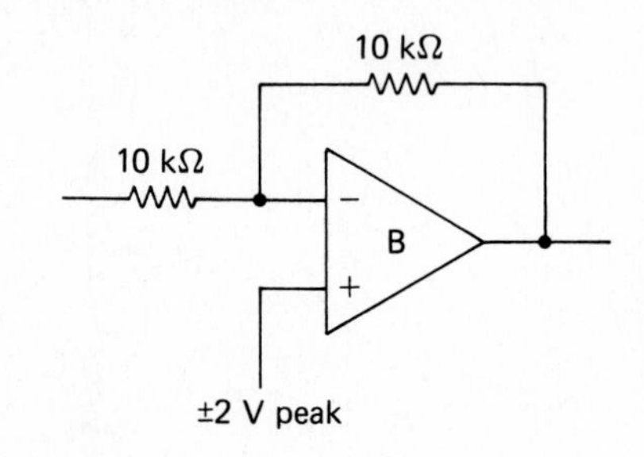

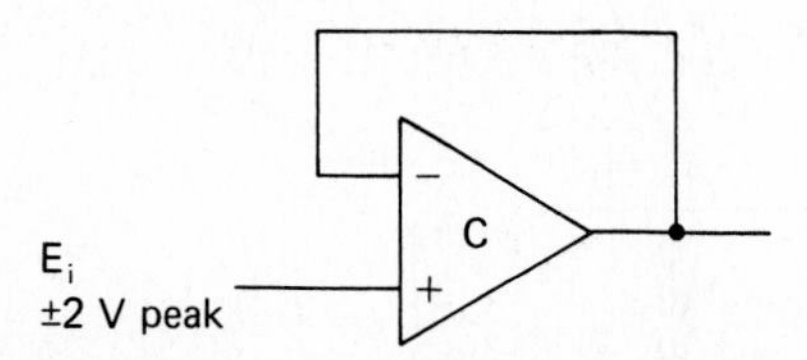

3-17. **(a)** $R_{i3} = 10\ \text{k}\Omega$, $R_{f3} = 50\ \text{k}\Omega$ **(b)** $R_{i1} = 50\ \text{k}\Omega$ **(c)** $R_{i2} = 16.67\ \text{k}\Omega$
3-19. Change Fig. 3-15(a) as follows: (1) Connect voltage followers (Fig. 3-8) between E_1 and its input and between E_2 and its input. (2) Change R_f to 50 kΩ. Then op amp B applies a gain of -5 to each channel. Thus, $V_o = (-E_1 \times -5) + E_2(-5) = 5E_1 - 5E_2 = 5(E_1 - E_2)$.

CHAPTER 4

4-3.

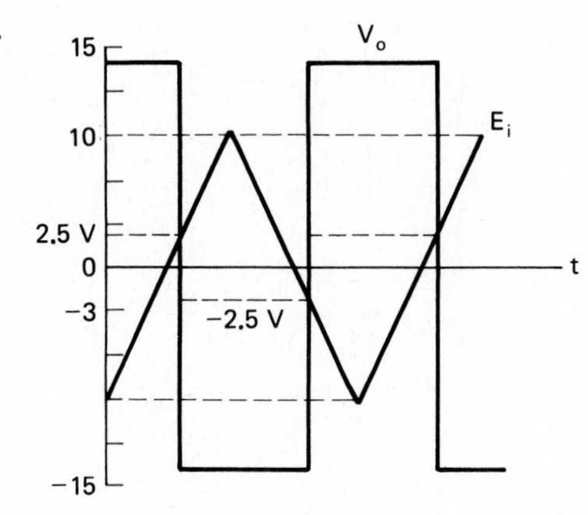

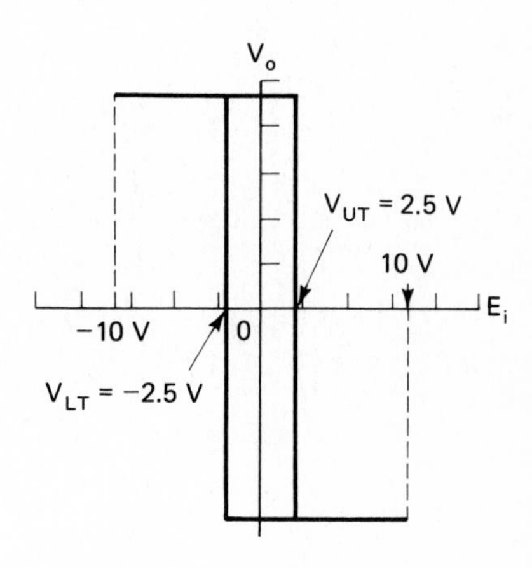

4-5. **(a)** 500 Hz **(b)** 10 V **(c)** 6 V **(d)** −6 V **(e)** 12 V
4-7. **(a)** $V_H = 1.5$ V, $V_{CTR} = -1.25$ V **(b)** $n - 20$ **(c)** −1.19 V **(d)** $R = 10$ kΩ, $nR - 200$ kΩ
4-9. $V_H = 1.5$ V, $V_{CTR} = 1.25$ V **(a)** $n = 17.3$ for $\pm V_{sat} = \pm 13$ V, $R = 10$ kΩ, $nR = 173$ kΩ
(b) $V_{ref} = -15$ V, $mR = 100$ kΩ **4-11.** $4.97 \text{ V} \simeq 5$ V **4-13.** $V_o = +5$ V **4-15.** 311

CHAPTER 5

5-1. **(a)** −1 mA **(b)** −2 V **5-5.** **(a)** 4.5 kΩ **(b)** 3.18 kΩ **(c)** 1.59 kΩ
5-7. **(a)** *pnp* **(b)** 100 mA **(c)** 5 V **5-9.** 500 Ω
5-11. **(a)** −0.2 mA, −1.0 V, −2 V **(b)** 0.2 mA, 1 V, 2 V **5-13.** $I_L = 0$, $V_L = 0$; $V_o = +5$ V
5-15. See Figs. 5-8 and 5-9. **5-17.** $I_L = 5$ mA **5-19.** $\theta = 64.2°$
5-23. $C = 1\ \mu$F, $R_i = 159$ kΩ, $R_{resistors} = 300$ kΩ
5-25. 100°C = 373 K. The AD590 outputs 373 μA. In Fig. 5-20(a), current through $R_F = 373 - 273\ \mu A = 100\ \mu A$ (R to left). $VR_F = 100\ \mu A \times 10\ k\Omega = 1000\ mV = V_o$,

$$V_o = \frac{10 \text{ mV}}{°\text{C}} \times 100\ °\text{C} = 1000 \text{ mV}.$$

CHAPTER 6

6-1. **(a)** 6.9 V **(b)** −6.9 V **6-3.** yes; $R_f = 5$ kΩ **6-7.** ±3 V; 1250 Hz
6-9. 9.4 V; 250 Hz **6-11.** **(a)** Switched gain amplifier **(b)** Amplifier B; $V_o = V_{ref}$ **(c)** With pin 9 at 1 V, V_o will be a sine wave identical to V_{ref}. With pin 9 at −1 V, V_o will be inverted from V_{ref}.
6-13. **(a)** 60° **(b)** 1.2 V **6-15.** Pick $C = 0.1\ \mu$F. At 0.5 Hz, $R_i = 500$ kΩ; at 50 Hz, $R_i = 5$ kΩ

CHAPTER 7

7-1. 3 V

7-3.

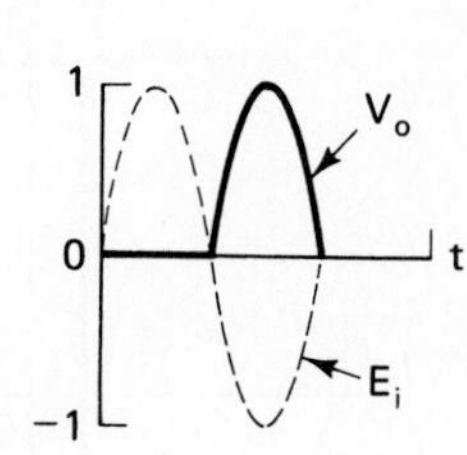

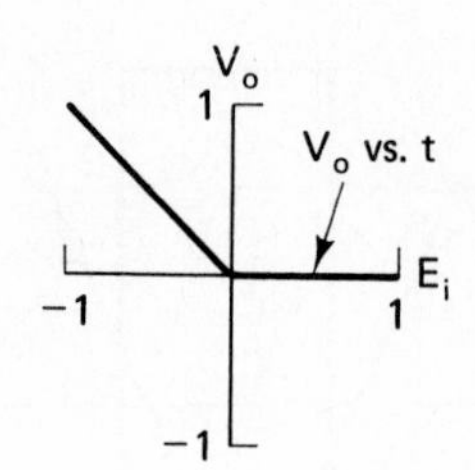

7-5. See Fig. 7-5. 7-9. See Fig. 7-14. 7-11. See Figs. 7-18(b) and 7-17.

CHAPTER 8

8-1. $V_o = -1$ V 8-3. $V_o = 0$ V 8-5. **(a)** −200 mV **(b)** 0 V
8-7. $V_o = -2\text{ V}\left(1 + \frac{2}{a}\right)$ 8-9. **(a)** 0.143 V **(b)** $a = 0.1$ 8-11. **(a)** $V_o = 5.0$ V **(b)** 2.5 V
8-13. **(a)** Down **(b)** 10 mA **(c)** 1 V **(d)** 11 V 8-15. **(a)** V_o goes positive **(b)** V_o decreases
8-17. Plot V_o versus temperature using data from Table 8-2. Choose E_i = positive and a reference at the low-temperature limit. 8-19. **(a)** $V_o = 25$ mV **(b)** $V_o = 50$ mV **(c)** $V_o = 100$ mV

CHAPTER 9

9-3. $I_{B-} = 0.2\ \mu A$ 9-5. $V_o = 2.2$ V 9-7. $V_o = -2.5$ mV 9-9. $V_{io} \cong 2$ mV
9-11. $R_c = 5\ k\Omega$ 9-13. $\Delta V_o = \pm 101$ mV
9-15. **(a)** $V_{io} = 1$ mV **(b)** $I_{B-} = 0.2\ \mu A$ **(c)** $I_{B+} \simeq 1\ \mu A$

CHAPTER 10

10-1. 200,000 10-3. 5 MHz 10-9. $A_{CL} = 990$
10-11. **(a)** $f_H = 100$ kHz **(b)** $A_{CL} = 70.7$ 10-13. $f_{max} = 15.92$ kHz 10-15. 6
10-17. Decrease

CHAPTER 11

11-5. $R = 7.2\ k\Omega$ 11-7. $|V_o| = 0.707$ at f_c. There is a 45° phase angle at f_c for each capacitor.
11-9. $f_c = 11.2$ kHz 11-11. $\omega_c = 25$ krad/s 11-13. $R = 8\ k\Omega$
11-15. $R_1 = 14\ k\Omega$, $R_2 = 7.07\ k\Omega$ 11-17. $R_3 = 6.35\ k\Omega$, $R_1 = 12.7\ k\Omega$, $R_2 = 3.17\ k\Omega$

11-19. **(a)** 10 Hz **(b)** 60 Hz **(c)** 6
11-21. 3000 Hz **11-23.** $Q = 0.35$
11-25. **(a)** Connect the bandpass filter to an inverting adder as in Fig. 11-15. **(b)** $f_L = 92$ Hz, $f_H = 177$ Hz

CHAPTER 12

12-1. **(a)** $V_o = 2.5$ V **(b)** $V_o = -2.5$ V **(c)** $V_o = -2.5$ V **(d)** $V_o = 2.5$ V
12-5. **(a)** 3.2 V **(b)** 3.2 V peak at 800 Hz **12-7.** $V_{odc} = 4.33$ V
12-9. Upper = 16 to 17 kHz; lower = 13 to 14 kHz **12-11.** Carrier is eliminated
12-13. 955 kHz

CHAPTER 13

13-3. 70 Hz **13-5.** 107 Hz **13-7.** 3.1 kΩ **13-9.** 6.95 ms
13-11. $1 \text{ ms} < t_{high} < 2 \text{ ms}$, $R_A = 15$ kΩ for $t_{high} = 1.65$ ms **13-13.** 170 ms
13-15. **(a)** 50 ms **(b)** 600 ms **(c)** 12.75 s **13-17.** 62.5 Hz

CHAPTER 14

14-3. **(a)** 1024 **(b)** 1.023 V
14-5. **(a)** 5 kΩ **(b)** 0.125 mA **(c)** 0.625 V/bit **(d)** $V_o = 0.625 \text{ V} \times D$ **(e)** 9.375 V
14-7. $V_o = 0.04$ V; 5.12 V = voltage across the feedback resistor
14-9. AD558—no; DAC-08—yes
14-11. **(a)** $V_o = 10 \text{ mV/bit} \times D$ **(b)** 1.28 V
14-13. **(a)** 83.33 ms **(b)** Reference integrate phase: $V_{ref} = -50$ mV, $T_2 = 41.65$ ms **(c)** 500 counts
14-15. Conversion
14-17. Ground pin 11
14-19. **(a)** 10 μS **(b)** 100,000 conversions **(c)** See Example 14-15.
14-21. 255

CHAPTER 15

15-1. 38 V **15-3.** **(a)** Decreases **(b)** Increases **15-5.** **(a)** 8 Ω **(b)** 29%
15-7. $V_L = 25$ V **15-9.** 6.1%
15-11. **(a)** $E_m = 35.3$ V **(b)** $V_{dcFL} = 29.3$ V **(c)** 20% **(d)** 1.42 V **(e)** 4.9% **(f)** 26.8 V

15-13. **(a)** 6.0 V **(b)** ΔV_o centered on $V_{dc} = 12$ V
(c) Waveshape

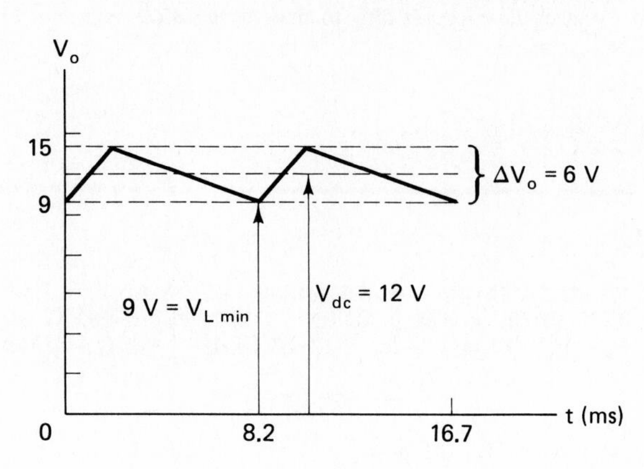

15-15. $R_2 = 4560\ \Omega$ **15-17.** **(a)** 1.2 V **(b)** −1.2 V **15-19.** 7.2 to 19.7 V
15-21. $R_2 = 1032\ \Omega$

APPENDIX 1

μA741 Frequency-Compensated Operational Amplifier*

*Courtesy of Fairchild Semiconductor, a Division of Fairchild Camera and Instrument Corporation.

Description

The μA741 is a high performance Monolithic Operational Amplifier constructed using the Fairchild Planar epitaxial process. It is intended for a wide range of analog applications. High common mode voltage range and absence of latch-up tendencies make the μA741 ideal for use as a voltage follower. The high gain and wide range of operating voltage provides superior performance in integrator, summing amplifier, and general feedback applications.

- **NO FREQUENCY COMPENSATION REQUIRED**
- **SHORT-CIRCUIT PROTECTION**
- **OFFSET VOLTAGE NULL CAPABILITY**
- **LARGE COMMON MODE AND DIFFERENTIAL VOLTAGE RANGES**
- **LOW POWER CONSUMPTION**
- **NO LATCH-UP**

Connection Diagram
10-Pin Flatpak

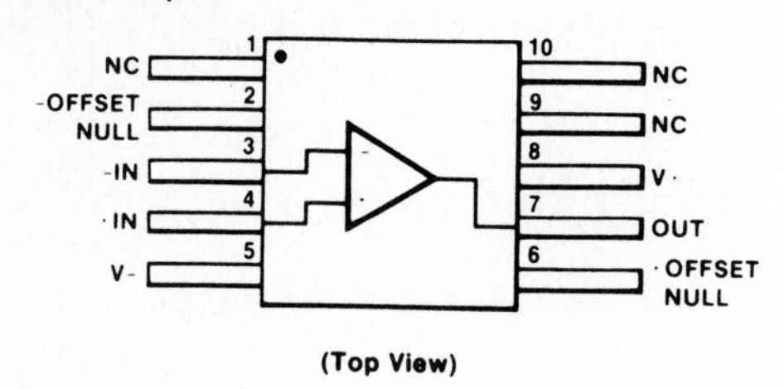

(Top View)

Order Information

Type	Package	Code	Part No.
μA741	Flatpak	3F	μA741FM
μA741A	Flatpak	3F	μA741AFM

Connection Diagram
8-Pin Metal Package

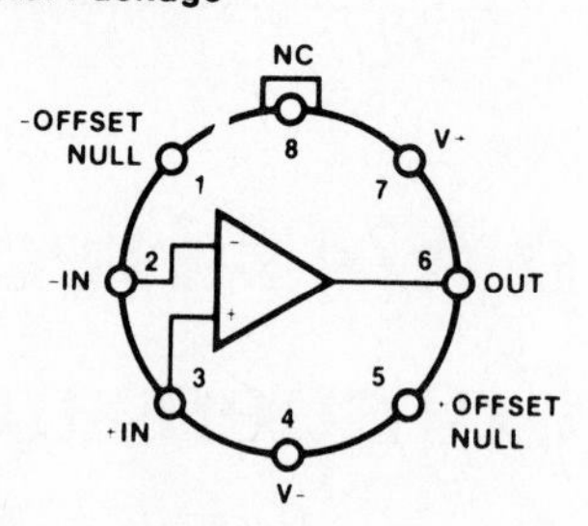

(Top View)

Pin 4 connected to case

Order Information

Type	Package	Code	Part No.
μA741	Metal	5W	μA741HM
μA741A	Metal	5W	μA741AHM
μA741C	Metal	5W	μA741HC
μA741E	Metal	5W	μA741EHC

Connection Diagram
8-Pin DIP

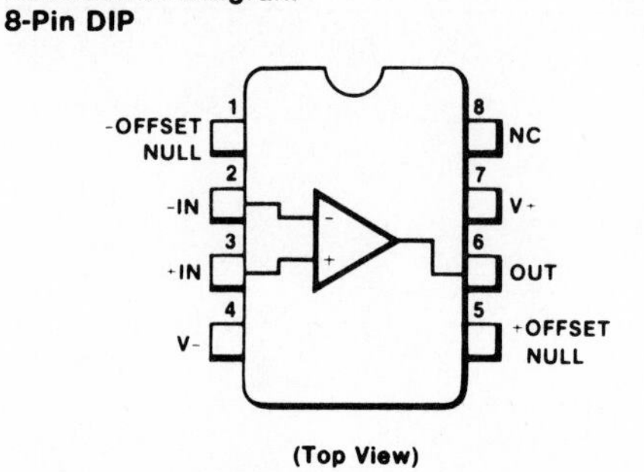

(Top View)

Order Information

Type	Package	Code	Part No.
μA741C	Molded DIP	9T	μA741TC
μA741C	Ceramic DIP	6T	μA741RC

Absolute Maximum Ratings

Supply Voltage	
μA741A, μA741, μA741E	±22 V
μA741C	±18 V
Internal Power Dissipation (Note 1)	
Metal Package	500 MW
DIP	310 mW
Flatpak	570 mW
Differential Input Voltage	±30 V
Input Voltage (Note 2)	±15 V
Storage Temperature Range	
Metal Package and Flatpak	−65°C to +150°C
DIP	−55°C to +125°C
Operating Temperature Range	
Military (μA741A, μA741)	−55°C to +125°C
Commercial (μA741E, μA741C)	0°C to +70°C
Pin Temperature (Soldering 60 s)	
Metal Package, Flatpak, and Ceramic DIP	300°C
Molded DIP (10 s)	260°C
Output Short Circuit Duration (Note 3)	Indefinite

Equivalent Circuit

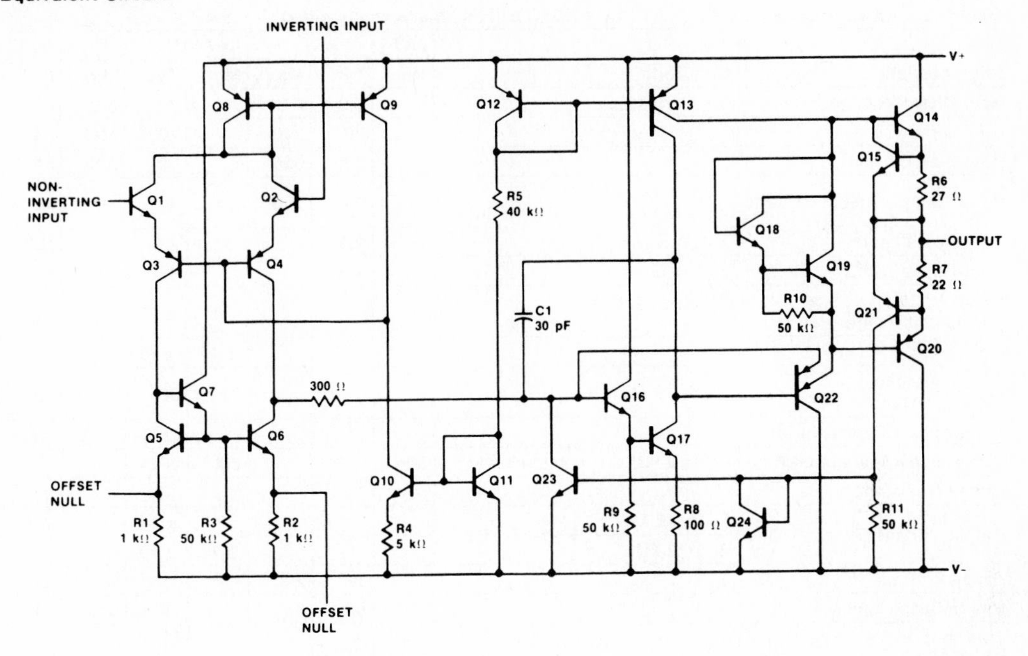

Notes

1. Rating applies to ambient temperatures up to 70°C. Above 70°C ambient derate linearly at 6.3 mW/°C for the metal package, 7.1 mW/°C for the flatpak, and 5.6 mW/°C for the DIP.
2. For supply voltages less than ± 15 V, the absolute maximum input voltage is equal to the supply voltage.
3. Short circuit may be to ground or either supply. Rating applies to +125°C case temperature or 75°C ambient temperature.

μA741 and μA741C
Electrical Characteristics $V_S = \pm 15$ V, $T_A = 25°C$ unless otherwise specified

Characteristic		Condition	μA741			μA741C			Unit
			Min	Typ	Max	Min	Typ	Max	
Input Offset Voltage		$R_S \leq 10$ kΩ		1.0	5.0		2.0	6.0	mV
Input Offset Current				20	200		20	200	nA
Input Bias Current				80	500		80	500	nA
Power Supply Rejection Ratio		$V_S = +10, -20$ $V_S = +20, -10$ V, $R_S = 50$ Ω		30	150		30	150	μV/V
Input Resistance			.3	2.0		.3	2.0		MΩ
Input Capacitance				1.4			1.4		pF
Offset Voltage Adjustment Range				± 15			± 15		mV
Input Voltage Range						± 12	± 13		V
Common Mode Rejection Ratio		$R_S \leq 10$ kΩ				70	90		dB
Output Short Circuit Current				25			25		mA
Large Signal Voltage Gain		$R_L \geq 2$ kΩ, $V_{OUT} = \pm 10$ V	50k	200k		20k	200k		
Output Resistance				75			75		Ω
Output Voltage Swing		$R_L \geq 10$ kΩ				± 12	± 14		V
		$R_L \geq 2$ kΩ				± 10	± 13		V
Supply Current				1.7	2.8		1.7	2.8	mA
Power Consumption				50	85		50	85	mW
Transient Response (Unity Gain)	Rise Time	$V_{IN} = 20$ mV, $R_L = 2$ kΩ, $C_L \leq 100$ pF		.3			.3		μs
	Overshoot			5.0			5.0		%
Bandwidth (Note 4)				1.0			1.0		MHz
Slew Rate		$R_L \geq 2$ kΩ		.5			.5		V/μs

Notes

4. Calculated value from $BW(MHz) = \dfrac{0.35}{\text{Rise Time } (\mu s)}$
5. All $V_{CC} = 15$ V for μA741 and μA741C.
6. Maximum supply current for all devices
   25°C = 2.8 mA
   125°C = 2.5 mA
   −55°C = 3.3 mA

μA741 and μA741C
Electrical Characteristics (Cont.) The following specifications apply over the range of $-55°C \leq T_A \leq 125°C$ for μA741, $0°C \leq T_A \leq 70°C$ for μA741C

Characteristic	Condition	μA741			μA741C			Unit
		Min	Typ	Max	Min	Typ	Max	
Input Offset Voltage							7.5	mV
	$R_S \leq 10\ k\Omega$		1.0	6.0				mV
Input Offset Current							300	nA
	$T_A = +125°C$		7.0	200				nA
	$T_A = -55°C$		85	500				nA
Input Bias Current							800	nA
	$T_A = +125°C$		.03	.5				μA
	$T_A = -55°C$		.3	1.5				μA
Input Voltage Range		±12	±13					V
Common Mode Rejection Ratio	$R_S \leq 10\ k\Omega$	70	90					dB
Adjustment for Input Offset Voltage			±15			±15		mV
Supply Voltage Rejection Ratio	$V_S = +10, -20$; $V_S = +20, -10\ V$, $R_S = 50\ \Omega$		30	150				μV/V
Output Voltage Swing	$R_L \geq 10\ k\Omega$	±12	±14					V
	$R_L \geq 2\ k\Omega$	±10	±13		±10	±13		V
Large Signal Voltage Gain	$R_L = 2\ k\Omega$, $V_{OUT} = \pm 10\ V$	25k			15k			
Supply Current	$T_A = +125°C$		1.5	2.5				mA
	$T_A = -55°C$		2.0	3.3				mA
Power Consumption	$T_A = +125°C$		45	75				mW
	$T_A = -55°C$		60	100				mW

Notes

4. Calculated value from $BW(MHz) = \frac{0.35}{\text{Rise Time } (\mu s)}$
5. All V_{CC} = 15 V for μA741 and μA741C.
6. Maximum supply current for all devices
   25°C = 2.8 mA
   125°C = 2.5 mA
   −55°C = 3.3 mA

μA741A and μA741E
Electrical Characteristics $V_S = \pm 15$ V, $T_A = 25°C$ unless otherwise specified.

Characteristic		Condition	μA741A/E Min	Typ	Max	Unit
Input Offset Voltage		$R_S \leq 50\ \Omega$		0.8	3.0	mV
Average Input Offset Voltage Drift					15	μV/°C
Input Offset Current				3.0	30	nA
Average Input Offset Current Drift					0.5	nA/°C
Input Bias Current				30	80	nA
Power Supply Rejection Ratio		$V_S = +10, -20$; $V_S = +20$ V, -10 V, $R_S = 50\ \Omega$		15	50	μV/V
Output Short Circuit Current			10	25	40	mA
Power Consumption		$V_S = \pm 20$ V		80	150	mW
Input Impedance		$V_S = \pm 20$ V	1.0	6.0		MΩ
Large Signal Voltage Gain		$V_S = \pm 20$ V, $R_L = 2$ kΩ, $V_{OUT} = \pm 15$ V	50	200		V/mV
Transient Response (Unity Gain)	Rise Time			0.25	0.8	μs
	Overshoot			6.0	20	%
Bandwidth (Note 4)			.437	1.5		MHz
Slew Rate (Unity Gain)		$V_{IN} = \pm 10$ V	0.3	0.7		V/μs

The following specifications apply over the range of $-55°C \leq T_A \leq 125°C$ for the 741A, and $0°C \leq T_A \leq 70°C$ for the 741E.

Characteristic	Condition			Min	Typ	Max	Unit
Input Offset Voltage						4.0	mV
Input Offset Current						70	nA
Input Bias Current						210	nA
Common Mode Rejection Ratio	$V_S = \pm 20$ V, $V_{IN} = \pm 15$ V, $R_S = 50\ \Omega$			80	95		dB
Adjustment For Input Offset Voltage	$V_S = \pm 20$ V			10			mV
Output Short Circuit Current				10		40	mA
Power Consumption	$V_S = \pm 20$ V	μA741A	−55°C			165	mW
			+125°C			135	mW
		μA741E				150	mW
Input Impedance	$V_S = \pm 20$ V			0.5			MΩ
Output Voltage Swing	$V_S = \pm 20$ V	$R_L = 10$ kΩ		±16			V
		$R_L = 2$ kΩ		±15			V
Large Signal Voltage Gain	$V_S = \pm 20$ V, $R_L = 2$ kΩ, $V_{OUT} = \pm 15$ V			32			V/mV V/mV
	$V_S = \pm 5$ V, $R_L = 2$ kΩ, $V_{OUT} = \pm 2$ V			10			V/mV

Notes

4. Calculated value from: $BW(MHz) = \frac{0.35}{\text{Rise Time } (\mu s)}$
5. All $V_{CC} = 15$ V for μA741 and μA741C.
6. Maximum supply current for all devices
   25°C = 2.8 mA
   125°C = 2.5 mA
   −55°C = 3.3 mA

Typical Performance Curves for μA741A and μA741

Open Loop Voltage Gain as a Function of Supply Voltage

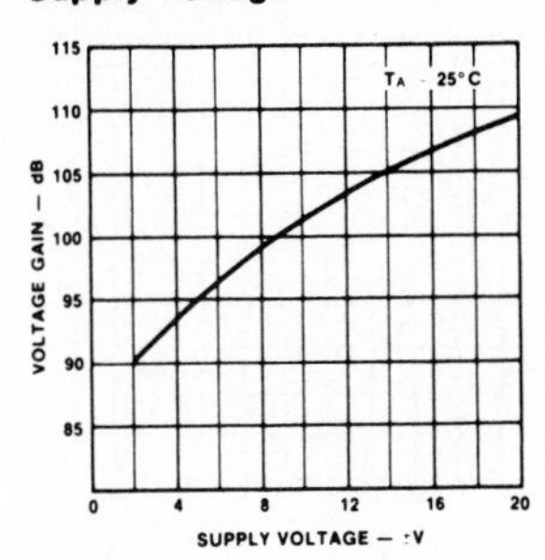

Output Voltage Swing as a Function of Supply Voltage

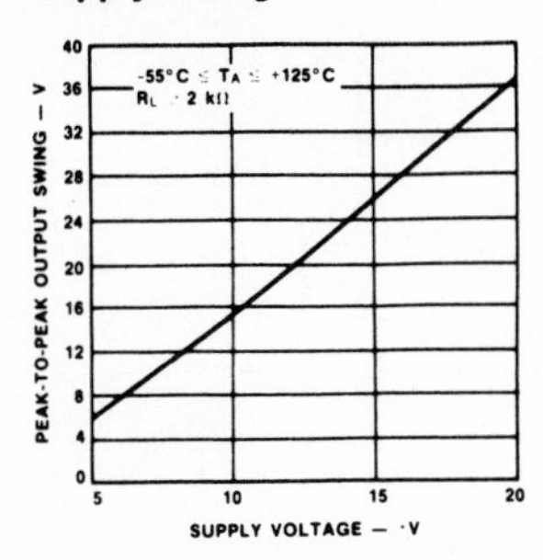

Input Common Mode Voltage as a Function of Supply Voltage

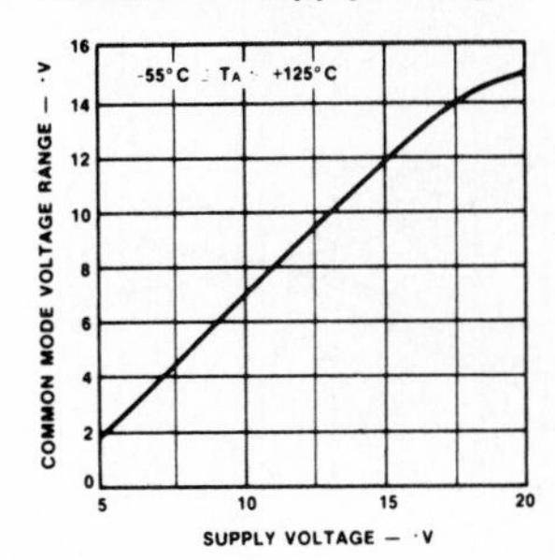

Typical Performance Curves for μA741E and μA741C

Open Loop Voltage Gain as a Function of Supply Voltage

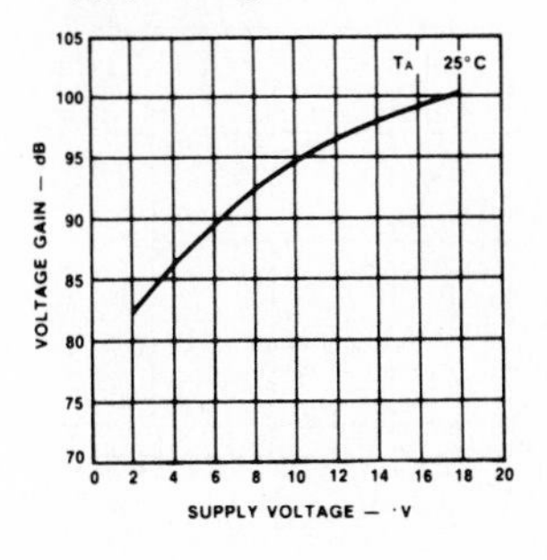

Output Voltage Swing as a Function of Supply Voltage

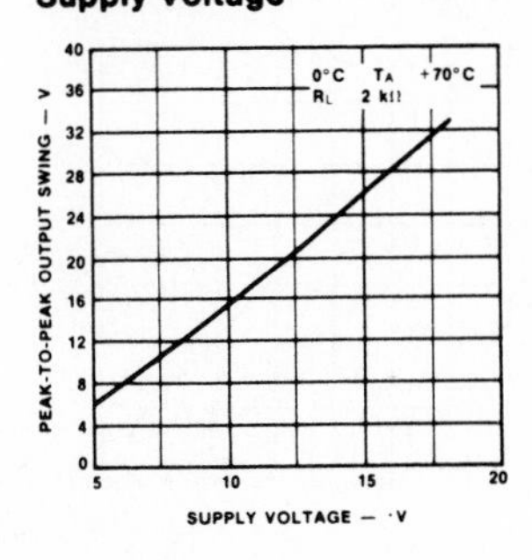

Input Common Mode Voltage Range as a Function of Supply Voltage

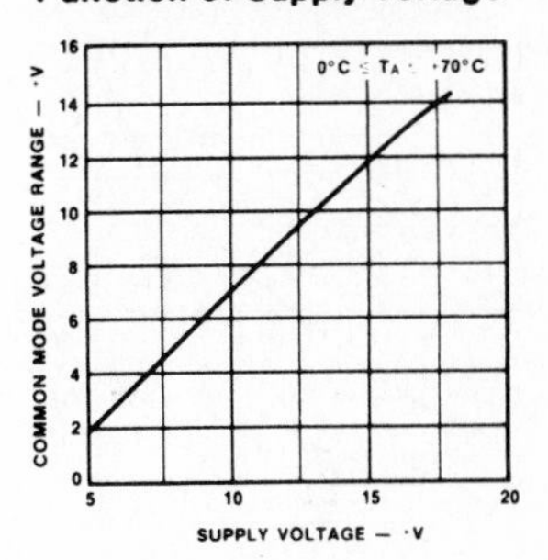

Transient Response

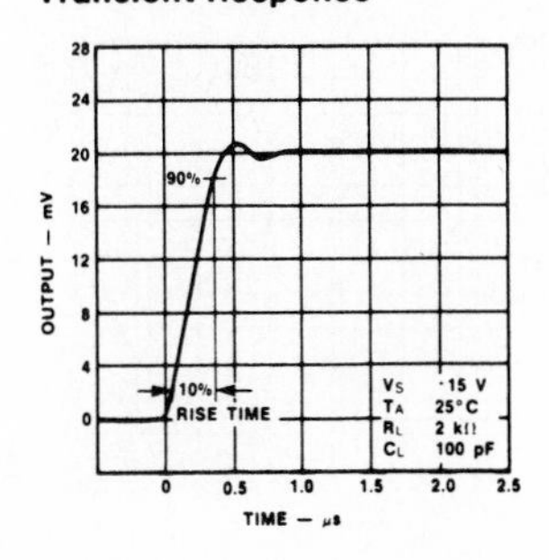

Transient Response Test Circuit

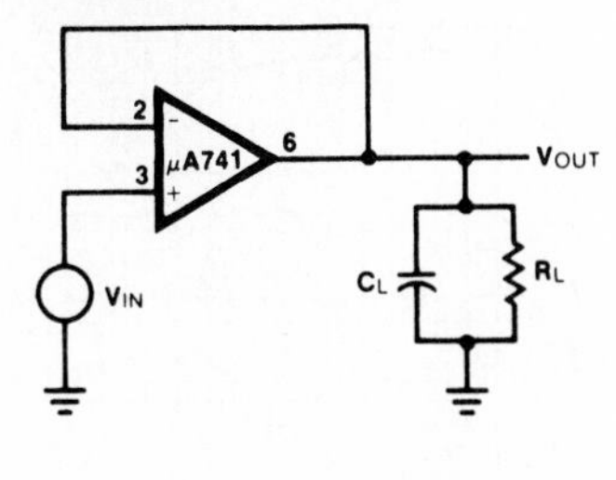

Common Mode Rejection Ratio as a Function of Frequency

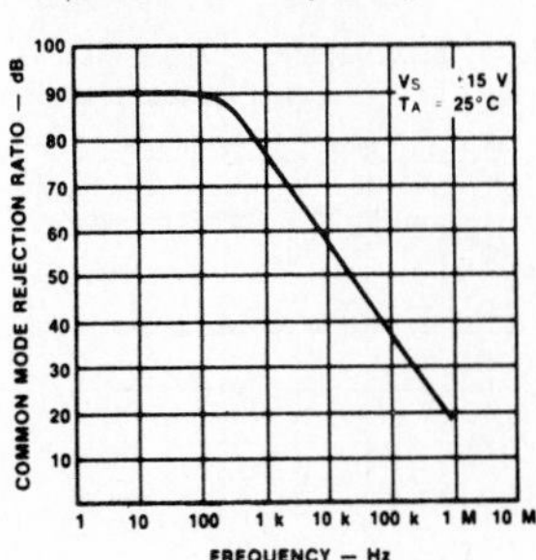

Typical Performance Curves for μA741E and μA741C (Cont.)

Frequency Characteristics as a Function of Supply Voltage

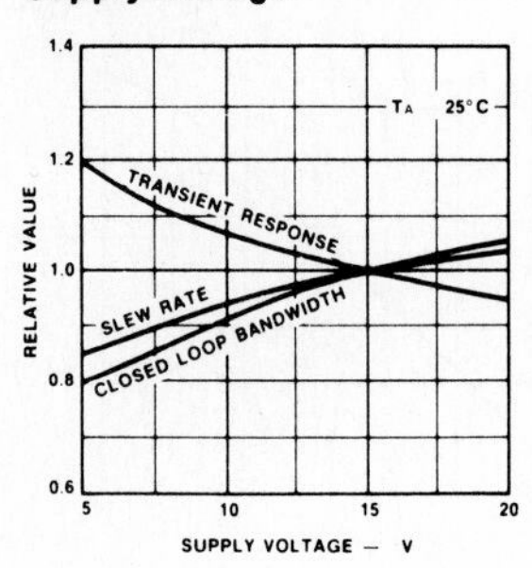

Voltage Offset Null Circuit

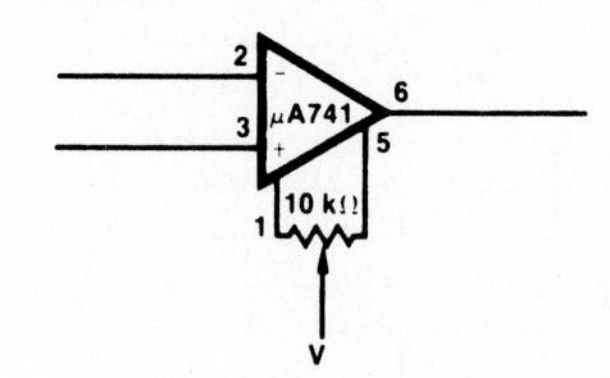

Voltage Follower Large Signal Pulse Response

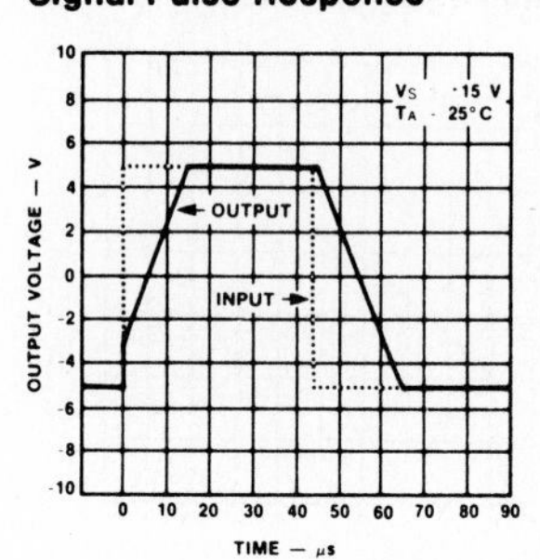

Typical Performance Curves for μA741A, μA741, μA741E and μA741C

Power Consumption as a Function of Supply Voltage

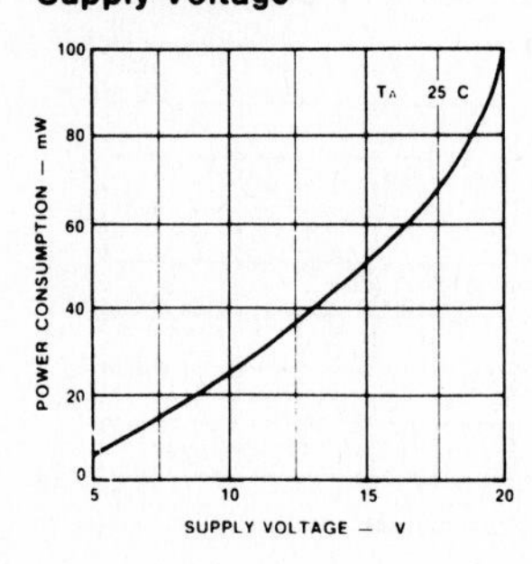

Open Loop Voltage Gain as a Function of Frequency

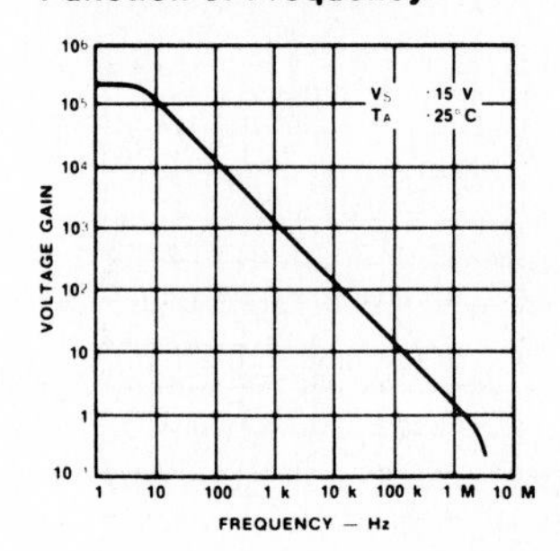

Open Loop Phase Response as a Function of Frequency

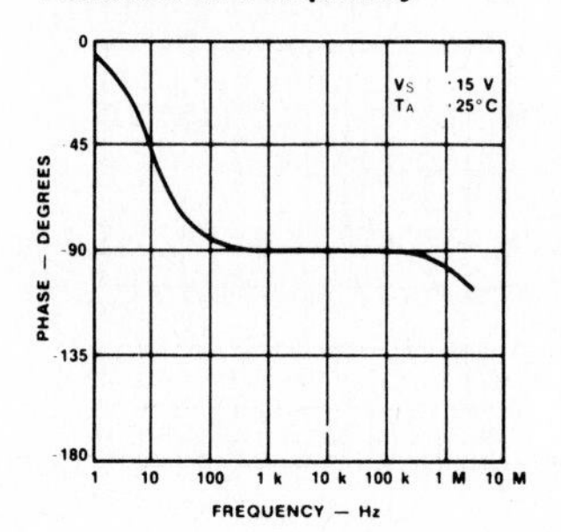

Input Offset Current as a Function of Supply Voltage

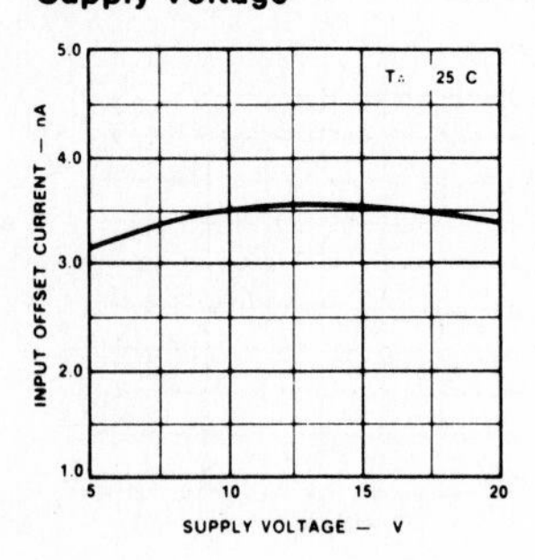

Input Resistance and Input Capacitance as a Function of Frequency

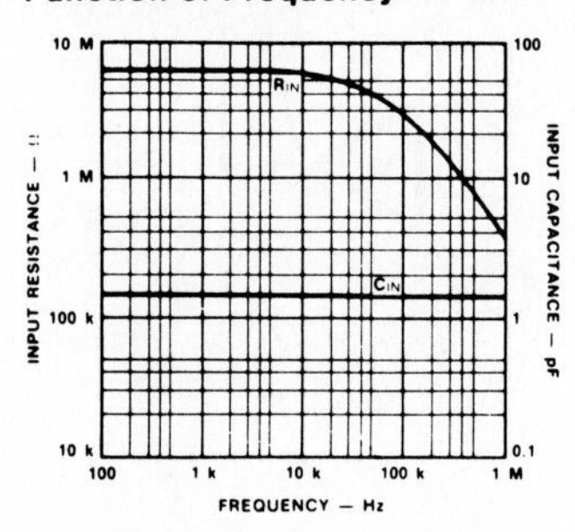

Output Resistance as a Function of Frequency

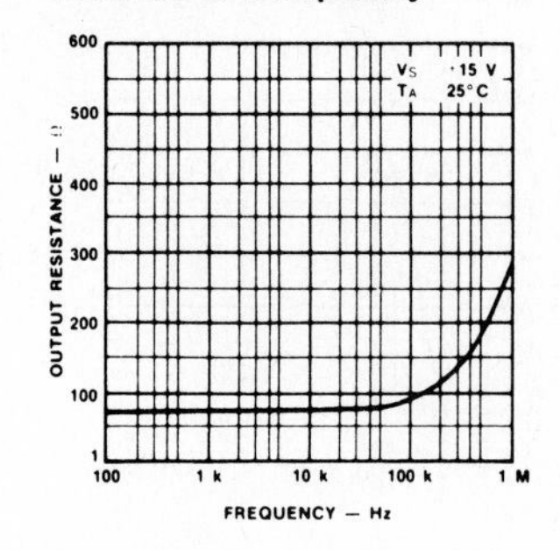

Typical Performance Curves for μA741A, μA741, μA741E and μA741C (Cont.)

Output Voltage Swing as a Function of Load Resistance

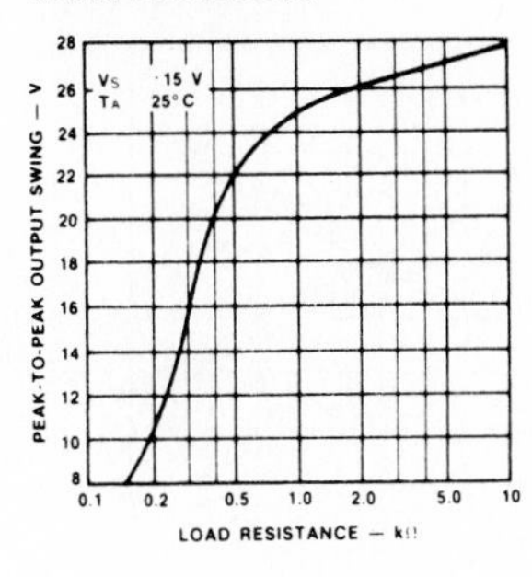

Output Voltage Swing as a Function of Frequency

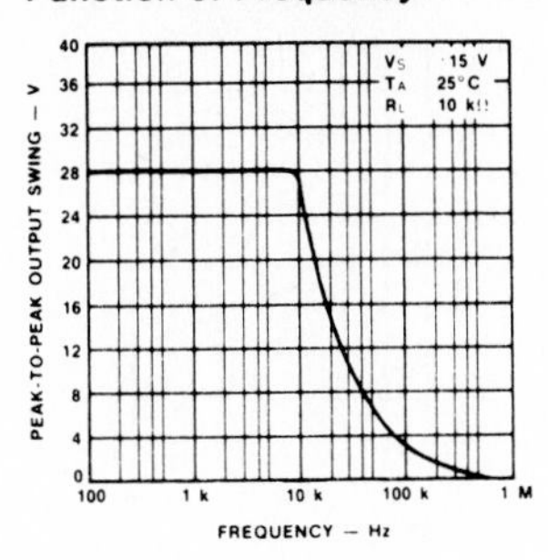

Absolute Maximum Power Dissipation as a Function of Ambient Temperature

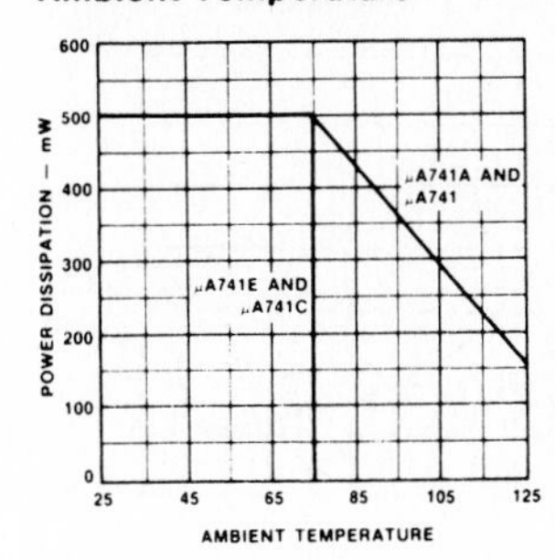

Input Noise Voltage as a Function of Frequency

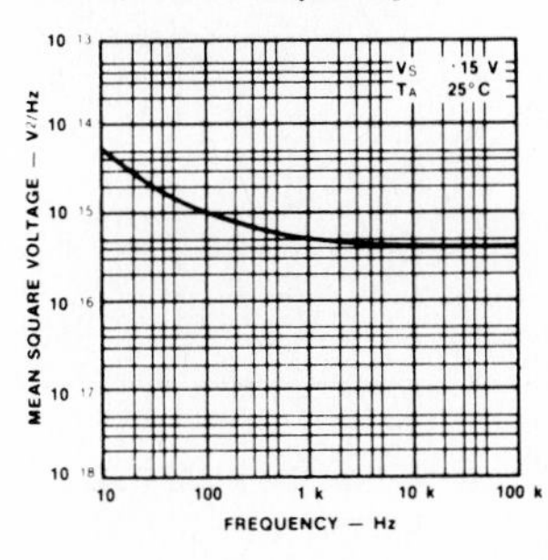

Input Noise Current as a Function of Frequency

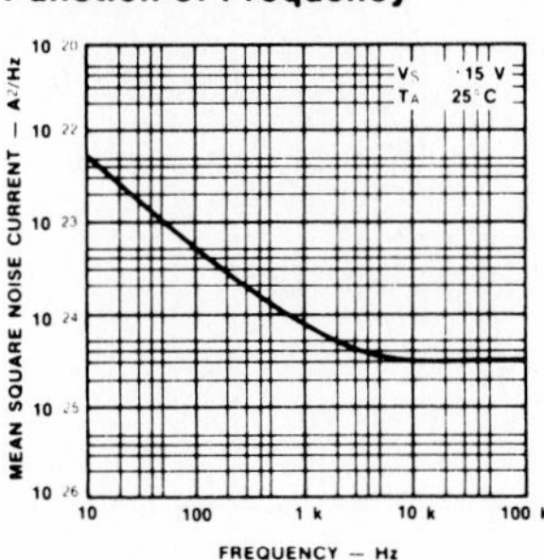

Broadband Noise for Various Bandwidths

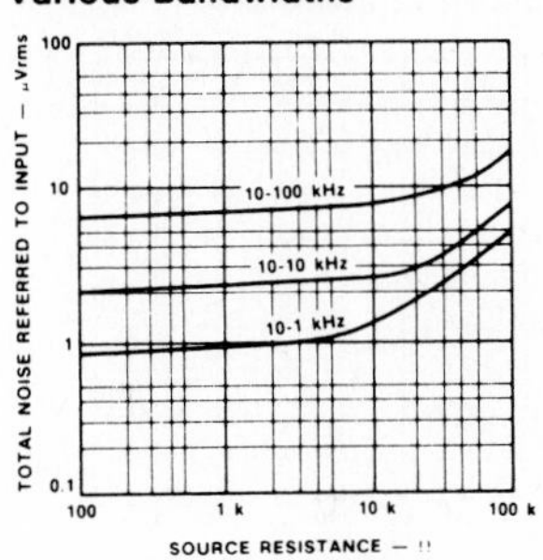

Typical Performance Curves for μA741A and μA741

Input Bias Current as a Function of Ambient Temperature

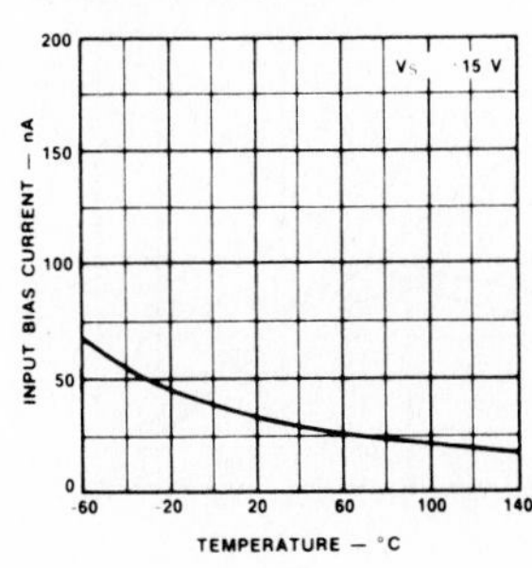

Input Resistance as a Function of Ambient Temperature

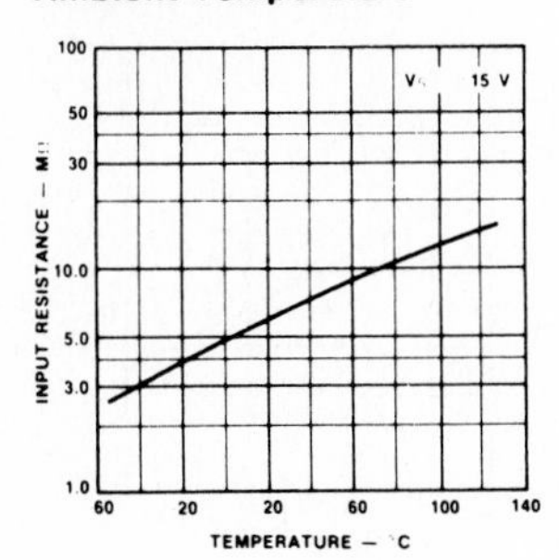

Output Short-Circuit Current as a Function of Ambient Temperature

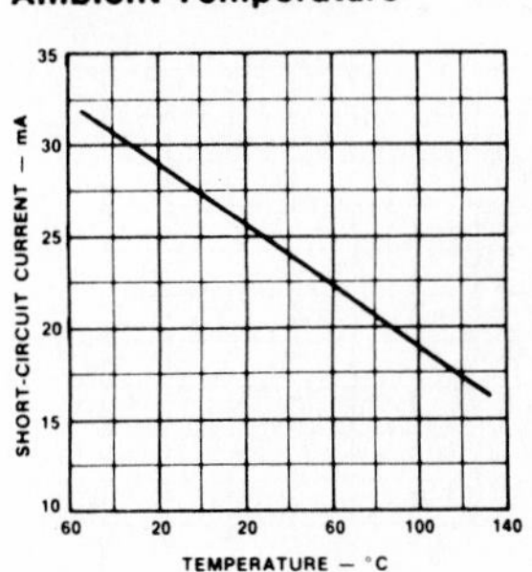

Typical Performance Curves for μA741A and μA741 (Cont.)

Input Offset Current as a Function of Ambient Temperature

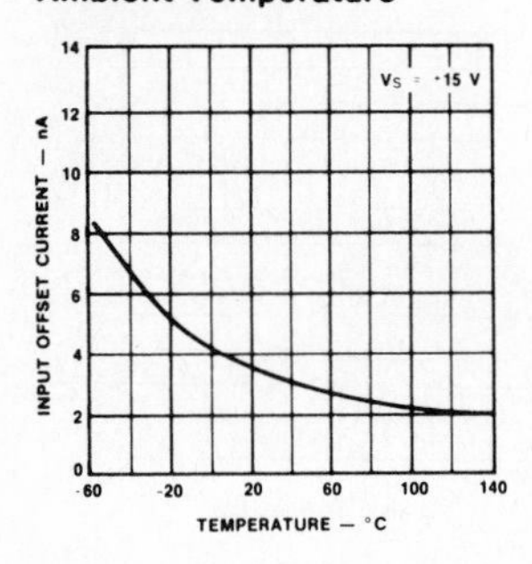

Power Consumption as a Function of Ambient Temperature

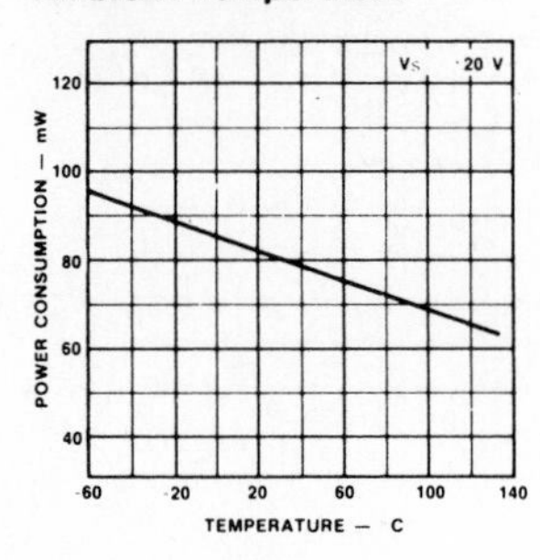

Frequency Characteristics as a Function of Ambient Temperature

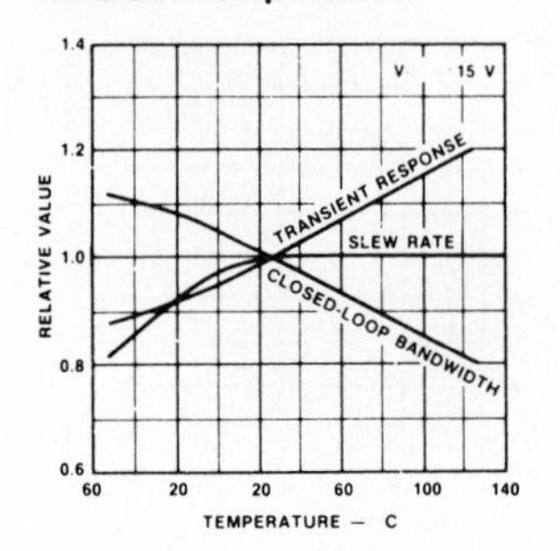

Typical Performance Curves for μA741E and μA741C

Input Bias Current as a Function of Ambient Temperature

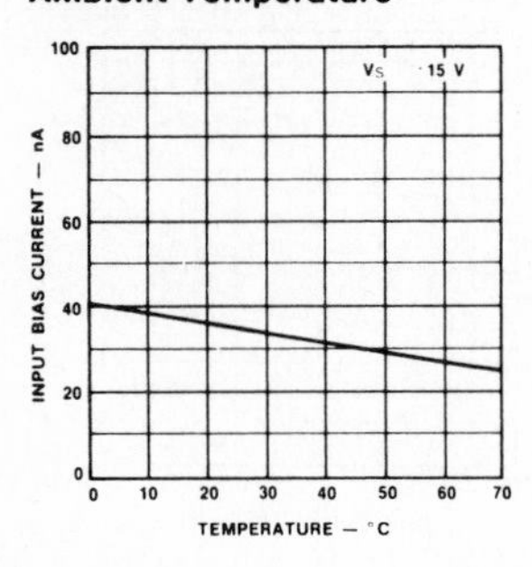

Input Resistance as a Function of Ambient Temperature

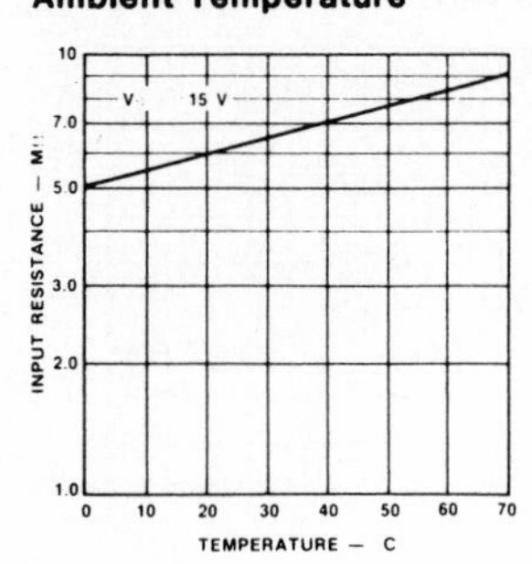

Input Offset Current as a Function of Ambient Temperature

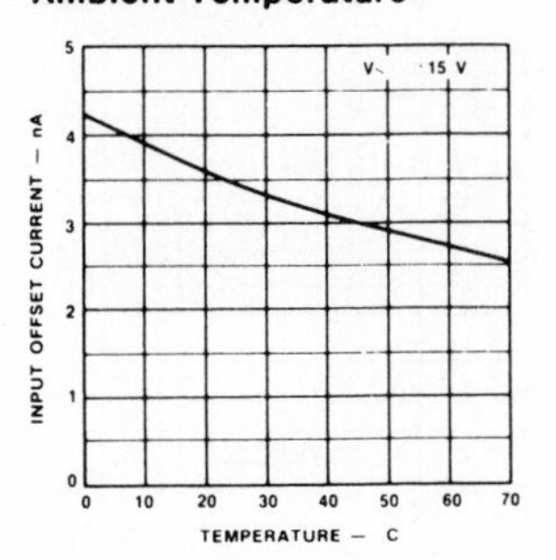

Power Consumption as a Function of Ambient Temperature

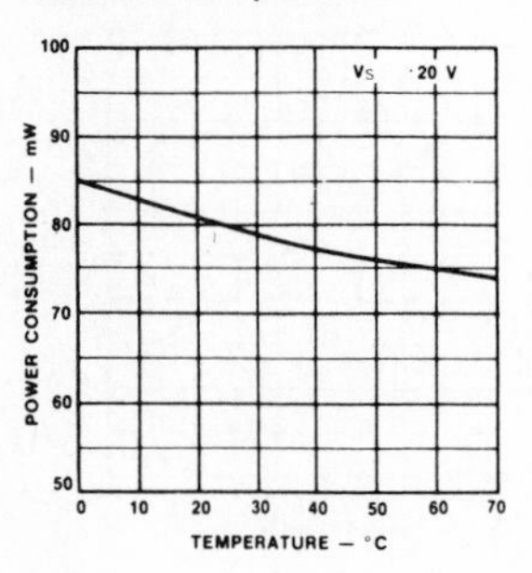

Output Short Circuit Current as a Function of Ambient Temperature

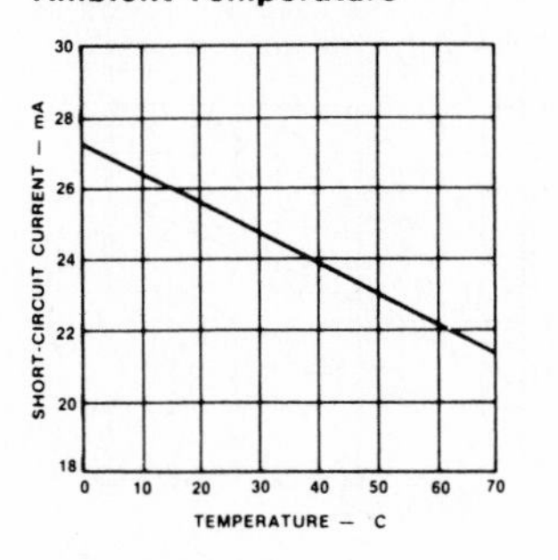

Frequency Characteristics as a Function of Ambient Temperature

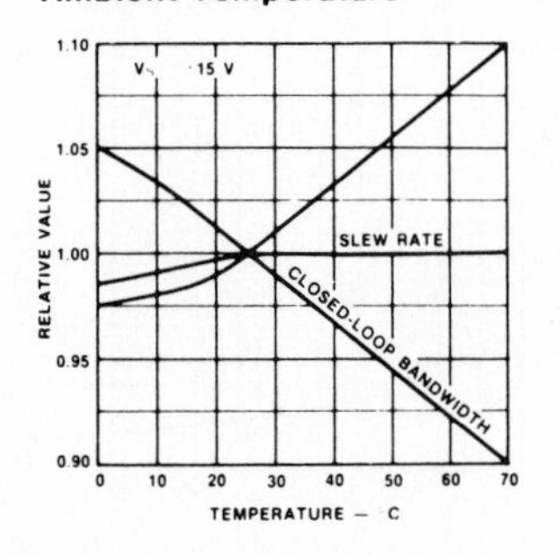

APPENDIX 2

LM301 Operational Amplifier*

*Courtesy of National Semiconductor Corporation.

General Description

The LM101A series are general purpose operational amplifiers which feature improved performance over industry standards like the LM709. Advanced processing techniques make possible an order of magnitude reduction in input currents, and a redesign of the biasing circuitry reduces the temperature drift of input current. Improved specifications include:

- Offset voltage 3 mV maximum over temperature (LM101A/LM201A)
- Input current 100 nA maximum over temperature (LM101A/LM201A)
- Offset current 20 nA maximum over temperature (LM101A/LM201A)
- Guaranteed drift characteristics
- Offsets guaranteed over entire common mode and supply voltage ranges
- Slew rate of 10V/μs as a summing amplifier

This amplifier offers many features which make its application nearly foolproof: overload protection on the input and output, no latch-up when the common mode range is exceeded, freedom from oscillations and compensation with a single 30 pF capacitor. It has advantages over internally compensated amplifiers in that the frequency compensation can be tailored to the particular application. For example, in low frequency circuits it can be overcompensated for increased stability margin. Or the compensation can be optimized to give more than a factor of ten improvement in high frequency performance for most applications.

In addition, the device provides better accuracy and lower noise in high impedance circuitry. The low input currents also make it particularly well suited for long interval integrators or timers, sample and hold circuits and low frequency waveform generators. Further, replacing circuits where matched transistor pairs buffer the inputs of conventional IC op amps, it can give lower offset voltage and drift at a lower cost.

The LM101A is guaranteed over a temperature range of −55°C to +125°C, the LM201A from −25°C to +85°C, and the LM301A from 0°C to 70°C.

Schematic ** and Connection Diagrams (Top Views)

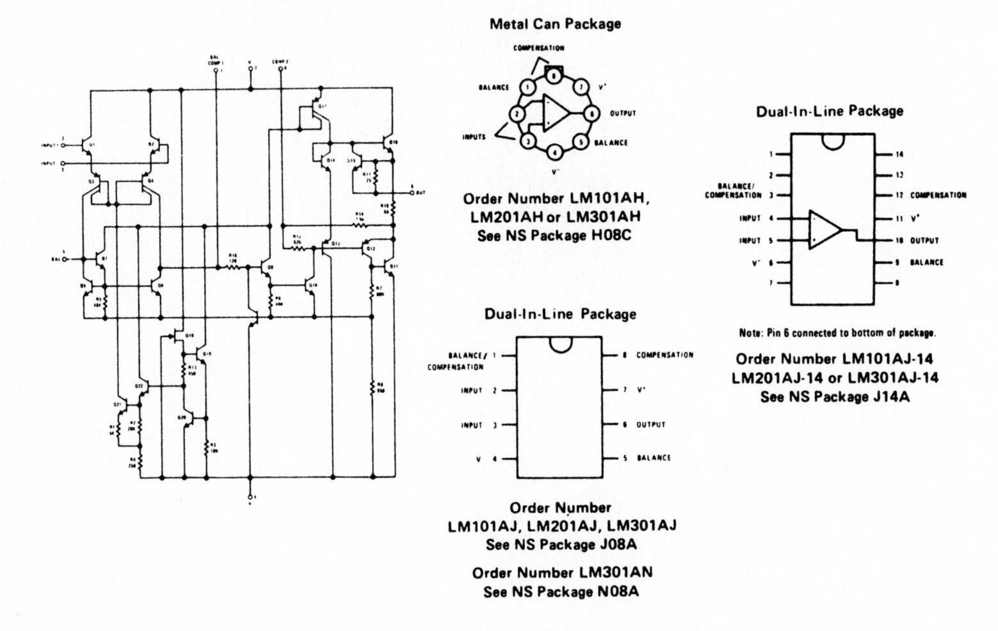

Order Number LM101AH, LM201AH or LM301AH
See NS Package H08C

Order Number LM101AJ, LM201AJ, LM301AJ
See NS Package J08A

Order Number LM301AN
See NS Package N08A

Note: Pin 6 connected to bottom of package.

Order Number LM101AJ-14
LM201AJ-14 or LM301AJ-14
See NS Package J14A

**Pin connections shown are for metal can.

Absolute Maximum Ratings

	LM101A/LM201A	LM301A
Supply Voltage	±22V	±18V
Power Dissipation (Note 1)	500 mW	500 mW
Differential Input Voltage	±30V	±30V
Input Voltage (Note 2)	±15V	±15V
Output Short Circuit Duration (Note 3)	Indefinite	Indefinite
Operating Temperature Range	−55°C to +125°C (LM101A) −25°C to +85°C (LM201A)	0°C to +70°C
Storage Temperature Range	−65°C to +150°C	−65°C to +150°C
Lead Temperature (Soldering, 10 seconds)	300°C	300°C

Electrical Characteristics (Note 4)

PARAMETER	CONDITIONS	LM101A/LM201A			LM301A			UNITS
		MIN	TYP	MAX	MIN	TYP	MAX	
Input Offset Voltage	$T_A = 25°C$							
LM101A, LM201A, LM301A	$R_S \leq 50\ k\Omega$		0.7	2.0		2.0	7.5	mV
Input Offset Current	$T_A = 25°C$		1.5	10		3.0	50	nA
Input Bias Current	$T_A = 25°C$		30	75		70	250	nA
Input Resistance	$T_A = 25°C$	1.5	4.0		0.5	2.0		MΩ
Supply Current	$T_A = 25°C$							
	$V_S = \pm 20V$		1.8	3.0				mA
	$V_S = \pm 15V$					1.8	3.0	mA
Large Signal Voltage Gain	$T_A = 25°C$, $V_S = \pm 15V$ $V_{OUT} = \pm 10V$, $R_L \geq 2\ k\Omega$	50	160		25	160		V/mV
Input Offset Voltage	$R_S \leq 50\ k\Omega$			3.0			10	mV
	$R_S \leq 10\ k\Omega$							mV
Average Temperature Coefficient of Input Offset Voltage	$R_S \leq 50\ k\Omega$		3.0	15		6.0	30	μV/°C
	$R_S \leq 10\ k\Omega$							μV/°C
Input Offset Current				20			70	nA
	$T_A = T_{MAX}$							nA
	$T_A = T_{MIN}$							nA
Average Temperature Coefficient of Input Offset Current	$25°C \leq T_A \leq T_{MAX}$		0.01	0.1		0.01	0.3	nA/°C
	$T_{MIN} \leq T_A \leq 25°C$		0.02	0.2		0.02	0.6	nA/°C
Input Bias Current				0.1			0.3	μA
Supply Current	$T_A = T_{MAX}$, $V_S = \pm 20V$		1.2	2.5				mA
Large Signal Voltage Gain	$V_S = \pm 15V$, $V_{OUT} = \pm 10V$, $R_L \geq 2k$	25			15			V/mV
Output Voltage Swing	$V_S = \pm 15V$							
	$R_L = 10\ k\Omega$	±12	±14		±12	±14		V
	$R_L = 2\ k\Omega$	±10	±13		±10	±13		V
Input Voltage Range	$V_S = \pm 20V$	±15						V
	$V_S = \pm 15V$		+15, −13		±12	+15, −13		V
Common-Mode Rejection Ratio	$R_S \leq 50\ k\Omega$	80	96		70	90		dB
	$R_S \leq 10\ k\Omega$							dB
Supply Voltage Rejection Ratio	$R_S \leq 50\ k\Omega$	80	96		70	96		dB
	$R_S \leq 10\ k\Omega$							dB

Note 1: The maximum junction temperature of the LM101A is 150°C, and that of the LM201A/LM301A is 100°C. For operating at elevated temperatures, devices in the TO-5 package must be derated based on a thermal resistance of 150°C/W, junction to ambient, or 45°C/W, junction to case. The thermal resistance of the dual-in-line package is 187°C/W, junction to ambient.

Note 2: For supply voltages less than ±15V, the absolute maximum input voltage is equal to the supply voltage.

Note 3: Continuous short circuit is allowed for case temperatures to 125°C and ambient temperatures to 75°C for LM101A/LM201A, and 70°C and 55°C respectively for LM301A.

Note 4: Unless otherwise specified, these specifications apply for C1 = 30 pF, $\pm 5V \leq V_S \leq \pm 20V$ and $-55°C \leq T_A \leq +125°C$ (LM101A), $\pm 5V \leq V_S \leq \pm 20V$ and $-25°C \leq T_A \leq +85°C$ (LM201A), $\pm 5V \leq V_S \leq \pm 15V$ and $0°C \leq T_A \leq +70°C$ (LM301A).

Guaranteed Performance Characteristics LM101A/LM201A

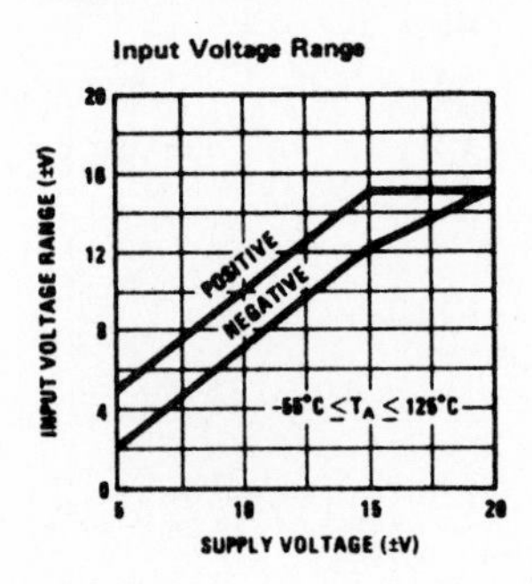

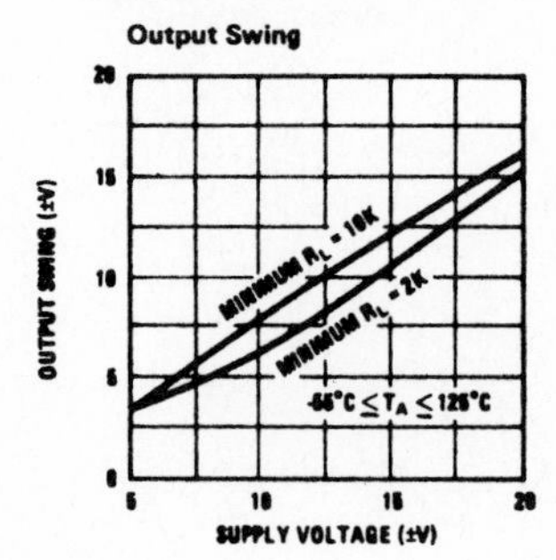

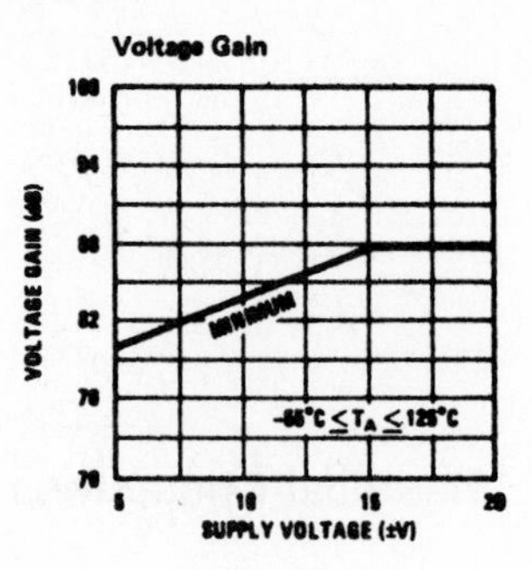

Guaranteed Performance Characteristics LM301A

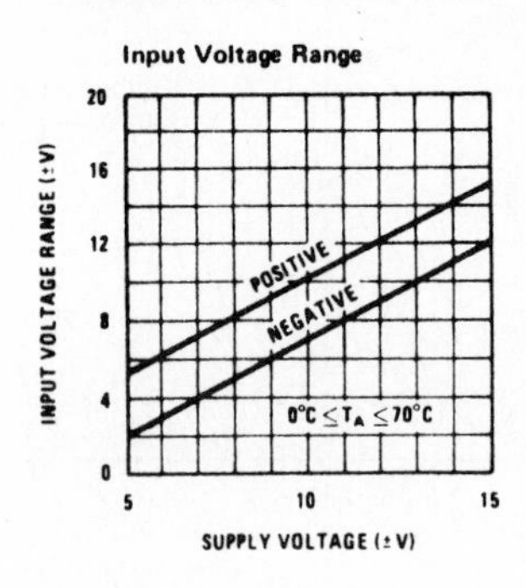

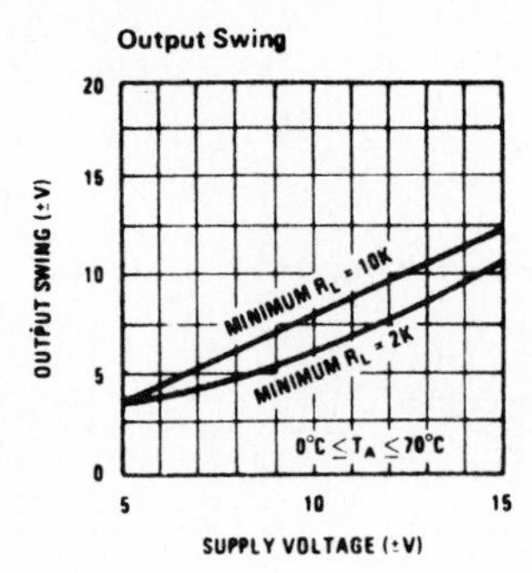

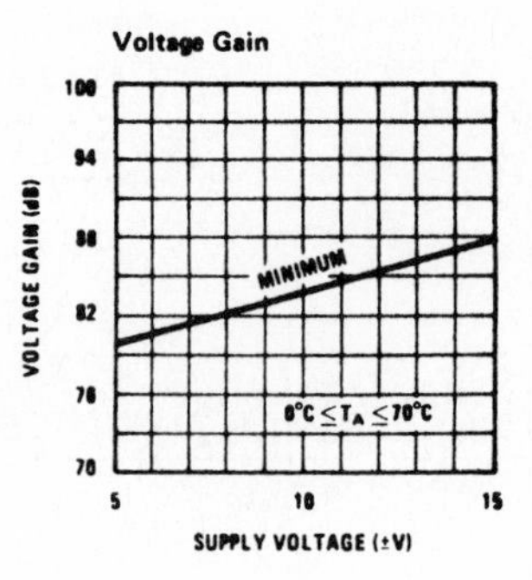

Typical Performance Characteristics

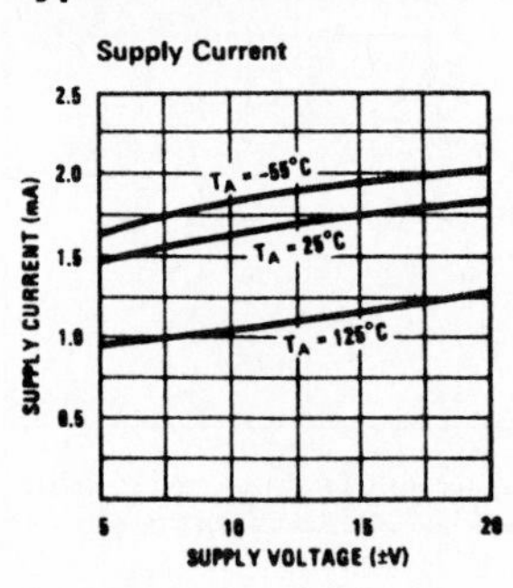

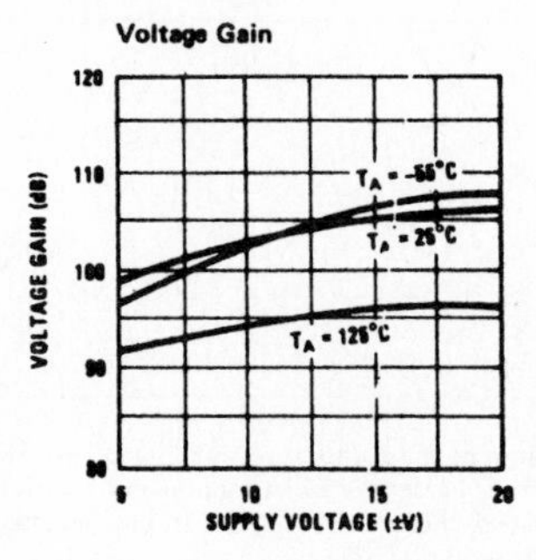

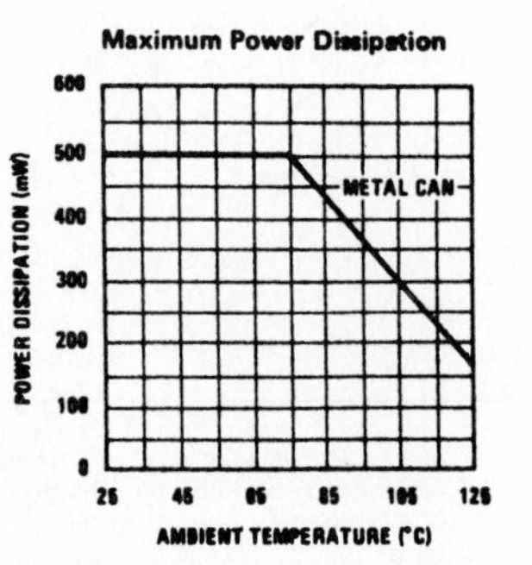

Typical Performance Characteristics (Continued)

Input Current, LM101A/ LM201A/LM301A

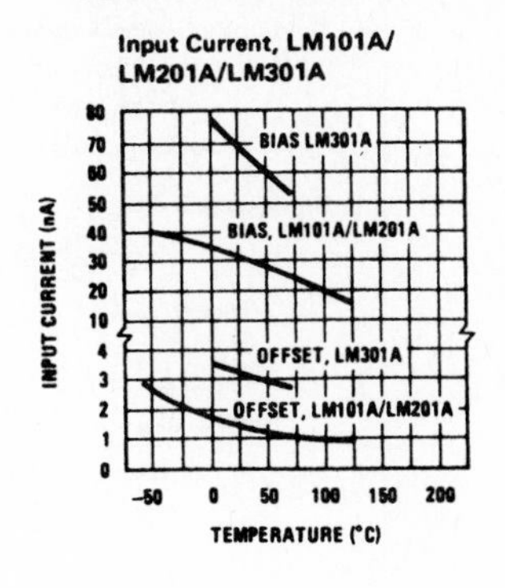

Current Limiting

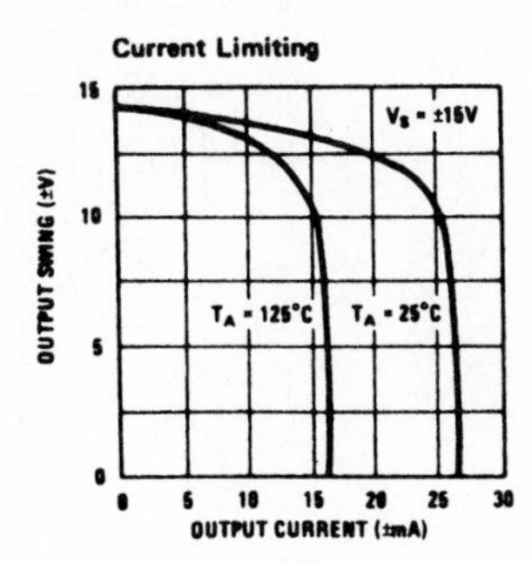

Input Noise Voltage

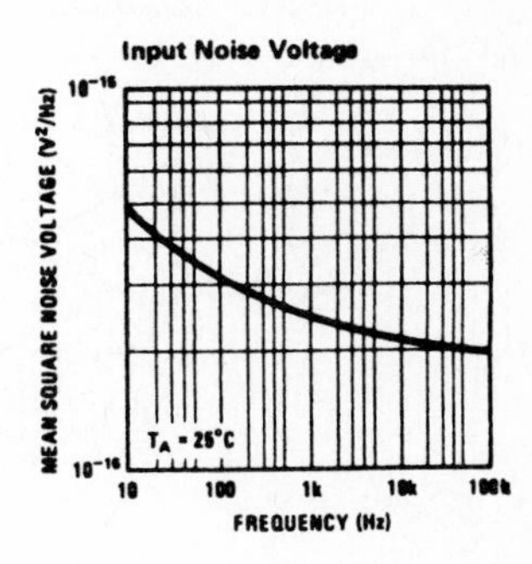

Input Noise Current

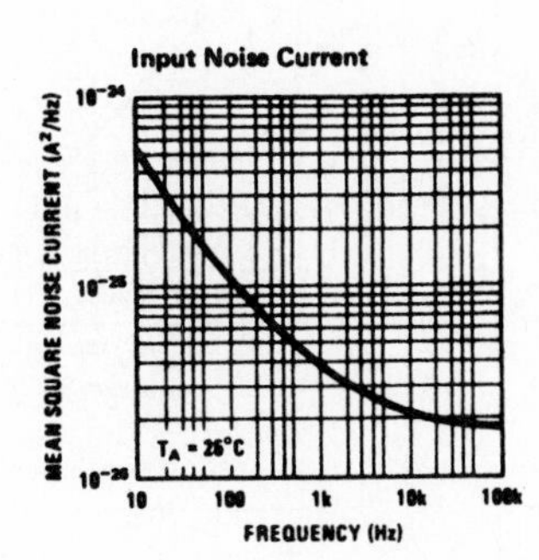

Common Mode Rejection

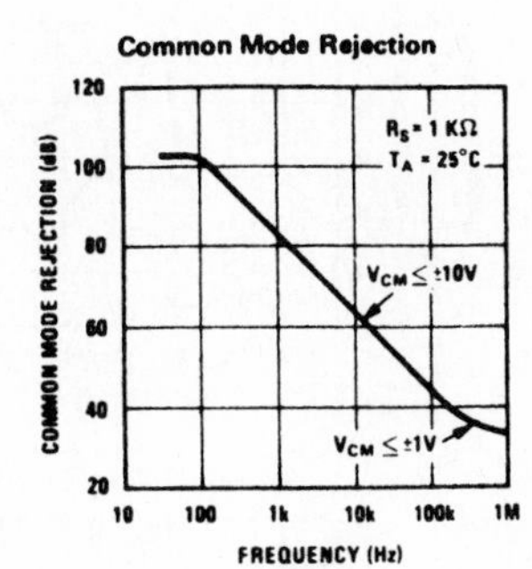

Power Supply Rejection

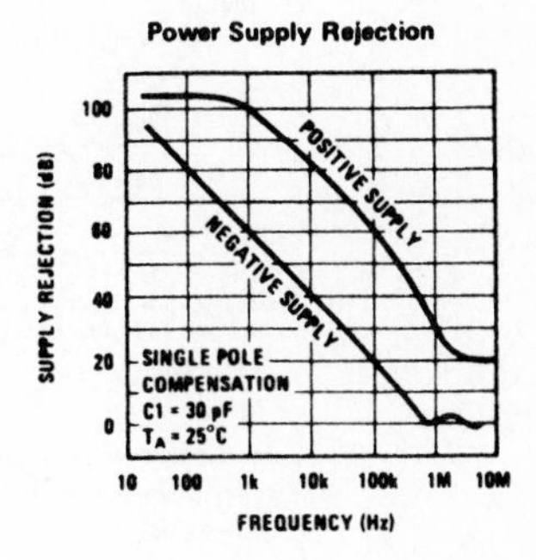

Closed Loop Output Impedance

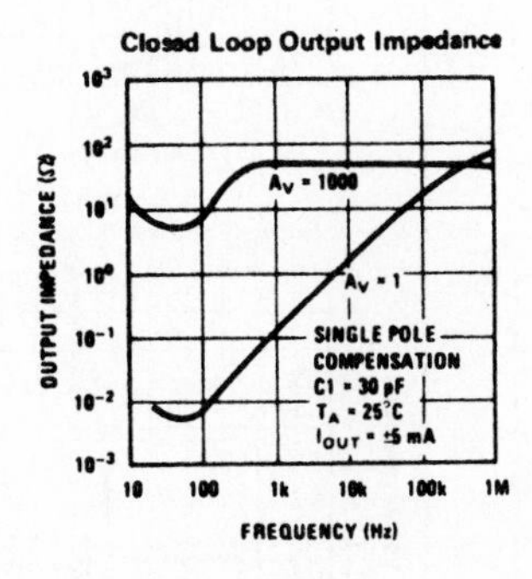

Typical Performance Characteristics for Various Compensation Circuits**

Single Pole Compensation

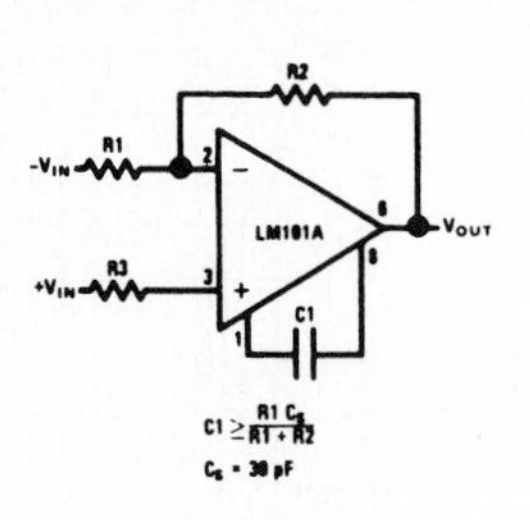

Two Pole Compensation

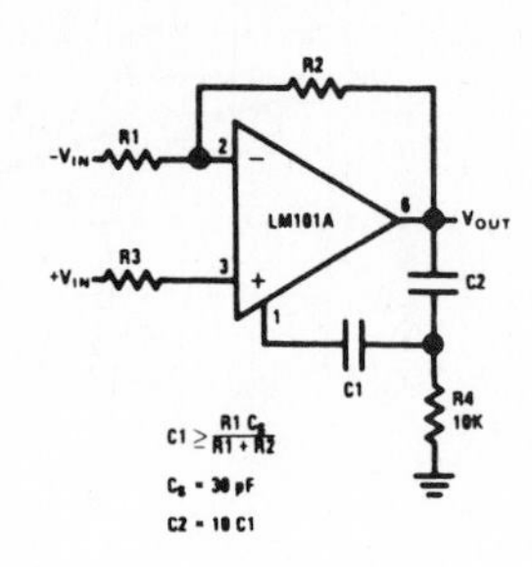

Feedforward Compensation

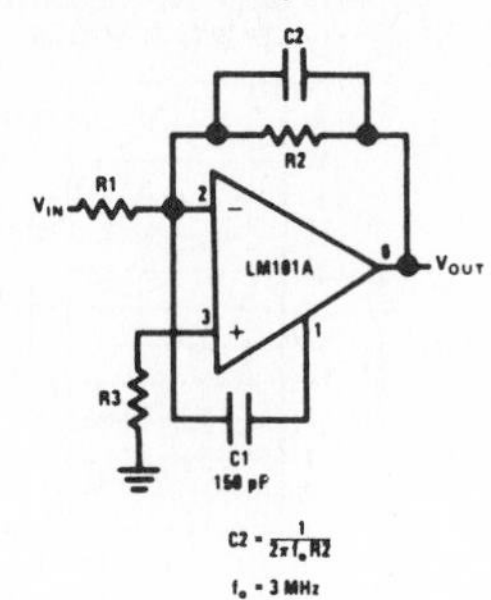

**Pin connections shown are for metal can.

Open Loop Frequency Response

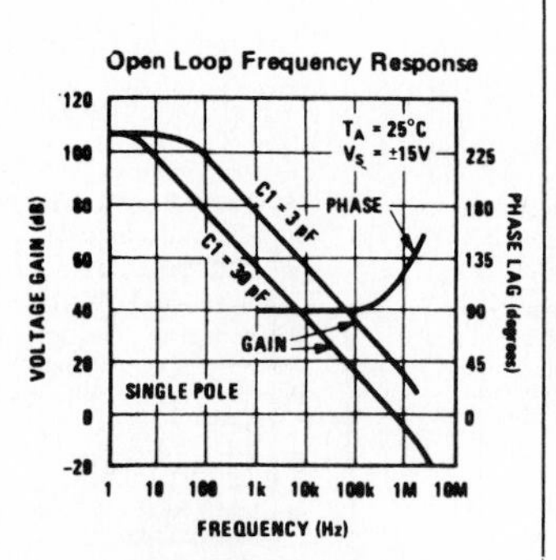

Open Loop Frequency Response

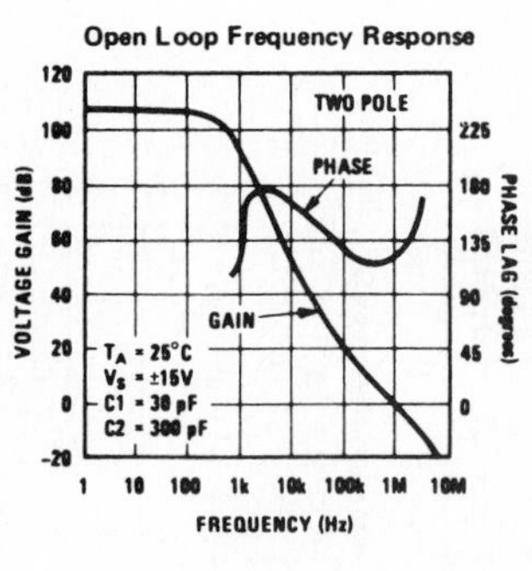

Open Loop Frequency Response

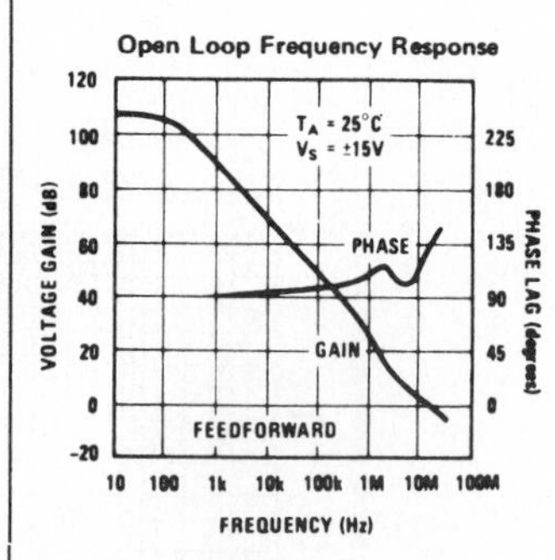

Large Signal Frequency Response

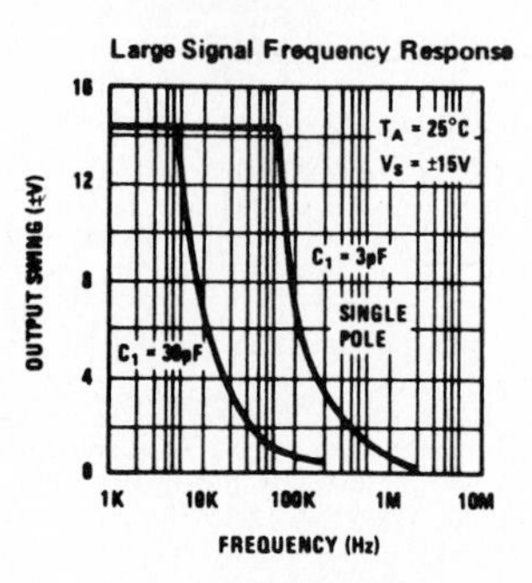

Large Signal Frequency Response

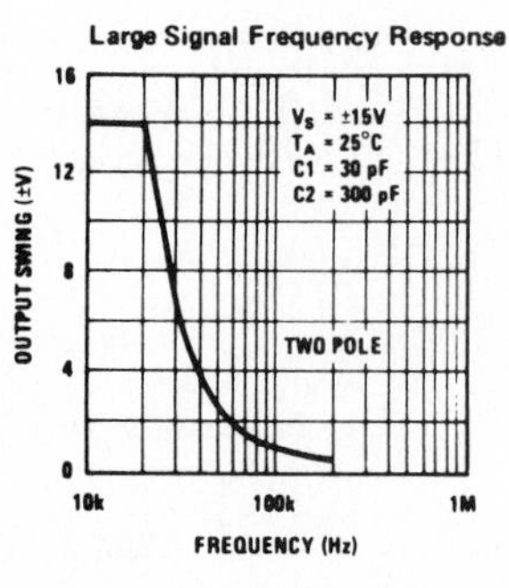

Large Signal Frequency Response

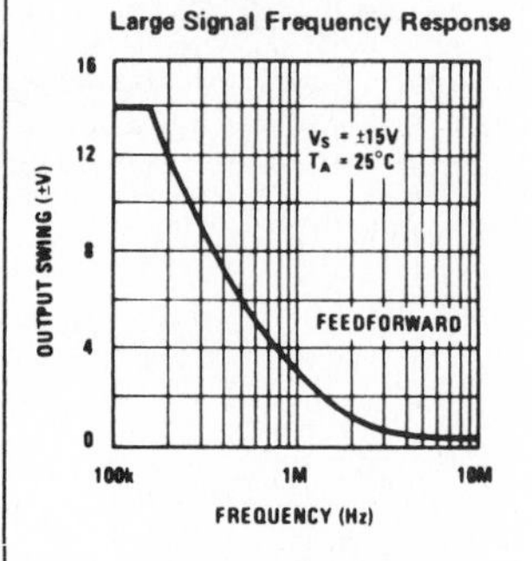

Voltage Follower Pulse Response

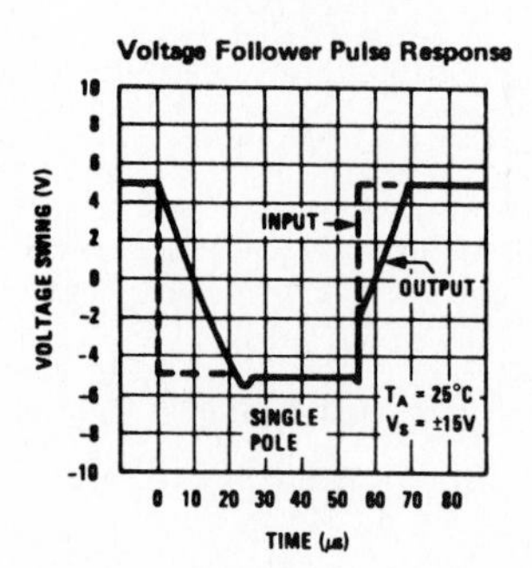

Voltage Follower Pulse Response

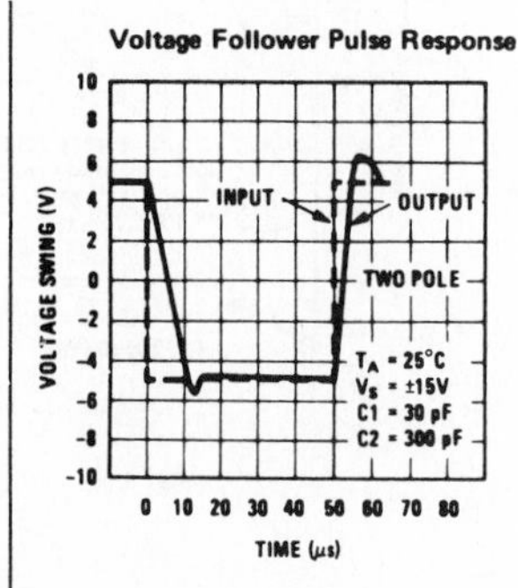

Inverter Pulse Response

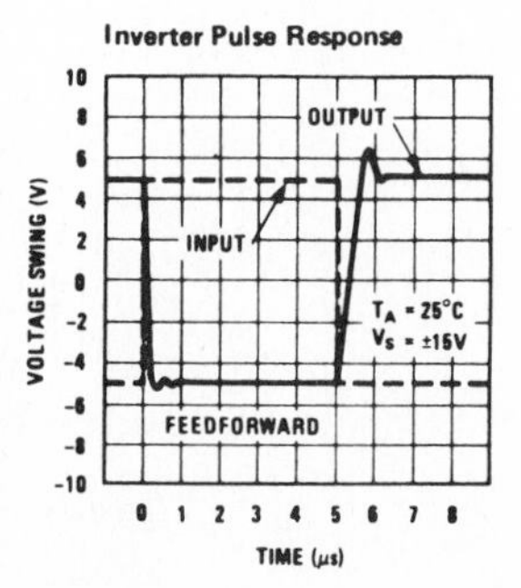

Typical Applications **

Variable Capacitance Multiplier

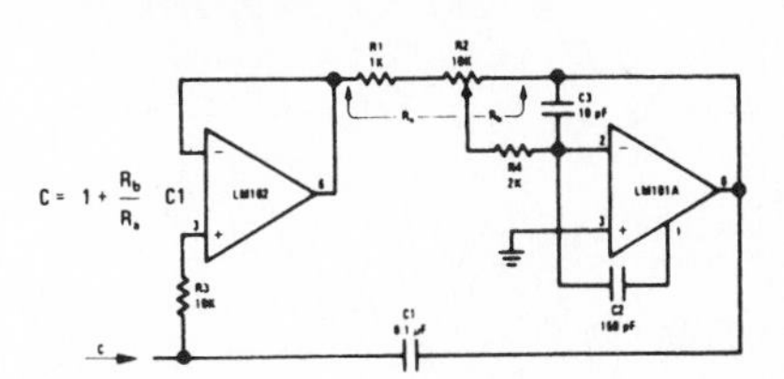

Simulated Inductor

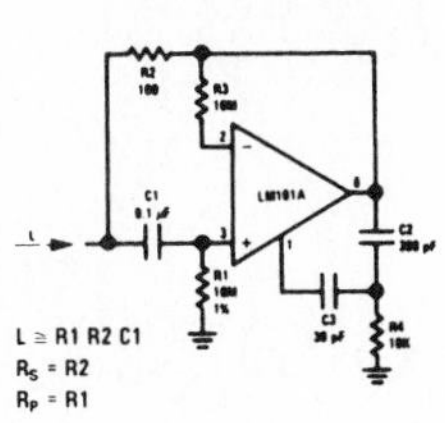

Fast Inverting Amplifier With High Input Impedance

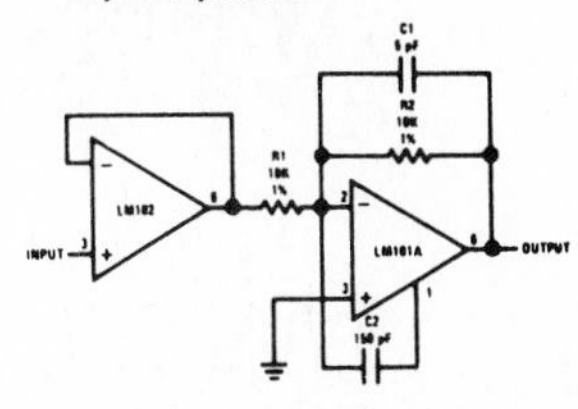

Inverting Amplifier with Balancing Circuit

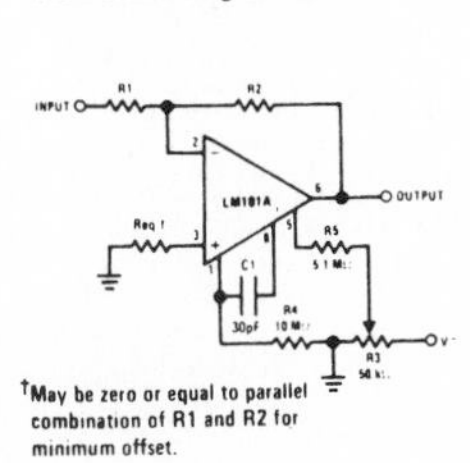

†May be zero or equal to parallel combination of R1 and R2 for minimum offset.

Sine Wave Oscillator

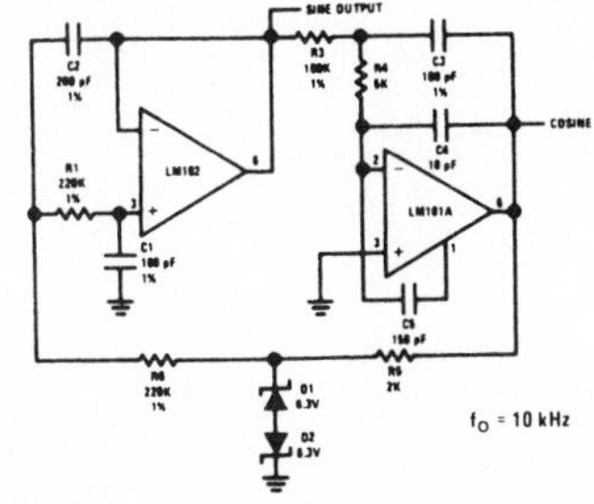

$f_O = 10$ kHz

Integrator with Bias Current Compensation

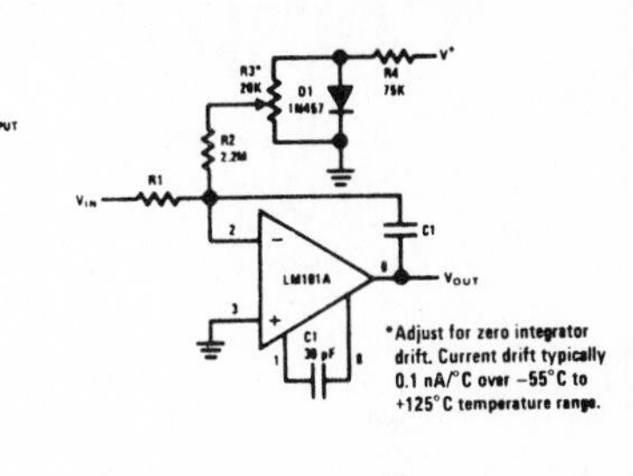

*Adjust for zero integrator drift. Current drift typically 0.1 nA/°C over −55°C to +125°C temperature range.

Application Hints **

Protecting Against Gross Fault Conditions

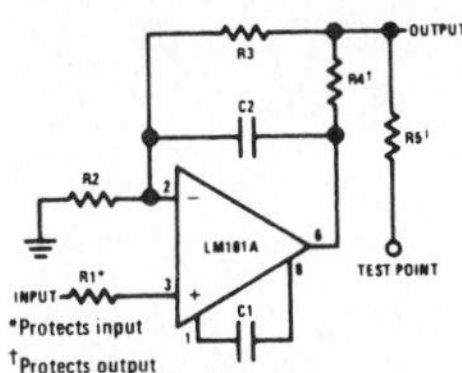

*Protects input

†Protects output

‡Protects output—not needed when R4 is used.

Compensating For Stray Input Capacitances Or Large Feedback Resistor

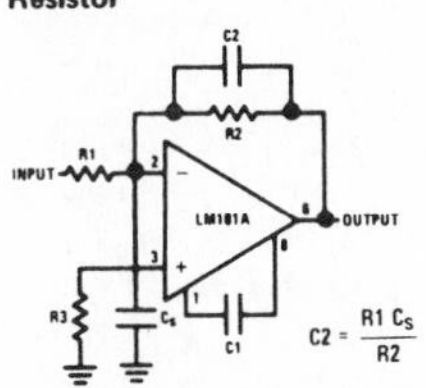

Isolating Large Capacitive Loads

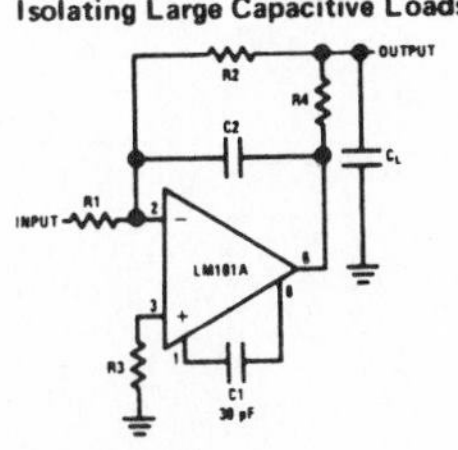

Although the LM101A is designed for trouble free operation, experience has indicated that it is wise to observe certain precautions given below to protect the devices from abnormal operating conditions. It might be pointed out that the advice given here is applicable to practically any IC op amp, although the exact reason why may differ with different devices.

When driving either input from a low-impedance source, a limiting resistor should be placed in series with the input lead to limit the peak instantaneous output current of the source to something less than 100 mA. This is especially important when the inputs go outside a piece of equipment where they could accidentally be connected to high voltage sources. Large capacitors on the input (greater than 0.1 μF) should be treated as a low source impedance and isolated with a resistor. Low impedance sources do not cause a problem unless their output voltage exceeds the supply voltage. However, the supplies go to zero when they are turned off, so the isolation is usually needed.

The output circuitry is protected against damage from shorts to ground. However, when the amplifier output is connected to a test point, it should be isolated by a limiting resistor, as test points frequently get shorted to bad places. Further, when the amplifier drives a load external to the equipment, it is also advisable to use some sort of limiting resistance to preclude mishaps.

Precautions should be taken to insure that the power supplies for the integrated circuit never become reversed—even under transient conditions. With reverse voltages greater than 1V, the IC will conduct excessive current, fuzing internal aluminum interconnects. If there is a possibility of this happening, clamp diodes with a high peak current rating should be installed on the supply lines. Reversal of the voltage between V^+ and V^- will always cause a problem, although reversals with respect to ground may also give difficulties in many circuits.

The minimum values given for the frequency compensation capacitor are stable only for source resistances less than 10 kΩ, stray capacitances on the summing junction less than 5 pF and capacitive loads smaller than 100 pF. If any of these conditions are not met, it becomes necessary to overcompensate the amplifier with a larger compensation capacitor. Alternately, lead capacitors can be used in the feedback network to negate the effect of stray capacitance and large feedback resistors or an RC network can be added to isolate capacitive loads.

Although the LM101A is relatively unaffected by supply bypassing, this cannot be ignored altogether. Generally it is necessary to bypass the supplies to ground at least once on every circuit card, and more bypass points may be required if more than five amplifiers are used. When feed-forward compensation is employed, however, it is advisable to bypass the supply leads of each amplifier with low inductance capacitors because of the higher frequencies involved.

**Pin connections shown are for metal can.

APPENDIX 3

LM311 Voltage Comparator*

*Courtesy of National Semiconductor Corporation.

General Description

The LM311 is a voltage comparator that has input currents more than a hundred times lower than devices like the LM306 or LM710C. It is also designed to operate over a wider range of supply voltages: from standard ±15V op amp supplies down to the single 5V supply used for IC logic. Its output is compatible with RTL, DTL and TTL as well as MOS circuits. Further, it can drive lamps or relays, switching voltages up to 40V at currents as high as 50 mA.

Features

- Operates from single 5V supply
- Maximum input current: 250 nA
- Maximum offset current: 50 nA
- Differential input voltage range: ±30V
- Power consumption: 135 mW at ±15V

Both the input and the output of the LM311 can be isolated from system ground, and the output can drive loads referred to ground, the positive supply or the negative supply. Offset balancing and strobe capability are provided and outputs can be wire OR'ed. Although slower than the LM306 and LM710C (200 ns response time vs 40 ns) the device is also much less prone to spurious oscillations. The LM311 has the same pin configuration as the LM306 and LM710C. See the "application hints" of the LM311 for application help.

Auxiliary Circuits**

**Note: Pin connections shown on schematic diagram and typical applications are for TO-5 package.

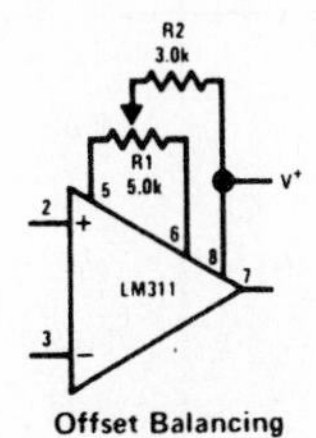

Offset Balancing

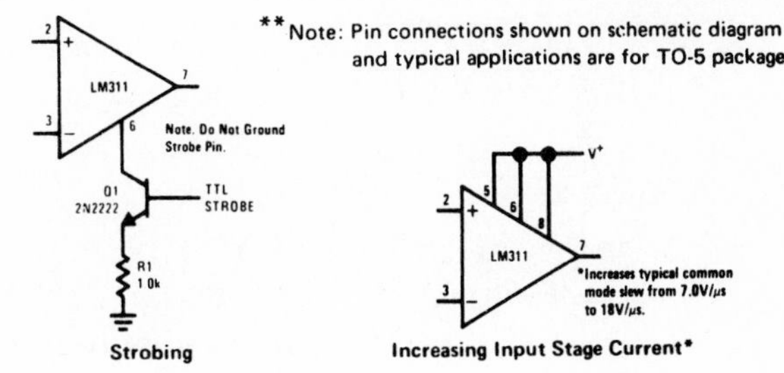

Strobing

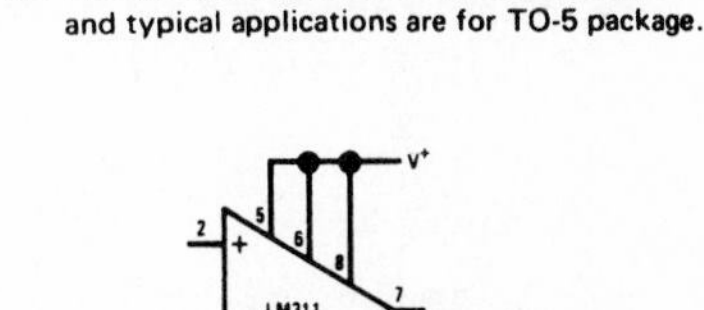

Increasing Input Stage Current*

Typical Applications**

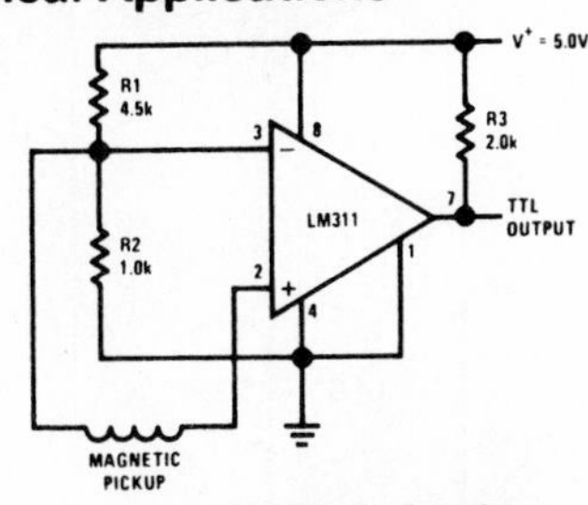

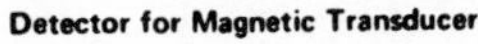

Detector for Magnetic Transducer

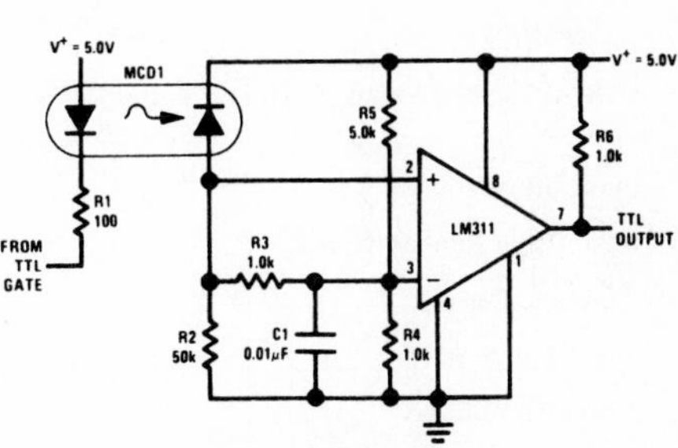

Digital Transmission Isolator

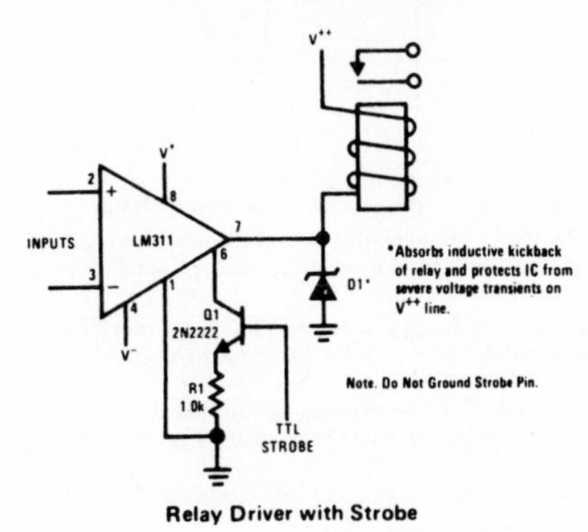

Relay Driver with Strobe

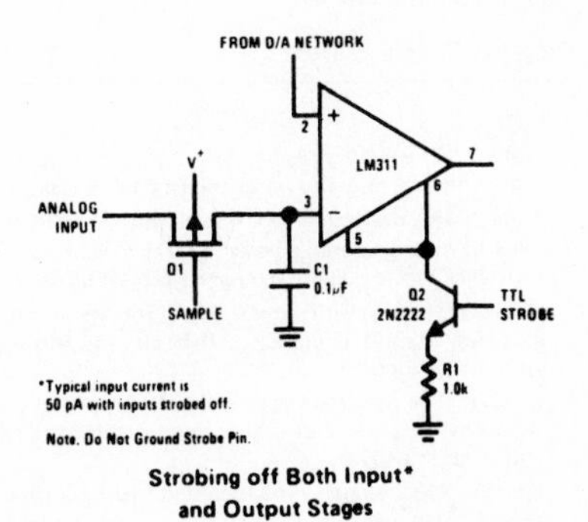

Strobing off Both Input* and Output Stages

Absolute Maximum Ratings

Total Supply Voltage (V_{84})	36V
Output to Negative Supply Voltage (V_{74})	40V
Ground to Negative Supply Voltage (V_{14})	30V
Differential Input Voltage	±30V
Input Voltage (Note 1)	±15V
Power Dissipation (Note 2)	500 mW
Output Short Circuit Duration	10 sec
Operating Temperature Range	0°C to 70°C
Storage Temperature Range	-65°C to 150°C
Lead Temperature (soldering, 10 sec)	300°C
Voltage at Strobe Pin	V^+–5V

Electrical Characteristics (Note 3)

PARAMETER	CONDITIONS	MIN	TYP	MAX	UNITS
Input Offset Voltage (Note 4)	T_A = 25°C, $R_S \leq$ 50k		2.0	7.5	mV
Input Offset Current (Note 4)	T_A = 25°C		6.0	50	nA
Input Bias Current	T_A = 25°C		100	250	nA
Voltage Gain	T_A = 25°C	40	200		V/mV
Response Time (Note 5)	T_A = 25°C		200		ns
Saturation Voltage	$V_{IN} \leq -10$ mV, I_{OUT} = 50 mA T_A = 25°C		0.75	1.5	V
Strobe ON Current	T_A = 25°C		3.0		mA
Output Leakage Current	$V_{IN} \geq$ 10 mV, V_{OUT} = 35V T_A = 25°C, I_{STROBE} = 3 mA		0.2	50	nA
Input Offset Voltage (Note 4)	$R_S \leq$ 50k			10	mV
Input Offset Current (Note 4)				70	nA
Input Bias Current				300	nA
Input Voltage Range		−14.5	13.8,−14.7	13.0	V
Saturation Voltage	$V^+ \geq$ 4.5V, V^- = 0 $V_{IN} \leq -10$ mV, $I_{SINK} \leq$ 8 mA		0.23	0.4	V
Positive Supply Current	T_A = 25°C		5.1	7.5	mA
Negative Supply Current	T_A = 25°C		4.1	5.0	mA

Note 1: This rating applies for ±15V supplies. The positive input voltage limit is 30V above the negative supply. The negative input voltage limit is equal to the negative supply voltage or 30V below the positive supply, whichever is less.

Note 2: The maximum junction temperature of the LM311 is 110°C. For operating at elevated temperatures, devices in the TO-5 package must be derated based on a thermal resistance of 150°C/W, junction to ambient, or 45°C/W, junction to case. The thermal resistance of the dual-in-line package is 100°C/W, junction to ambient.

Note 3: These specifications apply for V_S = ±15V and the Ground pin at ground, and 0°C < T_A < +70°C, unless otherwise specified. The offset voltage, offset current and bias current specifications apply for any supply voltage from a single 5V supply up to ±15V supplies.

Note 4: The offset voltages and offset currents given are the maximum values required to drive the output within a volt of either supply with 1 mA load. Thus, these parameters define an error band and take into account the worst-case effects of voltage gain and input impedance.

Note 5: The response time specified (see definitions) is for a 100 mV input step with 5 mV overdrive.

Note 6: Do not short the strobe pin to ground; it should be current driven at 3 to 5 mA.

Typical Performance Characteristics

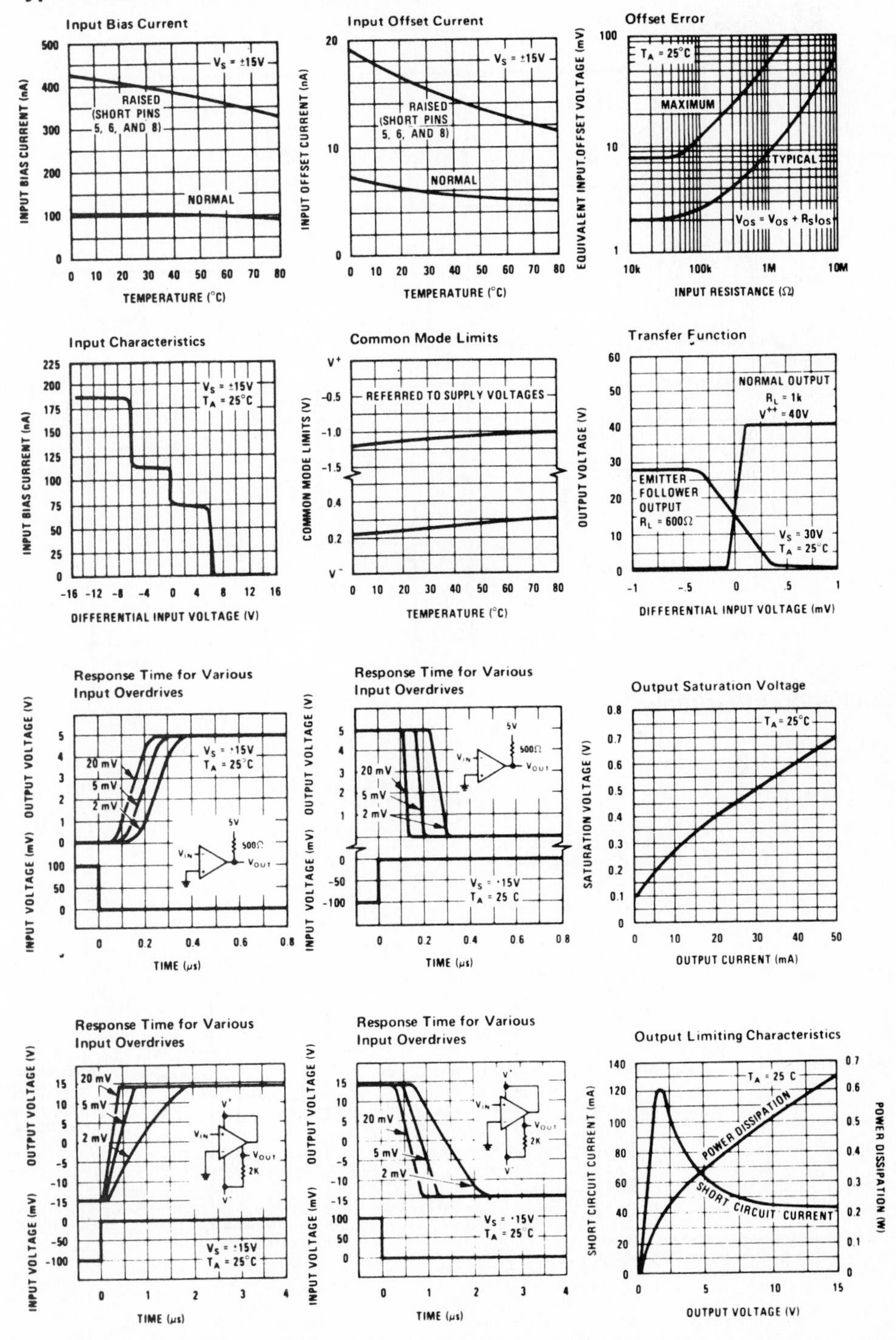

Schematic Diagram

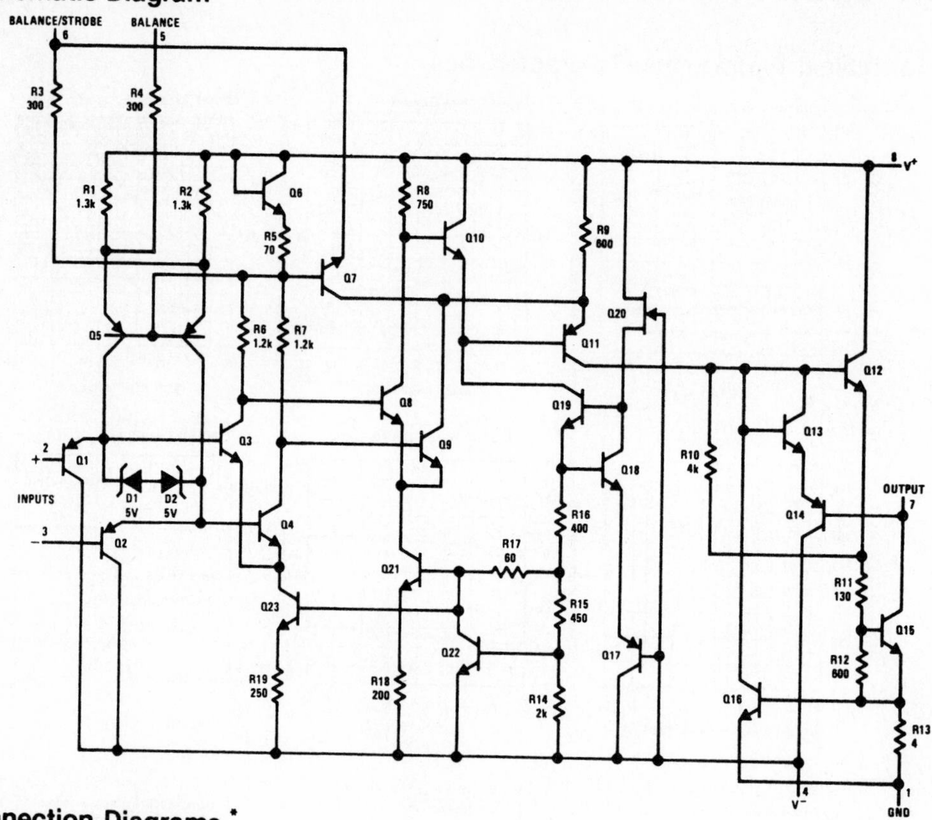

Connection Diagrams *

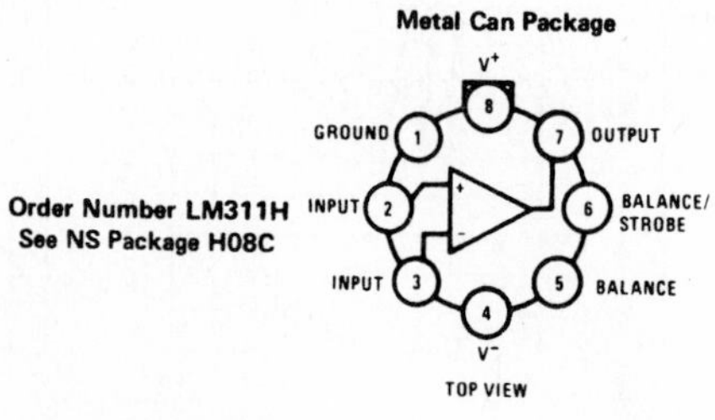

Order Number LM311H
See NS Package H08C

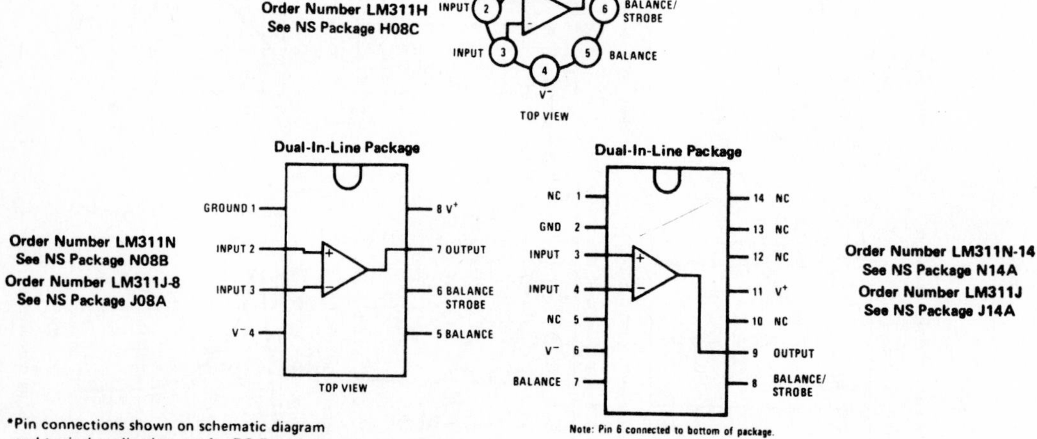

Order Number LM311N
See NS Package N08B
Order Number LM311J-8
See NS Package J08A

Order Number LM311N-14
See NS Package N14A
Order Number LM311J
See NS Package J14A

Note: Pin 6 connected to bottom of package.

*Pin connections shown on schematic diagram and typical applications are for TO-5 package.

Application Hints

CIRCUIT TECHNIQUES FOR AVOIDING OSCILLATIONS IN COMPARATOR APPLICATIONS

When a high-speed comparator such as the LM111 is used with fast input signals and low source impedances, the output response will normally be fast and stable, assuming that the power supplies have been bypassed (with 0.1 μF disc capacitors), and that the output signal is routed well away from the inputs (pins 2 and 3) and also away from pins 5 and 6.

However, when the input signal is a voltage ramp or a slow sine wave, or if the signal source impedance is high (1 kΩ to 100 kΩ), the comparator may burst into oscillation near the crossing-point. This is due to the high gain and wide bandwidth of comparators like the LM111. To avoid oscillation or instability in such a usage, several precautions are recommended, as shown in *Figure 1* below.

1. The trim pins (pins 5 and 6) act as unwanted auxiliary inputs. If these pins are not connected to a trim-pot, they should be shorted together. If they are connected to a trim-pot, a 0.01 μA capacitor C1 between pins 5 and 6 will minimize the susceptibility to AC coupling. A smaller capacitor is used if pin 5 is used for positive feedback as in *Figure 1*.

2. Certain sources will produce a cleaner comparator output waveform if a 100 pF to 1000 pF capacitor C2 is connected directly across the input pins.

3. When the signal source is applied through a resistive network, R_S, it is usually advantageous to choose an R_S' of substantially the same value, both for DC and for dynamic (AC) considerations. Carbon, tin-oxide, and metal-film resistors have all been used successfully in comparator input circuitry. Inductive wirewound resistors are not suitable.

4. When comparator circuits use input resistors (eg. summing resistors), their value and placement are particularly important. In all cases the body of the resistor should be close to the device or socket. In other words there should be very little lead length or printed-circuit foil run between comparator and resistor to radiate or pick up signals. The same applies to capacitors, pots, etc. For example, if R_S = 10 kΩ, as little as 5 inches of lead between the resistors and the input pins can result in oscillations that are very hard to damp. Twisting these input leads tightly is the only (second best) alternative to placing resistors close to the comparator.

5. Since feedback to almost any pin of a comparator can result in oscillation, the printed-circuit layout should be engineered thoughtfully. Preferably there should be a groundplane under the LM111 circuitry, for example, one side of a double-layer circuit card. Ground foil (or, positive supply or negative supply foil) should extend between the output and the inputs, to act as a guard. The foil connections for the inputs should be as small and compact as possible, and should be essentially surrounded by ground foil on all sides, to guard against capacitive coupling from any high-level signals (such as the output). If pins 5 and 6 are not used, they should be shorted together. If they are connected to a trim-pot, the trim-pot should be located, at most, a few inches away from the LM111, and the 0.01 μF capacitor should be installed. If this capacitor cannot be used, a shielding printed-circuit foil may be advisable between pins 6 and 7. The power supply bypass capacitors should be located within a couple inches of the LM111. (Some other comparators require the power-supply bypass to be located immediately adjacent to the comparator.)

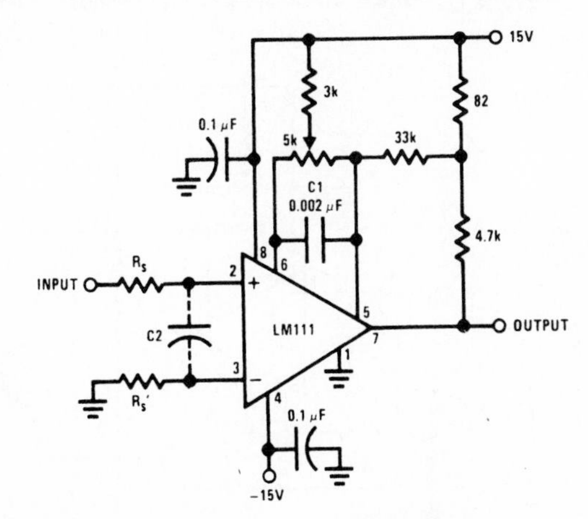

Pin connections shown are for LM111H in 8-lead TO-5 hermetic package

FIGURE 1. Improved Positive Feedback

Application Hints (Continued)

6. It is a standard procedure to use hysteresis (positive feedback) around a comparator, to prevent oscillation, and to avoid excessive noise on the output because the comparator is a good amplifier for its own noise. In the circuit of *Figure 2*, the feedback from the output to the positive input will cause about 3 mV of hysteresis. However, if R_S is larger than 100Ω, such as 50 kΩ, it would not be reasonable to simply increase the value of the positive feedback resistor above 510 kΩ. The circuit of *Figure 3* could be used, but it is rather awkward. See the notes in paragraph 7 below.

7. When both inputs of the LM111 are connected to active signals, or if a high-impedance signal is driving the positive input of the LM111 so that positive feedback would be disruptive, the circuit of *Figure 1* is ideal. The positive feedback is to pin 5 (one of the offset adjustment pins). It is sufficient to cause 1 to 2 mV hysteresis and sharp transitions with input triangle waves from a few Hz to hundreds of kHz. The positive-feedback signal across the 82Ω resistor swings 240 mV below the positive supply. This signal is centered around the nominal voltage at pin 5, so this feedback does not add to the V_{OS} of the comparator. As much as 8 mV of V_{OS} can be trimmed out, using the 5 kΩ pot and 3 kΩ resistor as shown.

8. These application notes apply specifically to the LM111, LM211, LM311, and LF111 families of comparators, and are applicable to all high-speed comparators in general, (with the exception that not all comparators have trim pins).

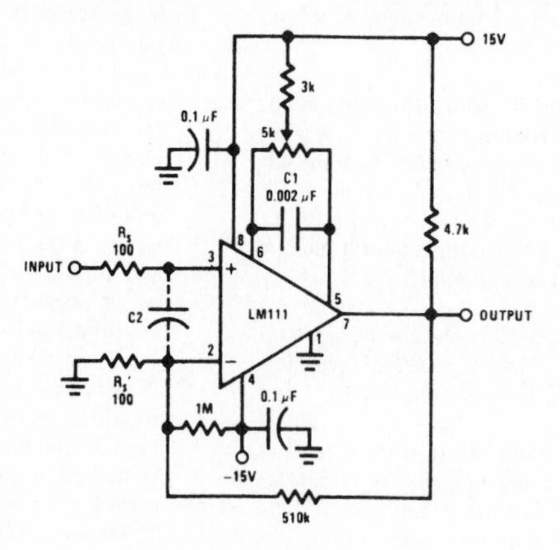

Pin connections shown are for LM111H in 8-lead TO-5 hermetic package

FIGURE 2. Conventional Positive Feedback

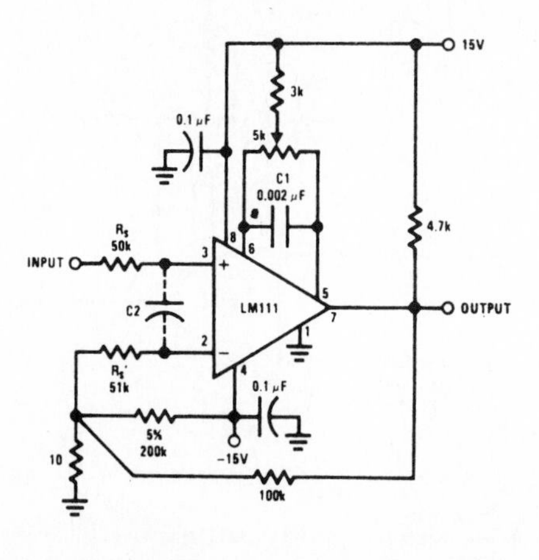

FIGURE 3. Positive Feedback With High Source Resistance

APPENDIX 4

Timer 555*

*Courtesy of Signetics Corporation, 811 East Arques Avenue, Sunnyvale, California, 94086, copyright 1974.

LINEAR INTEGRATED CIRCUITS

DESCRIPTION

The NE/SE 555 monolithic timing circuit is a highly stable controller capable of producing accurate time delays, or oscillation. Additional terminals are provided for triggering or resetting if desired. In the time delay mode of operation, the time is precisely controlled by one external resistor and capacitor. For a stable operation as an oscillator, the free running frequency and the duty cycle are both accurately controlled with two external resistors and one capacitor. The circuit may be triggered and reset on falling waveforms, and the output structure can source or sink up to 200mA or drive TTL circuits.

FEATURES

- TIMING FROM MICROSECONDS THROUGH HOURS
- OPERATES IN BOTH ASTABLE AND MONOSTABLE MODES
- ADJUSTABLE DUTY CYCLE
- HIGH CURRENT OUTPUT CAN SOURCE OR SINK 200mA
- OUTPUT CAN DRIVE TTL
- TEMPERATURE STABILITY OF 0.005% PER °C
- NORMALLY ON AND NORMALLY OFF OUTPUT

APPLICATIONS

PRECISION TIMING
PULSE GENERATION
SEQUENTIAL TIMING
TIME DELAY GENERATION
PULSE WIDTH MODULATION
PULSE POSITION MODULATION
MISSING PULSE DETECTOR

PIN CONFIGURATIONS (Top View)

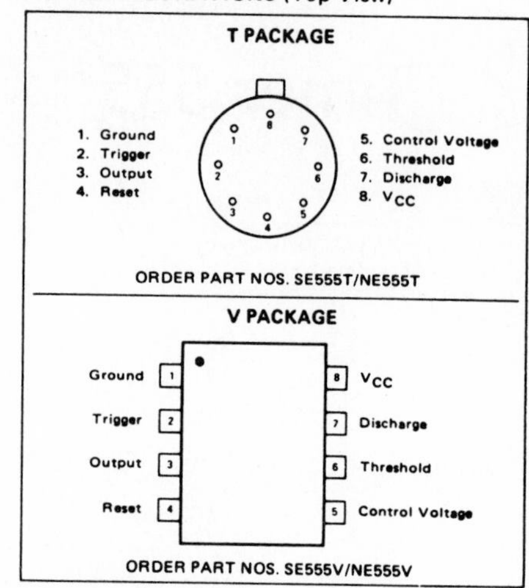

ABSOLUTE MAXIMUM RATINGS

Supply Voltage	+18V
Power Dissipation	600 mW
Operating Temperature Range	
NE555	0°C to +70°C
SE555	–55°C to +125°C
Storage Temperature Range	–65°C to +150°C
Lead Temperature (Soldering, 60 seconds)	+300°C

BLOCK DIAGRAM

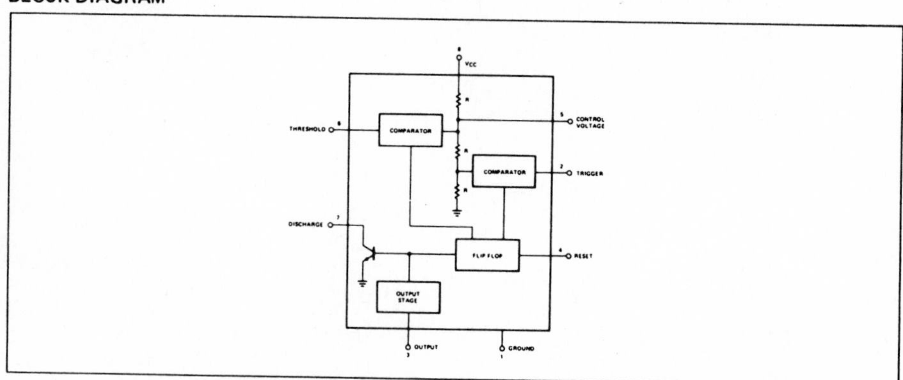

ELECTRICAL CHARACTERISTICS $T_A = 25°C$, V_{CC} = +5V to +15 unless otherwise specified

PARAMETER	TEST CONDITIONS	SE 555			NE 555			UNITS
		MIN	TYP	MAX	MIN	TYP	MAX	
Supply Voltage		4.5		18	4.5		16	V
Supply Current	V_{CC} = 5V $R_L = \infty$		3	5		3	6	mA
	V_{CC} = 15V $R_L = \infty$		10	12		10	15	mA
	Low State, Note 1							
Timing Error(Monostable)	R_A, R_B = 1KΩ to 100KΩ							
Initial Accuracy	C = 0.1 µF Note 2		0.5	2		1		%
Drift with Temperature			30	100		50		ppm/°C
Drift with Supply Voltage			0.05	0.2		0.1		%/Volt
Threshold Voltage			2/3			2/3		X V_{CC}
Trigger Voltage	V_{CC} = 15V	4.8	5	5.2		5		V
Timing Error(Astable)	V_{CC} = 5V	1.45	1.67	1.9		1.67		V
Trigger Current			0.5			0.5		µA
Reset Voltage		0.4	0.7	1.0	0.4	0.7	1.0	V
Reset Current			0.1			0.1		mA
Threshold Current	Note 3		0.1	.25		0.1	.25	µA
Control Voltage Level	V_{CC} = 15V	9.6	10	10.4	9.0	10	11	V
	V_{CC} = 5V	2.9	3.33	3.8	2.6	3.33	4	V
Output Voltage (low)	V_{CC} = 15V							
	I_{SINK} = 10mA		0.1	0.15		0.1	.25	V
	I_{SINK} = 50mA		0.4	0.5		0.4	.75	V
	I_{SINK} = 100mA		2.0	2.2		2.0	2.5	V
	I_{SINK} = 200mA		2.5			2.5		
	V_{CC} = 5V							
	I_{SINK} = 8mA		0.1	0.25				V
	I_{SINK} = 5mA					.25	.35	
Output Voltage Drop (low)								
	I_{SOURCE} = 200mA		12.5			12.5		
	V_{CC} = 15V							
	I_{SOURCE} = 100mA							
	V_{CC} = 15V	13.0	13.3		12.75	13.3		V
	V_{CC} = 5V	3.0	3.3		2.75	3.3		V
Rise Time of Output			100			100		nsec
Fall Time of Output			100			100		nsec

NOTES

1. Supply Current when output high typically 1mA less.
2. Tested at V_{CC} = 5V and V_{CC} = 15V
3. This will determine the maximum value of $R_A + R_B$. For 15V operation, the max total R = 20 megohm.

EQUIVALENT CIRCUIT (Shown for One Side Only)

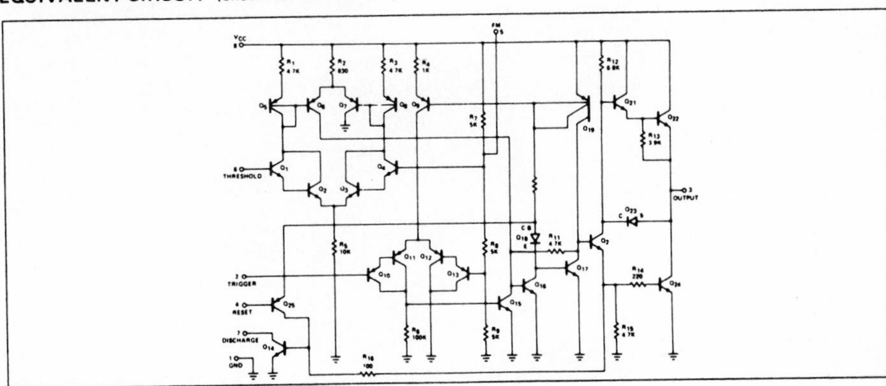

APPENDIX 5

LM117 3-Terminal Adjustable Regulator*

*Courtesy of National Semiconductor Corporation.

General Description

The LM117/LM217/LM317 are adjustable 3-terminal positive voltage regulators capable of supplying in excess of 1.5A over a 1.2V to 37V output range. They are exceptionally easy to use and require only two external resistors to set the output voltage. Further, both line and load regulation are better than standard fixed regulators. Also, the LM117 is packaged in standard transistor packages which are easily mounted and handled.

In addition to higher performance than fixed regulators, the LM117 series offers full overload protection available only in IC's. Included on the chip are current limit, thermal overload protection and safe area protection. All overload protection circuitry remains fully functional even if the adjustment terminal is disconnected.

Features

- Adjustable output down to 1.2V
- Guaranteed 1.5A output current
- Line regulation typically 0.01%/V
- Load regulation typically 0.1%
- Current limit constant with temperature
- **100% electrical burn-in**
- Eliminates the need to stock many voltages
- Standard 3-lead transistor package
- 80 dB ripple rejection

Normally, no capacitors are needed unless the device is situated far from the input filter capacitors in which case an input bypass is needed. An optional output capacitor can be added to improve transient response. The adjustment terminal can be bypassed to achieve very high ripple rejections ratios which are difficult to achieve with standard 3-terminal regulators.

Besides replacing fixed regulators, the LM117 is useful in a wide variety of other applications. Since the regulator is "floating" and sees only the input-to-output differential voltage, supplies of several hundred volts can be regulated as long as the maximum input to output differential is not exceeded.

Also, it makes an especially simple adjustable switching regulator, a programmable output regulator, or by connecting a fixed resistor between the adjustment and output, the LM117 can be used as a precision current regulator. Supplies with electronic shutdown can be achieved by clamping the adjustment terminal to ground which programs the output to 1.2V where most loads draw little current.

The LM117K, LM217K and LM317K are packaged in standard TO-3 transistor packages while the LM117H, LM217H and LM317H are packaged in a solid Kovar base TO-39 transistor package. The LM117 is rated for operation from −55°C to +150°C, the LM217 from −25°C to +150°C and the LM317 from 0°C to +125°C. The LM317T and LM317MP, rated for operation over a 0°C to +125°C range, are available in a TO-220 plastic package and a TO-202 package, respectively.

For applications requiring greater output current in excess of 3A and 5A, see LM150 series and LM138 series data sheets, respectively. For the negative complement, see LM137 series data sheet.

LM117 Series Packages and Power Capability

DEVICE	PACKAGE	RATED POWER DISSIPATION	DESIGN LOAD CURRENT
LM117	TO-3	20W	1.5A
LM217 LM317	TO-39	2W	0.5A
LM317T	TO-220	15W	1.5A
LM317M	TO-202	7.5W	0.5A
LM317LZ	TO-92	0.6W	0.1A

Typical Applications

1.2V–25V Adjustable Regulator

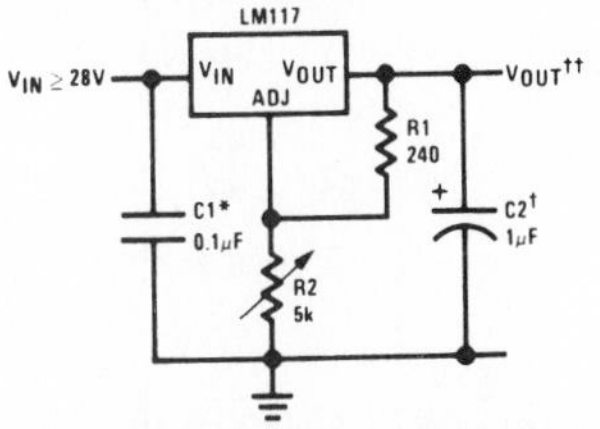

†Optional—improves transient response. Output capacitors in the range of 1 μF to 1000 μF of aluminum or tantalum electrolytic are commonly used to provide improved output impedance and rejection of transients.

*Needed if device is far from filter capacitors.

$$^{\dagger\dagger}V_{OUT} = 1.25V\left(1 + \frac{R2}{R1}\right)$$

Digitally Selected Outputs

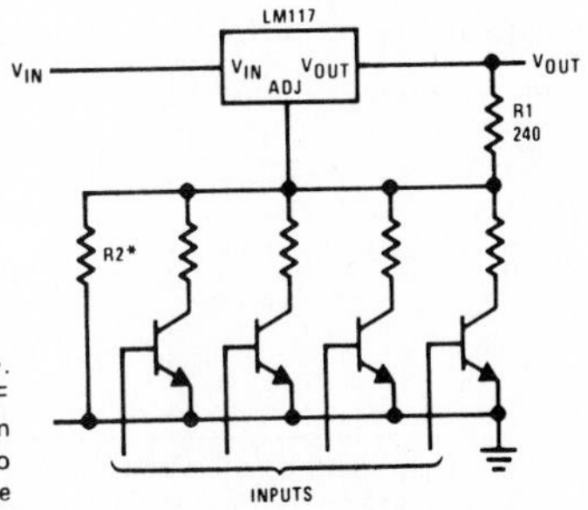

*Sets maximum V_{OUT}

5V Logic Regulator with Electronic Shutdown*

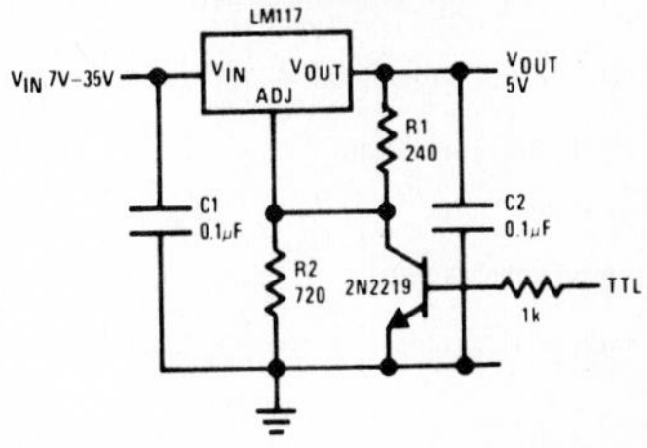

*Min output ≈ 1.2V

Absolute Maximum Ratings

Power Dissipation	Internally limited
Input–Output Voltage Differential	40V
Operating Junction Temperature Range	
LM117	−55°C to +150°C
LM217	−25°C to +150°C
LM317	0°C to +125°C
Storage Temperature	−65°C to +150°C
Lead Temperature (Soldering, 10 seconds)	300°C

Preconditioning

Burn-In in Thermal Limit **100% All Devices**

Electrical Characteristics (Note 1)

PARAMETER	CONDITIONS	LM117/217			LM317			UNITS
		MIN	TYP	MAX	MIN	TYP	MAX	
Line Regulation	T_A = 25°C, 3V ≤ V_{IN} − V_{OUT} ≤ 40V (Note 2)		0.01	0.02		0.01	0.04	%/V
Load Regulation	T_A = 25°C, 10 mA ≤ I_{OUT} ≤ I_{MAX}							
	V_{OUT} ≤ 5V, (Note 2)		5	15		5	25	mV
	V_{OUT} ≥ 5V, (Note 2)		0.1	0.3		0.1	0.5	%
Thermal Regulation	T_A = 25°C, 20 ms Pulse		0.03	0.07		0.04	0.07	%/W
Adjustment Pin Current			50	100		50	100	μA
Adjustment Pin Current Change	10 mA ≤ I_L ≤ I_{MAX} 3V ≤ (V_{IN}−V_{OUT}) ≤ 40V		0.2	5		0.2	5	μA
Reference Voltage	3V ≤ (V_{IN}−V_{OUT}) ≤ 40V, (Note 3) 10 mA ≤ I_{OUT} ≤ I_{MAX}, P ≤ P_{MAX}	1.20	1.25	1.30	1.20	1.25	1.30	V
Line Regulation	3V ≤ V_{IN} − V_{OUT} ≤ 40V, (Note 2)		0.02	0.05		0.02	0.07	%/V
Load Regulation	10 mA ≤ I_{OUT} ≤ I_{MAX}, (Note 2)							
	V_{OUT} ≤ 5V		20	50		20	70	mV
	V_{OUT} ≥ 5V		0.3	1		0.3	1.5	%
Temperature Stability	T_{MIN} ≤ T_j ≤ T_{MAX}		1			1		%
Minimum Load Current	V_{IN}−V_{OUT} = 40V		3.5	5		3.5	10	mA
Current Limit	V_{IN}−V_{OUT} ≤ 15V							
	K and T Package	1.5	2.2		1.5	2.2		A
	H and P Package	0.5	0.8		0.5	0.8		A
	V_{IN}−V_{OUT} = 40V, T_j = +25°C							
	K and T Package	0.30	0.4		0.15	0.4		A
	H and P Package	0.15	0.07		0.075	0.07		A
RMS Output Noise, % of V_{OUT}	T_A = 25°C, 10 Hz ≤ f ≤ 10 kHz		0.003			0.003		%
Ripple Rejection Ratio	V_{OUT} = 10V, f = 120 Hz		65			65		dB
	C_{ADJ} = 10μF	66	80		66	80		dB
Long-Term Stability	T_A = 125°C		0.3	1		0.3	1	%
Thermal Resistance, Junction to Case	H Package		12	15		12	15	°C/W
	K Package		2.3	3		2.3	3	°C/W
	T Package					4		°C/W
	P Package					12		°C/W

Note 1: Unless otherwise specified, these specifications apply −55°C ≤ T_j ≤ +150°C for the LM117, −25°C ≤ T_j ≤ +150°C for the LM217, and 0°C ≤ T_j ≤ +125°C for the LM317; V_{IN} − V_{OUT} = 5V; and I_{OUT} = 0.1A for the TO-39 and TO-202 packages and I_{OUT} = 0.5A for the TO-3 and TO-220 packages. Although power dissipation is internally limited, these specifications are applicable for power dissipations of 2W for the TO-39 and TO-202, and 20W for the TO-3 and TO-220. I_{MAX} is 1.5A for the TO-3 and TO-220 packages and 0.5A for the TO-39 and TO-202 packages.

Note 2: Regulation is measured at constant junction temperature, using pulse testing with a low duty cycle. Changes in output voltage due to heating effects are covered under the specification for thermal regulation.

Note 3: Selected devices with tightened tolerance reference voltage available.

Application Hints

In operation, the LM117 develops a nominal 1.25V reference voltage, V_{REF}, between the output and adjustment terminal. The reference voltage is impressed across program resistor R1 and, since the voltage is constant, a constant current I_1 then flows through the output set resistor R2, giving an output voltage of

$$V_{OUT} = V_{REF}\left(1 + \frac{R2}{R1}\right) + I_{ADJ}R2$$

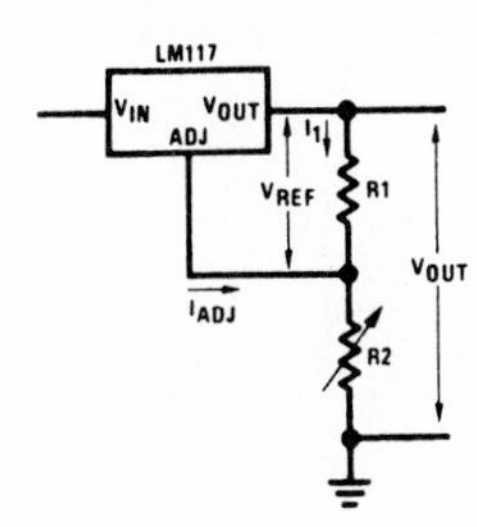

FIGURE 1.

Since the 100μA current from the adjustment terminal represents an error term, the LM117 was designed to minimize I_{ADJ} and make it very constant with line and load changes. To do this, all quiescent operating current is returned to the output establishing a minimum load current requirement. If there is insufficient load on the output, the output will rise.

External Capacitors

An input bypass capacitor is recommended. A 0.1μF disc or 1μF solid tantalum on the input is suitable input bypassing for almost all applications. The device is more sensitive to the absence of input bypassing when adjustment or output capacitors are used but the above values will eliminate the possibility of problems.

The adjustment terminal can be bypassed to ground on the LM117 to improve ripple rejection. This bypass capacitor prevents ripple from being amplified as the output voltage is increased. With a 10μF bypass capacitor 80 dB ripple rejection is obtainable at any output level. Increases over 10μF do not appreciably improve the ripple rejection at frequencies above 120 Hz. If the bypass capacitor is used, it is sometimes necessary to include protection diodes to prevent the capacitor from discharging through internal low current paths and damaging the device.

In general, the best type of capacitors to use are solid tantalum. Solid tantalum capacitors have low impedance even at high frequencies. Depending upon capacitor construction, it takes about 25μF in aluminum electrolytic to equal 1μF solid tantalum at high frequencies. Ceramic capacitors are also good at high frequencies; but some types have a large decrease in capacitance at frequencies around 0.5 MHz. For this reason, 0.01μF disc may seem to work better than a 0.1μF disc as a bypass.

Although the LM117 is stable with no output capacitors, like any feedback circuit, certain values of external capacitance can cause excessive ringing. This occurs with values between 500 pF and 5000 pF. A 1μF solid tantalum (or 25μF aluminum electrolytic) on the output swamps this effect and insures stability.

Load Regulation

The LM117 is capable of providing extremely good load regulation but a few precautions are needed to obtain maximum performance. The current set resistor connected between the adjustment terminal and the output terminal (usually 240Ω) should be tied directly to the output of the regulator rather than near the load. This eliminates line drops from appearing effectvely in series with the reference and degrading regulation. For example, a 15V regulator with 0.05Ω resistance between the regulator and load will have a load regulation due to line resistance of 0.05Ω x I_L. If the set resistor is connected near the load the effective line resistance will be 0.05Ω (1 + R2/R1) or in this case, 11.5 times worse.

Figure 2 shows the effect of resistance between the regulator and 240Ω set resistor.

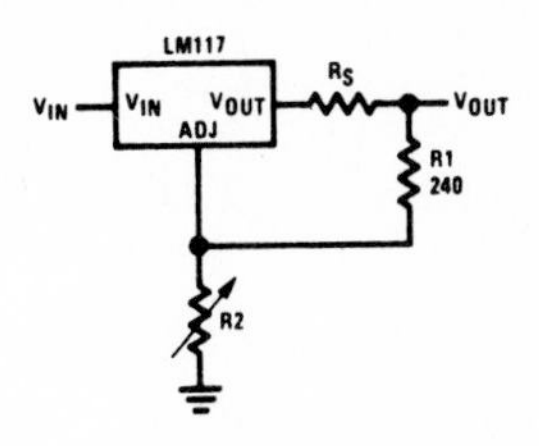

FIGURE 2. Regulator with Line Resistance in Output Lead

With the TO-3 package, it is easy to minimize the resistance from the case to the set resistor, by using two separate leads to the case. However, with the TO-5 package, care should be taken to minimize the wire length of the output lead. The ground of R2 can be returned near the ground of the load to provide remote ground sensing and improve load regulation.

Protection Diodes

When external capacitors are used with *any* IC regulator it is sometimes necessary to add protection diodes to prevent the capacitors from discharging through low current points into the regulator. Most 10μF capacitors have low enough internal series resistance to deliver 20A spikes when shorted. Although the surge is short, there is enough energy to damage parts of the IC.

When an output capacitor is connected to a regulator and the input is shorted, the output capacitor will discharge into the output of the regulator. The discharge

Application Hints (cont'd.)

current depends on the value of the capacitor, the output voltage of the regulator, and the rate of decrease of V_{IN}. In the LM117, this discharge path is through a large junction that is able to sustain 15A surge with no problem. This is not true of other types of positive regulators. For output capacitors of 25µF or less, there is no need to use diodes.

The bypass capacitor on the adjustment terminal can discharge through a low current junction. Discharge occurs when *either* the input or output is shorted. Internal to the LM117 is a 50Ω resistor which limits the peak discharge current. No protection is needed for output voltages of 25V or less and 10µF capacitance. *Figure 3* shows an LM117 with protection diodes included for use with outputs greater than 25V and high values of output capacitance.

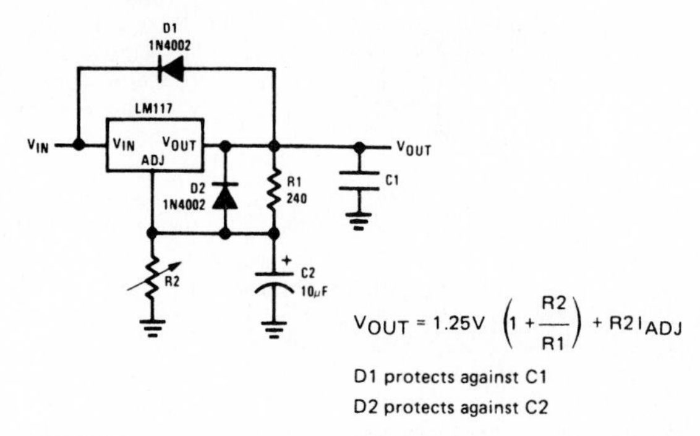

FIGURE 3. Regulator with Protection Diodes

Schematic Diagram

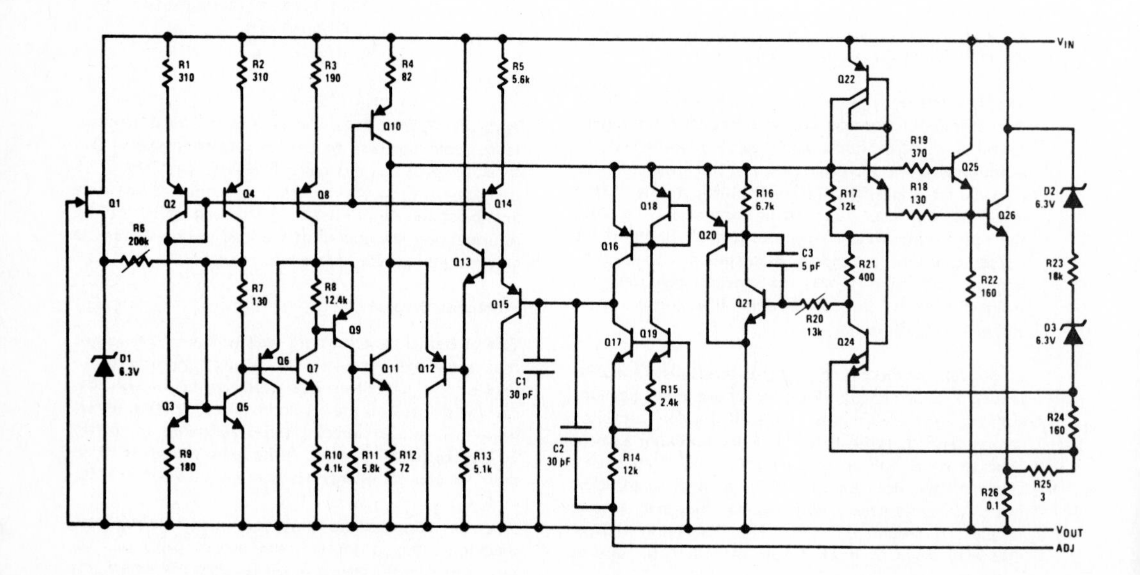

Typical Applications

AC Voltage Regulator

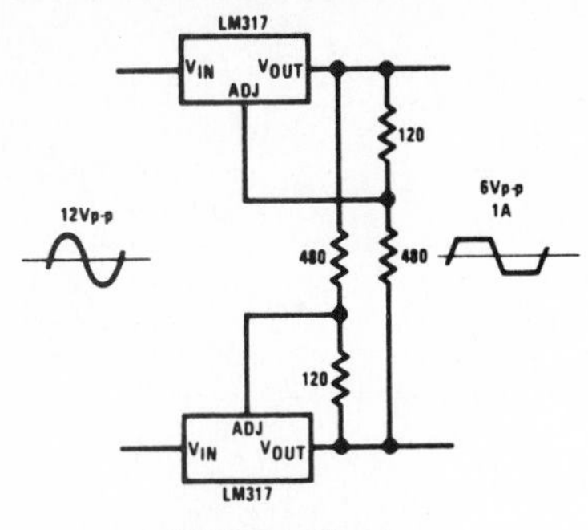

Adjustable 4A Regulator

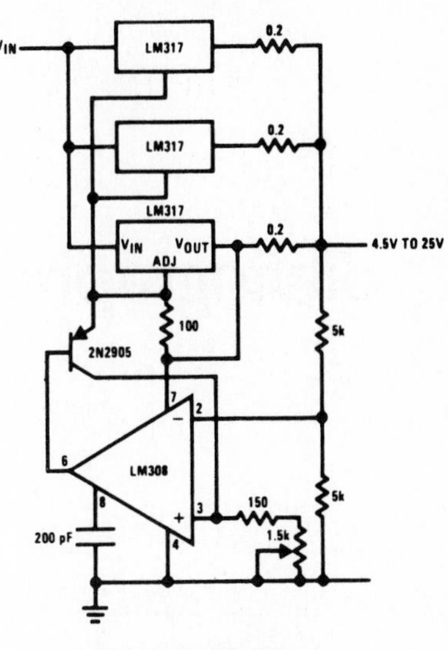

12V Battery Charger

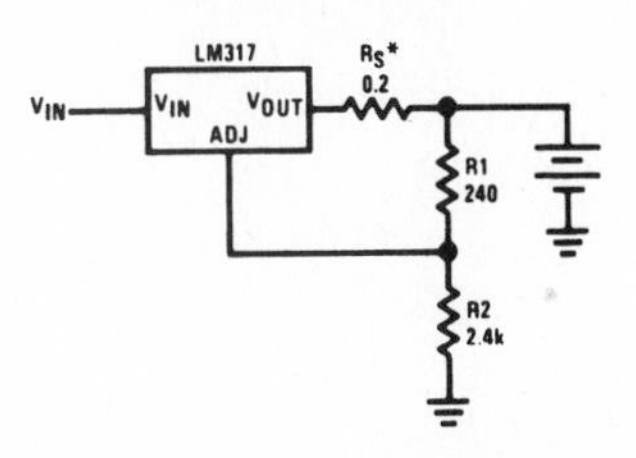

*R_S—sets output impedance of charger $Z_{OUT} = R_S\left(1 + \frac{R2}{R1}\right)$ Use of R_S allows low charging rates with fully charged battery.

50 mA Constant Current Battery Charger

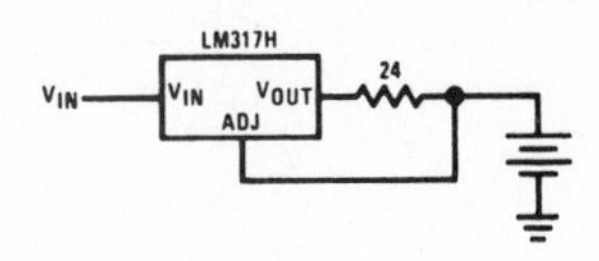

Current Limited 6V Charger

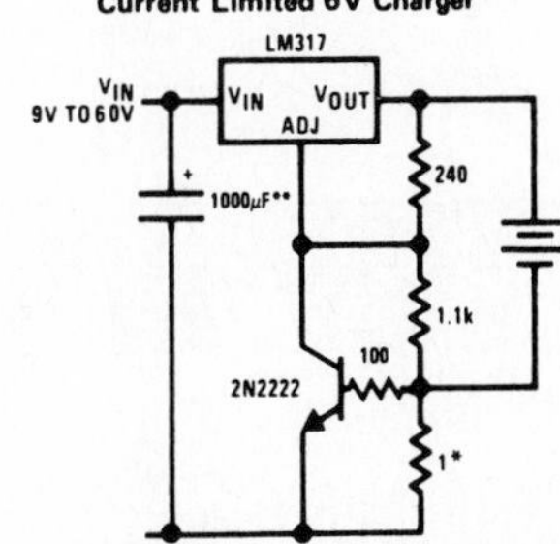

*Sets peak current (0.6A for 1Ω)

**The 1000 µF is recommended to filter out input transients

Connection Diagrams

(TO-3 STEEL)
Metal Can Package

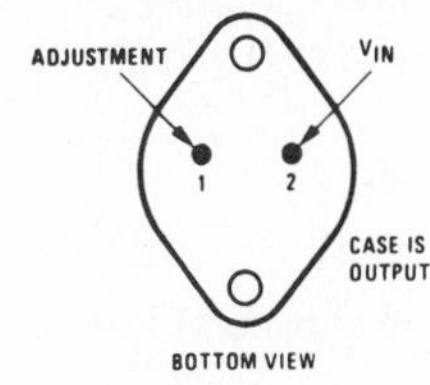

Order Number:
LM117K STEEL
LM217K STEEL
LM317K STEEL
See Package K02A

(TO-39)
Metal Can Package

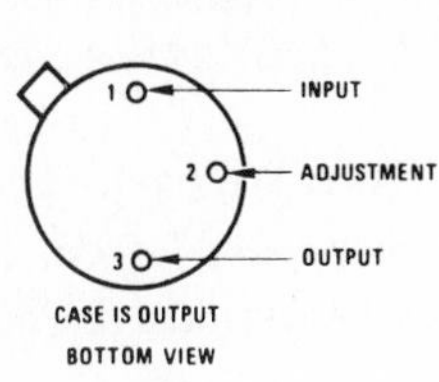

Order Number:
LM117H
LM217H
LM317H
See Package H03A

(TO-220)
Plastic Package

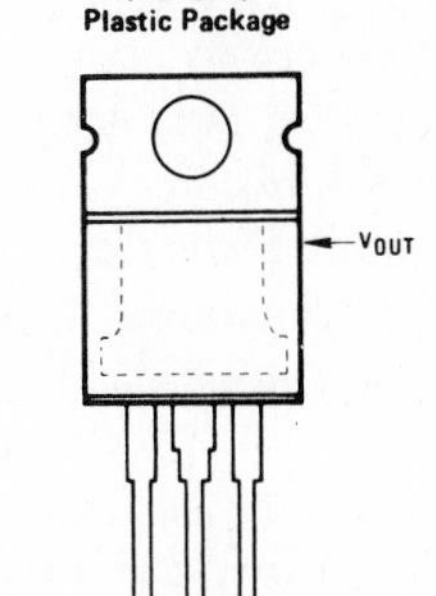

Order Number:
LM317T
See Package T03B

(TO-202)
Plastic Package

V_{OUT}
1 3 2
ADJ
V_{IN}
V_{OUT}
FRONT VIEW

Order Number:
LM317MP
See Package P03A
Tab Formed Devices
LM317MP TB
See Package P03E

ANALOG DEVICES, INC., *Analog-Digital Conversion Handbook* (1972), *Analog-Digital Conversion Notes* (1977), *Data Acquisition Products Catalog* (1984), *Non-linear Circuits Handbook* (1979), *Data Acquisition Products Catalog Supplement* (1984), Analog Devices Inc., Norwood, Mass.

BLH ELECTRONICS, *Strain Gages, SR4,* BLH, Waltham, Mass. (1979).

BURR-BROWN INTERNATIONAL, *Product Data Book*, Tucson, Ariz. (1984).

CLAYTON, G. G., *Operational Amplifiers,* Butterworth & Company (Publishers) Ltd., London (1971).

COUGHLIN, ROBERT F., and VILLANUCCI, ROBERT F., *Introductory Operational Amplifiers and Linear ICs, Theory, and Experimentation,* Prentice-Hall, Inc., Englewood Cliffs, N.J. (1990).

DRISCOLL, FREDERICK F., *6800/68000 Microprocessors,* Breton Publishers, Boston (1987).

DRISCOLL, FREDERICK F., and COUGHLIN, ROBERT F., *Solid State Devices and Applications,* Prentice-Hall, Inc., Englewood Cliffs, N.J. (1975).

FAIRCHILD INSTRUMENT AND CAMERA CORPORATION, *Linear Division Products* (1982), *Voltage Regulator Handbook* (1974), Fairchild, Mountain View, Calif.

FREDERIKSEN, THOMAS M., *Intuitive IC Op Amps*, National Semiconductor Technology Series, National Semiconductor Corporation, Santa Clara, Calif. (1984).

HARRIS CORPORATION, *Linear and Data Acquisition Products*, Harris, Melbourne, Fla. (1980).

MORRISON, R., *Grounding and Shielding Techniques in Instrumentation,* John Wiley & Sons, Inc., New York (1967).

MOTOROLA SEMICONDUCTOR PRODUCTS, INC., *More Value Out of Integrated Operational Amplifier Data Sheets*, AN-273A, Motorola, Phoenix, Ariz. (1970).

NATIONAL SEMICONDUCTOR CORPORATION, *Linear Databook* (1988), *Linear Applications Handbook* (1988), *Special Functions Databook* (1979), *Voltage Regulator Handbook* (1980), *Audio Handbook* (no date), *Linear Supplement Databook* (1988), National Semiconductor Corporation, Santa Clara, Calif.

PHILBRICK RESEARCHES, INC., *Applications Manual for Computing Amplifiers for Modeling, Measuring, Manipulating and Much Else*, Nimrod Press, Inc., Boston (1966).

PRECISION MONOLITHICS, INC., *Linear and Conversion IC Products*, Precision Monolithics, Santa Clara, Calif. (1982).

SHEINGOLD, DANIEL H., *Transducer Interfacing Handbook*, Analog Devices, Inc., Norwood, Mass. (1980).

SIGNETICS CORPORATION, *Data Manual,* Signetics, Sunnyvale, Calif. (1977).

SMITH, J. I., *Modern Operational Circuit Design,* John Wiley & Sons, Inc., New York (1971).

VILLANUCCI, R., ET AL., *Electronic Techniques: Shop Practices and Construction,* 4th ed., Prentice-Hall, Inc., Englewood Cliffs, N.J. (1990).

Index